D0838553

Restraint and Handling of Wild and Domestic Animals

MURRAY E. FOWLER

Restraint and Handling of Wild and Domestic Animals

Iowa State University Press, Ames

This book is gratefully dedicated to my wife, Audrey, for her tireless effort in copyreading the complete work. She has been an intimate participant in this work since its inception.

MURRAY E. FOWLER is Professor and Chairman of the Department of Medicine, School of Veterinary Medicine, University of California, Davis, where he is also Chief of the Zoological Medicine Service of the teaching hospital. He received the B.S. degree from Utah State University and the D.V.M. degree from Iowa State University. He has written extensively for both professional journals and nontechnical publications. He is a Diplomate of the American Board of Toxicology and Diplomate of the American College of Veterinary Internal Medicine. He is a past president of the American Association of Zoo Veterinarians.

Composed by Iowa State University Press
Printed in the United States of America

First edition, 1978
Second printing, 1979
Third printing, 1981
Fourth printing, 1983
Fifth printing, 1985
Sixth printing, 1987
Seventh printing, 1989

Library of Congress Cataloging in Publication Data

Fowler, Murray E.
Restraint and handling of wild and domestic animals.

Includes index.
1. Animal immobilization. I. Title.
QL62.5.F68 636.08′3 78-5036
ISBN 0-8138-1890-7

CONTENTS

Preface v
Acknowledgments vi

Part 1. General Concepts

1. Introduction 3
2. Tools of Restraint 6
3. Rope Work 17
4. Chemical Restraint 35
5. Stress 53
6. Thermoregulation 63
7. Medical Problems during Restraint 73

Part 2. Domestic Animals

8. Horses, Donkeys, Mules . . 93
9. Cattle and Other Domestic Bovids 113
10. Sheep and Goats 131
11. Swine 139
12. Dogs 148
13. Cats 156
14. Laboratory Animals 161
15. Poultry and Waterfowl . . . 171

Part 3. Wild Animals

16. Introduction 181
17. Monotremes and Marsupials 183
18. Small Mammals 189
19. Carnivores 201
20. Primates 214
21. Marine Mammals 223
22. Elephants 231
23. Hoofed Stock 240
24. Birds 262
25. Reptiles 286
26. Amphibians and Fish 311

Appendixes 317
Index 325

PREFACE

THE original intent in this book was to deal only with wild animal restraint. However, upon deliberation, it was realized that fundamental principles of restraint apply to both domestic and wild animals, so it was decided to include both groups to present a more comprehensive picture of the subject.

The objectives of this book are to collect under one cover discussions and illustrations of the principles of animal restraint and handling and to describe some restraint practices for diverse species of vertebrate wild and domestic animals. Heretofore no single source has offered information for handling such diverse animals as a 2.5 g hummingbird and an elephant weighing 5,000-6,000 kg. It is hoped that this book will satisfy that need for all who handle animals—particularly veterinarians; animal husbandmen; wildlife biologists; personnel of zoos, research, and humane society facilities; and any others who deal with animals.

Government regulatory agencies require humane treatment and proper care and handling of all animals in captivity. It is legally necessary for those maintaining wildlife to provide adequate restraint facilities and personnel trained in satisfactory handling techniques to prevent or minimize injuries.

Restraint and handling techniques for domestic animals have long been well documented and described. Although the most recent text[1] was written in 1954, the principles outlined in that excellent publication are still valid.

Wild animal restraint and handling techniques are not as well known nor as widely publicized except in those notorious instances when inhumane and torturous methods used in capture and transport attract the attention of the news media. Some people feel that all wildlife should be returned to the native habitat and left to live and die undisturbed by human beings. This attitude is naive in this day and time. Wild animals have become an integral part of society and will continually be handled. It behooves us to know and use techniques safe for both animal and handler.

The need for understanding restraint principles, particularly for wild animals, is exemplified by the statement of an experienced zoo veterinarian in a recent publication: "It is all very well to plan an operation on a tiger, but the problem that arises is how to catch the beast, and once having caught it, how safely to secure it. Nor is this difficulty restricted to the tiger, it applies in a lesser or greater degree to every type of wild animal in captivity. Not one of them will cooperate in your well-meaning efforts to help them, and no such thing as gratitude exists in their primitive makeup."[2]

This book is not, nor is it meant to be, an exhaustive encyclopedia on animal restraint. The author is well aware

1. Leahy, J. R., and Barrow, P. 1954. Animal Restraint. Ithaca, N.Y.: Comstock.

2. Graham-Jones, O. 1973. First Catch Your Tiger. New York: Taplinger.

that certain individual researchers or biologists may favor one or more techniques or special tools not mentioned here. It is impossible for any individual to acquire a personal knowledge of all possible combinations of restraint and handling procedures for every species or even for groups of species. However, the techniques presented have proved successful in the hands of experienced individuals and should serve as guides for anyone faced with similar problems.

It is only through an enlightened understanding of restraint principles that humane handling with the least amount of stress will be possible for any animal. It is hoped that by bringing all this information together in one source, more people will be able to share in saving wild animals for posterity.

ACKNOWLEDGMENTS

IN addition to my own practice, I have been privileged to visit many zoos on the North American continent, in Europe, England, Australia, New Zealand, and Africa. I have also enjoyed opportunities to see programs in operation at primate centers throughout the United States, at wildlife research facilities, and at many other installations where wild animals are handled routinely.

Special thanks are due to the staffs of these facilities. Wherever I have visited, the welcome has been warm and the conversation candid and enlightening. Veterinarians, directors, curators, and keepers have all enthusiastically shared experiences, frustrations, and successful techniques.

I was afforded the unique opportunity of spending a sabbatical year at the San Diego Zoo developing autotutorial programs on restraint and handling of wild animals. I am indebted to the Zoological Society of San Diego and zoo personnel for financial, physical, and moral support for that project, which also provided valuable material for this book.

Animal care personnel of all institutions—those persons most intimately involved in day-to-day animal handling—have been a storehouse of philosophies, ideas, and gadgets. I recognize their respect and concern for the animals in their charge. These are often the compassionate individuals who have developed techniques for the safety and well-being of both animal and handler.

I make no pretense of being the originator of all the techniques described herein. In many instances the originator of the technique is not known; in others, the technique is the result of the work of many. I am, nevertheless, grateful to all those who willingly described and demonstrated techniques. I have used, or seen used, most of the techniques described. Where my experience was lacking I have gleaned from the literature or the personal experiences of others.

This work could not have been accomplished without the help of numerous people who read and criticized those parts of the manuscript pertaining to their particular area of expertise. Nonetheless, errors and omissions are my responsibility.

Thanks are also due to my staff for preparation of the manuscript and illustrations. These include Tamara Sturak, Mary Watkins, Margaret Hardy, Carol Rowe, Velma Ayres, Terry Schulz, and Basil Knox. The photographs and drawings were made by me or done under my direction unless otherwise noted.

Restraint and Handling of Wild and Domestic Animals

It is incumbent upon a person who takes the responsibility of manipulating an animal's life to be concerned for its feelings . . .

1 General Concepts

1 INTRODUCTION

RESTRAINT varies from confinement in an unnatural enclosure to complete restriction of muscular activity or immobilization (hypokinesia). Both physical and chemical restraint are now practiced. Anciently only physical restraint was utilized. Just when man learned of chemical immobilization (poison arrows) is not known, but it antedates recorded history.

Only within the past two decades have some of the physiological effects of restricted movement been studied and appreciated. For centuries, extended bed rest for ill or postsurgical human patients was practiced—to the detriment of the patient. Now it is known that many deleterious effects result from this type of immobility (see Chapter 5). Solitary confinement is known to be extremely devastating for a human being. Similar confinement of social animals produces severe psychological stress.

Restraint practices evolved with the domestication of animals for food, fiber, labor, sport, and companionship. Domestication necessitated special husbandry practices. As people began to minister to animals' needs, they found it necessary to restrict activity by placing them in enclosures. If animals resisted when wounds were treated or medication administered, it was necessary to further restrain them. Trial and error combined with the shared experiences of fellow human beings ultimately produced satisfactory practices.

A person who undertakes to restrict an animal's activity or restrain the animal is assuming a responsibility that should not be considered lightly. Each restraint incident has some effect on the behavior, life, or other activities of an animal. From a humane and moral standpoint, the minimum amount of restraint consistent with accomplishing the task should be used. This should become a maxim for persons who must restrain animals.

Each time it is proposed to restrain an animal, the following questions should first be asked: Why must this animal be restrained? What procedure will produce the greatest gain with the least hazard? When will it be most desirable to restrain the animal? Who is the most qualified to accomplish the task in the least amount of time and with the least stress to the animal? What location would be best for the planned restraint procedure?

WHY RESTRAINT

Everyone agrees that domestic animals require transporting, medicating, and handling. Some contend that all wild animals should be free-living, without human interference. This philosophy seems naive in the present time.

Wild animals kept in captivity require special husbandry practices. They must be transported, housed, and fed. If they become ill, they must be examined and treated.

Free-living animals may have to be translocated, as was necessary when the Kariba dam was built in Southern Rhodesia. Diseases in wildlife populations must be monitored, since some have far-reaching consequences for the health of domestic livestock and human beings. Many wild populations are managed. As far as wild animals are concerned, any captive situation involves some form of restraint.

GENERAL CONCEPTS

Four basic factors should be considered when selecting a restraint technique: (1) Will it be safe for the person who must handle the animal? (2) Does it provide maximum safety for the animal? (3) Will it be possible to accomplish the intended procedure by utilizing the suggested restraint method? (4) Can constant observation and attention be given the animal following restraint until it is fully recovered from the physical or chemical effects? Once these four factors are evaluated, a suitable technique can be selected.

Many wild animals can inflict serious, if not fatal, injury. The first concern when dealing with wild animals should be the safety of human beings. To think otherwise is foolhardy, and those who grandstand or show off by manipulating dangerous animals without benefit of proper restraint are potentially injuring themselves or bystanders. Those who own or have administrative responsibility for wild animals must recognize that the animal, no matter how valuable, cannot be handled in such a way as to jeopardize the safety of those who must work around it. Techniques are known that when properly used can safeguard both animal and operator.

It is desirable to build proper facilities into areas where wild animals must be kept so that these handling procedures can be safely carried out. It is foolish to pay thousands of dollars for a zoo specimen if facilities are not available in which to handle or restrain the animal for prophylactic measures or treatment of disease or injury.

Certain wildlife populations have become so depleted they are near extinction. We should not practice on these species. It is not economically feasible, nor is there sufficient animal life for each person to gain through personal experience the intimate knowledge of various behavioral patterns and characteristics to enable them to develop expertise in the successful use of restraint procedures. Therefore we must learn from the experiences of others who have dealt extensively with one species or family of animals and trade on their knowledge of the more successful techniques.

To be successful in working with animals, one must understand their behavioral characteristics and the aspects of their psychological makeup that will allow for provision of their best interests. Successful restraint operators must understand and have a working acquaintance with the tools of restraint. They must understand the use of voice, manual restraint, and chemical restraint. Special restraint devices and their application should be thoroughly understood. These are explained in the text, with a major emphasis on physical restraint methods. It has been my experience that an operator really understanding what can be done with physical restraint can build upon this information to carry out more successful chemical immobilization—if it is indicated.

The general principles of chemical restraint will be outlined and specific tables presented to give current usage of chemical restraint agents in various classes of animals. There is a marked swing toward the use of chemical restraint when working with wild animals. Pharmaceutical companies are carrying out research on newer and better restraint agents. This has led to the marketing of new products on a continuing basis. This ongoing activity may lead to the false assumption that applying physical restraint techniques is no longer necessary. Nothing could be further from the truth.

Just as the indiscriminate use of antibiotics may cloud test results and cause the inefficient clinician to make an inaccurate diagnosis, indiscriminate chemical restraint can likewise produce clinical aberrations and is often hazardous to the animal.

Chemical restraint is an extremely important adjunct to physical restraint practices, particularly in regard to wildlife. However, it is far from universally ideal and cannot replace special squeeze cages and other specially arranged facilities for wild animals, which allow them to be approached without imposing undue stress or hazard. Those who work extensively with wild animals know that no single chemical or group of chemical restraint agents fulfills all the safety and efficacy requirements to qualify for universal application.

The decision whether to use chemical or physical restraint is based on the skill of the handlers, facilities available, and the psychological and physical needs of the species to be restrained. No formula can be given. If in doubt, some one who has had experience should be consulted.

WHEN TO RESTRAIN

One does not always have a choice of times when restraint should be carried out. Emergencies must be dealt with immediately. In the majority of instances, however, planning can be done.

Environmental Considerations

Thermoregulation is a critical factor in many restraint procedures. Hyperthermia and, more rarely, hypothermia are common sequelae. Heat is always generated with muscle activity. During hotter months of the year, select a time of day when ambient temperatures are moderate. Special cooling mechanisms such as fans may be required. Place restrained animals in the shade to avoid radiant heat gain. Conversely, use the sun's heat if the weather is cool. Avoid handling when the humidity is 70–90%. Cooling is difficult under such circumstances.

Take advantage of light and dark. Diurnal animals may best be handled at night when they are less able to visually accommodate. Nocturnal species may be more easily handled under bright lights.

Behavioral Aspects

An animal's response to restraint varies with the stage of life. A tiger cub grasped by the loose skin at the back of the neck will curl up just as a domestic kitten does. Such a reaction is not forthcoming with adults.

A female in estrus or with offspring at her side reacts differently than at other times. Males near conspecific estrus females may be aggressive.

Male cervids (deer, elk, caribou) go into rut in the fall of the year. By this time the antlers are stripped of velvet and are no longer sensitive. Now the antlers are weapons. Although a handler may safely enter an enclosure of cervids during the spring or summer, it may be hazardous to do so during the rutting season.

Hierarchial Status

Most social animals establish a pecking order. A person trying to catch one animal in an enclosure may be attacked by other members of the group. Dominant male primates are especially prone to guard their band. I have seen similar responses in domestic swine and Malayan otters.

Animals removed from a hierarchial group for too long a time may not be accepted back into the group. At the very least they will have lost a favored position and must win a place in the order.

Infants removed from the dam and kept separated for more than a few hours may be rejected upon reuniting. Species vary greatly in this behavioral response. An infant Philippine macaque was accepted back by the mother after a 3-month separation. Some species may reject the infant if it has human scent on it. A further hazard of hours-long separation occurs if the dam has engorged mammary glands. The hungry infant may overeat and suffer from indigestion.

Health Status

Recently transported animals are poor restraint risks. Transporting in crates, trucks, and planes is a stressful event. The longer the journey, the more stress. The method of handling and type of accommodations used in transport are also important. Allow the animal time to acclimate to a new environment if possible before carrying out additional restraint.

Sick domestic animals are routinely handled for examination and treatment. It may be more difficult to evaluate the health status of wild animals. Standard techniques of measuring body temperature or evaluating heart and respiratory rate may yield meaningless results because of excitement. Even though a captive wild animal may exhibit some signs of a disease, it may be prudent not to handle it. The following incidents illustrate two such cases.

A nine-year-old child wrote a letter to the president of the United States following a visit to a small zoo. She told him the yak had long hair and long toenails and asked why the zoo didn't give it a haircut and trim its toenails. The letter was answered in an admirable way by a zoo director who explained that the long hair was normal and that it might be more dangerous to catch the yak than to let it be slightly uncomfortable with the long toenails.

In another situation a bison had dermatitis. A decision was made to catch it to check the lesion. The animal died of overexertion during the process.

Deciding when to intervene is difficult. Clinical experience may be the governing factor.

Territoriality

Domestic animals differ in response to handling depending on where they are. A veterinarian attempting to handle a dog in the owner's home will find a more defiant individual than if the same dog is placed in the strange environment of a hospital examining room. Cattle, horses, swine, and sheep likewise respond differently in their own corral or pen than if in a strange place. An animal can sometimes establish its territory rather quickly. A dog placed in a hospital cage may defend it as "home" within a few hours. After removal from the cage the dog may become more docile.

Many wild animals are highly territorial. In order to work on such animals they must be moved to a new enclosure.

HUMANE CONSIDERATIONS

It is incumbent upon a person who takes the responsibility of manipulating an animal's life to be concerned for its feelings, the infliction of pain, and the psychological upsets that may occur from such manipulation. One must, however, be able to be objective about such manipulations and realize that the manipulation is for the best interests of the animal. Some feel that to restrict an animal's activity in any way is immoral and inhumane. At the opposite extreme is the person who has a total disregard for the life of animals.

Pain is a natural phenomenon that assists an animal to remove itself from danger in response to noxious influences. *No animal is exempt from experiencing pain.* Pain is relative; individual persons and animals experience pain in varying degrees in response to the same stimulus. Pain can become so intense, however, that an animal may die from shock induced by pain. We should not minimize the effect of pain, nor should we overemphasize it. Some persons cannot cope with pain in themselves, their children, or their pets.

Working as a medical technologist while a student in veterinary school, I frequently saw mothers bring children into the laboratory for a blood count, telling them, "This isn't going to hurt." Nonsense, it does hurt. Why not face the fact and learn to cope with it? We all experience numerous painful stimuli every day. We live through it and so do animals.

Sensitive people do not like to inflict pain. Veterinarians and others who manipulate animals are morally and ethically obligated to minimize pain in the animals they handle. The animal under restraint is incapable of escaping from pain. The handler must perceive the feelings of the animal and take appropriate steps to alleviate pain.

Some of the tools used in restraint practices involve the infliction of mild pain to divert the animal's attention from other manipulative procedures. The equine twitch is an example. The chain is placed over the nose of the horse and twisted down, causing a certain degree of pain. If the horse is preoccupied with the mild pain of the nose, nonpainful manipulative procedures can be carried out elsewhere on the body.

Every restraint procedure should be preceded by an evaluation as to whether or not the procedure will result in the greatest good for that animal. Animals have rights, and they have feelings. People should not look upon animals as machines to be manipulated at will.

Albert Schweitzer was one of the foremost proponents of the concept of reverence for life[6]. Human beings may have supreme power over other forms of life on this earth, but unless they recognize a dependence upon other life forms and have an appreciation for their position in the scheme of things, they will fail to develop an attitude that will result in humane care for animals under their charge. Persons who seek to work in animal restraint would do well to read some of the literature of the humane movement so they might become more empathetic in their approach to procedures that involve the infliction of pain and understand the emotional trauma associated with restraint [1-7].

Plan each restraint episode in detail. Anticipate potential problems. Provide equipment and facilities commensurate with the procedure. Time is crucial—get the job done fast. Follow through with observation and care until the animal is back to normal. If you lack experience in handling a given species, ask someone who does have the experience.

Remember: (1) Safety to the handler; (2) Safety to the animal; (3) Will it do the job? (4) Get the animal back to normal.

REFERENCES

1. Caras, R. 1970. Death As a Way of Life. Boston: Little, Brown.
2. Carson, G. 1972. Men, Beasts and Gods. New York: Charles Scribner's Sons.
3. Dembeck, H. 1965. Animals and Men. Garden City, N.Y.: Natural History Press. (Trans. from German)
4. Hume, C. W. 1957. The Status of Animals in the Christian Religion. London: United Federation for Animal Welfare.
5. Scheffer, V. B. 1974. A Voice for Wildlife. New York: Charles Scribner's Sons.
6. Schweitzer, A. 1965. The Teaching of Reverence for Life. New York: Holt, Rinehart and Winston.
7. Thoreau, H. D. 1965. Walden. New York: Harper & Row.

2 TOOLS OF RESTRAINT

ALTHOUGH some instances of tool use have been described in nonhuman vertebrates, only man has developed a high degree of skill in the use of tools. Every vocation, profession, or activity in which man engages requires the use of tools. The animal restrainer must become acquainted with a wide variety of tools used to handle animals safely, humanely, and effectively.

Tools may make a job easier or more efficient. The degree of skill attained by the restrainer is directly proportional to the degree of proficiency achieved in the use of tools of the trade. Tools must be kept in good repair; the art and practice of their use must be kept toned up.

Restraint levels can vary from the level achieved by arousing the subordinate feelings of an animal by voice and/or force of personality to the level of complete physical or chemical immobilization (hypokinesia). The tools used in effecting a given degree of restraint vary greatly. Some tools may be desirable for dealing with one species and be contraindicated when working with another. Success in the art of restraint requires both experience and study to know when it is appropriate to use a specific type of restraint. Inappropriate use of certain techniques may be not only unwise but dangerous to animal or human being.

When skilled animal restrainers are asked to share their secrets of success in working with animals, they can seldom give a detailed description of techniques. They have learned and habituated various means of restraint and sometimes do not even recognize the use of a system of techniques and tools. Undoubtedly the use of the tools of restraint has become second nature or instinctive to them.

For ease of discussion, the tools have been placed into seven categories: (1) psychological restraint—understanding a certain biological characteristic enables more satisfactory manipulation of a given animal; (2) diminishing sense perceptions of animals; (3) confinement; (4) lending added strength to or extension of the arms; (5) physical barriers—used to protect us or allow closer scrutiny of animals; (6) physical force—used to subdue animals; and (7) chemical restraint—used to sedate, immobilize, or anesthetize animals.

PSYCHOLOGICAL RESTRAINT

The successful restrainer must know a given species' particular behavioral patterns. For instance, to handle swine with a snout rope one must know that it is the nature of the pig to pull back when the upper jaw is grasped. The same technique would be dangerously unsuitable for handling a carnivore, because a carnivore would attack instead of pulling back on the rope.

Each species exhibits its own behavioral pattern, its own degree of nervousness, and other unique traits. Knowledge of these patterns enables restrainers to counteract or incorporate them into restraint practices.

Voice is an important tool, frequently overlooked by animal handlers because of its simplicity. Emotional states are reflected in the voice.

Both domestic and wild animals readily perceive fear or lack of confidence. Some scientists believe that a frightened person may actually exude odorous substances, which can be smelled by animals. Others believe that persons betray fear to animals through voice or other behavior, and that animals will not perceive fear hidden by self-confident behavior and voice control.

Students sometimes struggle for many minutes to halter a horse which whirls away each time the head is approached. Another person walks confidently into the box stall, speaks to the horse in a firm tone, then walks up to the animal and places the halter. This is extremely frustrating to students who can see essentially no difference between their mode of approach to the horse and that of the skilled person. They failed because the horse perceived their uncertainty.

Perhaps an excellent teaching tool could be developed by making an audiovisual presentation revealing the differences between the sound of the students' voices and their attitudes in approaching the animal and the voice and attitude of the skilled restrainer.

Voice differences have been graphically demonstrated to me in a slightly different situation. I was anesthetizing an African puff adder while making a television teaching tape to demonstrate the restraint technique. Both video and audio recordings were made during the actual procedure. After the procedure was completed I listened to the playback. At the point in the procedure when I grasped the animal by hand at the back of the head, after pinning it, my voice jumped—almost half an octave higher in pitch. This was a graphic illustration of an altered emotional pattern being reflected in the voice. Obviously I was somewhat concerned as I grasped this poisonous snake. I did not recognize the change of voice at the time, but it was clearly heard on the playback.

Such subtle changes affect animal behavior in a given situation and signify confidence or lack of it. Perhaps the best advice that can be given is that a handler who lacks confidence in either self or procedure should remain silent.

Other mannerisms of the restrainer also reflect emotional state. Timidity when approaching the animal, the way the hands are held, quickness or slowness in using the hands, and general stance indicate confidence or lack of confidence to the animal.

Be sure animals are aware of your approach. As a boy I came alongside an old cultivating horse and threw a burlap sack over its back, intending to jump on and ride. The startled horse kicked out and flattened me. This was not so much a matter of restraint as failure to make contact with the animal. Contact may be by voice or through sight, but an unstartled animal is easier to manipulate. If the principle of surprise must be utilized to catch the animal, be prepared to cope with the results of fright.

Both domestic and wild animals can be trained to permit the carrying out of certain manipulative procedures. Approaching a 5-year-old stallion that has run free on pasture or range since birth is much more difficult than approaching a 5-year-old stallion accustomed to people and trained for riding.

Likewise, wild animals can be trained to perform various acts or allow certain procedures to be carried out. Usually they cannot be trained to allow any procedure inflicting even minimal degrees of pain. Sometimes even this inhibition can be overridden under certain circumstances, at least to the extent of injecting medications or sedative agents to properly trained animals. A killer whale can be trained to lay its flukes on the bank at the side of a pool and lie quietly while a blood sample is withdrawn from a vein.

With wild animals, it is important to recognize that the training may involve establishing dominance over the animal by the trainer. This is a complex behavioral phenomenon, and it is unlikely that a casual person who comes in to manipulate that animal can acquire such dominance in a short time. Thus it is usually necessary for the trainer to perform the manipulative procedure for the clinician or veterinarian who must carry out an examination or make injections.

Hypnosis has been practiced on human beings for many years. Even surgery has been performed on individuals under hypnosis. The same technique has been effectively applied to animals. Many species of animals can be hypnotized. For instance, a chicken blindfolded and placed on its back will lie quietly in that position for a long period. Crocodilians can be manipulated in the same way, relaxing and entering a hypnotic state if placed on the back and stroked on the belly for a moment or two. Animals that "play dead," such as the opossum, enter a state essentially hypnotic. There is reason to believe that a horse may likewise become semihypnotized when the twitch is placed on its nose.

Many tales indicate that various animals hypnotize their prey when capturing food. It is unlikely that these states are bonafide representations of true hypnosis. It may be that entering this torpid state is a phenomenon that permits prey species to become free of the final pain of death when captured by a predator. It is not uncommon for a yet unharmed animal, chased by a predator, to give up and seemingly accept death without struggle. A zebra chased and grabbed by a lion will usually give up without a fight although in many instances the zebra may be fully capable of striking and killing the lion. Instead of doing so, the zebra becomes semicomatose, accepting the inevitable.

This response is utilized by those capturing such wild animals as zebras, giraffe, and some antelope species. For a few minutes immediately following capture by roping, they appear to be in a hypnotic state and can be approached and placed in crates without their kicking or striking. Those same animals, released from the crate into a holding pen, cannot be approached without dire consequence to the unwise person who makes that attempt.

Self-confidence is perhaps the single most important attribute that can be developed by the restrainer. This confidence can be acquired by experience, though some individuals seem to possess such ability almost innately. Some handlers develop the ability to manipulate or handle only one species or group of animals. Others handle many species with ease.

I am acquainted with one individual who possesses a phenomenal ability to work with the large wild felids. He had not worked with these animals extensively prior to a few years ago when, as an adult, he began to acquire an interest in some of the cats. I have seen him enter an enclosure containing mixed species of large, adult, untrained wild cats, including tigers and lions. These cats would wait in line to place their forepaws upon his shoulders and lick his face. He has such a degree of rapport, I am told, that he has entered an enclosure containing a half-dozen adult male African lions to successfully quell a fight.

This man has absolute confidence in his ability to work with these cats. There is no evidence of fear-mastery or dominance over the cats. He has studied their behavior sufficiently to know how to respond to the animals and how to get along with them, although many of the cats were adult when he acquired them. He has also taken lions known to be vicious toward other persons, studied them for a time, and safely entered an enclosure with them, feeling perfectly at ease.

To some this may appear a foolhardy and hazardous undertaking. Certainly it would be foolhardy for a person lacking the great confidence and behavioral skills of this individual to enter such an enclosure. Nonetheless it vividly illustrates what can be accomplished by someone with confidence and skill.

The successful restrainer must acquire detailed knowledge of the anatomy and physiology of the species to be manipulated, including the distance the limbs can reach to kick or strike. It is important to know the degree of agility and speed of the species in question. Techniques such as the use of a half-hitch chest rope to cast bovine species make use of a physiological response. The importance of gaining as much knowledge as possible of the biology and physiology of any species to be restrained cannot be overstated.

The significance of the physiological and behavioral phenomena of social and flight distances must be understood. All animals, including human beings, live with certain social interactions. These interactions involve both intimate and casual relationships. Social distances are inherent in the evolutionary development of a species. Social distances are precise for a given species and cannot be encroached upon without adverse effects. The general relationship of social distance is illustrated in Figure 2.1.

Domestic animals are less intensely affected by the lack of sufficient social distance than are wild species. The process of domestication necessitated that animals allow closer social contact with other individuals of the same species and with many other species.

Wild animals are habituated to social interactions and respond to violations of social distance in a prescribed manner. The usual response of the animal is to fight or flee. For example, a gazelle approached by a cheetah, though obviously aware of the cheetah's presence, stands

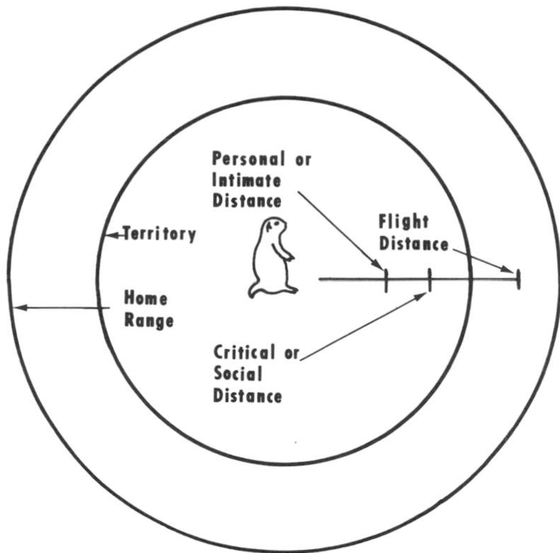

FIG. 2.1. Social interactions of animals.

quietly. But as soon as the cheetah approaches within a narrowly defined distance, the gazelle explodes into flight. This narrowly defined distance is "flight distance."

Flight distance varies among species according to agility, speed, and other behavioral traits and possibly the speed and agility of the enemy. A skilled keeper at a zoo gave me a vivid illustration of how an understanding of flight distance may be applied in restraint. He entered a cage to net some monkeys. As he approached the animals, he showed me that he could come within a certain distance without disturbing them. He then described and illustrated that if he moved one foot another few inches closer, the monkeys were startled and ran from him. This certain distance was the flight distance for that particular species in that situation.

He showed me that by using a long-handled net, he could reach out and catch one of the monkeys without startling it because the animal did not recognize the net as violating the flight distance. Thus by understanding this basic phenomenon, he was able to capture the animal without undue stress for the animal or himself.

Understanding flight and social requirements of various species is an important management tool for zoos and other institutions maintaining captive wild animals. Individuals of the antelope species in particular can be placed under continual severe stress if unable to maintain required social distances from other members of the group, from human beings, or from predator species. Sight barriers may meet these requirements as well as actual distance barriers. Flight distance can be modified by training. Furthermore, flight distance for animals raised in captivity differs from that of animals captured and brought into captivity as adults.

The response of an animal to a violation of flight distance is usually explosive; the animal either flees or attacks at full capacity. A wild animal may fling itself against a wall or into other barriers without regard for the consequences, once the flight response is initiated. An animal with no means of escape may attack without regard for its own safety.

Weapons Used against Man

All animals have both defensive and offensive mechanisms enabling them to cope with encounters with enemies. In most restraint situations, the restrainer is the enemy, and the response of the animal to the manipulative procedure involves one or another of the mechanisms used by that animal to cope with danger. Thus the restrainer, in addition to understanding behavior, should know the defense and offense mechanisms operating in that species in order to modify or counter the effects of such responses.

Defense mechanisms may involve a display or demonstration of one sort or another, which warns the responsive handler that the animal intends to protect itself. Anatomical structures used for defense and offense include claws, talons, feet and legs, teeth, bills or beaks of birds, special glands that exude scent, and the body itself.

Any animal with teeth and/or the ability to open the mouth widely enough to grasp some part of the restrainer's body is capable of biting. Not all who are capable will readily bite. All the carnivores, however, are prone to use the teeth, particularly the large canine teeth, to protect themselves and/or obtain food. The bite of many carnivores is serious and may be fatal. Birds, although not possessing teeth, are capable of biting or pecking. Some of the larger birds, such as the macaw, are able to crush bones with their heavy beaks, and large raptorial species (hawks, eagles) can severely tear tissue. Smaller birds can also inflict serious wounds. Birds with straight bills, such as fish-eating cranes and storks, may peck the eyes of the handler unless handled carefully.

Some animals, particularly invertebrate species, have special stingers with which to defend themselves, which may also be used in gathering food. Bees, wasps, some coelenterates, some marine cones, mollusks, fish, and other species have developed stinging structures that inflict pain and can cause illness or even fatalities in handlers.

Animals possessing horns or antlers can seriously injure by goring. Horns and antlers are used for display in combat with one another and in defense against enemies. Therefore it may be necessary to protect the horns from injury, as well as to prevent injury to animal and handler.

Some animals without sharp horns or antlers are capable of using the head as a battering ram to severely bruise or crush a handler against the wall. The giraffe is particularly prone to butt in defense or offense. Wild sheep also use their heavy horns and heads as battering rams, as do domestic goats and large domestic sheep. In fact, all horned animals, even if dehorned or deantlered, are capable of crushing a person. Serious injury can result from failure to understand this characteristic.

Large animals such as the hippopotamus, elephant, and rhinoceros can cause serious damage by crushing a handler against walls or posts. Large constrictor snakes do not crush the bones of victims, but kill by suffocation. By throwing their coils around the body, snakes can cause serious injury or death. Even a small constrictor is dangerous if the coils become wrapped around the neck of a handler.

The teeth of most herbivores are not adapted to biting, nor is this a usual fighting technique for these species. Nonetheless these animals may become prone to bite when placed in a captive situation. Deer may reach out and grab a handler if frustrated or frightened. The hippopotamus, which is a grazer, bites in both offense and defense. The teeth of the hippo are formidable, and many persons have succumbed to a hippopotamus bite. Wallaroos often bite.

A few herbivorous species have large canines used for fighting. The camel is a formidable adversary and is often a vicious biter. Likewise the small muntjac deer has enlarged canine teeth, used primarily in intraspecific fighting but which can be used in defense against handlers.

Hoofed animals are capable of kicking, the only defense mechanism of some species. The response may be reflexive and is often elicited simply by touching the animal anywhere on the body. A knowledge of the length of the leg and the direction of the kick is important in such cases. The horse usually kicks straight backward. However, a few individuals kick forward and outward in a manner similar to the kick of the domestic bovine, referred to as "cow-kicking" by horsemen. As indicated, the cow does kick forward and out, so the most dangerous position for a handler may be just in front of the hind leg.

Novices may believe they can jump away when an animal initiates a kick, but experience will teach that this is not possible. The strength of some animals is phenomenal, and one must keep this in mind at all times. A camel can kick a 10 cm × 10 cm (4 in. × 4 in.) support for a building and break it in two.

Front limbs primarily strike or paw in defense. Many species, including South American camelids (llamas, alpacas), camels, giraffes, and equines, are prone to strike or paw.

Some species, such as the shark, have very rough skin surfaces used to rub against an enemy, inflicting serious abrasions. The handler who does not recognize that the surface is rough may be injured when manipulating sharks, certain lizards, and pangolins.

Poisonous snakes and lizards and some poisonous mammals are capable of envenomating enemies or prey with potent toxins. Handling such species requires the use of highly specialized techniques and should be restricted to those who are fully qualified to do so by experience and inclination.

Some animals utilize the technique of spraying the enemy with urine or other substances. The octopus emits an inky fluid to hide itself as it escapes. Some primates, and other species such as the chinchilla, may urinate on the person who is trying to capture it. Such urination may also occur as an anger phenomenon or be used to delineate territory. Defecation may fit into the same category.

Numerous species have scent glands which produce materials objectionable to people. The skunk is the most noteworthy in this regard, but many other species, including carnivores and reptiles, have such glands. The musk gland is usually associated with the anus, and the material is often discharged under excitement. The scent glands sometimes serve purposes other than defense.

Spitting is a means of defense for some species. The expectorant may be composed of saliva, regurgitated stomach contents, or a specialized venom. Some cobra species are notorious for accurately projecting venom for distances up to 3 m (10 ft). Camelids and some apes are spitters. In one unique instance a shark in an aquarium surfaced and spit water on the author.

Regurgitation may occur in response to fright, but it is often a direct response to handling. Camels and llamas may deliberately spew foul-smelling material from the rumen on the handler. Cranes, storks, and pelicans may emit crop contents. Wolves and other carnivores may regurgitate as a stress response.

Although an elephant may use the tusks to gore or the trunk to grasp and fling an offender, it primarily tramples the enemy. Any large heavy mammal is capable of placing someone under its feet and trampling him. Fatalities from elephants, camels, rhinoceros, hippopotamuses, and other large animals have occurred in zoos.

Carnivores and other species may defend by clawing. Claws, whether sharp or dull, can inflict serious injury. Perhaps the worst injury I have received while manipulating animals was caused when a giant anteater drove its two blunted claws into the bone of my wrist. Clawing may result in infected scratches, or severe slashes transecting muscles, skin, blood vessels, and nerves, possibly incapacitating the handler permanently. In addition, the claws can grasp and pull a person into close contact within reach of teeth and strong forelimbs to bite and/or squeeze.

In short, the whole spectrum of the animal kingdom possesses abilities for self-protection. The restrainer must acquire knowledge of these mechanisms and be able to counter them in the restraint procedure. There are safe places to stand next to domestic animals. There are proper distances to recognize in working with animals and many ways to counter offensive and defensive mechanisms. Some specific mechanisms possessed by various animal groups will be described in the appropriate sections.

DIMINISHING SENSE PERCEPTIONS

Reducing or eliminating an animal's visual communication with its environment is an important restraint technique. A parakeet experiences less stress when placed in a darkened room before it is grasped for examination and/or medication. Blindfolding the domestic horse may make it possible to introduce it into a new environment, such as a trailer or a new stall, without engendering fright. Obviously it is impossible to blindfold most wild animals until the animal is already in hand. However, one can frequently place animals in a darkened environment.

If a herd of flighty and nervous black buck antelope are placed in a darkened room, the keeper can usually enter and grasp one animal without causing the pandemonium that develops if such an attempt is made from a herd in a lighted enclosure. It is important to recognize that manipulation of animals in such a restricted environment is somewhat hazardous if the herd includes males with horns.

This technique is contraindicated for species that possess excellent nighttime vision; in a darkened enclosure nocturnal species may well have better eyesight than the handler. It is therefore obvious that a detailed knowledge of the behavior and biology of a species is necessary before attempting any manipulation.

It may be necessary to blindfold large, flightless birds such as the ostrich and emu before they can be approached. Special devices can be constructed to place a blindfold over the head of such an individual (see Chapter 24). Most wild animals cannot be blindfolded until after capture. However, subsequent to capture, much stress can be relieved if the animal is blindfolded. A blindfolded animal may lie quietly for a long period while nonpainful manipulations are carried out. Sedation and anesthesia are required for painful procedures. Sedated animals handled in sunlight should be blindfolded to prevent damage to the retina by direct rays of the sun on an eye that cannot accommodate properly.

Sound is important in restraint. The importance of tone and quality of the voice as a restraint technique has already been described. Conversely, excessive sounds of people talking, motors, noisy vehicles, and other strange noises may seriously upset a wild animal. Restraint is easier to achieve if sounds can be dampened and harsh tones of voice eliminated or diminished in proximity to the animal. Cotton plugs in the animal's ears may suffice; however, it is extremely important that they be removed before the animal is released.

The skilled handler of domestic animals can accomplish much by proper use of the hands on the animal. Soothing, by stroking in the proper direction in the proper areas of the body, can be very valuable. Placing a hand firmly on the neck or shoulder of a horse and stroking it elicits desirable responses, while a lighter touch in the flank area may induce kicking. Most trained lions and tigers will frequently rub up against an individual; if one recognizes this as a friendly gesture, much can be accomplished. However, the handler who perceives it as a threat or is frightened will be unable to take advantage of this behavior in restraint.

Most untrained wild animals respond negatively to the touch of a person and institute defense mechanisms in response. Once such an animal is in hand, stress on the animal will be diminished if touching is kept at a minimum.

Cooling diminishes an animal's ability to respond to stimuli, particularly with poikilothermic species. My first experience in handling a large snake involved treating a large python for tail rot. I experienced much trepidation until I arrived at the ranch and found that the animal had been placed in a walk-in refrigerator some two hours previously. The animal was torpid and easily manipulated.

Cooling has been used to sedate nonvenomous species of snakes and lizards for surgery; however, it is insufficient for immobilizing poisonous species and its use for other species may be questionable. The potential hazard of the development of respiratory infections following prolonged cooling must be recognized. Sedative techniques are available to replace cooling as a technique. However, hypothermia as a surgical technique is used in certain types of human surgery, and it can be satisfactory for animal restraint if careful monitoring and care of the animal during the procedure and proper rewarming afterward are instituted.

CONFINEMENT

Confinement is a tool of restraint, but the acceptable degree of confinement may vary considerably, depending on the species and the situation. To the free-living wild animal, being placed in a large fenced-in area represents confinement, resulting in a certain degree of stress on the adult wild animal. Confinement can be progressively intensified by smaller enclosures. In a zoo situation this may be in an alleyway; for a domestic animal it may be confinement in a stall or shed. Close confinement makes it easier to evaluate clinical signs. The closest and most stressing confinement is that requiring an animal to be placed into a special cage, such as a transfer cage (Fig. 2.2) in a zoo, a special night box or bedroom, a shipping crate, or one of the many different types of squeeze cages (Figs. 2.3, 2.4).

FIG. 2.2. Transfer cage, mounted on an overhead track, designed to fit between two rows of cages. Used to move animals from one cage to another.

FIG. 2.3. Home-constructed squeeze cage for small mammals.

Squeeze cages are an extremely valuable restraint tool for wild animals. It is important to recognize that no squeeze cage can be adapted for universal use. Animals vary in both anatomical conformation and physiological requirements; the design of the squeeze cage must accommodate these to be safe and useful for carrying out various procedures. Squeeze cages designed for use in particular wild species will be described under those groups. Commercial squeeze chutes are available for domestic sheep, cattle, and swine.

Confinement can likewise be carried out by the use of special bags. The cat bag is useful for handling domestic species and can be adapted for use with many different species of small wild mammals. The Clifton quarantine station of the U.S. Department of Agriculture has

FIG. 2.4. Squeeze cage for large felids.

developed a special bag for use in manipulating various species of ungulates processed through the quarantine station (Fig. 2.5). Similar bags can be constructed for handling almost any kind of animal. If such bags are not available, burlap or jute sacks can be used to protect many species of birds and animals, as well as the handler. A limb

FIG. 2.5. Special restraint bag for handling antelope. (Courtesy, USDA Animal Import Center, Clifton, N.J.)

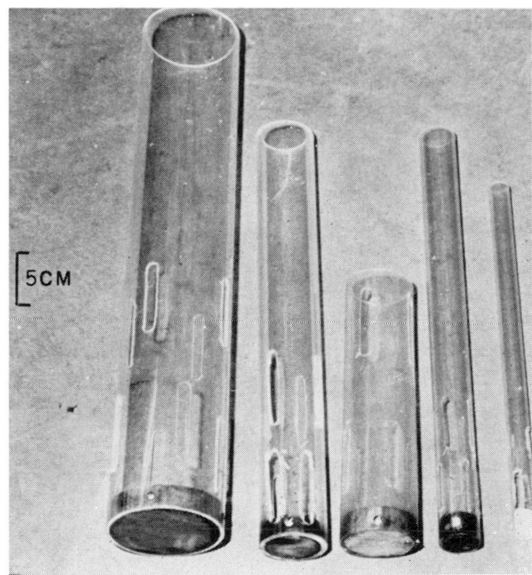

FIG. 2.6. Plastic tubes used for restraint.

or a wing can be extended from a hole in the side or from the partially opened top.

Towels or other flat cloths can be used to wrap animals for short manipulative procedures. This technique is frequently used with parrots, domestic cats, and the young of many carnivorous species. Birds are also placed in stockinettes or women's nylon hose.

Small mammals, birds, and reptiles can be restrained for radiographic studies, anesthesia, or other mild manipulative procedures by inserting them into plastic tubes (Fig. 2.6).

Complete restraint of an animal, usually under sedation, can be carried out with a restraint board (Fig. 2.7). This is routinely used with birds, snakes, and small primates. Such a board may consist of a Plexiglas sheet to which the animal is fastened with adhesive or masking tape, or the board can be equipped with velcro straps to hold the animal. The techniques are similar to those used in a hospital to restrain an infant needing specialized intravenous medication or restriction of movement.

Well-trained domestic cattle and horses can be placed into stocks for examination and manipulative procedures.

FIG. 2.7. Bird taped to a plastic restraint board.

This is the common method used to restrict movement, medicate, carry out dentistry, or examine a horse.

Confinement can also be accomplished with the use of ropes, cables, or wire panels. Gregarious species such as domestic sheep or cattle can be herded into a restricted chute area and the handler walk among the animals to urge them into the chute to examine, medicate, drench, or carry out other desired procedures.

EXTENSION OF ARMS

Ropes are an excellent means of extending the arm. Details of rope use are found in Chapter 3.

Snares are used to capture and restrain animals in a variety of situations. A snare is an important tool, but used carelessly, it can cause unnecessary pain or suffocate an animal. Commercial snares are usually designed with swivels for more humane and effective manipulation. An excellent snare is produced by the Ketch-all Company of San Diego (Fig. 2.8). It is a quick-release snare that permits the animal to twist without being suffocated. Homemade snares can be constructed from either metal or plastic pipe, using rope or cable. Snares used in obstetrics for extraction of the fetus can be adapted for use in animal handling. Special snares are made for handling swine.

FIG. 2.8. Commercial cable snare "Ketch-all" pole.

A special combination of a snare and a rope can be made by placing the rope loop at the end of a long bamboo pole, lending rigidity to the rope for placement of the loop. As soon as the neck is surrounded, the pole is loosed, and one has the animal by the rope. This technique is used in capturing free-living wild ungulates. The animal is pursued in a 4-wheel drive vehicle and captured, the pole is removed, and the animal is brought to a stop by snubbing it to some part of the vehicle.

Snares are hazardous in the hand of untrained persons. The animal can be suffocated by careless use. When grasped, some animals immediately begin to twist. This is natural behavior for carnivores like the wild felids and for crocodilians. If the handler does not compensate by countermanipulating the snare, the cable twists up, causing the animal to strangle.

It is desirable to include one front leg through the snare,

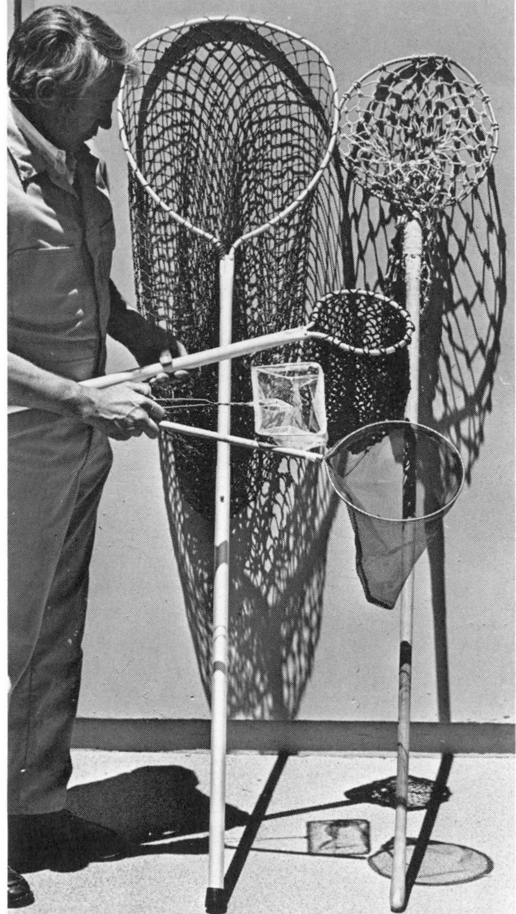

FIG. 2.9. Nets.

but it is not always possible to catch the animal in the best position. Catching the animal around the abdomen allows too much mobility of the head, making it possible for the animal to injure the handler. Animals having great dexterity of the forelimbs are not suitable candidates for use of the snare. They push the snare away, preventing application. This is true of some carnivores, such as the raccoon, and most primates. As soon as it is possible to grasp the animal, remove the snare and apply other methods of restraint to minimize the possibility of strangulation.

Nets are important tools for animal restraint. They come in all sizes and shapes (Figs. 2.9, 2.10), from those used to capture tiny insects to the very large cargo net used to restrain a musk-ox. Keep them readily available and in good repair. Obtain a variety of sizes to provide the right net for manipulating a wide range of species. By placing a net on the animal, many manipulative procedures such as injection with sedatives, medication, examination, or obtaining samples for laboratory work can be carried out.

Hoop nets with long handles can be directed at an animal. It is important for the handler to recognize that the hoop edge may injure the animal. It is better to allow the animal to enter the net rather than swing the net at the animal and possibly bang or crush it with the hoop. The net should be of sufficient depth to allow the hoop to be twisted, incarcerating the animal in the bottom of the net. Too shallow a net may permit the captured animal to

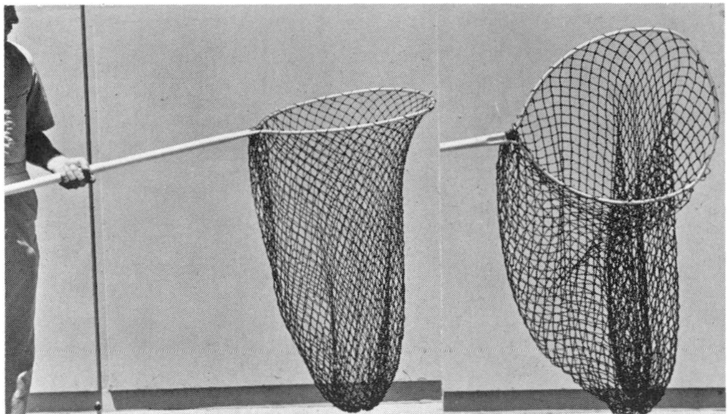

FIG. 2.10. A net must be deep enough to allow closure with the animal in the bottom of the net.

climb back out of the net. If a net is shallow, immediately placing the hoop against the ground will further restrict the animal and prevent escape. An animal in a net may retain too much mobility. Pressing the animal with a broom, shovel, or stick can further restrict movement.

Birds with talons or animals with claws are difficult to handle in large mesh nets. They may poke their limbs through the netting or cling to it, making it extremely difficult to extract them from the net. It is better to utilize a smooth net for some birds, perhaps one made from the plastic sacking now being used for livestock and other animal feeds. If the mesh of a net is too large, the animal may force its head through the mesh and strangle before it can be released. The size of the mesh should correspond to the size of the animal to be netted.

Rectangular nets can be placed in the path of various types of animals. As the animal runs toward it, the net is extended. Since a net is not usually recognized as a barrier, the animal will run into it. The net can then be dropped over it, and the animal will entangle itself. Handlers can then grasp the animal and proceed with further restraint. This technique has been successfully used on animals of various sizes, up to the size of the musk-ox. The size of the mesh and strength of the rope must be commensurate with the species being manipulated.

Nets can be used to lift animals from precarious spots, such as out of moats into which they have fallen, or transport them for short distances via helicopters or airplanes.

It is important to know the characteristics of the materials with which a net is constructed. Nylon, cotton, and manila are all used and each will withstand different degrees of stretch and wear. Carnivorous species are apt to chew at netting and may effect escape by chewing holes. The net should be inspected before each use for flaws that may allow the animal to escape at an inopportune time and place or to grasp and injure the handler.

Very fine nets called mist nets are used to capture small birds and bats. Cannon nets are sometimes used to capture animals. The animal is baited to a selected area and a folded net is shot over the top of it to entrap it.

Another technique is to suspend a net over an area such as a salt lick or feeding area. Animals are enticed beneath the net, which is then dropped to entangle them.

Special tongs have been developed for working with various species of animals, including swine and certain of the canids, such as the fox. A vise tong is used to grasp the animal at the neck, much as is done with a snare. The tong doesn't completely encircle the neck but clamps behind the back of the head. Tongs are usually used only to obtain initial hold of the animal; other means are then used to apply further restraint.

Nose tongs are widely used for handling domestic bovine species (see Fig. 9.7). One may grip the nose with the fingers, like improvised nose tongs (see Fig. 9.5).

Bulls, particularly dairy bulls, usually have a ring placed through the nose to allow for safer manipulation or handling (see Fig. 9.9). A special bull lead, which is a pipe with a hook on the end of it, can be used to grasp the ring to guide the animal, allowing the handler to stay away from the animal.

The snake hook, in its various forms, is utilized in handling all species of reptiles. Descriptions are found in Chapter 25.

Trained wild animals, particularly the large cats or bears, may be restrained to some extent with the use of chains. The chains either snap into heavy collars on the neck or encircle the neck and snap into a link. For short manipulative procedures, particularly if it is necessary to sedate the individual, the chain can be placed around the neck by the trainer and wrapped around a post or other solid object. The animal can be grasped by the tail and the injection quickly given.

PHYSICAL BARRIERS

Physical barriers can be used to protect handler and animal or to allow the handler closer proximity without alarming the animal.

Shields are important tools of restraint (Figs. 2.11, 2.12). They may consist of plywood sheets or be constructed with handles on the back to be held by the manipulator. Shields may allow close approach to the animal or can be used between two transfer cages having

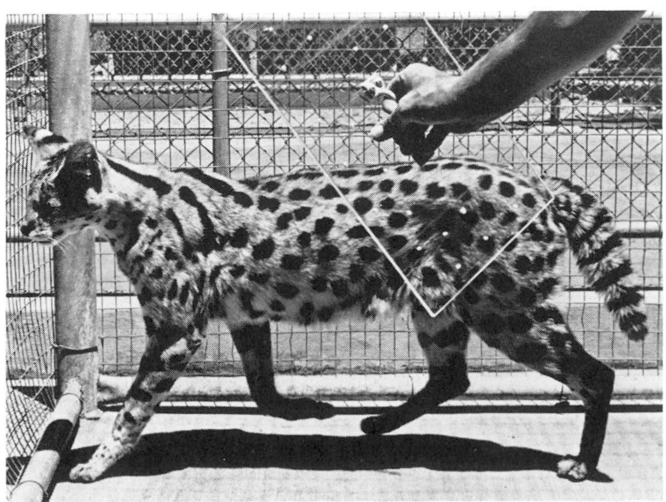

FIG. 2.11. Plastic shield used to ward off attack by a small mammal or bird while maintaining visual contact.

FIG. 2.12. Plywood shield used to move a wapiti.

swinging doors instead of guillotine doors. Plastic shields allow the handler vision without exposing the head. They are useful in handling large nonvenomous reptiles, some of the smaller mammals, and some birds.

A head screen offers protection from extremely agile animals that may become aggressive (see Fig. 20.7). The head screen must be made of small enough mesh that birds cannot peck through it or animals cannot reach through to scratch the face. These are especially useful in handling hornbills, cranes, storks, and certain species of primates.

A blanket can be used to shield the animal from the handler. It will also protect the handler from legs, horns, or antlers. A small antelope can frequently be captured by allowing it to jump into a blanket, completely enclosing it, and holding it in the blanket. Small mattresses can be similarly used. A mattress is more solid than a blanket and can be used to press the animal against a wall to inject it with a sedative or to carefully grasp it for additional restraint. A mattress is also valuable to cushion the body and protect the eyes of a restrained animal from trauma caused by thrashing on the ground.

Bales of hay or straw can be used as physical barriers for working on or around animals. Rectal examinations or perineal surgery of mares or stallions can be performed using these devices to prevent injury to the manipulator from kicking. Also, they are soft and do not traumatize the legs of an animal that kicks. Such barriers should be high enough that the animal cannot kick over the top of them.

Wire panels or solid gates can be used to squeeze animals against the walls of buildings or fences. It is important to prevent the animal from sticking a leg through the mesh or slats of such panels. Fractures of the limb are common from slatted panels.

Opaque plastic sheeting is excellent for use as a physical barrier to direct animals to proceed in a desired direction (see Chapter 23). Animals recognize an opaque plastic sheet as a barrier, while they may not recognize a wire or wooden fence as such. Thus animals can be directed into loading crates or into chutes with opaque plastic sheeting in a manner heretofore impossible. Plastic sheeting has its greatest application when moving hoofed animals.

Giraffelike species do not respond well to the barrier type of manipulative procedure. They look over the top or reach the head underneath to look, instead of moving away from the barrier.

PHYSICAL FORCE

The hands are used in most manipulative procedures; the wise restrainer takes every precaution to protect them. The hands can be used alone to grasp an animal. The restrainer must know where and how to grasp the animal to protect himself and to accomplish the restraint required.

The pressure required varies with the species. Handling a 50 g parakeet is indeed different than holding onto a 12 kg macaque. The amount of force applied must be appropriate to the species. Suffocation may result from the application of too great a pressure. Limbs or ribs may be fractured by applying too much force.

The greatest protection for the hands is detailed knowledge of the animal. Many people use gloves, which are an important tool of restraint. Gloves vary from thin cotton gloves used to handle small rodents to heavy, double-layered, coarse leather gloves used to handle large primates (Fig. 2.13). Leather welder's gloves are excellent for general use. The thicker and heavier the glove, the less the ability of the handler to determine how tightly he is grasping an animal or to feel the response of the animal. Because of this, many handlers refuse to use gloves.

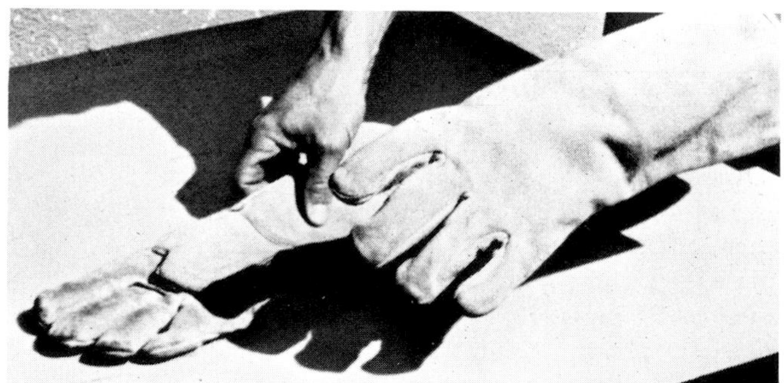

FIG. 2.13. Gauntleted leather welder's gloves.

Carnivorous species can likely bite through the thickest gloves available, so a glove is not an absolute guarantee of protection from biting. Preferably, leather gloves should be loose on the hand so that if an animal bites, the digit can slip sideways and be missed as the canine teeth penetrate the leather.

Gloves do not protect from crushing by powerful jaws. A tiger can crush the bones of the hand, or a large macaw can fracture finger bones without breaking the skin of a gloved hand.

Chain-mail gloves used either alone or within a leather glove offer more protection against the tearing effects of large canine teeth, particularly in regard to primates. Such gloves are useful as restraint aides, but they do decrease tactile discrimination and do not entirely eliminate the crushing effects of the bites of animals with strong jaws.

A whip is another tool used to exert physical force on both domestic and wild animals. In the hands of a good trainer who has established mastery over the animal, they can be effective. Unfortunately, the average individual uses a whip only as a means of punishment, and the whip usually becomes a vicious weapon, useless as a restraint tool. Whips are not recommended.

Electric shock prods have application in persuading reluctant domestic stock to enter chutes and passageways (Fig. 2.14). Be sure that chutes and passageways so used are sturdy enough to contain an excited animal. Electric shock prods are primarily used against wild animals only as a means of defense. An electric shock prod made it possible for keepers and handlers to enter the cage of a vicious male Philippine macaque; before its use, the animal would immediately attack anyone who entered the enclosure. The use of the stock prod to encourage a wild animal to move in one direction or another is usually unsuccessful and stimulates angry aggression.

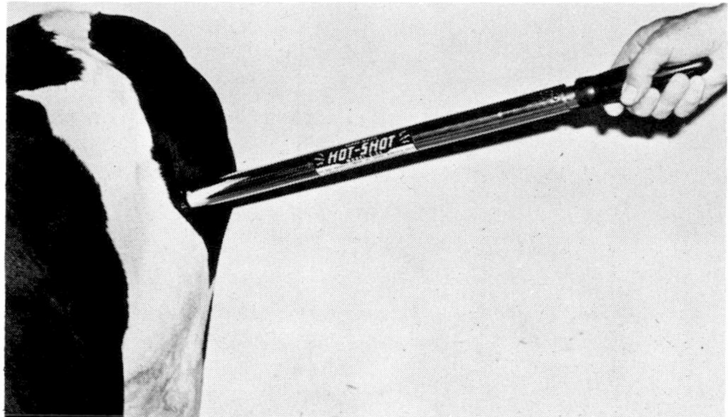

FIG. 2.14. Electroshocker.

Other means of frightening or encouraging an animal to move in a given direction include using a rolled up newspaper, a scoop shovel, or a house broom. Using a broom to persuade a horse to enter a trailer is a time-tested technique, much better than using a whip. All these devices inflict minimal pain, but the noise of the slap encourages the desired response. These techniques are useful in handling both domestic and wild animals.

Other tools for applying physical force include poles or bars to additionally restrict animals in cages or to press on a netted animal to hold the head down for a short time while injections are being made.

Carbon dioxide-charged fire extinguishers have been utilized to encourage or frighten animals to move out of a den or into another enclosure (Fig. 2.15). Apparently the sound and/or the fog resulting from discharge of the fire extinguishers frighten them into moving away from the source of annoyance. This technique is not infallible but has been successful. It may also be used as a defensive weapon for manipulating such animals should they escape.

CHEMICAL RESTRAINT

Chemical restraint has been the single most important contribution to the art of animal handling that has occurred in recent years. It enables one to manipulate some species of animals that heretofore simply could not be handled. Many different agents are used for chemical restraint. None of them are satisfactory in all cases. Each has its indications and limitations and each must be used with judicious understanding of what it can and cannot do. Perhaps the greatest evil of the chemical restraint era is for the novice to assume that all that is required to solve all restraint problems is a drug, a syringe, and some method to inject it into the animal. Such is emphatically not the case.

FIG. 2.15. Carbon dioxide fire extinguisher can be used to frighten an animal to move into another enclosure.

Chemical restraint is such an important tool for restraint that a special chapter is devoted to it.

SPECIAL TECHNIQUES

Slings are not necessarily restraint tools, but are adjuncts to the proper care and management of animals unable to maintain the upright position because of injury or illness. Slinging may also be necessary to extract an animal from a precarious position, such as from a moat or, in the case of free-living animals, a bog.

The rope sling described in Chapter 3 is adaptable to most species of quadrupeds. The size of the rope should vary with the size of the animal. Other slings can be purchased commercially, or they can be constructed if one understands the basic anatomy and physiology of the animal. Slings are commonly used for horses and cattle with injuries that necessitate resting one or more limbs. Birds can also be slung, as depicted in Figure 24.20. Bird slings usually must be improvised.

A speculum is used to hold the mouth open for oral examination, dental surgery, or gastric intubation. Specula may be elaborate commercial metal devices (Fig. 2.16) or they can be constructed from doweling (Fig. 2.17). A set of hardwood dowels 25 cm (10 in.) long of the following dimensions will accommodate birds, reptiles, and mammals up to 10 kg (22 lb): 25 mm (1 in.) with a 13 mm (0.5

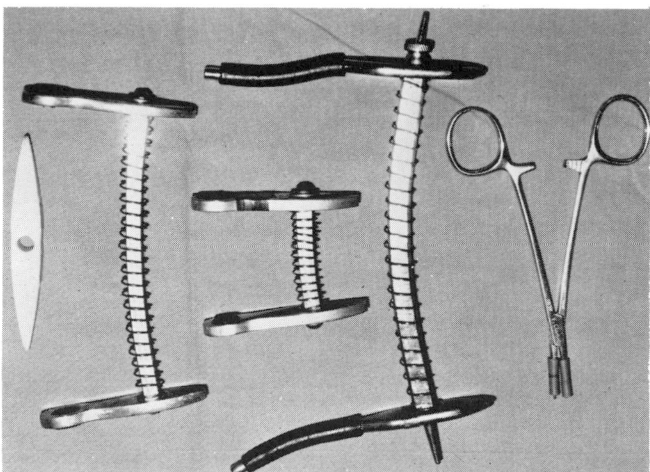

FIG. 2.16. Dental specula for small mammals.

in.) diameter hole; 19 mm (0.75 in.) with a 13 mm (0.5 in.) diameter hole; 13 mm (0.5 in.) with a 6 mm (0.25 in.) diameter hole; 6 mm (0.25 in.) with a 3 mm (0.12 in.) diameter hole. Old broom handles serve the need for larger specula. The beveled end of the dowel is gently inserted between the lips and teeth to open the mouth.

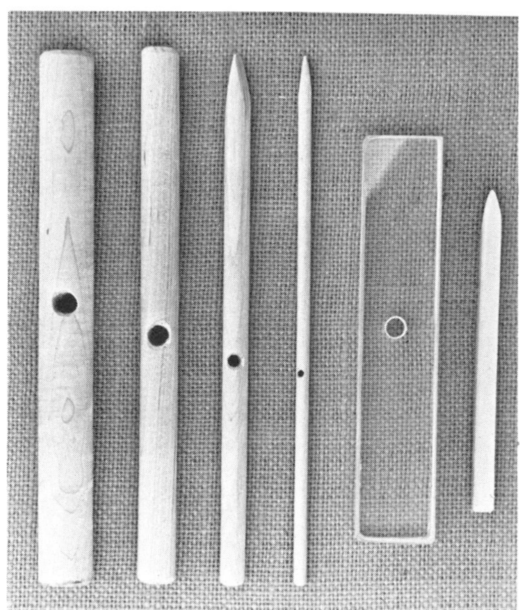

FIG. 2.17. Hardwood dowels and plastic used for specula.

Clear plastic can be used to allow a better view of oral structures, but plastic is harder and may be more traumatic to the teeth than wood. The size of the plastic speculum illustrated is 35 mm (1.5 in.) × 15 mm (0.65 in.) × 20 cm (8 in.) with a 13 mm (0.5 in.) hole.

To transfer an animal from a swinging-door cage to another cage, procure a shield to cover the swinging door. Slowly open the door of the cage containing the animal just wide enough to allow insertion of the solid shield behind it. When the shield is in place, open the door (Fig. 2.18). Place the new cage with the opening closely against the shield. Remove the shield, allowing the animal to enter the new cage (Fig. 2.19).

Obviously this technique is highly effective only for small to medium-sized primates, carnivores, rodents, and other small mammals not strong enough to push away the shield.

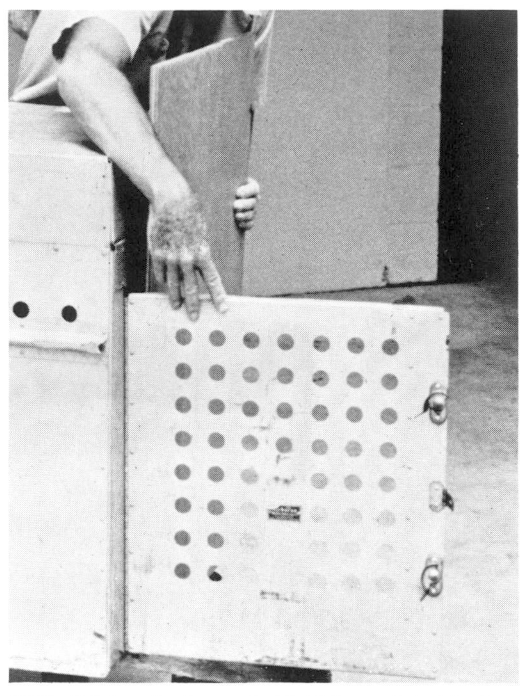

FIG. 2.18. A plywood shield is used to block the opening of a swinging-door cage until the cage can be pushed up against the door of another cage.

FIG. 2.19. Removal of the shield to allow passage of animal to a new cage.

3 ROPE WORK

ROPE was one of man's earliest tools. Prehistoric man first used vines but soon began to fashion rope from the fibers of bark, cotton, hides, hair, coconuts, and silk. Braided rope of animal hair was known in southwest Asia prior to 4000 B.C. Every culture has used rope in one form or another to hoist, haul, tie, secure, hunt, fish, build, explore, bridge, sail, and catch [7].

The Egyptians used rope made from papyrus and camel hair to build pyramids and to hunt hippopotamuses in the Nile. The Incas built rope bridges across the gorges of the Andes. Mayans used rope to haul stones for building temples. North American Indians fashioned ropes from bark and horsehair. Tribes near the ocean hunted whale with ropes of cedar. By the fourteenth century a rope-making guild was formed in England [7]. The western and southwestern areas of the United States were settled by a sturdy lot of cowboys proficient in the use of ropes.

Rope is a basic tool required for many manipulative procedures on wild and domestic animals. Even though drugs and other devices are often used in restraint practice, fundamental knots, hitches, and rope techniques have wide application.

CONSTRUCTION OF A ROPE [6,7,8]

The fundamental unit of all cordage is the *yarn* or thread. Plant or synthetic fibers are straightened by combing, then drawn into a small tube and twisted. Individual fibers are overlapped and interlocked during the twisting process to bind and give strength. The hardness of the twist is determined by the number of twists per foot, which varies from 8 to 22. Two or more yarns twisted together in opposite directions to that forming the yarn produces a *strand*. The final *rope* is made by twisting three or more strands together.

FIG. 3.1. Soft fiber ropes: **A.** Linen (flax). **B.** New Zealand flax. **C.** Braided cotton (sash cord). **D.** Twisted cotton (7/16 in.). **E.** Hemp.

The twist of the individual strand is opposite to the twist given the strands as they are laid together. The twists are directed so they turn in toward each other to securely bind and prevent untwisting in the final rope. Three or more ropes can be twisted together to form a rope *cable.* Instead of twisting, strands of soft fibers such as cotton can be machine-braided into a rope. Such ropes are easier to handle than ordinary rope and annoying twisting is less likely to occur.

Select the proper type and size of rope for the job (Tables 3.1-3.4; Figs. 3.1-3.3). Care for rope as for a precision instrument and it will provide long, useful service. Keep rope clean. Dragging it through dirt and feces or over rough gritty surfaces allows abrasive particles to work into the rope and damage fibers. If the rope becomes soiled, wash it in plain water and dry thoroughly before storing to prevent fungal decay. Avoid using soaps and detergents [8].

FIG. 3.2. Hard fiber ropes: **A.** Horsehair. **B.** Sisal (7/16 in.). **C.** Manila (7/16 in.). **D.** Tarred Manila (7/16 in.). **E.** Maguey.

Protect ropes from acids, alkalies, oils, paints, and other agents not chemically neutral. Rinse rope that has been soaked with urine.

Prevent kinks, which cause permanent damage and weakening of the rope. Do not tie knots in hard-twist ropes. Minimize sudden strains or jerking as they may break a rope otherwise strong enough to handle the load.

TABLE 3.1. Soft fibers used for ropes

Soft Fibers	Common Name of Plant	Scientific Name of Plant	Geographical Source of Fiber	Part of Plant Used for Rope	Advantages	Disadvantages
Cotton	Cotton	*Gossypium* spp.	Worldwide	Fibers attached to seed	Soft, flexible, least likely to cause rope burns, inexpensive, excellent for hobbles	Weak, subject to abrasion, water damage and fungal deterioration
Linen	Flax	*Linum usitatissimum*	Northern Europe	Leaves	Strong	Expensive
Hemp	Marijuana, hemp	*Cannabis sativa*	Middle East, China, Italy, South America	Stems, leaves	Very soft and pliable, used for twine, excellent for hobbles	Weak
Jute	Jute	*Corchorus* sp.	Malaysia, India	Stems	Used primarily for burlap and twine	Weak, makes poor rope

TABLE 3.2. Hard fibers used for ropes

Common Name of Fiber	Common Name of Plant	Scientific Name of Plant	Geographical Source of Plant	Part of Plant Used	Advantages	Disadvantages
Manila	Abaca	*Musa textilis*	Philippines, Central America	Sheathing of leaf stalks	One of strongest natural fibers known	Fibers are hydroscopic; unless treated, rope becomes unmanageable when wet
Sisal	Agave, sisal	*Agave sisalana*	Mexico, Central America, Africa, Hawaii, Indonesia	Fibers stripped from long, pulpy leaves	Inexpensive, used for twines and ropes where strength not important	80% as strong as manila
Maguey	Maguey sisal	*Agave cantala*	India, Java	Same as sisal	Used for special trick and fancy ropes	Similar to sisal
Mexican maguey henequen	Maguey	*Agave* spp.	Mexico	Same as sisal	Used for twine and cheap rope	Similar to sisal
New Zealand hemp	New Zealand flax	*Phormium tenas*	New Zealand	Leaves	Softer, more flexible than manila	Lacks strength
Coir or sennit	Coconut	*Cocos nucifera*	Tropical Pacific	Husk of fruit	Highly water resistant	Availability

TABLE 3.3. Other fibers used for ropes

Name of Fiber	Source of Fiber	Where Used	Advantage	Disadvantage
Nylon	Synthetic, from chemicals derived chiefly from petroleum and natural gas products	Worldwide	Strongest rope available, resists moisture and fungus, only safe rope for handling large bovids and equids	Highly elastic, difficult to take up slack in slings and casting ropes, inflammable, causes rope burns easily
Dacron	Synthetic polyester fiber	Worldwide	Similar to nylon, slightly weaker	Similar to nylon
Polypropylene	Synthetic fiber, produced by polymerization of propylene, a by-product of crude oil refining	Worldwide	Water resistant, will float, resistant to acids and alkalies	Will melt with heat, should not be used where friction is expected
Leather or rawhide	Cured leather or rawhide of many different animals	Mexico. Many cultures have used leather for rope throughout history	Rope is nicely balanced for throwing	Extreme variability in quality and strength, used only with dally system
Horsehair and other animal hair	Tail and mane hairs of horses, etc.	Southwestern United States, other areas of world	Can be fashioned into ornamental ropes, has little use in restraint	Harsh on the hands, causes rope burns easily

TABLE 3.4. Strength comparison of rope

Diameter	Diameter	Breaking Strength* (approximate)													
		Manila		Sisal		Cotton		Nylon		Dacron		Polypropylene		Wire cable†	
(in.)	*(mm)*	*(lb)*	*(kg)*	*(lb)*	*(kg)*	*(lb)*	*(kg)*	*(lb)*	*(kg)*	*(lb)*	*(kg)*	*(lb)*	*(kg)*	*(lb)*	*(kg)*
3/16	4.8	450	204	360	164	250	114	1,110	504	1,050	477	800	363	2,120	962
1/4	6.4	600	272	480	218	420	191	1,850	840	1,750	795	1,350	613	4,100	1,861
3/8	9.4	1,350	613	1,080	490	890	404	4,000	1,820	3,600	1,634	2,650	1,203		
1/2	12.7	2,650	1,200	2,120	960	1,450	658	7,100	3,220	6,100	2,770	4,200	1,907	17,820	8,090
5/8	15.7	4,400	2,000	3,520	1,600	2,150	976	10,500	4,770	9,000	4,086	5,700	2,588		
3/4	19.1	5,400	2,450	4,320	1,960	3,100	1,408	14,200	6,470	12,500	5,675	8,200	3,723	35,480	16,108
1	25.4	9,000	4,100	7,200	3,270	5,100	2,315	24,600	11,170	20,000	9,800	14,000	6,356	62,400	28,330
1 1/2	38.1	18,500	8,400	14,800	6,720			55,000	25,000	36,000	16,344	29,700	14,484		
2	50.8	31,000	14,100	24,800	11,260			91,000	41,315	61,500	27,921	53,000	24,062		

*A safe working load is double the breaking strength.
†Ordinary flexible steel cable (6 strands, 19 wires, 1 fibre core).

FIG. 3.3. Synthetic ropes: **A.** Nylon (3/8 in.). **B.** Polypropylene (5/16 in.). **C.** Terylene (3/8 in.).

A short glossary of terms is provided at the end of the chapter to acquaint the reader with the terminology used when handling rope.

BASIC ROPE WORK

Figure 3.4A illustrates specific terms used to refer to the parts of a rope. The standing part "w" is the segment not being used; the length "u" is a bight or bend that will be used in procedures or to construct knots; and "v" is the end of the rope used to thread through knots, etc. Figure 3.4B illustrates a loop or half hitch, a fundamental step in building some knots. Figure 3.4C shows an overhand knot.

To prevent unraveling, use one of the methods illustrated in Figures 3.5–3.8. Burn nylon rope at the tip, melting the strands together. This prevents unraveling but does not form a bulge at the end of the rope. The bulge may actually be desirable.

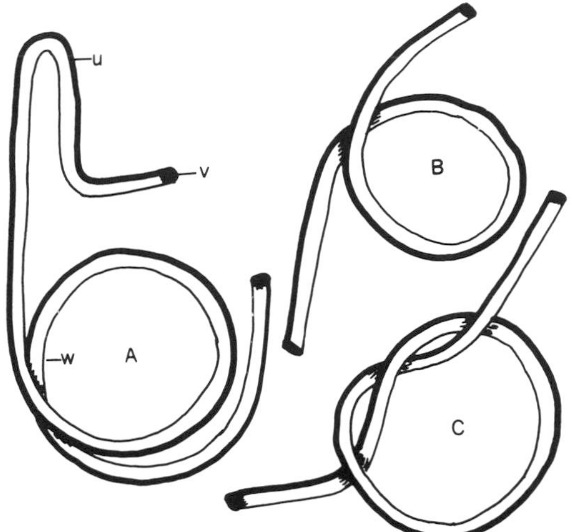

FIG. 3.4. **A.** Parts of a rope: (u) bend or bight; (v) end or running part; (w) standing part. **B.** Loop or half hitch. **C.** Overhand knot.

SPLICING ROPE [2,4,6]

Splicing may occasionally be necessary to join or repair ropes. Various types of splices are available (Fig. 3.9). In general, short splices add bulk but little rope is wasted in the splice. A long splice is less bulky but requires a longer segment of rope. A long splice may be necessary if the rope must pass through a block and tackle. The eye splice and back splice have special uses in the construction of rope halters.

A short splice is made by unraveling a sufficient length of the ends of both segments to be joined that at least three over and unders may be carried out. For a ½ in. rope, this should amount to approximately 6 in. The unraveled ends are placed together as illustrated in Figure 3.10A. The strands of one segment are anchored to the other rope with a piece of string or masking tape. Interlace one loose strand over the adjacent strand and under the following strand. Each strand, in turn, is handled in this manner until all of the strands are interlaced over and under three complete times. When one side is completed, the string or tape is removed and the same procedure followed on the opposite side. A completed splice is illustrated in Figure

FIG. 3.5. Methods of preventing unraveling: **A.** Overhand knot. **B.** Double crown knot. **C.** Whipping. **D.** Simple crown. **E.** Burned end of nylon rope. **F.** Back splice.

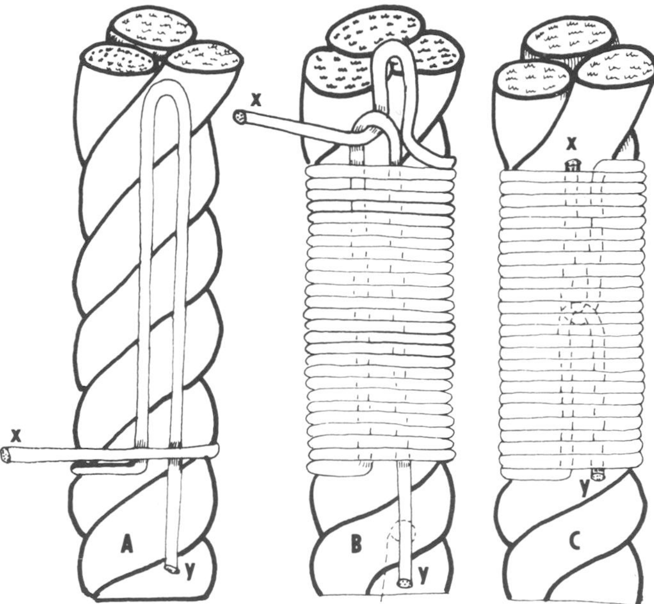

FIG. 3.6. Whipping a rope.

3.11. Remember that a splice is about 80% as strong as unspliced rope.

A long splice is begun in the same manner as a short splice except that longer strands must be unraveled. To splice a ½ in. rope, at least 1 ft should be used. Place the two ropes together, intermeshing the strands as illustrated in Figure 3.10A. The splice is continued by unwrapping one strand while intertwining a corresponding strand from the opposite rope in the place of that strand, as illustrated in Figure 3.10C. This unraveling and relaying of the strands is continued until the relaid strand is used up. Tie an overhand knot in these two strands as shown in Figure 3.10D. At the center of the splice, unravel a strand in the opposite direction, laying the opposite strand in the open track. These two strands are finished like the first two. The third set of strands, left in the center of the splice, are tied in place with an overhand knot. Cut the ends of the strands, leaving tiny projections to prevent knots from loosening. The completed splice is illustrated in Figure 3.9B.

A back splice is used as a stopper knot and is begun by unraveling the rope and interweaving the strands as illustrated in Figure 3.8A,B. The interweavings are tightened and the basic splicing method of carrying one strand over and under adjacent strands is followed. Again, this procedure is repeated in triplicate. The completed knot is illustrated in Figure 3.9C.

The eye splice is begun by unraveling the strands and laying the strands across the rope as illustrated in Figure

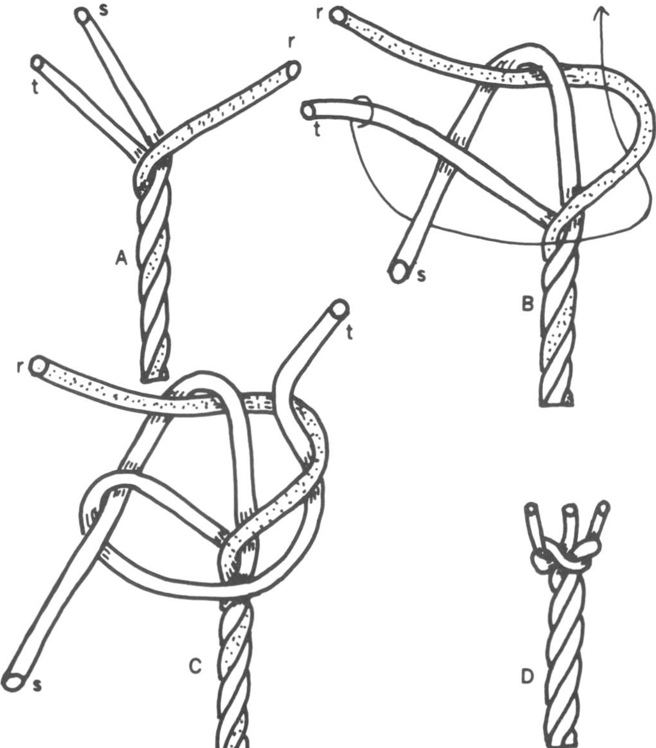

FIG. 3.7. Wall knot.

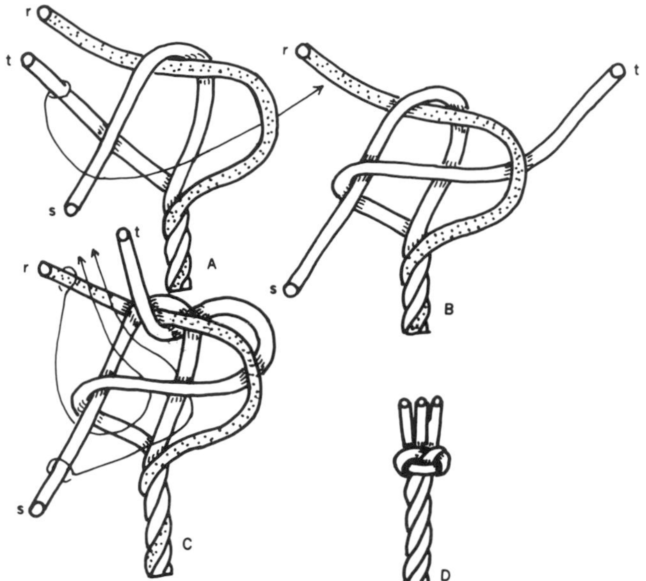

FIG. 3.8. Matthew Walker knot.

FIG. 3.9. Splices: **A.** Short splice. **B.** Long splice. **C.** Back splice. **D.** Eye splice.

FIG. 3.11. Making a short splice.

3.12. The knot is completed by inserting each strand in turn under and over the subsequent strand. When all three strands have been laced through a corresponding strand on the standing part, all strands are pulled tight. The basic splicing over-under procedure is repeated until the splice is completed (Fig. 3.9D).

HANKING A ROPE

A rope must be kept coiled or secured in some manner to prevent tangling. One method is to hank the rope (Fig. 3.13). First coil the rope. The size of the coil depends on the length and diameter of the rope. To complete the hanking, wrap one end around the coils as shown in Figure 3.13A. Secure the ends as illustrated in Figure 3.13B or C. In B, a loop is grasped at x, put through the coils, and brought back over the top. The standing part (the end) is then pulled tight. A variation is to bring the loop beneath a previous wrap.

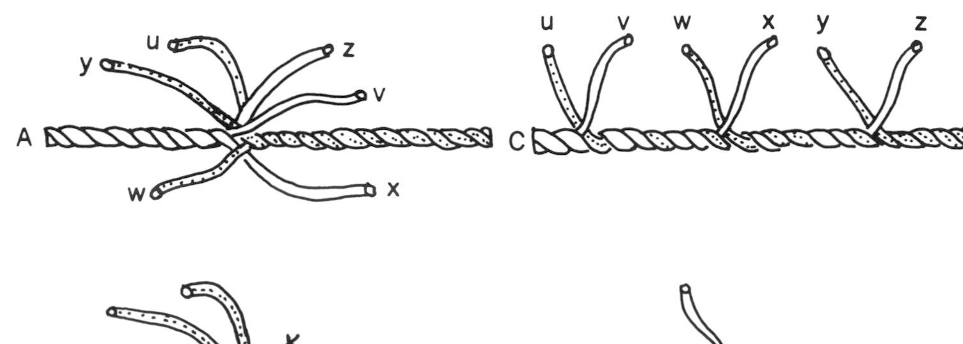

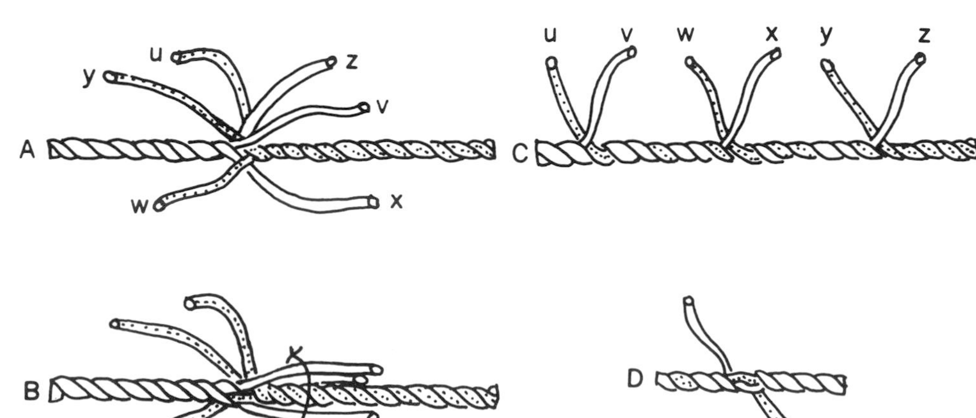

FIG. 3.10. Construction of short and long splices.

FIG. 3.12. Starting the eye splice. Continue by inserting each free strand alternately under and over subsequent strands in the standing segment.

Long, heavy ropes, block and tackles, and electrical cord can be stored in a crocheted loop (Fig. 3.14). Lariats and other hard-twist ropes should be coiled only.

KNOTS [2,3,4]

Variations of the square knot are used in many aspects of rope work. It is the basic knot of surgery and has wide application in restraint, especially for securing crates and cages. The basic knot is tied as illustrated in Figure 3.15A–C. In the completed knot, both strands of each loop are parallel and project through the opposite loop on the same side.

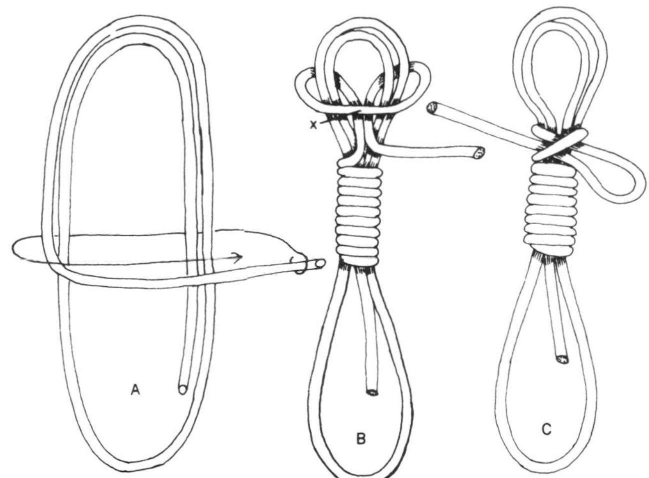

FIG. 3.13. Hanking a rope: **A.** Basic coil. **B.** Securing the end. **C.** Alternate method of securing end.

A single or double bowknot (used to tie shoelaces) is a variation of the square knot. Bowknots are often used in restraint because they can be quickly untied.

The square knot must not be used where linear tension is applied to the knot. When tension is improperly applied (Fig. 3.16), it becomes a slipknot and is dangerous to use on a loop around the neck or leg of an animal. If the knot becomes a slipknot, the animal may strangle, or if on the leg, gangrene may develop.

A sheet bend is used to join two ropes of unequal size. This knot is tied by forming a bight in the end of the larger rope and interlacing the smaller rope (Fig. 3.16C). A variation of this knot is used as the tail tie (see Fig. 3.33).

The bowline is the universal knot of animal restaint. It is the basis for many specialized knots and hitches. Temporary rope halters, casting ropes, slings, breeding hobbles, and sidelines all require the bowline. The advantage of the bowline is that it is secure, yet can be easily untied despite excessive tightening.

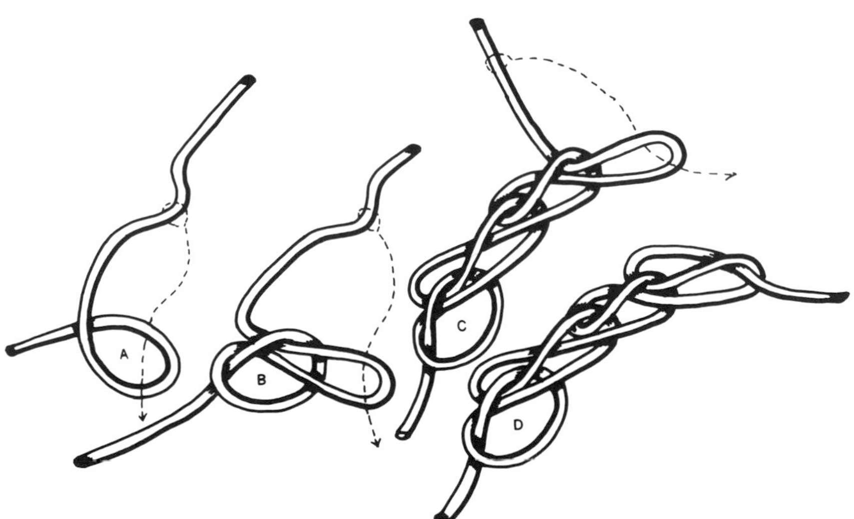

FIG. 3.14. Crocheting loop for storing large rope, block and tackle, and electrical cord.

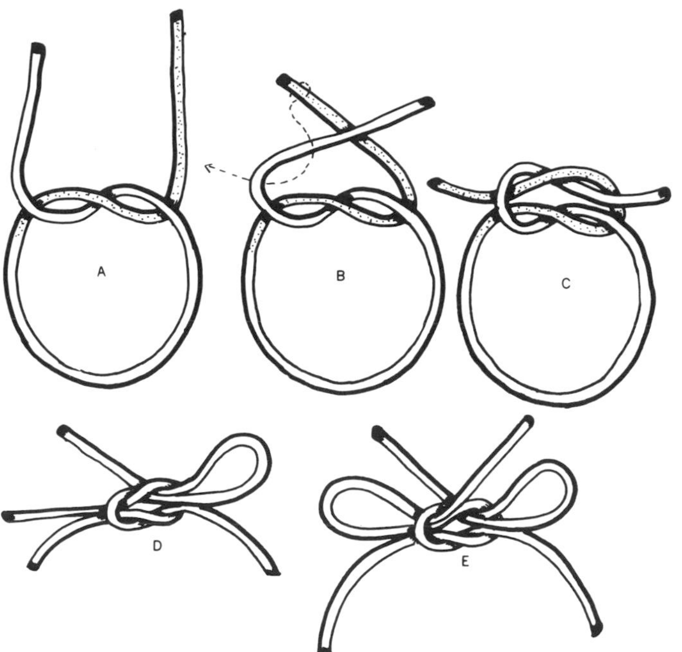

FIG. 3.15. **A, B, C.** Basic square knot. **D.** Single bowknot. **E.** Double bowknot.

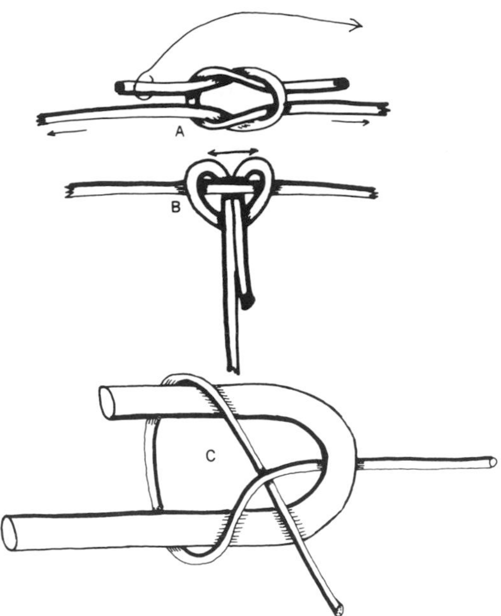

FIG. 3.16. **A, B.** Square knot converted to a slipknot by linear tension. **C.** Sheet bend.

There are ten or more variations of the bowline, each claimed by its adherents to be easier to tie or better for a particular purpose than the basic knot. However, shortcut methods are usually not adaptable to all situations. Time and effort spent learning the shortcut would be better spent in really understanding and becoming proficient in tying the basic knot.

The knot is begun by forming a loop in the standing part of the rope, leaving an end long enough to encircle the object being secured (Fig. 3.17A). The end is then inserted through the loop (Fig. 3.17B,C). The final knot should be tightened carefully. If a mistake is made in the direction of threading the end through the loop, the knot can still be tied by encircling the other segment of the standing rope. In this instance the direction of pull will be across the knot, but this is usually of little consequence (Fig. 3.17D,E). The same configuration is basic to the tail tie and the sheet bend.

The clove hitch (Fig. 3.18) is frequently used to begin other procedures. This knot can be tied around a leg or post (Fig. 3.19). This hitch is used around the hock to form a temporary breeding hobble. Encircle the object, bringing the end above the standing part to a half hitch (Fig. 3.19A). Continue around the object in the same direction

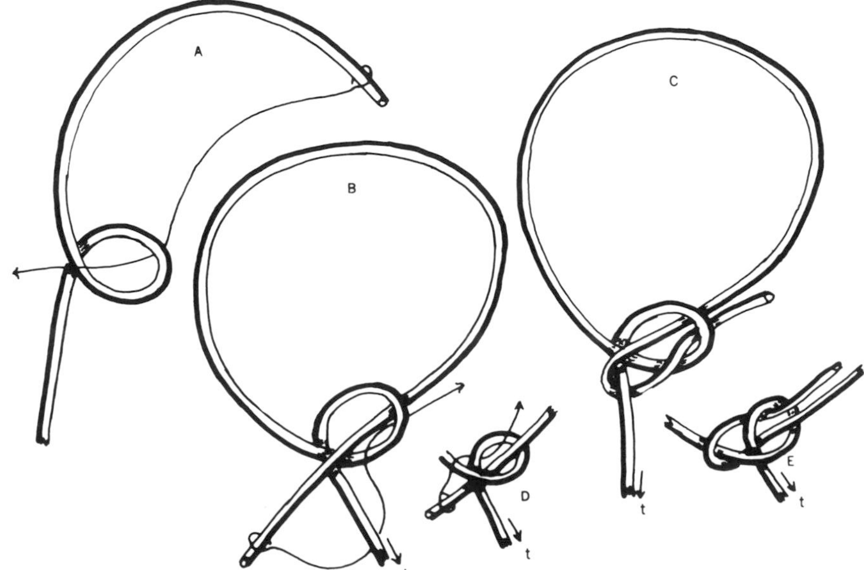

FIG. 3.17. Basic bowline.

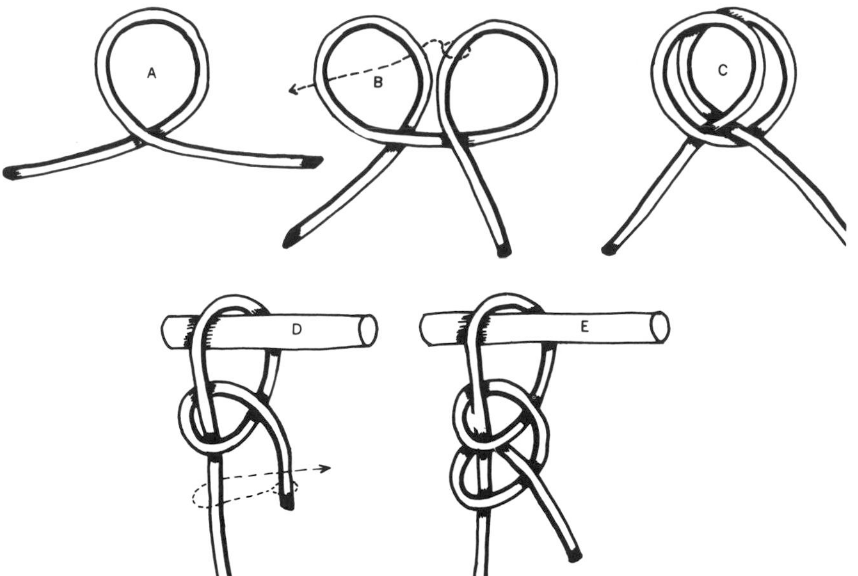

FIG. 3.18. **A, B, C.** Basic clove hitch. **D, E.** Clove hitch around same rope, also called a double half hitch.

and thread the end above the second loop (Fig. 3.19B). The clove hitch is not a secure knot. Tension applied on the standing end of the rope, either intermittently or continuously, may slacken the loops and free the animal.

The halter tie has numerous applications in animal restraint, only one of which is to secure an animal to a post, fence, or ring. Tying this knot should become second nature to anyone wishing to become proficient in animal handling. The knot is tied by wrapping the end of the rope around a post or ring, then forming a loop, and laying it over the standing part of the rope as in Figure 3.20A. The end of the rope is then grasped and brought beneath the formed loop and the standing part of the rope, continuing through the loop (Fig. 3.20B).

This is the basic knot. Carry the standing end of the rope through the loop as illustrated in Figure 3.20C to make it less likely that accidentally pulling on the standing end of the rope will release the halter. This knot must be tied close to the object to which the animal is anchored. A major advantage of using the halter tie is that the series of loops allow easy release of the knot. Even if an animal pulls back on the rope, a quick tug on the standing part, after the end has been removed from the loop, releases the knot. This knot is used for forming breeding hobbles, casting ropes, slings, and sidelines.

A honda (Fig. 3.21) forms a small loop through which the standing part of the rope may be passed to form a larger loop for securing or catching an animal. There are many ways of fashioning a honda. To tie the honda knot, a wall or overhand knot is tied in the end of the rope. Then an overhand knot is tied (Fig. 3.22A). The distance from x to y will be approximately two and a half times the length of the final loop. The knot is finished by gently pulling on the standing end of the rope until the loop is tightened and the knot is secured. If tied and secured properly, the standing part emerges from the middle of the loop, as depicted in Figure 3.22C.

A very simple honda can be fashioned by doubling the end of the rope and tying an overhand knot in the doubled rope (Fig. 3.22D). A disadvantage of this easily formed knot is that the honda comes off at approximately a 45° angle from the rope, producing an imbalance on the rope end.

The weight of a metal honda causes some loss of balance from the rope and is a potential hazard if the honda strikes an eye or the body of an animal. Nevertheless, a quick-release honda is valuable when working with wild animals since the honda can be released by pulling the latch (Fig. 3.23).

THROWING OR TOSSING A ROPE [1,5]

Any type of rope can be thrown at an animal, but it takes little experience to recognize that hard-twist ropes such as manila or nylon are much more efficient for throw-

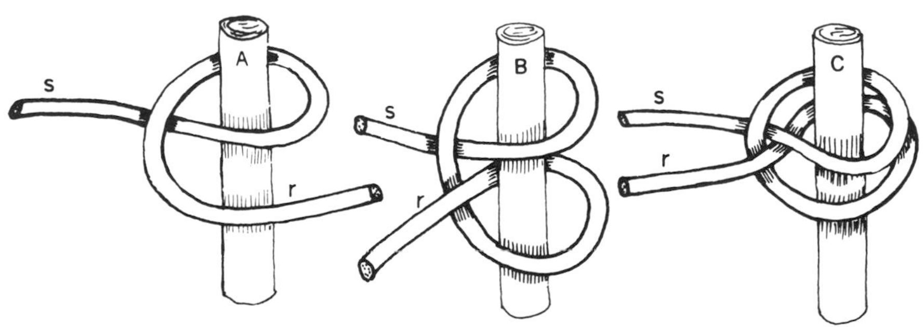

FIG. 3.19. Clove hitch tied around an object.

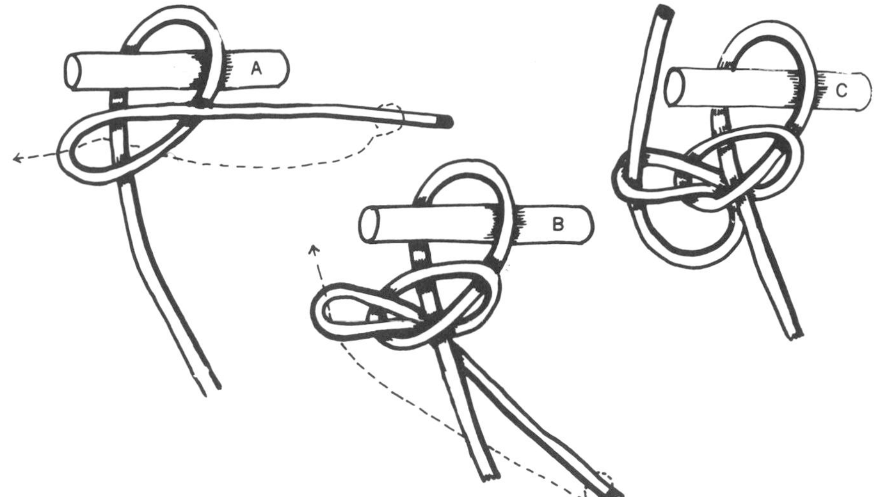

FIG. 3.20. Halter tie.

ing than softer ropes. Hard-twist ropes have better balance and the loop stays open better.

A novice may become frustrated by the amount of practice necessary to acquire proficiency in roping, but persistence yields dividends. Although it is possible to catch most domestic and wild animals without being able to toss a rope, most handlers will find that it is desirable to have mastered this art.

Roping an animal to establish a first contact is an art that was highly developed by old-time cowboys of the southwest and western United States [1]. Their feats are legendary. Roping styles varied from region to region. Vestiges of the glory of the past are now found primarily in the sport of rodeo.

Roping has its greatest application when handling cattle. Nonetheless, horses, sheep, and even swine can be caught under proper circumstances. Wild species are so much faster at dodging the loop than domestics, greater proficiency is required to capture them.

Roping is not without hazard to the animal and the operator. If the rope is used injudiciously and a tightened loop is left around the neck too long, the animal strangles. Bruises or lacerations of the skin or the cornea of the eye may occur when struck by metal hondas. Rope burns are also potential injuries. Animals frightened by roping may injure themselves by jumping against or over fences or walls.

I once roped a weanling foal in a wooded pasture. The loop had settled low on the neck, and as the slack was jerked, the head and neck were drawn to the side, pulling the foal off balance. At that moment the foal ran into a low branch of a tree and fractured its spine.

A properly coiled rope lies smooth and is flexible. As the coil is formed the rope may require twisting or untwisting to conform to the natural twist or lay of the rope. If a coil kinks and fails to lie open and smooth against other coils, it is an indication that the rope must be twisted one way or the other.

A right-handed roper can coil the rope in either hand. When coiling in the right hand, first build a loop (Fig. 3.24). Then grasp the standing part of the rope with the left thumb and forefinger and form a coil of the desired diameter. As the coil is brought to the right hand, remove twists or kinks by rolling the rope one way or the other between left thumb and forefinger until the coil is smooth. Continue the same motion until the entire rope is coiled.

FIG. 3.21. **A.** Honda knot. **B.** Quick release honda. **C.** Brass honda in an eye splice. **D.** Galvanized metal honda. **E.** Eye splice.

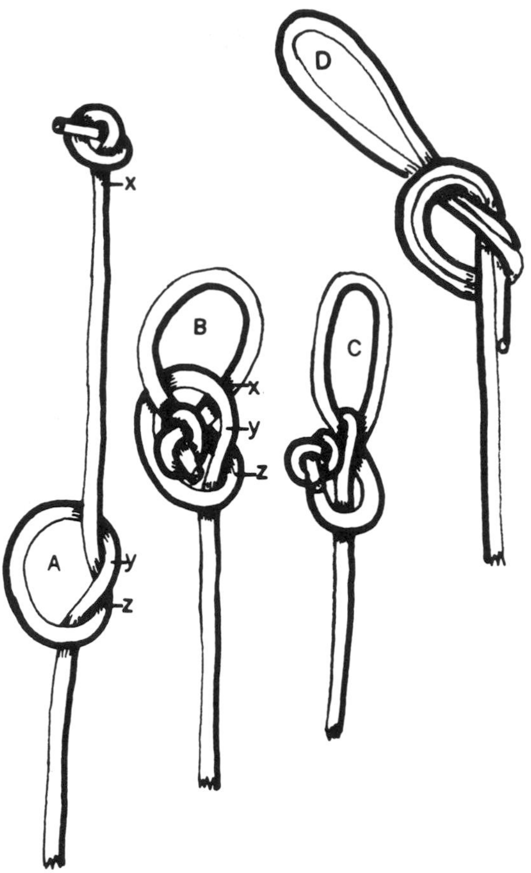

FIG. 3.22. Honda knots: **A, B, C.** Basic knot. **D.** Quick honda knot.

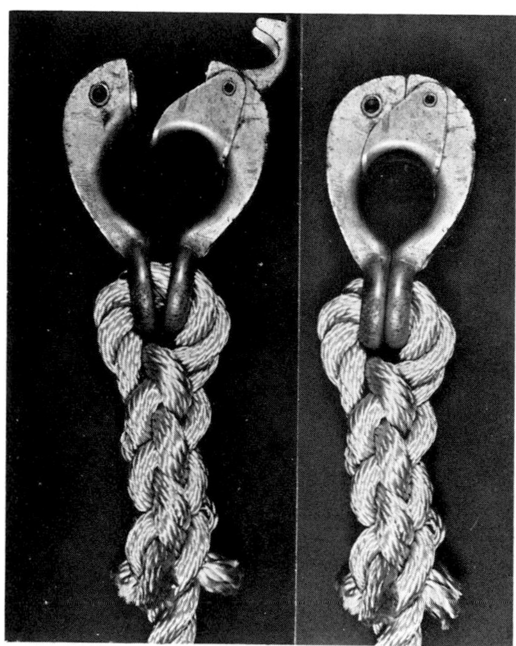

FIG. 3.23. Quick-release honda.

FIG. 3.24. Right-handed roper coiling a rope to his right hand. Appropriate twist is made with left thumb and forefinger to make the coil lie properly.

To coil into the left hand, grasp the end of the rope in the left hand and grasp the standing part with the right thumb and forefinger. Bring the coil over, untwisting as necessary, and place it in the left hand (Fig. 3.25). Notice the direction of the coiling. This is important. Continue coiling until the honda is reached.

As the rope is coiled a specific twist is built into each coil. If one simply grasps the honda and pulls out a loop without untwisting the rope, kinks will form in the loop that *cannot be shaken out* (Fig. 3.26). The loop may be

FIG. 3.25. Right-handed roper coiling a rope to his left hand.

FIG. 3.26. Twisted loop. Twist cannot be shaken out. Remove twist by rotating the honda end in the appropriate direction.

FIG. 3.27. Lariat held properly for throwing.

untwisted after forming or the kink may be prevented by feeding the honda backward around each coil until the loop is the desired size. Because of the built-in twist, a left-handed roper cannot use a rope coiled by a right-handed roper, or vice versa.

Two methods of throwing a rope are pertinent to animal restraint: the *drag toss* and the *swing toss.* The drag toss is less likely to frighten the animal than the swing, but less speed is generated for the throw and the animal may more easily dodge the loop. The techniques described and illustrated are for a right-handed roper. The loop is grasped so that the honda hangs approximately half way down the loop (Fig. 3.27) and is carried behind the body at the side (Fig. 3.28). The coil is held loosely in the opposite hand with the end of the rope held firmly between thumb and forefinger. At the appropriate moment, the loop is brought forward as the arm is thrown toward the head of the animal. The loop is opened by a quick forward thrust of the wrist (Fig. 3.29). As it settles around the animal the loop is tightened by drawing the unused coils back sharply. This is called "jerking the slack."

The swing toss is used when the rope must be thrown a greater distance or a fast-moving animal must be caught. Hold the rope as for the drag toss. The loop is kept open by

FIG. 3.28. Initial position for drag-toss method of roping. Wrist is bent back and to the right.

twisting with the proper wrist action. The hand must rotate in a circle at the wrist. To practice opening a loop, hold your arm in front of your body. Keep the arm steady and rotate your hand. For the right-handed person, this is performed by bending the hand to the right as the wrist is flexed back. Then the hand is bent toward the left while a quick snap of the wrist upward completes the rotation. Practice this motion with a small loop held waist high.

FIG. 3.29. Drag-toss throw. Arm is brought forward and directed at the object to be roped. Hand is brought forward and to the left to flip the loop open.

FIG. 3.31. Swing-throw release. Right hand follows through to the object being roped. Coils play off the fingers of the left hand.

Keep the arm stationary. You should learn to swing an open loop indefinitely, using wrist action only. The quick snap of the wrist upward is the key to maintaining an open loop. When the wrist action is mastered, arm movement can be added to develop more momentum. The wider the arc the more momentum acquired. The rapidity of the swing also affects momentum.

Timing of the throw and release is critical. Practice throwing at a stationary object. Swing the rope over the head (Fig. 3.30). As the loop is rotated forward, direct the arm toward the object. When the arm is at maximum stretch, open the hand and let the loop fly (Fig. 3.31). Throwing a rope is much like throwing a baseball. The rope will go in the dirction the arm is pointed. Follow-through with the arm is of prime importance.

The coils in the opposite hand are allowed to play out freely from a partially opened hand. As the loop drops over the object the slack is jerked. This should be done each time the rope is thrown, to establish a habit. Key factors for the swing and toss are:

FIG. 3.30. Initial position of swing-throw method of roping.

1. No twists in the loop
2. Wrist action to keep the loop open
3. Direct the arm at the target
4. Follow through
5. Jerk the slack immediately

A third method of roping used by professional animal capturers is by means of a loop hung from the end of a long bamboo pole. The loop is attached to the pole by a light string that will break easily. The animal is usually pursued in a vehicle. The roper stands in the back. When the vehicle comes alongside the animal, the roper places the loop over the head, the pole breaks away, and the vehicle slows down to stop the animal.

Catching the animal is only half the problem. The animal has to be stopped and subdued before it strangles itself or injures the roper. Obviously this should be planned for before roping begins.

A large animal must be dallied to a post or ring. The rope must be long enough to reach the anchor. To stop a smaller animal, the roper presses the rope across the body or legs while standing in a braced position (Fig. 3.32). Be prepared to be jerked off balance. A light pair of gloves will protect against rope burns.

Domestic animals are easily subdued from this position since they habitually pull back and stand until grabbed. Wild animals are unpredictable. They are as likely to jump forward or to the side as to pull back. If a dally can be taken and the animal snubbed up before contact is made, injuries are likely to be fewer.

If two ropers of sufficient skill are available, the animal can be stretched by simultaneously roping the neck and the hind legs and pulling in opposite directions. This method may be the only way to subdue larger bovids in order to trim feet or collect laboratory samples.

The tail tie is frequently used to prevent the tail from swishing and striking the operator, but it may also be used in certain instances to restrain an animal or to lift it. When the tail tie is used to lift an animal, be sure the particular species can be lifted by the tail. In equine species the tail

FIG. 3.32. Bracing for stopping a roped animal.

can be used to support the hindquarters. In bovine species the tail will break.

The knot is tied by bending the switch of the tail back on itself, forming a bight. The rope used to complete the knot is brought through the bight, circled behind both segments and brought forward across the bight, and looped underneath the previously placed strand (Fig. 3.33). The basic configuration of this knot is similar to that of the bowline. Thus the knot is secure and is also easily untied, no matter what degree of tension is placed on it. Its disadvantage is that if tension is not constantly maintained, the knot may loosen and fall off. The bight or bend in the tail can only be made when there is sufficient hair to form the loop. Do not bend and tie the coccygeal vertebra.

A trucker's hitch can be tied either at the end of a rope or at any point along a rope to remove slack. It can also be used to stretch the legs of a cast animal. Since the end is never threaded through a loop, this hitch is easily untied and tension on the hitch is adjustable. If the animal moves, slack can be taken up or released. Tension must be constantly applied to the hitch to keep it tied. This hitch is tied by passing the rope around a post or hook and bending the rope back on itself. Form another bend in the standing segment and loop the running end over it (Fig. 3.34A,B). Form a half hitch in the standing segment and place it over the bend. If there is to be excessive tension over a prolonged period, a double half hitch may be used. Pulling on the free end places tension on the whole hitch (Fig. 3.34C).

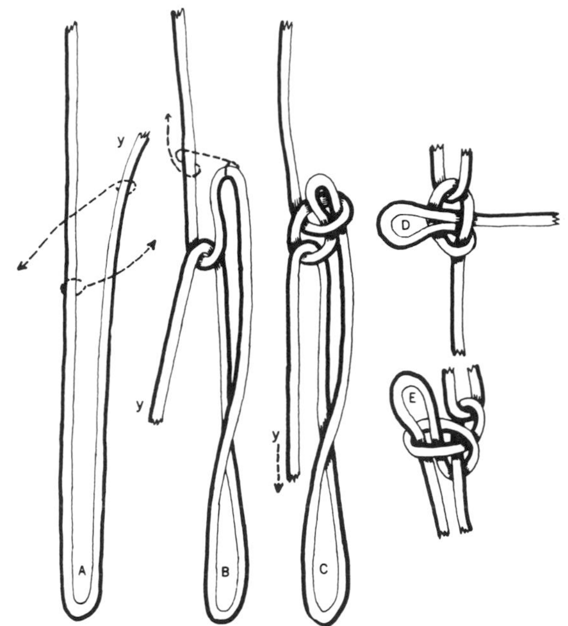

FIG. 3.34. Trucker's hitch.

When an animal must be stretched, a foot hitch is desired that will sufficiently secure the feet, yet be easily untied (Fig. 3.35). First a clove hitch or a honda knot is placed on one of the legs. Next both legs are encircled with the rope. Then a loop of the standing part of the rope is brought between the legs and placed over both feet. The standing part is then anchored to a post or ring. A trucker's hitch can be used to hold the tension satisfactorily. As the animal struggles, the knot is tightened, but when the animal relaxes, the knot will loosen. The release of the knot is accomplished by simply taking the loop back over the feet, pulling on the loop, and unwrapping. No secure knot is used in this procedure.

The anchor hitch is used to secure a part of a complex roping procedure, allowing the standing part to continue on or change direction. Suppose a rope is used to lash an animal to a board or table. If you wish to anchor the rope to take up slack or change direction, apply the anchor hitch (Fig. 3.36). To tie an anchor hitch, bring the standing part of the rope around either another segment of rope or around any handy object such as a post. The slack is

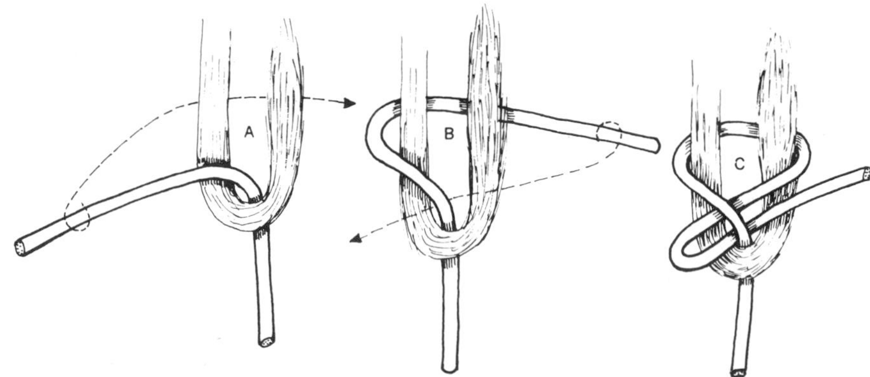

FIG. 3.33. Tail tie.

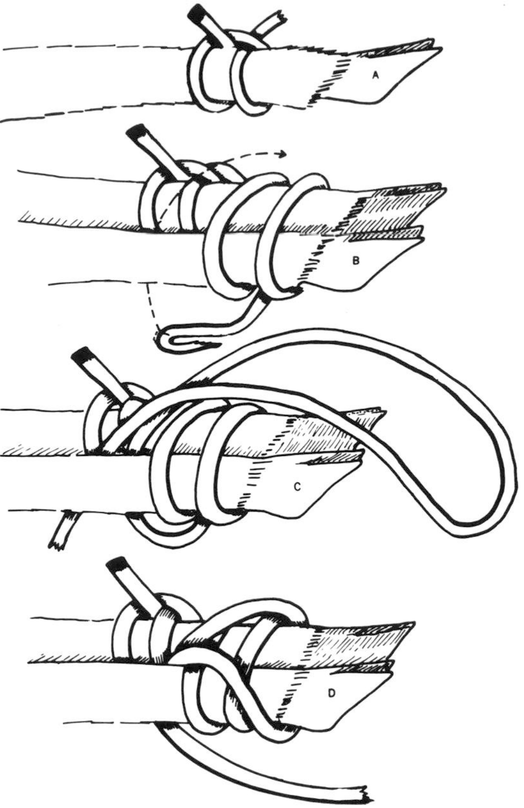

FIG. 3.35. Foot hitch: Start with clove hitch around one leg. **A.** Wrap both legs two times. **B.** Form loop in the standing part and anchor as in **C** and **D.**

taken up and an overhand knot tied, using a loop instead of the end of the rope (Fig. 3.36B). In order to continue using the standing part of the rope, a half hitch must be placed over the loop. Then the standing part can proceed in any direction and produce tension in a different manner. The loop can also be used as a pulley or a fulcrum on which to tie other knots. This knot is easily undone by releasing the half hitch and pulling on the standing part of the rope.

ROPE HALTER

Rope halters are used to lead animals, tie them up, secure the head to operating tables during surgery, or steady the head when manipulating under chemical restraint. Excellent rope halters are available commercially for most domestic animals. Temporary rope halters (Fig. 3.37) can be adapted to any species. The size of the rope should be varied to suit the strength of the animal.

No animal should be left tied with any of these halters without supervision, as the nose loop may slacken and fall from the nose, freeing the animal or, worse, strangling it.

A more permanent rope halter is easily constructed (Figs. 3.38, 3.39). The rope size and length of the nose piece can be varied to make a halter suitable for either a small domestic calf or a bull elk.

BLOCK AND TACKLE USAGE

A block and tackle (Fig. 3.40) provides a mechanical advantage for slinging animals, lifting crates, or casting an animal such as an elephant. Various types of block and tackle are represented in Figures 3.41 and 3.42. One sheave does not provide any mechanical advantage but changes the direction of the pull. Multiple sheaves provide a significant mechanical advantage. Multiple sheaved blocks can easily become tangled if not stored properly. The crocheting loop illustrated in Figures 3.14, 3.43, and 3.44 is excellent for this purpose. Extend the block and tackle to the maximum and proceed as illustrated.

Crates are usually moved manually or with a forklift. Large crates for rhinoceros or hippopotamuses must be moved with a crane. Be certain the cables used are of sufficient size to accommodate the load. The angle of attachment of the guy wire to the load is important in distributing the load (Fig. 3.45). Work with a large safety margin.

An excellent sling can be built with rope. The diameter and length of the rope used are dependent on the size of the animal. A 50 ft, ½ in. nylon rope is suitable to lift a 100 lb horse. The sling is illustrated as applied to a standing horse. In practice it can be put on a recumbent animal as well.

First form a neck loop in the doubled rope and a bowline

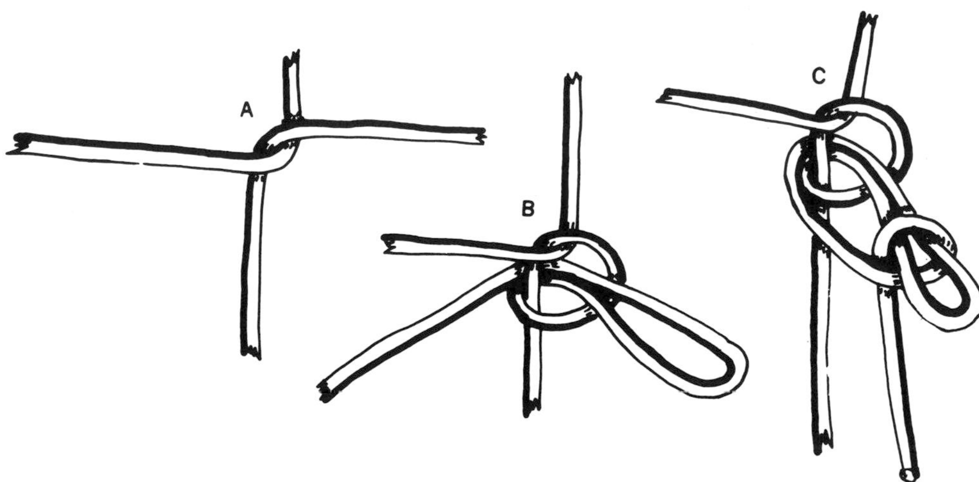

FIG. 3.36. Anchor hitch.

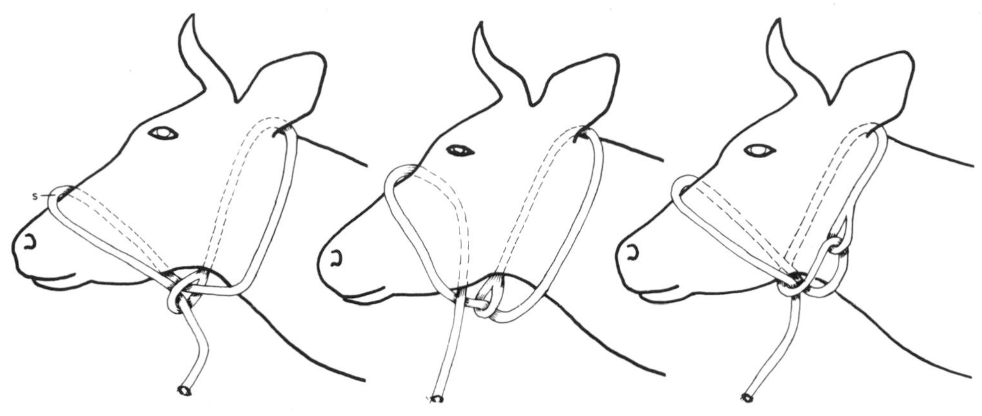

FIG. 3.37. Temporary rope halters.

FIG. 3.38. Rope halter: B goes over the nose, D behind the ears, E under the chin. Distance A-B-C must vary with the size of the animal. Loop at A is formed first, then an eye splice is made at C.

FIG. 3.39. Rope halter cheek loop: First establish location of this loop on the rope. Allow enough distance to construct the eye splice. Form the loop by running long segment of the rope under one strand of the short segment (**A**). Complete the loop by inserting short segment through long segment and pulling it tight (**B,C**).

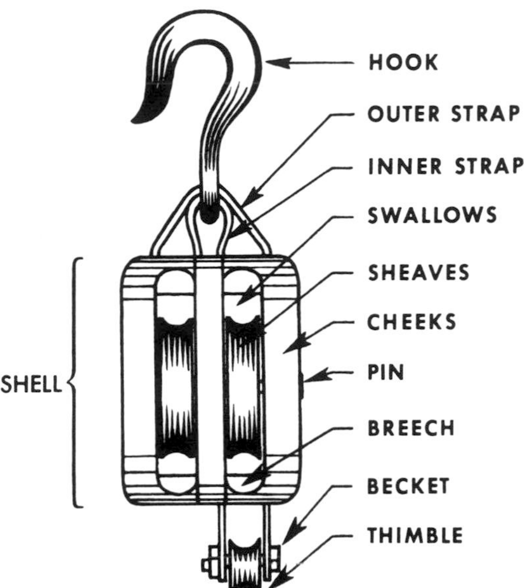

FIG. 3.40. Terminology used for a block. (Courtesy, Tubbs Cordage Company)

at the base of the chest. Bring the running segment between the front legs and back through the corresponding side of the neck loop (Fig. 3.46). Cross the ropes over the back, then bring the ropes between the hind legs. Be certain not to cross ropes underneath the animal. The ropes should pass on either side of the tail and underneath all the ropes on the back (Fig. 3.47). Take all slack out of the ropes and complete the sling by doubling the running ends back and tying with a halter tie (Fig. 3.48). An attempt should be made to place the lifting site over the approximate center of gravity of the animal. Even if this is not done the animal can still be lifted, since each leg is in its own sling loop. The animal cannot slip out of the sling.

Rope can be used to hobble an animal to prevent it from kicking, striking, or wandering away (Fig. 3.49). I have seen this technique used to prevent a camel cow from kicking at a newborn calf. Soft cotton rope is the most suitable for hobbles, but other fibers may be used.

A trained animal can be hobbled with a short rope. A longer rope is necessary with dangerous species. Make the tie with a bowknot so it can be released quickly if the animal fights too hard or falls down. Do not leave the ends long enough to step on lest the knot be released.

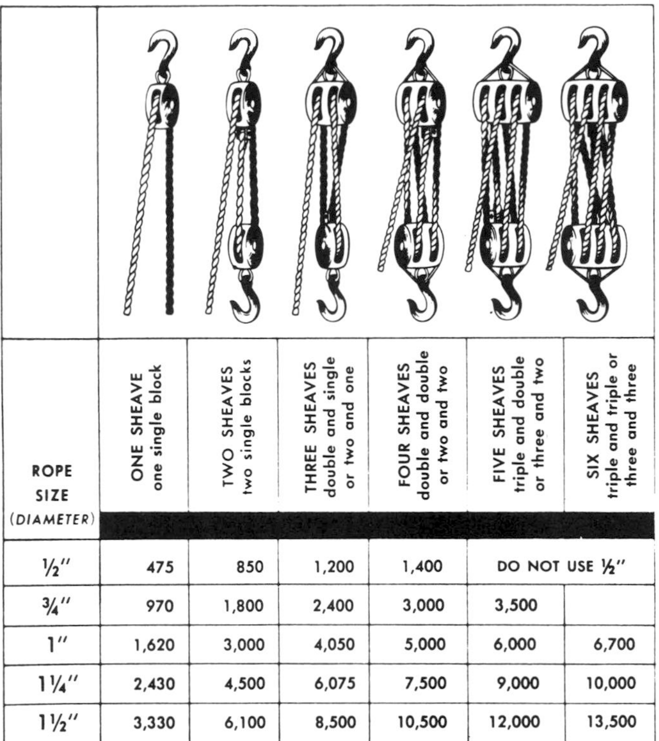

ROPE SIZE (DIAMETER)	ONE SHEAVE one single block	TWO SHEAVES two single blocks	THREE SHEAVES double and single or two and one	FOUR SHEAVES double and double or two and two	FIVE SHEAVES triple and double or three and two	SIX SHEAVES triple and triple or three and three
½"	475	850	1,200	1,400	DO NOT USE ½"	
¾"	970	1,800	2,400	3,000	3,500	
1"	1,620	3,000	4,050	5,000	6,000	6,700
1¼"	2,430	4,500	6,075	7,500	9,000	10,000
1½"	3,330	6,100	8,500	10,500	12,000	13,500

FIG. 3.41. Safe loads when used with a given rope size and type of block and tackle. (Courtesy, Tubbs Cordage Company)

FIG. 3.42. Snatch block. Used to change direction of pull in a rope without threading rope through the block.

FIG. 3.43. Storage of block and tackle: Beginning of crocheting loop *(left)*. Continuation of crocheting loop *(right)*.

FIG. 3.44. Block and tackle ready for storage.

ROPE AND CORDAGE TERMS

Bend: A form of knot to fasten two ropes together.

Bight: The part of a rope that is curved, looped, or bent—the working part of thc linc.

Binder twine: Single oiled or treated yarn, usually sisal, used for binding sheaves in harvesting; generally in 5 or 8 lb balls.

Block: The framework into which sheaves or pulleys are fitted, over which rope may be led to reduce power necessary for lifting. Blocks are designated as single, double, etc.

Cord: Two or more yarns twisted together much the same as a strand, used to tie bundles and whip ropes.

Dally: Placing tension on a rope by wrapping it around an object without securing the rope. The friction produced provides the anchoring necessary but also allows for some slackening if the tension is likely to be greater than the strength of the rope.

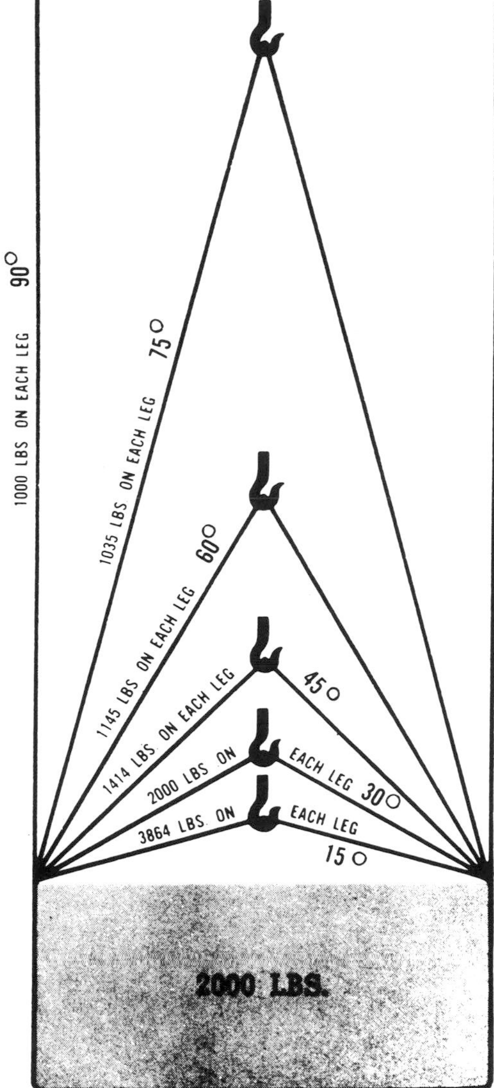

FIG. 3.45. Weights borne by supporting ropes when lifting crates. All weights indicate the load borne by each leg of sling at the various angles of lift. (Courtesy, Tubbs Cordage Company)

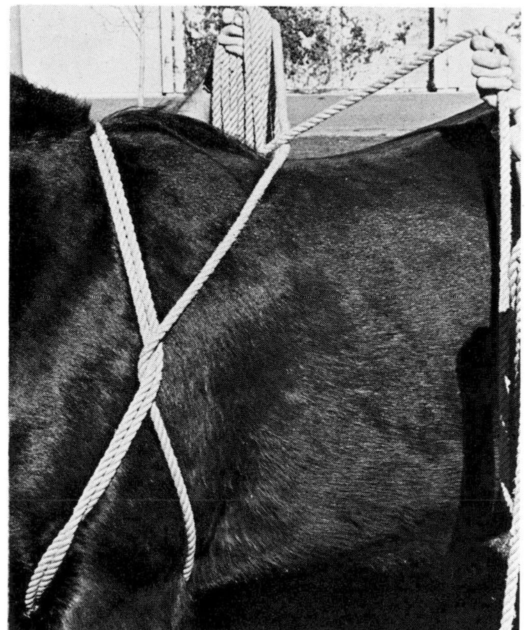

FIG. 3.46. Beginning a rope sling.

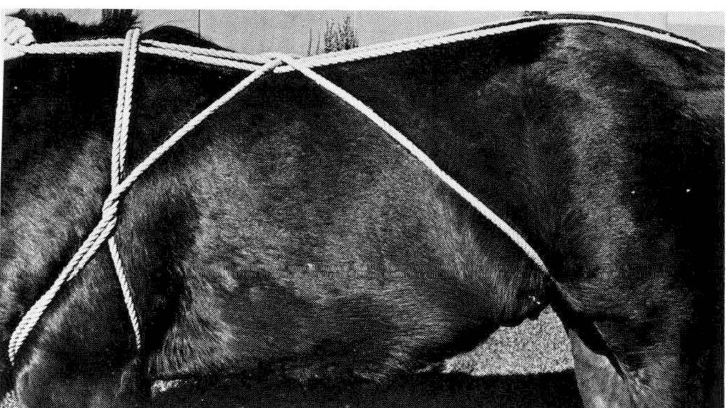

FIG. 3.47. Continuation of rope sling.

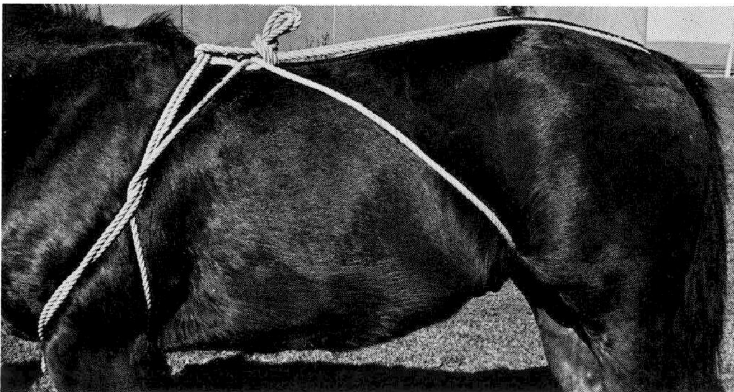

FIG. 3.48. Completed rope sling.

Heaving line: A small rope weighted at one end, that is thrown across the water to assist in moving a larger rope.

Hitch: Types of knots used for making a rope fast to an object, usually for a temporary purpose.

Lariat: In general usage, this is any rope used to form a loop and throw at an animal to catch it. Specifically, it is a rope constructed with an extremely tight or hard twist. The fiber is either manila or nylon. The hard twist provides more "body" to the rope and helps keep the loop open. The lariat is used only for catching and holding an animal. Knots are difficult to tie in a lariat because of the stiffness of the rope. In fact, tying knots in such a rope will put kinks in it and minimize its usefulness as a lariat.

Lay: A term used to designate the amount of turn or twist put in a rope, such as soft, medium, or hard lay.

Laying rope: The operation of twisting together three or more strands into rope.

Line: The term used by seamen for rope.

Snatch block: A special block with one side of the shell capable of being opened to allow a cable or rope to be placed in the block without having to thread it. Useful in

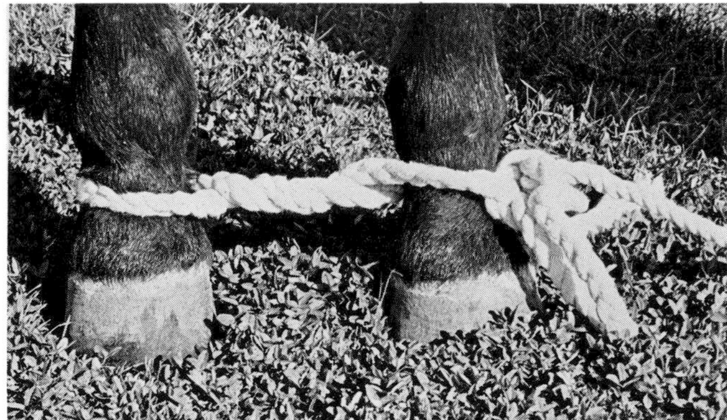

FIG. 3.49. Temporary rope hobble constructed of braided cotton rope.

restraint to change the direction of pull in inconvenient situations (Fig. 3.42).

Splicing: To unite ends of ropes by interweaving the strands.

Standing part: The principal part of the rope as compared with the working part.

Tackle: A combination of rope and blocks for the purpose of decreasing the power necessary in lifting or moving loads.

Whipping: Binding the strands at the ends of the rope to prevent unraveling.

Twine: A simple yarn, usually made from sisal and not tightly twisted.

REFERENCES

1. French, W. M. 1940. Ropes and roping. Cattleman 26(May):17–30. (excellent presentation of techniques of roping animals)
2. Gibson, C. E. 1953. Handbook of Knots and Splices. New York: Emerson Books.
3. Graumont, R., and Hensel, J. 1952. Encyclopedia of Knots and Fancy Rope Work, 4th ed. Cambridge, Md.: Cornell Maritime Press.
4. Leahy, J. R., and Barrow, P. 1953. Restraint of Animals, 2nd ed., pp. 6–37. Ithaca, N.Y.: Cornell Campus Store.
5. Mason, B. S. 1940. Roping. New York: A. S. Barnes. (trick and fancy)
6. McCalmont, J. R. 1943. Care and use of rope on the farm. USDA, Farmer's Bull. 1931.
7. Some Brief Facts of Life about Rope. Its Origins, Development, Uses and Characteristics. (N.D.) Auburn, N.Y.: Columbian Rope.
8. Tubbs Technical Reference on Rope. (N.D.) San Francisco: Tubbs Cordage.

4 CHEMICAL RESTRAINT

THE LAST decade has seen a phenomenal increase in the use of drugs for restraint and immobilization, particularly by those dealing with wild animals. Commonly used drugs now permit manipulative procedures that were heretofore impossible. The lives of many animals have been spared by the judicious use of drugs that minimize stress and trauma.

Although many of these agents have been widely used only during the past 10-15 years, such drugs are not new. Anesthetic agents have been used for many years. Primitive hunters treated the tips of arrows with poisonous plant and animal extracts. The art of posion arrow hunting was most successful in Africa and South America. Poisoned darts are still used by poachers in some areas of Africa.

The concentration of toxic agent in the poison mixtures was extremely variable, depending on the hunter's source and the method used to prepare the material. Analyses of some arrows found in Africa yielded as much as 5 g of ouabain (cardioactive glycoside) on one tip. The adult human lethal dose is 0.002 g.

The identity of plants and animals used for making poisonous extracts was a well-kept secret for decades. Even after the species were known, the techniques of modern chemistry were required to solve the riddles of the toxic principles.

South American hunters used plants containing curare (*Strychnos toxifera* and *Chondrodendron tomentosum*). In some areas hunters also had access to frogs (*Dendrobates* spp.) producing cardioactive glycosides in skin glands.

In Africa plants containing potent cardioactive glycosides were used (*Acocanthera* spp., *Strophanthus* spp.). Snake venoms and animal extracts were added to the concoctions in some instances. Small animals and birds injected with poisoned arrows died in as short a time as 15 minutes. Elephants sometimes took days to die.

Curare and its various derivatives were among the earliest chemical restraint agents used at the beginning of the present era [10]. Curare has many deficiencies as a restraint agent. It neither sedates nor anesthetizes the animal but merely prevents muscular activity. Therefore it is hazardous to use, since alarm responses, psychogenic stimulation, and other detrimental physiological neural functions continue unabated. Furthermore, the therapeutic index of curare is narrow, with respiratory arrest common. Unless rapid assistance in breathing is provided, the animal is likely to die.

Curare is rarely used for chemical immobilization today, yet the techniques for drug delivery and the experience gained through working with this drug served as the foundation for modern chemical restraint practices.

This chapter is not meant to be a definitive work or a technical reference on the use of chemical agents in restraint. Books on the development and current usage of restraint agents are presently available and should be consulted for details [8,10,12,16,18]. I am not aware of a current complete bibliography of the literature dealing with chemical restraint. However, I have in my own files over 600 citations on this subject. Articles on new drugs appear weekly. Only review or general papers and books will be used as references in this chapter. Definitions of some terms are given at the end of the chapter.

Much credit should be given to the pioneers in the development of chemical restraint agents and delivery systems [4,10,13,14]. Today the clinician and biologist have drugs and delivery systems to fit almost any need. Some fail to realize the endless toil, frustration, expense, and persistence that was necessary to reach the present state of the art.

Even more important, one need not experiment or take needless risks with valuable animals by trying one drug after another. Collectively, there are persons with immobilization experience in nearly every class of vertebrate. We cannot morally, ethically, or economically gain all this experience ourselves, particularly at the expense of the lives of animals. When faced with a unique immobilization problem, help should be sought from the literature or from veterinarians at large zoos.

THE IDEAL RESTRAINT DRUG

The search for a drug that will meet all the requirements of effective and safe chemical restraint continues. That ideal drug has not yet been found. However, certain drugs meet most of the qualifications for individual species.

An ideal drug should have a high therapeutic index (TI) (TI = lethal dose/effective dose). A high TI allows a margin for error in estimating body weight or for individual variations in physiological response. In order to increase the TI, many chemical restraint agents are combined with other agents. Combination often decreases the required dose of each while increasing the effectiveness of immobilization. Therefore the ideal restraint drug should be physically and chemically compatible with other useful drugs.

The majority of chemical restraint agents are administered intramuscularly and should not irritate the muscle. Some agents sting or cause transient localized pain upon injection but cause no damage. Most of the presently used agents meet this qualification. Even though a solution may not be irritating in itself, liquid injected with excessive pressure may tear muscle fibers. This can occur if charges higher than necessary are used to expel the fluid from the syringe. Muscle and skin can be bruised by the impact of the syringe.

A short induction period is desirable. Movies and television have fostered the belief that darted animals fall immediately. Chemical restraint agents presently available require 10-20 minutes following intramuscular injection before immobilization is effected. This time lapse allows some of the more fleet-footed animals to run for miles if unconfined. This is a serious drawback when working with free-living wild animals, since after the animal is injected it may escape into the bush and be lost to the restrainer. Furthermore, the unassisted animal may die from respiratory depression or be killed by a predator while under the influence of the drug.

The ideal drug should have an available antidote. The antidote reverses the effects of the drug, preventing death from respiratory arrest or other problems that may arise during the course of immobilization. Antidotes are available for etorphine, fentanyl (the narcotic component of Innovar-Vet) and *d*-tubocurarine.

Solution stability is important. Many restraint drugs are used in situations where refrigeration is not available. The ideal drug should remain stable in solution for long periods. Unrefrigerated succinylcholine chloride in solution has a very short shelf life.

Finally, it is important that the effective dose of the drug is low enough to allow its use in the small-volume syringes necessary for dart injection.

DELIVERY SYSTEMS

The first challenge facing the person using chemical agents for immobilization and restraint is to administer the drug agent in a site that allows absorption. The most satisfactory technique varies from species to species and from animal to animal, according to size, distance from the operator, ability to partially confine the animal, operator skill, and effectiveness of available equipment.

Oral

Twenty years ago I tried oral chemical restraint for the first time in order to treat a young 40 lb semitame chimpanzee with a diseased tooth. The effective dose of phenobarbital was carefully calculated and a tablet hidden in a piece of banana. The chimpanzee readily accepted the banana and began eating it greedily. I noted that the piece containing the tablet had been taken into the mouth and gloated over the fact that soon I would have the animal in hand. As I continued to watch, the animal began chewing more carefully. In a moment the tablet appeared on her lips. She reached up, picked the tablet out, and flipped it away.

Numerous subsequent experiences have taught me the futility of depending on oral medication to sedate wild animals. Aside from the lack of acceptance of such drugs by many species of wild animals, the effectiveness of oral medication is often minimal, since many chemical restraint agents are either unabsorbed or destroyed in the digestive tract.

However, sometimes oral administration is the only available choice. Phencyclidine hydrochloride is effective when administered orally in dosages two to three times the normal parenteral dose. Some primates can be provoked to come to the side of a cage and scream at a strange person, allowing medication to be squirted into the mouth. Though some of the drug may be lost, enough is usually ingested to effect immobilization or sedation.

A useful technique for the sedation of an obstreperous horse follows: Isolate the horse from all water for a period of 24 hours. Dissolve 30–60 g of chloral hydrate in 4 L of water and offer it to the animal as the sole source of water. Chloral hydrate tastes bitter and may be rejected for a time. If the animal refuses the water, remove it, then reoffer the solution 4 or 5 hours later. Within an hour of the time the water is consumed the animal will be sedated sufficiently to allow application of other means of restraint.

Drugs can be placed in the food of carnivores and other species to quiet or immobilize them. However, since other delivery systems now available are more satisfactory, oral administration is generally used only as a last resort.

Hand-held Syringe

Intramuscular injections can be given very quickly with a syringe held in the hand. Veterinarians accustomed to aspirating before injection must recognize that this is not possible when administering chemical restraint agents to wild animals.

A metal or plastic syringe will not break if knocked from the hand or if an animal kicks, strikes, or pushes it into the side of a cage. Use a large-gauge needle to deliver the liquid in the syringe quickly. Tighten the needle securely on the hub so the pressure buildup by rapid injection will not blow the needle from the syringe. A luer lock is desirable.

Hand-held syringes and quick intramuscular and subcutaneous injections are commonly used for vaccinating sheep, swine, and beef cattle in chutes. Similar techniques can be used to administer drugs quickly to animals held in stalls or stanchions before they become overly excited. While standing at one side, reach across and give the injection on the opposite side of the animal's body. Most animals kick out with the leg on the side pricked by the needle.

The technique is as follows: Grasp the syringe as indicated in Figure 4.1. Be prepared to press the plunger at the same time the needle is thrust through the skin. Many animals that have been previously injected are wary. Conceal the syringe when approaching primates or carnivores. If the animal is in a cage, wait for it to present a suitable muscular area near the side of the cage. Quickly jab the syringe and needle through the skin and at the same time make the injection. The animal will jump away, but if the thrust has been properly made, the medication will be injected before the animal can react.

Projected Syringes or Darts

Modern chemical restraint requires equipment capable of projecting a syringe some distance and discharging the contents upon impact. The first suitable weapons were developed by Jack Crockford and co-workers [4]. Palmer Cap-Chur equipment is the standard equipment used today in the United States. Other companies in New Zealand and Africa are now competing for the world market.

There are three types of Palmer projectors (Fig. 4.2). The short-range projector (pistol) is a modified pellet gun powered by compressed carbon dioxide. The range is 13.7 m (15 yd). The long-range projector (rifle) is also powered by compressed carbon dioxide. The range is 32 m (35 yd). The extra long range projector is powered by per-

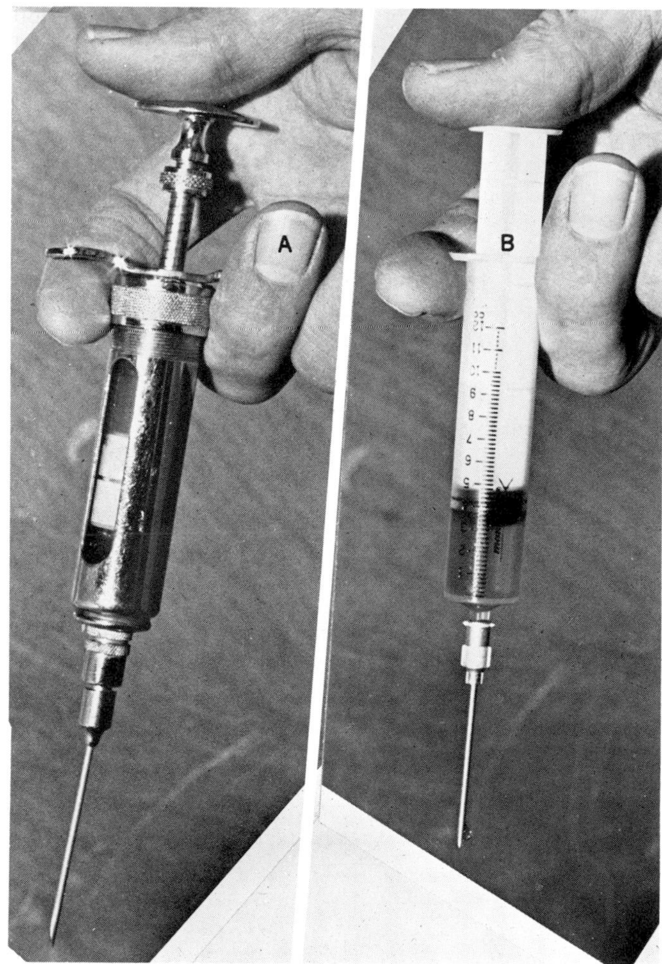

FIG. 4.1. Hand-held syringes: **A.** Metal and glass. **B.** Plastic.

cussion caps. The strength of the caps varies according to the distance from the target. The maximum range is 73 m (80 yd).

Both the long-range rifle and the short-range pistol are loaded with a filled syringe. The breech (mechanism) is closed and the weapon is cocked, sighted, and fired. When using the extra long range projector, the syringe is placed in the barrel with a special adaptor inserted behind it. A .22 blank charge is placed in the adaptor. The breech is closed and the weapon is cocked, sighted, and fired. Detailed instructions on the use of this equipment are included in the instruction manual supplied by the company.

The syringe itself is made up of the components illustrated in Figure 4.3. The syringe charge works on the principle illustrated in Figure 4.4. When the syringe strikes the target, pressure against the hub of the syringe forces a small weight in the back of the charge forward against a tiny spring. The sharp tip of the weight penetrates a seal, setting off the charge and driving the plunger forward to discharge the liquid from the syringe. If the charge is improperly placed in the neoprene plunger, the discharge will occur when the projectile is expelled from the weapon.

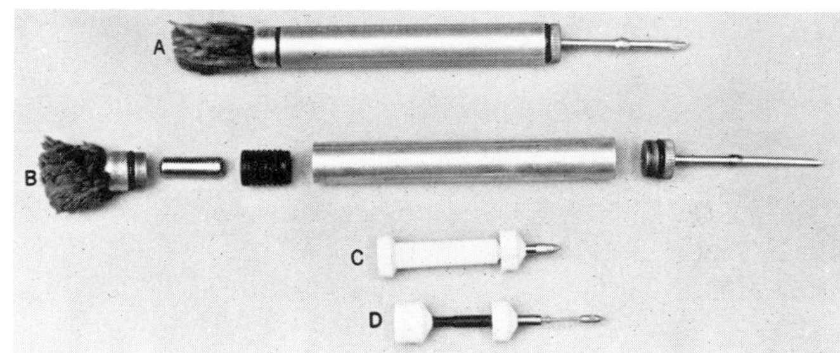

FIG. 4.3. **A, B.** Palmer Cap-Chur syringe makeup. **C.** Pneu-dart ready to be fired. **D.** Pneu-dart after ejection of powder.

The Paxarm pistol and syringe (Fig. 4.5) are powered by compressed air. The air is injected into a compartment behind the plunger after the syringe is loaded. Pressure can be varied according to the volume of solution to be injected.

Blowgun

Variations of the blowgun and poisoned darts have been used for both warfare and food gathering for centuries in

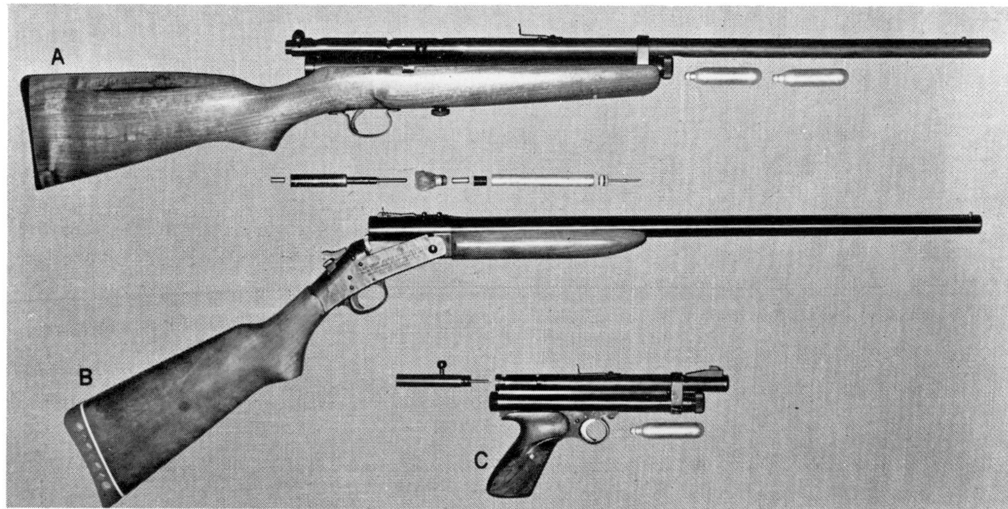

FIG. 4.2. Palmer Cap-Chur projectors: **A.** Long range, carbon dioxide-powered. **B.** Extra long range, powder cap-charged. **C.** Short range, carbon dioxide-powered.

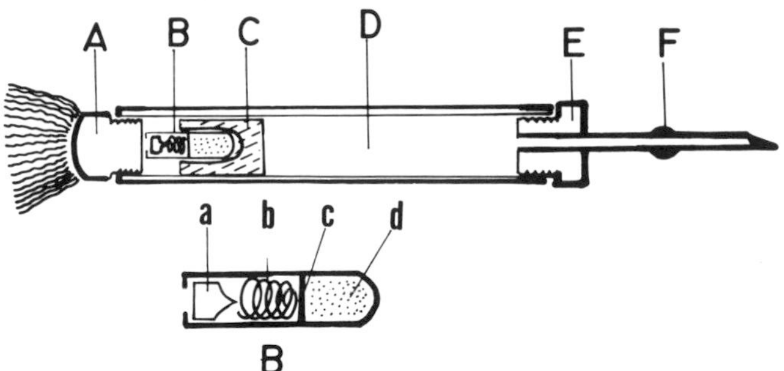

FIG. 4.4. Diagram of the mechanism of a Cap-Chur syringe: **A.** Tailpiece. **B.** Ejection charge. **C.** Plunger. **D.** Chamber for medication. **E.** Nosepiece hub. **F.** Collared needle: (a) sharpened weight, (b) coiled spring to keep weight separated from (c) diaphragm to contain (d) charge.

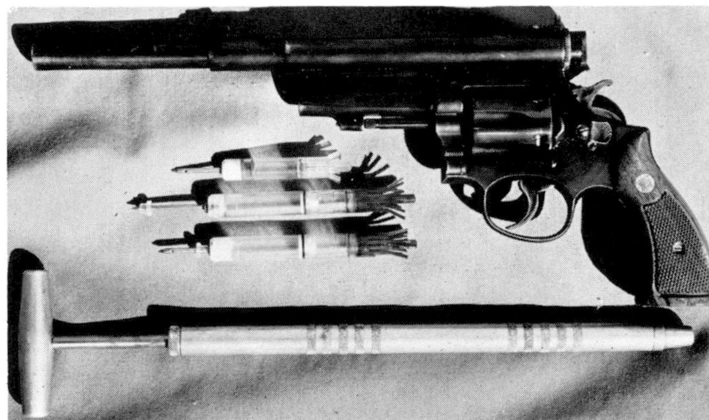

FIG. 4.5. Paxarm revolver, syringes, and pump.

primitive cultures (Fig. 4.6). The blowgun is now becoming popular with animal restrainers since it offers certain advantages over other delivery systems.

A major advantage of the blowgun is silent projection with less trauma on impact. It is adaptable for use on small animals, easily sighted, and has no mechanical parts requiring maintenance. The only disadvantages of the blowgun are its length and short range. It is unwieldy in confined spaces and the range is limited to approximately 13.7 m (15 yd). Blowguns can be purchased commercially, but aluminum electrical conduit, copper, stainless steel, plastic, or other types of tubing of appropriate diameter are satisfactory (Fig. 4.7). The inside of the tube must be smooth, polished if necessary, to minimize friction as the syringe is blown through it. A mouthpiece on the tube permits development of greater propulsion pressure.

The length of the blowgun is determined by the distance required to project the syringe. This varies from 1 to 2 m (3 to 6 ft). A longer tube permits greater accuracy. The maximum range varies with the length of the tubing and the skill of the operator, but 13.7 m (15 yd) is average (Fig. 4.8). Various commerical and home-designed syringes are used with blowguns (Figs. 4.9-4.11) [13].

Crossbow

The crossbow has been adapted for use with various types of syringes (Fig. 4.12). Crossbows are quite accurate and silent [16]. Tension can be adjusted according to target distance. The disadvantages of crossbows are that they are bulky, difficult to manipulate in a restricted space, and require considerable strength to tense the bowstring.

Stick Syringe

Various homemade and commercial stick syringes act as extensions of the hand for administering drugs to dangerous animals (Fig. 4.13). All work on the principle of

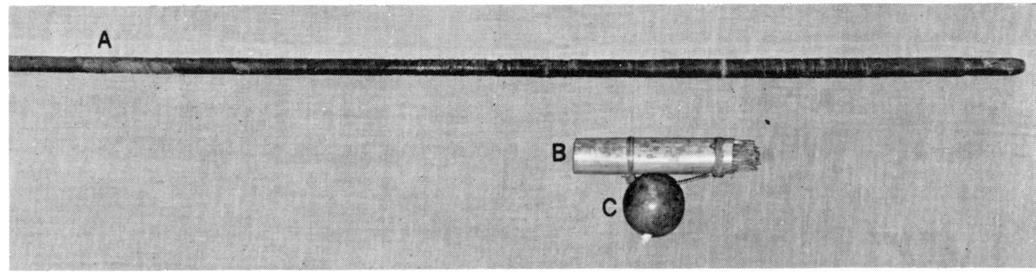

FIG. 4.6. Blowgun used by Equadorian Indian in the early 1900s: **A.** 2 m blowgun made by wrapping two pieces of hollowed bamboo with plant fiber yarn. **B.** Quiver to hold arrows. **C.** Hollow dried gourd. Tailpiece is made by inserting nontipped end into a cottony plant fiber and twisting as one would make a cotton-tipped applicator stick.

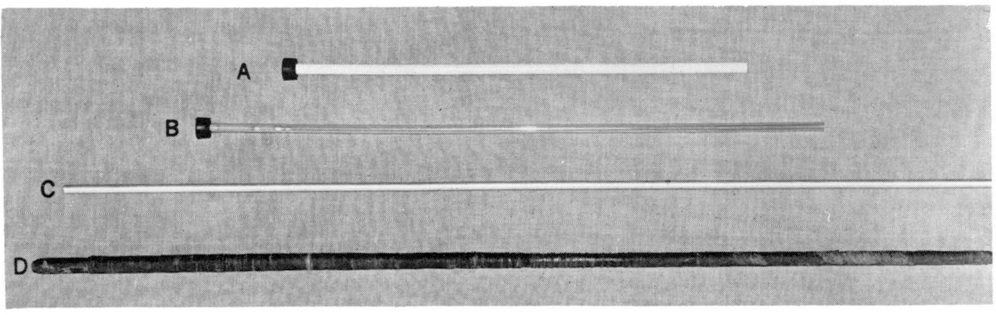

FIG. 4.7. Tubing used for blowguns: **A.** Polyvinyl chloride plastic. **B.** Plexiglas. **C.** Aluminum electrical conduit tubing. **D.** Bamboo tubing.

FIG. 4.8. A blowgun can be used for chemical immobilization or medication.

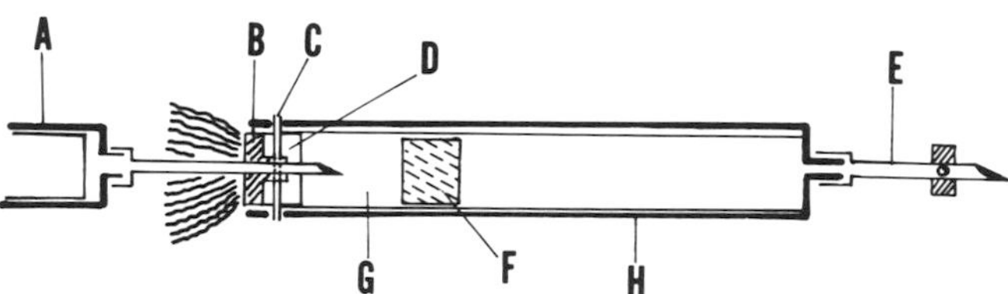

FIG. 4.9. Haigh-designed syringe for use in blowgun: **A.** Standard syringe. **B.** Glue or silicone to attach yarn to rubber plunger. **C.** Disposable needle, ends broken off. **D.** Plunger from a disposable syringe. **E.** Special needle, plugged at the tip and holes bored in the side. **F.** Another syringe plunger. **G.** Chamber for injecting air or butane as ejection propellant. **H.** Disposable plastic syringe, 3 ml.

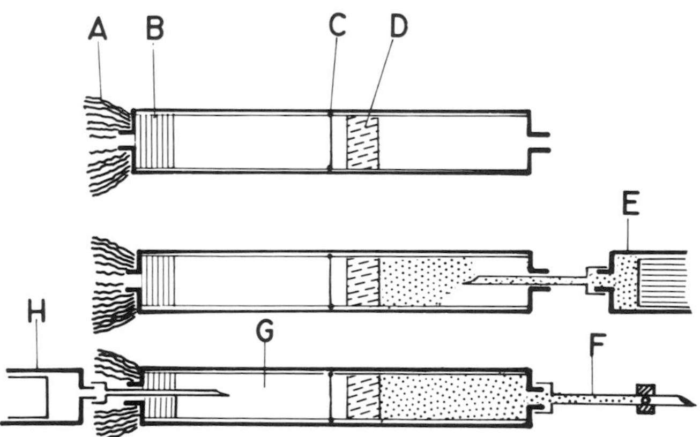

FIG. 4.10. Reudi-designed syringe for use in blowgun: **A.** Yarn glued onto tip of syringe. **B.** Silicone poured seal. **C.** Leads where two 3 ml syringes have been cut and bonded together. **D.** Syringe plunger. **E.** Standard syringe used to load medication. **F.** Special needle. **G.** Air chamber. **H.** Standard syringe to charge the chamber.

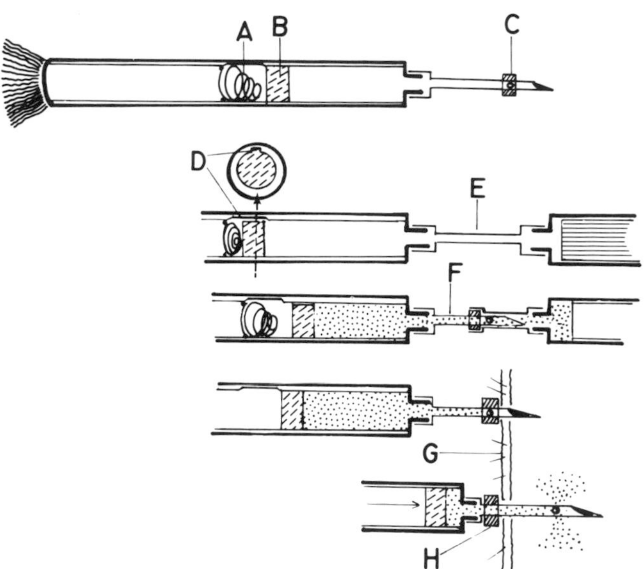

FIG. 4.11. Telinject blowgun syringe: **A.** Spring to keep plunger forward. **B.** Plunger. **C.** Collared side-hole needle. **D.** Cutaway showing groove on inside of the barrel of the syringe. When air pressure is applied to plunger, it collapses the spring, and plunger allows passage of air behind the syringe. This charges the syringe. **E.** Double-hubbed needle to charge syringe and load it. **F.** Loading the syringe. **G.** Skin. **H.** Collar pushed back, allowing ejection of liquid.

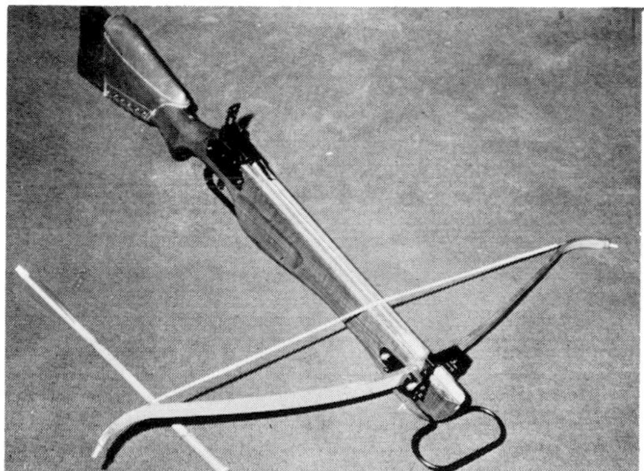

FIG. 4.12. Crossbow.

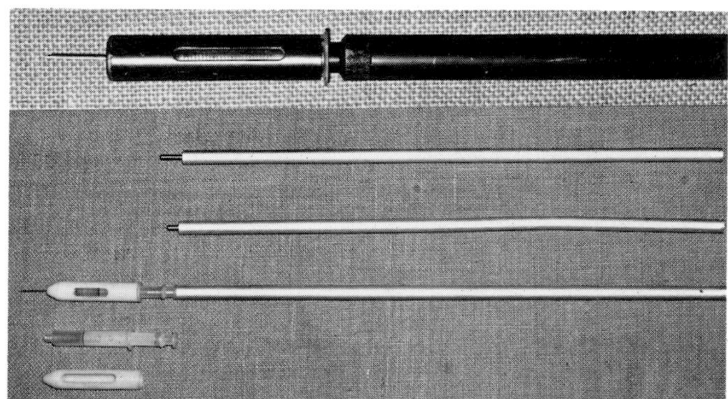

FIG. 4.13. Different types of pole or stick syringes.

injection immediately upon insertion of the needle. A quick jab is necessary to effect administration. The operator must maintain pressure against the animal until all the material has been injected or until the animal jumps away. If the animal jumps away before all the material is injected, a second injection is necessary. Sharp, large-bore needles should be used for easy insertion and quick expulsion of the drug. In many instances the animal reaches around and bites at the syringe, damaging it (Fig. 4.14). Occasionally an animal bites off the syringe and may swallow it. Hoofed animals are likely to kick out, and the quick movement may bend the needle. The needle may break off, but most of these devices have protecting shields around the needle to enclose and support the hub.

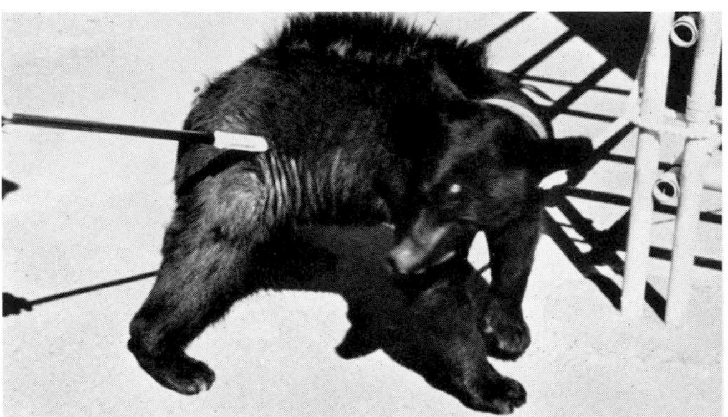

FIG. 4.14. Chemical immobilization utilizing a pole syringe.

FACTORS TO CONSIDER

When preparing for chemical immobilization, the following should be considered: (1) species, (2) physiological status of the animal, including general health, age, sex, and lactation, (3) physical condition, and (4) emotional status.

Species Differences

No presently available chemical restraint agent is equally effective and safe for use with all 45,000 vertebrate species. Many of these drugs were used on every available species during the testing phase of production and early clinical usage. Experience has verified that most agents have strict species limitations. The restrainer must establish the suitability of a particular restraint agent for a given species before using it.

Physiological Factors

Age is a factor. Young animals may respond differently to a certain chemical restraint agent than do adults of the same species. Similarly, aged animals may suffer from a diminished capacity for detoxification because of kidney or liver degeneration.

The sex of an animal rarely affects response to chemical restraint. However, it is possible that pregnancy or lactation may modify the action of a particular restraint agent. Fortunately, the agents available at the present time are not likely to cause abortion. Etorphine, ketamine, phencyclidine, and fentanyl all cross the placental barrier and may sedate the fetus. When these are used prior to cesarean section, the young may require special attention to initiate breathing after delivery. In the case of etorphine and fentanyl, the antidote must be used on the young as well as on the mother to effect reversal.

Physical Condition

Dosage of chemical restraint drugs is based on body weight. It is important to develop the ability to accurately estimate body weight. Persons experienced with domestic animals find it difficult at first to accurately estimate weights of wild species. Small species such as cage birds or rodents must be accurately weighed prior to chemical immobilization. Reference to the literature may or may not be helpful in establishing average weights, since not all references are accurate.

Animals requiring immobilization may suffer from varying degrees of malnutrition. Emaciation indicates that the animal is probably in negative nitrogen balance. This affects many metabolic activities and may modify the effects of the chemical restraint agent. Obese animals are likely to be in poor condition and also present problems of weight estimation to determine drug dosage.

Free-living wild animals must maintain excellent physical condition to obtain food and escape predators. Captive-born or captive-reared wild animals often lack the optimum physical condition of their free-living wild cousins. Captive animals often react adversely to exercise stress. Cardiovascular and pulmonary systems are usually not prepared for maximum output.

Torpidity is a means of escaping intolerable environmental conditions. Many species reduce metabolic activity during times of heat or cold stress. Metabolic pathways change, and response to chemical restraint agents changes at the same time. The inactive state may be seasonal or it may be triggered by changes in environmental conditions. In some instances it may be brought on by emotional responses, as in the case of the opossum, which becomes torpid or hypnotic from extreme fright or excitement. It is important for the animal restrainer to be able to recognize the torpid state of certain species and adjust the use of chemical restraint agents to accommodate the restricted metabolic activities of the torpid animal.

Emotional Status

The most important modifying factor to be considered in chemical restraint is the emotional status of the animal at the time of injection. Injecting a drug into an animal in a state of alarm with high catecholamine and cortisol release may produce effects opposite to those occurring in a normal, quiet animal. Emotions such as excitement, fear, and rage all produce the alarm response. Excitement may cause aimless running, which in turn produces acidosis, sensitizing the cardiac muscle to the effects of catecholamines. The end result may be an episode of ventricular fibrillation.

It is not always possible to immobilize an animal under ideal conditions. However, the closer the approach to the ideal, the safer the immobilization procedure. It may be desirable to place the animal in a small enclosure and allow time for the emotional state to stabilize before a drug is administered. If it is impossible to quiet the animal before injection, the restrainer must recognize the hazards and be prepared to counteract any adverse responses. Equipment for providing supplemental oxygen and sodium bicarbonate solutions to correct metabolic acidosis should be on hand.

ADVERSE EFFECTS OF RESTRAINT

Successful immobilization of a wild animal is an art. Many factors are involved. The operator must consider not only which equipment to use and the animal's condition, but personal ability as well. If the operator is not skilled in the use of the chosen weapon, it will be difficult to hit the target.

Other causes for failure of the immobilization procedure can be categorized in three major areas: (1) equipment failure, (2) operator fault, and (3) miscellaneous conditions over which the operator has little control.

Equipment Failure

High impact darts may break at the needle hub, allowing the needle to imbed deeply in the animal's body. I have seen a needle pass completely through a small deer.

Modified commercial needles or needles fabricated by the user must be scrutinized closely (Fig. 4.15). The shaft of the needle should be inserted through the nosepiece or hub and flanged slightly before soldering.

A syringe charge may fail to explode. This may be due to a faulty charge or insufficient force of impact to set off the charge.

The propelling charge may be insufficient to carry the dart to the animal. This may be a result of an incorrect estimation of the distance, but also may be due to faulty projection. Carbon dioxide cartridges may be nearly empty, resulting in decreased propulsion. An improper powder charge may have been selected. In the case of blowguns, insufficient breath may have been used.

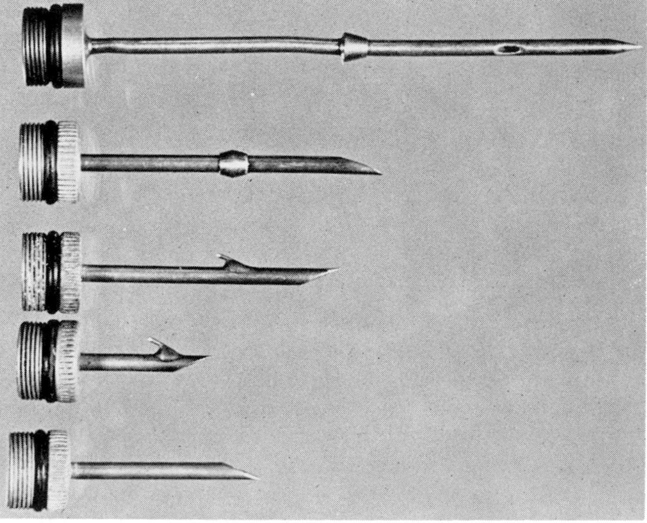

FIG. 4.15. Needles used with Palmer Cap-Chur syringes: Top needle was specially constructed for use in elephants. Tip was closed and pointed and holes drilled in the side. Various collars and barbs aid in retention of needle in the tissue. Barbed needles must be removed surgically.

Large-bore needles may cut a plug while traveling through the dermis, preventing discharge of the contents. Although rare, this can be prevented by plugging the tip of the needle and drilling holes through the side. Medication is then discharged through the side rather than out the tip. This technique is especially useful when dealing with elephants, rhinos, or other thick-skinned animals.

The syringe normally flies on an arched path to the target. If the syringe is improperly loaded or if the tailpiece is malfunctioning, the syringe may wobble in its course. Fabric tailpieces should be trimmed so they are symmetrical; they should not be stiff from contamination with moisture or exudates. If the syringe is out of balance, it may hit the animal broadside instead of at the needle. Air in the drug chamber may change the balance and hence the ballistics of the syringe. When loading syringes, excess air should be removed from the drug chamber and water added (physiological saline or 5% dextrose solution) to completely fill the chamber.

Operator Fault

The operator must be sure all equipment is in proper repair, clean, and lubricated. Needles should be inspected for plugs. The commonest operator fault is missing the target entirely or making an injection at an inappropriate site.

Restraint drugs are not absorbed equally well or at the same rate from all tissues. Injection into fat deposits, fascia, or nerves or into intraperitoneal, intrathoracic, or intradermal sites can result in prolonged induction or failure of effect. Intravascular injection is indicated by accelerated induction and may result in an overdose.

Injection at the wrong site can also result in the needle striking a bone. Impact from a syringe and needle may fracture the femur in small ungulates.

With sufficient impact, the whole syringe may be driven through a thin-skinned animal. This is likely to happen if too high a charge is used or if distance is misjudged. Small antelopes and deer have thin skin which is easily traumatized, especially when Palmer equipment is used. Such trauma is not only esthetically undesirable, but it also results in decreased absorption of the drug. The delivery of a syringe from a blowgun is much more gentle.

In some capture operations, a chase becomes mandatory. When this occurs, more misses can be expected. So far as is practical, do not attempt to shoot at a moving target. The operation of most discharge mechanisms requires a certain degree of impact, necessitating that the needle enter the skin at a perpendicular angle. If the needle strikes at too acute an angle, the syringe may fail to discharge the contents or may glance off the animal, or the force of impact may bend the needle or break it off.

Primates and some carnivores may bite the syringe or reach around and grab it with their hands or claws. Again, the needle may break off. It is extremely difficult to retrieve a broken needle because muscular activity moves it in the tissue. Radiographs taken from at least two angles are required to locate the needle prior to surgical removal.

The syringe charge may explode prematurely during projection and discharge all of the material before the dart reaches the animal. With Palmer syringes, this is an indication that the charge was inserted backward into the syringe and that the tiny percussion point was driven into the charge at the time of projection.

Sometimes infection develops at the injection site. It is impossible to clean the skin or apply antiseptic. Therefore the needle travels through dirty hair and skin, possibly carrying surface bacteria into the muscle. Syringes, needles, and all paraphernalia used to load syringes should be clean and sterile to minimize the chances of postinjection infection. The administration of parenteral antibiotics is not warranted unless the dosage can be maintained for 3-5 days. Wound infection occurs more frequently when there is extensive contusion at the impact site.

The dart occasionally falls from the animal before all of the contents have been discharged. The needle bore may have been too small for the volume to be discharged, the charge may have been too weak, or the impact was so great that the needle bounced out before discharge was completed. Collars have been attached to many needles to impede the release of the syringe, allowing complete discharge of the contents.

Although chemical restraint drugs can be given via other parenteral routes, they are primarily prepared for intramuscular administration. Any muscle mass is a suitable site. In practice, the muscles of the upper hind leg, rump, and shoulder are preferred sites because they present the largest target area. Figures 4.16-4.18 illustrate the preferred sites on various groups of animals.

The neck is an undesirable target site. The cervical vertebrae lie close to the surface in the midportion of the neck; the trachea, esophagus, and major blood vessels traverse the lower neck; and if the injection is made into the heavy fascial bands (nuchal ligaments) in the upper portion of the neck, no absorption will take place.

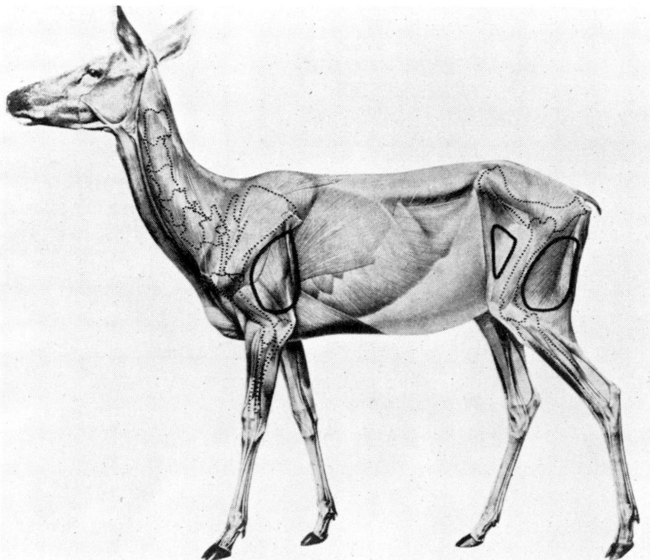

FIG. 4.16. Suitable injection sites in a deer. (Modification of an illustration in W. Ellenberger. 1956. An Atlas of Animal Anatomy for Artists, rev. ed. New York: Dover)

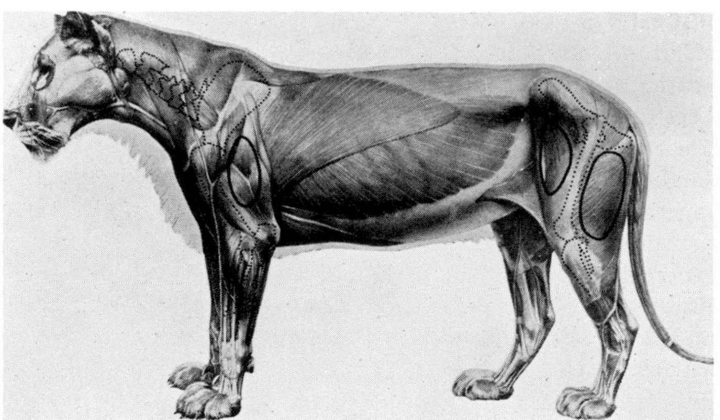

FIG. 4.17. Suitable injection sites in a lion. (Modification of an illustration in W. Ellenberger. 1956. An Atlas of Animal Anatomy for Artists, rev. ed. New York: Dover)

FIG. 4.18. Injection sites on medium-sized mammals.

Primates are exceedingly difficult to immobilize because of the difficulty in obtaining access to a muscle of sufficient size to provide a desirable target. A chimpanzee that has been injected previously may sit in a corner, legs crossed and arms folded in such a manner that no muscle is readily available for injection. If the animal is not sitting in a corner, it may be ranting and raving, running around the cage, or throwing food particles or feces at the operator, preventing a still shot. Successful immobilization of primates requires patience and quick, accurate response to opportunity.

Miscellaneous Conditions

Climatic conditions affect the functioning of equipment and the flight path of the dart. Wind can have a marked effect on trajectory. Sometimes it is necessary to select a different time because of inability to cope with wind. Warm weather increases gas efficiency, adding to the range of the carbon dioxide projector. Conversely, cold weather decreases the range.

A constant concern in using chemical immobilizers is the possibility that the partially drugged animal may stumble and fall into precarious positions. A sudden fall may result in a fracture or contusions. Incisor teeth have been fractured from a fall forward onto the mouth. A tympanitic stomach or loop of intestine can rupture from suddenly increased pressure exerted on the abdomen by a fall.

Prevent access to pools, ponds, and moats when using chemical restraint. Partially sedated animals may stumble into them and drown. Primates frequently climb up on bars, trees, or other fixtures and injure themselves by falling as the drug takes effect. Above all, have the proper equipment available to administer oxygen or to give other emergency treatment. Endotracheal intubation is difficult to achieve in some species but is made possible with suitable laryngoscopes (Fig. 4.19). An endotracheal tube allows positive pressure resuscitation; if cuffed, it prevents aspiration of regurgitated ingesta. Some special resuscitation devices are inexpensive and highly effective (Figs. 4.20, 4.21).

INDIVIDUAL DRUG DESCRIPTION

Only drugs commercially available in the United States will be discussed. (Companies supplying these agents are listed in the Appendix.) Some drugs prohibited in the United States are used in other countries [11, 12]; they are not available to practitioners in the United States because of various manufacturing, distributing, or legal problems.

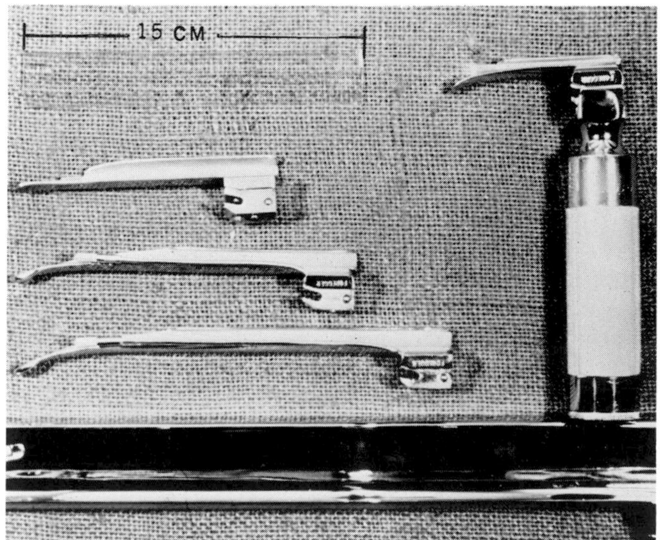

FIG. 4.19. Laryngoscopes for endotracheal intubation.

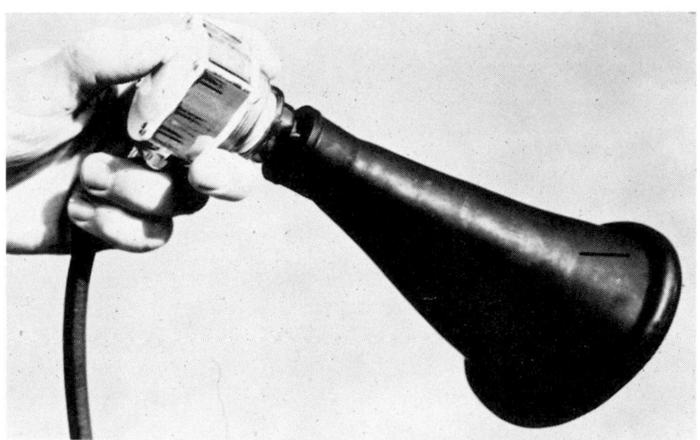

FIG. 4.20. Elder valve for resuscitation during restraint. It must be attached to compressed air or oxygen.

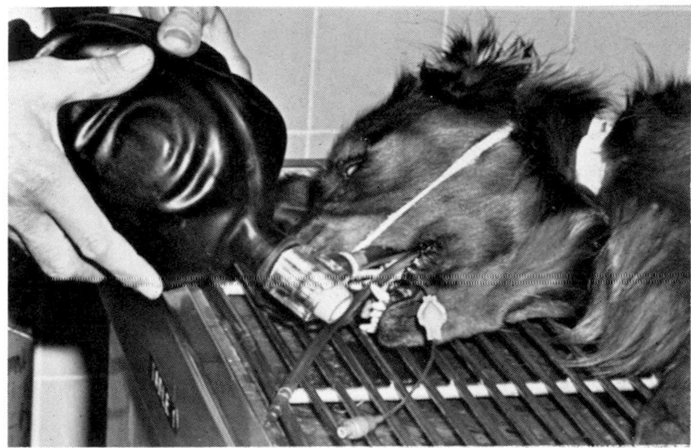

FIG. 4.21. Ambu resuscitation bag and valve.

Research is rapidly advancing the field of chemical immobilization. Many drugs are undergoing tests for efficacy and suitability as immobilizing agents. Some of these drugs will ultimately become available to the practitioner, but it would be unwise to discuss experimental drugs in a general text. This text discusses the following drugs as chemical immobilizing agents: etorphine, fentanyl-droperidol, ketamine, phencyclidine, succinylcholine, tiletamine-zolazepam, and xylazine (Table 4.1). Table 4.2 gives the names of drugs used to modify the effects of some of these agents.

Although it is possible to predict the pharmacological effect of a given drug from experimental reactions observed, it is an entirely different matter to predict the specific effects of the drug under field conditions. The physiological, physical, and emotional condition of the animal alters the effect of any drug.

TABLE 4.1. Immobilizing drugs

Generic Name	Trade Name and Company	How Supplied	Class of Agent
Etorphine	M99—D-M Pharmaceutical	20 ml vial, 1 mg/ml	Narcotic
Fentanyl-Droperidol	Innovar-Vet—Pitman-Moore	20 ml vial 0.4 mg/ml fentanyl 20 mg/ml droperidol	Narcotic, tranquilizer
Ketamine hydrochloride	Vetalar—Parke, Davis Ketalar—Parke, Davis Ketaject—Bristol	100 or 10 ml vials 100 mg/ml, 50 mg/ml, 20 mg/ml	Cyclohexamine
Phencyclidine hydrochloride	Sernylan—Bio-Ceutics	10 ml, 100 mg/ml	Cyclohexamine
Succinylcholine chloride	Quelicin—Abbott Sucostrin—Squibb Anectine—Burroughs Wellcome	10 ml, 20 mg/ml 10 ml, 100 mg 12 ml, 20 mg/ml	Neuromuscular blocking agent
Xylazine	Rompun—Chemagro	50 ml, 100 mg/ml 20 ml, 20 mg/ml	1-3-Thiazine derivative

TABLE 4.2. Drugs used to modify effects of immobilizing agents

Generic Name	Trade Name and Company	How Supplied	Class of Agent
Atropine sulfate	Atropine—Lilly, Fort Dodge	Small animal 20 ml, 0.4 mg/ml Large animal 100 ml, 2 mg/ml	Parasympatholytic
Acepromazine maleate	Acepromazine—Ayerst Acetylpromazine—Boots (England)	50 ml, 10 mg/ml 2 mg/ml	Phenothiazine
Diazepam	Tranimal—Roche (England) Valium—Roche (USA)	10 ml, 5 mg/ml	Benzodiazepine, tranquilizer
Diprenorphine	M50-50—D-M Pharmaceutical	20 ml, 2 mg/ml	Narcotic antagonist
Naloxone hydrochloride	Narcan—Endo	1 ml vial, 0.4 mg/ml	Narcotic antagonist
Nalorphine hydrochloride	Nalline—Merck	10 ml vial, 5 mg/ml	Narcotic antagonist

IMMOBILIZING AGENTS

Etorphine Hydrochloride (M99, Immobilon) [1,11,12,17]

CHEMISTRY. The formula is tetrahydro-7α-(1-hydroxy-1-methylbutyl)-6,14 endoethenooripavine hydrochloride. It is a synthetic derivative of one of the opium alkaloids (thebaine). Etorphine is a powder, readily soluble in slightly acidified water. The solution as supplied by the manufacturer is stable and can be stored at room temperature for years.

PHARMACOLOGY. Etorphine is a highly potent analgesic. It has up to 10,000 times the analgesic potency of morphine hydrochloride. Etorphine produces pharmacological effects similar to those of morphine, namely depression of the respiratory and cough centers, decreased gastrointestinal motility, and behavioral changes. Stimulation or depression of the central nervous system (CNS) varies with the species. The elephant is depressed. Eland and nyala may be stimulated to aimless walking or running. Aggressive animals usually become tractable. In addition, blood pressure is elevated, accompanied by tachycardia [11].

INDICATIONS. Etorphine has been tested in a wide range of species. It is particularly useful for immobilizing large ungulates such as the elephant, rhinoceros, or hippopotamus. It has been given to most species of artiodactylids with varying degrees of efficacy and safety.

ADMINISTRATION. Etorphine is readily absorbed from an intramuscular site. An elephant has been maintained in surgical anesthesia for as long as 4 hours by intermittent intravenous injections (1 mg every 15 minutes) or by continuous drip injection (4 mg etorphine hydrochloride in 250 ml of 0.9% sodium chloride solution, given at 1 drop per second or 1 mg every 15 minutes) [6].

Onset of anesthesia occurs 10–20 minutes after intramuscular injection. If no antidote is administered, recovery is slow, up to 3 hours. When the antidote is injected, the animal becomes ambulatory within 4–10 minutes.

Do not mix with atropine, since atropine reduces the solubility of etorphine.

SIDE EFFECTS. The excitement noted in the horse and cat with the use of morphine may also be observed following administration of etorphine. This effect is dose related. Minimal doses for a species may result in excitement, tremors, and convulsions. Aimless walking or running has been observed, continuing until the animal is physically restrained (Fig. 4.22).

Inhibition of respiratory centers may directly or indirectly influence blood gases and acid-base balance.

Reflex vomiting, a common disturbing side effect of morphine administration, is rare with etorphine. However, passive regurgitation is not unusual, particularly with prolonged immobilization with etorphine, probably as a result of total relaxation of the cardiac sphincter of the stomach coupled with abdominal pressure.

As with most other agents, fatal hyperthermia may occur with etorphine. An animal that is chased or highly excited at the time of immobilization experiences a tremendous buildup of body heat. Body insulation and central depression of thermoregulatory systems may inhibit heat dissipation. Smaller animals are more prone than larger species to thermal stress following the use of any chemical restraint agent.

Etorphine is extremely dangerous to human beings. If

FIG. 4.22. This eland was in the aimless walking stage of immobilization with etorphine. The animal was roped and subdued in lieu of further drugging.

injected accidentally, seek medical help immediately. Administer the antidote (diprenorphine) or naloxone intravenously. Equipment for artificial respiration should be available to deal with possible respiratory arrest.

Etorphine is readily absorbed through mucous membranes and may be absorbed through the intact skin. Avoid inhalation, ingestion, or contamination of the skin, particularly of the hands, which might touch the mouth.

ANTIDOTE. Diprenorphine (M50-50) is a specific antidote developed for etorphine. The standard dose is double the amount of etorphine injected. If diprenorphine is unavailable, naloxone may be used.

Fentanyl-Droperidol (Innovar-Vet, Sublimaze [fentanyl]-Inapsine [droperidol]) [11,12].

CHEMISTRY. Fentanyl is a morphine derivative with the formula *N*-(1-phenethyl-4-piperidyl)propionanilide. Droperidol is a tranquilizer with the formula 1-[1-[3-(*p*-fluorobenzoyl)propyl]-1,2,3,6-tetrahydro-4-pyridl]-2-benzimidazolinone.

PHARMACOLOGY. Fentanyl has a pharmacological spectrum similar to that of morphine, but it is as much as 180 times more potent than morphine as an analgesic. It does not produce emesis in most species except man. Droperidol produces a reduced responsiveness to environmental stimuli similar to that of many other tranquilizers. The combination of fentanyl and droperidol may produce a slight decrease in blood pressure due to direct vasodilatation or to α-adrenergic blockade.

INDICATIONS. This combination is satisfactory as a sedative, analgesic, and anesthetic agent in the dog. It is also acceptable for administration to a wide range of wild species, particularly carnivores, nonhuman primates, and various small mammals. It is particularly desirable for short procedures because the effects of fentanyl can be quickly reversed and the animal becomes ambulatory within minutes.

ADMINISTRATION. Intramuscular injection of an immobilizing dose is effective within 10–15 minutes. Analgesia is maintained for approximately 40 minutes. The tranquilizing effect may last for several hours. Usually, animals are ambulatory within minutes after injection of the antidote.

SIDE EFFECTS AND PRECAUTIONS. The combination of fentanyl and droperidol potentiates the action of barbiturates. It is not recommended for use in cesarean sections because it sedates the fetus.

ANTIDOTE. All the effects of fentanyl can be reversed by the administration of naloxone hydrochloride (0.006 mg/kg IV or IM). When the antagonist is used, recovery from immobilization is immediate. Tranquilization persists for 1–2 hours.

Ketamine Hydrochloride (Ketaset, Ketalar, Vetalar, Ketaject, Ketanest) [3,9,10,11,12]

CHEMISTRY. The formula is 2-(o-chlorophenyl)-2-(methylamino)-cyclohexanone hydrochloride. Ketamine is a derivative of phencyclidine hydrochloride. It is a white crystalline powder, readily soluble in water.

PHARMACOLOGY. Ketamine is a nonbarbiturate dissociative anesthetic agent. The animal usually retains normal pharyngeal-laryngeal reflexes. This desirable effect minimizes inhalation of food or ingesta near the glottis. However, endotracheal intubation is difficult when ketamine is the only agent used.

Ketamine does not produce skeletal muscle relaxation. Catatonia is common. Profound analgesia is rapidly produced, although analgesia of the visceral peritoneum may be less than optimal. Excessive salivation can be alleviated with atropine.

Many animals experience transitory pain upon injection of the solution. The effect on blood pressure varies from species to species. In the dog and man, blood pressure is elevated. In the rhesus monkey, blood pressure is depressed. Nystagmus may be noted during induction.

Ketamine crosses the placenta in all species. Anesthetic effects are noted in the fetus when ketamine is used as a sedative for cesarean section or dystocia. Ketamine is not known to produce abortion when used on pregnant animals.

Ketamine is detoxified in the liver. Metabolites are excreted via the urine.

Ketamine produces a fixed expression in the eyes [12].

The eyelids stay open, yet the cornea usually remains moist. Occasionally corneal ulceration may result from prolonged exposure. Palpebral reflexes persist. Since swallowing reflexes are usually unaffected, excessive saliva is swallowed as usual.

Induction is characterized by ataxia. The animal lies down. A characteristic licking motion is called "serpentine tongue" [12]. The animal becomes insensitive to external stimulation. Lateral nystagmus appears, then disappears with increased depth of anesthesia.

INDICATIONS. Ketamine has U.S. Food and Drug Administration (FDA) clearance for use in man, domestic cats, and nonhuman primates. However, it has also been safely and effectively used for anesthesia of numerous other species of wild animals. It is particularly effective in wild carnivores, reptiles, and birds but is not suitable for most ungulates.

ADMINISTRATION. Ketamine is supplied as a solution of 20 mg/ml, 50 mg/ml, and 100 mg/ml concentrations. It can be administered orally or parenterally. In domestic cats the intravenous and intramuscular routes are used. In wild species either intramuscular or subcutaneous routes are acceptable. Immobilization dose variation is tremendous (2-50 mg/kg). When used as an anesthetic, intravenous redosage is usually required except for very short procedures. Specific doses are tabulated with each animal group.

Parenteral injections take effect within 3-5 minutes. Complete immobilization is produced within 5-10 minutes. The duration of effect varies with the species and the dosage administered. The operator usually has from 15 to 30 minutes to complete a task if ketamine is not augmented with other CNS depressant drugs.

Recovery is usually smooth. An animal is usually ambulatory within 1 hour; however, periods of up to 5 hours are not rare. Some felids may show slight depression for 24 hours following anesthesia.

SIDE EFFECTS AND PRECAUTIONS. Ketamine causes tonic clonic convulsions in a small percentage of domestic and wild felids. Other carnivores are similarly affected. Primates are less commonly affected. Convulsive effects can be obviated by administering diazepam (0.25-0.5 mg/kg) with the ketamine. Acepromazine maleate is often used in combination with ketamine. Acepromazine decreases catatonia and provides better muscle relaxation. Theoretically acepromazine lowers the threshold for convulsive stimulation, thus increasing the likelihood of convulsions [5,7]. However, in practice acepromazine seems to minimize the development of convulsions.

Once convulsions have begun, intravenous administration of diazepam or pentobarbital sodium may be required to control seizures. I have observed two typical cases of mania in lion cubs given ketamine alone. Convulsive seizures began, which gave way to wild thrashing, vocalization, and compulsive running. The cubs had to be physically restrained to prevent head trauma. Repeated intravenous injections of diazepam over a period of 2 hours were necessary to control the episodes.

Prolonged apnea is sometimes observed in large wild felids. It may be necessary to assist respiration. Endotracheal intubation is highly desirable. Salivation may be profuse but is readily controlled with atropine.

Accidental oral ingestion or injection can produce anesthesia or immobilization in man. Hospitalization is required. Respiration should be monitored and assistance given if necessary. Ketamine is known to produce hallucinations in man. It is difficult to establish the occurrence of hallucinations in animals, but some primates and felids behave strangely during recovery. They may vocalize and appear to be frightened.

Ketamine may be contraindicated in animals with elevated blood pressure. Do not mix ketamine and barbiturates in the same syringe; they are chemically incompatible and will precipitate. Ketamine is pharmacologically compatible with other anesthetic and CNS depressant drugs. Recovery time is prolonged when barbiturates or narcotic agents are used in addition to ketamine.

Ketamine should not be used as the sole anesthetic for reduction of fracture or luxations. Neither should it be used singly for diagnostic or surgical procedures of the pharynx, larynx, or bronchial tube.

Hyperthermia is a side effect of ketamine anesthesia. Catatonia causes heat production. Hyperthermia is very likely to occur if seizures appear. Swine and marine mammals with heavy insulation cannot dissipate heat rapidly and are particularly subject to hyperthermia.

Hypothermia is not a common side effect of ketamine, but as with any anesthetic, prolonged deep ketamine anesthesia in a cold environment may result in hypothermia.

ANTIDOTE. There is no known clinical antidote for ketamine.

Phencyclidine Hydrochloride (Sernylan, Sernyl) [3,11,12]

CHEMISTRY. The formula is 1-(1-phenylcyclohexyl)piperidine hydrochloride. This was one of the first cyclohexanone compounds to be used as an immobilizing agent.

PHARMACOLOGY. Phencyclidine directly affects the CNS. The exact mode of action is not understood. Phencyclidine and its derivatives are called "dissociative agents," that is, they disrupt communications between various sections of the CNS, with combined stimulation and depression. The predominant manifestation is determined by species and dosage. This agent is a cataleptic. Muscle relaxation is poor. In some situations the excessive muscle tone produced can be detrimental. Human subjects experience muscle soreness following anesthesia with phencyclidine.

Generally, the pharmacological action is similar to that of ketamine hydrochloride.

INDICATIONS. Phencyclidine hydrochloride has been used extensively for nonhuman primate, carnivore, and human immobilization and anesthesia. It is not presently used in human medicine because of its hallucinatory qualities; it is approved by the FDA for veterinary use in primates only.

Newer analogs such as ketamine hydrochloride have supplanted the general use of phencyclidine hydrochloride except when small volumes are required to dart large carnivores and primates. It is also a desirable agent when prolonged immobilization is necessary.

ADMINISTRATION. Dosage varies from 0.5 to 2 mg/kg. The drug can be given orally, intramuscularly, or subcutaneously.

Phencyclidine is slower in onset than ketamine; 15-25 minutes is usual. Immobilization lasts from 1 to 2 hours, but the animal continues to be depressed for 5-48 hours.

SIDE EFFECTS AND PRECAUTIONS. Disorientation (erratic behavior and excitement) is produced in many species. Convulsive seizures are commoner with phencyclidine than with ketamine. Seizures are rare in primates, most common in canids, and occasional in felids. Excessive salivation can be controlled with atropine.

The same precautions are necessary as with the use of ketamine. Catatonia or seizures may produce hyperthermia.

ANTIDOTE. There is no known antidote.

Succinylcholine Chloride (Sucostrin, Anectine, Quelicin, Scoline) [8,11,12,15]

CHEMISTRY. Succinylcholine chloride is the dicholine chloride ester of succinic acid. It is a white, odorless powder, sensitive to light, and readily soluble in water. It is easily hydrolyzed in an alkaline solution; therefore, it is chemically incompatible with barbiturates if mixed in the same syringe. Other similar drugs (suxamethonium, decamethonium) have been used in North America and other countries.

PHARMACOLOGY. Succinylcholine chloride acts by depolarizing the motor endplate and disrupting impulse transmission to the skeletal muscles. There is *no analgesia* or *anesthesia*. The animal retains consciousness and can be affected by auditory, visual, and psychological stimulation.

The animal may exhibit all the signs of alarm except those related to skeletal muscle response. Thus tachycardia and elevated blood pressure are potential hazards of use.

Succinylcholine and related drugs are detoxified by plasma cholinesterase or pseudocholinesterase. Animals vary tremendously in the levels of this enzyme present in the plasma. This may account for the wide variation in effective dosage and persistence of action of the drug in various species. Ruminants have a low pseudocholinesterase level and thus experience prolonged succinylcholine action. The pattern of muscle paralysis proceeds from the head to the neck, shoulders, limbs, abdomen, thorax (respiratory), and finally to the diaphragm. Artificial respiration equipment must always be available when working with this drug.

The first visible reaction to the drug is muscle fasciculation, described as very painful by human volunteers. The plasma potassium level rises, which can have a marked action on cardiac muscle funtion [8].

INDICATIONS. Succinylcholine is useful in providing additional muscle relaxation in the anesthetized surgical patient. Anesthesia must be carefully monitored lest succinylcholine mask pain reflexes and the surgery become inhumane.

Succinylcholine chloride has been widely used in equine practice to immobilize stallions for castration. Its use unaccompanied by suitable anesthesia is barbaric.

The availability of tranquilizing agents has reduced the need for, and the use of, this drug in equine practice. However, it is a rapidly acting, effective immobilizing agent. Some fatalities have occurred with its use in the horse, especially when organic phosphate anthelmintics were administered within the previous 30 days.

Succinylcholine chloride and similar agents have been used extensively in wild animal immobilization in the past, but newer agents have supplanted them. Succinylcholine has a marked variation in dosage and requires extensive experience for effective use. Solutions are unstable and quickly lose potency, further complicating determination of the proper dosage. This agent is still useful for rapid immobilization.

ADMINISTRATION. The dosage range is tremendous and is listed in appropriate animal groups. The chemical, supplied as a powder in a stoppered vial, is reconstituted as used. The solution must be kept cool, preferably refrigerated, and free from light exposure.

Intravenous administration results in immobilization within 1-2 minutes. Intramuscular injections require 3-10 minutes. The duration of immobilization is variable, depending on the concentration of plasma cholinesterases. In the horse the effect lasts for 3-5 minutes. A calf remained in flaccid paralysis for 45 minutes from a single injection of succinylcholine chloride. The calf required artificial respiration throughout this period.

Recovery is rapid and complete. Although there is no CNS effect, animals are less aggressive for a few minutes after recovery. Animals appear to be bewildered by the whole experience. Psychological stress is associated with being unable to fight or flee—the normal alarm response.

SIDE EFFECTS AND PRECAUTIONS. Cessation of respiration is a natural result of succinylcholine chloride immobilization. Procedures should not be attempted without resuscitation equipment available. Elevated blood pressure with ruptured aneurysms (weakened arteries) has occurred in the horse.

Prolonged effects, due to cholinesterase inhibitors in the plasma, are a sequel to immobilization induced shortly after administering organic phosphate anthelmintics.

This drug should never be used without thorough knowledge of the dosage for the species to be worked on. Be prepared to assist respiration.

ANTIDOTE. There is no specific antidote.

Tiletamine Hydrochloride-Zolazepam Hydrochloride (Tilazol) [3]

Tiletamine was previously known as CI-634 and zolazepam as CI 716 and flupyrapon. The combination is known as CI 744.

CHEMISTRY. Tiletamine hydrochloride is a cyclohexanone dissociative agent related to ketamine hydrochloride and phencyclidine hydrochloride. The chemical formula is 2-ethylamino-2-(2-thienyl)cyclohexanone hydrochloride. It is soluble in water in up to 30% concentrations.

Zolazepam hydrochloride is a nonphenothiazine pyrazolodiazepinone tranquilizer with the formula 4-(o-fluorophenyl)-6,8-dihydro-1,3,8-trimethylpyrazolo-[3,4-e][1,4]diazepine-7(IH)-1 monohydrochloride.

PHARMACOLOGY. The combination capitalizes on the desirable characteristics of each while minimizing the side effects. Tiletamine hydrochloride used alone produces analgesia and cataleptoid anesthesia plus convulsive seizures in some animals. The combination with zolazepam eliminates these undesirable effects.

The combination is classified as a general anesthetic. However, the eyelids usually remain open and the following reflexes persist, even during deep anesthesia: corneal, palpebral, laryngeal, pharyngeal, pedal, and pinnal. Since reflexes are unaffected, they cannot indicate planes of anesthesia—the classical differentiation technique. Instead, the following phases are observed: (1) induction—the time interval from injection of teletamine-zolazepam to the onset of surgical anesthesia; (2) surgical anesthesia—the time from induction to the beginning of emergence from anesthesia; (3) emergence—the time from the end of surgical anesthesia until the animal has returned to the preanesthetic condition; (4) return of righting reflex—the animal can right itself to sternal recumbency; and (5) recovery—the time from first standing position until completely normal.

INDICATIONS. Tiletamine hydrochloride-zolazepam hydrochloride is used for chemical immobilization and surgical anesthesia in a wide variety of carnivores, artiodactylids, birds, reptiles, and amphibians.

ADMINISTRATION. The combination can be given orally or parenterally. The dosage is dependent on the degree of anesthesia required and the species of animal.

Onset occurs within 5-12 minutes after intramuscular injection. The onset following oral administration is less predictable but effects should be observed in 15-30 minutes.

The duration of analgesia and anesthesia is directly proportional to the dose administered. The effect may last for 1 hour at low doses and up to 5 hours with higher doses.

SIDE EFFECTS AND PRECAUTIONS. Excessive salivation is common but can be controlled by the administration of atropine sulfate. Rare untoward reactions include muscle rigidity, tremors, vocalization, rough recovery, vomiting, apnea, cyanosis, and tachycardia.

ANTIDOTE. There is no known antidote for tiletamine hydrochloride-zolazepam hydrochloride.

Xylazine (Rompun, BAY 1470) [2,11,12]

CHEMISTRY. The formula is 5,6-dihydro-2-(2,6-xylidino)-4H-1,3-thiazine hydrochloride. It is a colorless crystal with a bitter taste, readily soluble in water and stable in solution.

PHARMACOLOGY. Xylazine is a nonnarcotic sedative, analgesic, and muscle relaxant. These effects are mediated by CNS depression.

Animals appear to be sleeping. Stimulation during the induction stage may prevent optimum sedation. When subjected to a too-rapid approach, a seemingly sedated animal may rouse explosively, jeopardizing the safety of the operator.

INDICATIONS. Xylazine is used as a mild sedative in the horse. Higher doses effect immobilization. It is a popular immobilizing agent, used either singly or in combination with other drugs for a wide variety of species. See the animal groups for specific indications.

ADMINISTRATION. Xylazine may be given intravenously or intramuscularly. There is wide species variation in the optimum dosage. In the horse, the dose is 1 mg/kg (IV) and 2 mg/kg (IM).

Immobilization occurs within 3-5 minutes following intravenous injection or 10-15 minutes after intramuscular injection. Analgesia lasts from 15 to 30 minutes, but the sleeplike state is maintained for 1-2 hours. Painful procedures should not be performed after 30 minutes.

SIDE EFFECTS AND PRECAUTIONS. Occasionally muscle tremors, bradycardia, and partial A-V block occur with standard doses. Atropine should be given to dogs and cats to prevent cardiac effects. Explosive response to stimuli, particularly to auditory stimuli, may cause operator injury.

Xylazine produces an additive effect when combined with tranquilizers and barbiturates. In such instances, reduce dosages and exercise caution. The analgesic effect is variable. The depth of analgesia should be ascertained before clinical diagnostic or surgical procedures are begun. A mildly sedated animal may make effective use of defensive mechanisms if pain is inflicted. Analgesia of the distal extremities of the horse is variable.

Xylazine is not cleared for use in food-producing animals.

ANTIDOTE. There is no known antidote.

Atropine Sulfate

CHEMISTRY. Atropine sulfate is a white crystalline powder, stable in aqueous solution.

PHARMACOLOGY. Atropine is a parasympatholytic drug with action equivalent to blockage of the parasympathetic autonomic nervous system. It decreases salivation, sweating, gut motility, bladder tone, gastric secretions, and respiratory secretions. Vagal blockage produces tachycardia. Mydriasis occurs.

INDICATIONS. Atropine diminishes excessive secretions induced by ketamine and phencyclidine. It is also commonly used as a preanesthetic medication to prevent reflex vagal stimulation of the heart (cholinergic bradycardia) during induction.

ADMINISTRATION. Atropine can be given orally or parenterally at dosages of 0.04 mg/kg. Atropinization occurs within 1-15 minutes, depending on the route of administration.

SIDE EFFECTS AND PRECAUTIONS. Animals that dissipate excess heat and moisture by sweating may develop hyperthermia because of sweat inhibition. Protect dilated pupils from direct sunlight to prevent retinal damage. Atropine is contraindicated for patients with glaucoma.

ANTIDOTE. Parasympathomimetic drugs may aid in counteracting atropine effects, but atropine is difficult to reverse.

TRANQUILIZERS

Tranquilizers are often used to calm aggressive domestic animals. Many classes and brands of these agents are available. The pharmacology and clinical use of tranquilizers are important but are outside the purview of this book. The reader is referred to textbooks on anesthesia and pharmacology for detailed discussions of tranquilizers.

Tranquilizers used in combination with immobilizing agents are discussed here.

Acepromazine Maleate (acetylpromazine maleate)

CHEMISTRY. The formula for acepromazine maleate is 10-[3-(dimethylamino)propyl]phenothiazin-2-yl-methyl ketone maleate. It is an acetyl derivative of promazine. It is a yellow crystalline powder.

PHARMACOLOGY. Acepromazine maleate is a potent tranquilizing agent which depresses the CNS. It produces muscular relaxation and reduces spontaneous activity. It exhibits antiemetic, hypotensive, and hypothermic properties [11].

INDICATIONS. Acepromazine maleate is rarely used singly for immobilization purposes but rather in combination with etorphine, ketamine, or phencyclidine. Its muscle-relaxing characteristic is of particular value when used with ketamine and phencyclidine. Acepromazine maleate is used extensively in domestic dogs, cats, and horses as an antiemetic, to inhibit pruritis, and to control intractable animals for examination and minor surgical procedures.

ADMINISTRATION. This agent may be injected intravenously, intramuscularly, or subcutaneously. A tablet form is available for oral administration, but the effects of oral administration are somewhat unpredictable. Dosage varies with species. The parent compound, promazine, is available in a granulated form for inclusion in the diet and is more suitable than acepromazine maleate for this route of administration.

Acepromazine maleate is commonly used as a preanesthetic agent. Since it potentiates the action of barbiturates, dosage of both the barbiturate and acepromazine maleate should be decreased.

When given intravenously, effects are noted within 1-3 minutes. Intramuscularly, it takes 15-25 minutes for full effect. Oral effects appear within 30-60 minutes. The general dosage for domestic animals is:

dog—0.5-1 mg/kg
cat—1-2 mg/kg
horse—0.04-0.08 mg/kg

SIDE EFFECTS AND PRECAUTIONS. Caution should be exercised when using acepromazine in combination with other hypotensive agents. Occasionally, instead of producing CNS depression, it acts as a stimulant and hyperexcitability ensues [5].

Do not use acepromazine to control convulsions caused by organic phosphate insecticide poisoning.

ANTIDOTE. There is no known antidote.

Diazepam (Valium, Tranimal, Tranimul)

CHEMISTRY. The formula is 7-chloro-1,3-dihydro-1-methyl-5-phenyl-2*H*-1,4-benzodiazepin-2-one. It is a colorless crystalline compound and has limited stability in solution.

PHARMACOLOGY. Diazepam acts on the thalamus and hypothalamus, inducing calm behavior. It has no peripheral autonomic blocking action like some other tranquilizers. Transient ataxia may develop with higher doses as muscle relaxation progresses. Spinal reflexes are blocked. Diazepam is an effective anticonvulsant.

INDICATIONS. Diazepam prevents the convulsive effects of ketamine and phencyclidine. If injected intravenously, it effectively controls convulsive seizures in progress. It can also be used as preanesthetic medication to calm an excited animal.

ADMINISTRATION. Diazepam can be given orally, intravenously, or intramuscularly. Oral administration is not recommended for immobilization procedures. The dosage varies from 1 to 3.5 mg/kg, depending on the species and the degree of excitement at the time of injection. Onset is within 1-2 minutes when given intravenously. If given intramuscularly it takes 15-30 minutes, depending on the dose. Diazepam is metabolized slowly in the normal liver. Usually the clinical effects disappear within 60-90 minutes.

SIDE EFFECTS AND PRECAUTIONS. Venous thrombosis and phlebitis at the injection site are complications of use.

Diazepam is chemically incompatible with most other

immobilizing agents and should not be mixed with them in the same syringe nor with intravenous solutions. Some pain is associated with intramuscular injection, and a transient inflammatory reaction may develop at the site. Diazepam is contraindicated in patients suspected of having glaucoma.

ANTIDOTE. There is no known antidote.

Diprenorphine

CHEMISTRY. The formula is *N*-(cyclopropylmethyl)6,7,8,14-tetrahydro-7-a-(1-hydroxy-1-methyethyl)-6,14-endoethanonororipavine hydrochloride.

PHARMACOLOGY. Diprenorphine is a narcotic antagonist used to reverse the effects of etorphine.

INDICATIONS. Diprenorphine acts as a depressant on the CNS and if used in excessive dosages may complicate recovery.

ADMINISTRATION. Diprenorphine is injected intravenously if possible, otherwise, intramuscularly. The recommended dose is double the injected dose of etorphine. When injected intravenously, reversal usually occurs within 1-4 minutes. Intramuscular injection requires 15-25 minutes for reversal effects.

SIDE EFFECTS. No adverse effects should be noted unless an overdose is administered.

ANTIDOTE. Naloxone is the antidote for diprenorphine.

Nalorphine Hydrochloride (Nalline) Naloxone Hydrochloride (Narcan) [12]

CHEMISTRY. The formula of nalorphine is *N*-allynormorphine. The formula of naloxone is *N*-allyl-noroxymorphone hydrochloride, a synthetic congener of oxymorphone.

PHARMACOLOGY. Nalorphine has a structure similar to that of morphine and competes with morphine and other narcotics at receptor sites.

Nalorphine has activity similar to that of naloxone, except that nalorphine may produce agonistic effects similar to those of morphine.

Naloxone is a narcotic antagonist. In the absence of narcotics or agonistic effects of other narcotic antagonists, it exhibits essentially no pharmacological activity.

INDICATIONS. These two drugs are used solely as antagonists to narcotic immobilizing agents. Naloxone is the preferred drug, but either may be used to reverse the effects of fentanyl. Nalorphine has been used to reverse etorphine, but naloxone or diprenorphine are much preferred. In case of accidental human injection of etorphine, physicians may not have access to diprenorphine, so it is important to recognize that naloxone is effective.

ADMINISTRATION. Intravenous, intramuscular, and subcutaneous routes for injection are used. Effects of nalorphine are noted within 1-3 minutes when injected intravenously. For intramuscular injection, 10-20 minutes are required. The dose of nalorphine is approximately 0.25 mg/kg. Induction time for naloxone is within 2-3 minutes when injected intramuscularly. The dose of naloxone is 0.006 mg/kg.

ANTIDOTE. Occasionally nalorphine produces agonistic effects in an animal. Naloxone will reverse these effects.

MISCELLANEOUS CONSIDERATIONS

Combinations of immobilizing agents are commonly used in zoo practice. None of these are cleared by the FDA. There is, however, a decided advantage in using combinations that allow dosage reduction.

Skill and experience are prerequisites to successful combination of immobilizing agents. No clear-cut guidelines can be given.

Curare (*d*-tubocurarine), succinylcholine chloride, and nicotine alkaloid are drugs used in the early development of chemical restraint. Both *d*-tubocurarine and nicotine alkaloid have been supplanted with safer and more effective drugs.

Animal control officers face a dilemma. Nicotine alkaloid is the only drug readily available for lay use, yet the TI is very low. Wide experience with use of this drug in a species such as the dog allows the development of sufficient skill to keep mortality rates low. When attempts are made to use nicotine alkaloid in wild species or in domestic animals other than the dog, mortality rates rise to unacceptable levels.

Despite the tremendous historical importance of these drugs in the developmental stages of chemical restraint, neither curare nor nicotine can be recommended as suitable immobilizing agents.

LEGAL ASPECTS OF USING IMMOBILIZING DRUGS IN THE UNITED STATES [17]

Few of the available immobilizing agents have been given FDA clearance for use in other than specific domestic species. Drug companies are unable to justify the expense of carrying out the extensive testing necessary to license a drug for use on wild animals. However, immobilizing agents must be used by the zoo veterinarian if proper health care is to be given.

Immobilizing drugs are potent and dangerous. None can be purchased or used without prescription on the order of a licensed veterinarian. Additionally, some are classified as controlled drugs by the Drug Enforcement Administration of the federal government. Examples are listed in Table 4.3. Drugs classified as controlled substances are controlled by both federal and state law. Use of controlled substances requires a special federal registration (BND number).

The federal law is encompassed in the Controlled

TABLE 4.3. Legal status of restraint agents in the United States

Controlled Substances	Special Schedule II	Schedule II	Schedule III	Schedule IV	Prescription	Prescription Vet Only
d-Tubocurarine					+	
Etorphine	+					
Fentanyl-Droperidol		+				
Ketamine					+	
Nicotine salicylate					+	
Phencyclidine			+			
Succinylcholine					+	
Xylazine						+
Atropine					+	
Acepromazine					+	
Diazepam				+		
Diprenorphine	+					

Substances Act of 1970. The act is administered by the Drug Enforcement Administration branch of the Department of Justice. The federal law serves as a model for state law, but states may enact more stringent regulations. In any case, the stricter law always prevails.

The Controlled Substances Act of 1970 places dangerous drugs in categories known as schedules and specifies regulations for their possession, use, and dispensing.

Schedule I drugs (heroin, LSD, and marijuana) have no accepted medical use in the United States and are outlawed.

Schedule II drugs are potent narcotics and short-acting barbiturates. This group also includes the amphetamines. Five restraint drugs fall into this schedule (Table 4.3). Etorphine and diprenorphine are in a special category within Schedule II. These drugs have been removed from Schedule I and placed in Schedule II to allow clinical usage, but security must be consistent with schedule I regulations.

Etorphine and diprenorphine can be purchased only by those having a valid Drug Enforcement Administration registration. Furthermore, such persons must comply with the following guidelines and be specifically investigated and approved by the Drug Enforcement Administration.

1. Federal law restricts this drug to use by or on the order of a licensed veterinarian.

2. Distribution of the product will be limited to veterinarians engaged in zoo and exotic animal practice, wildlife management programs, and research. Individuals who are nonveterinarians and desire to work with the aforementioned drugs are required to obtain the substance through a licensed and approved veterinarian, along with the official DEA 222C Narcotics Order form. The veterinarian must name the individual for whose use the substance is being obtained on the order form.

3. All registrants desiring to handle M99 are required to use a safe or steel cabinet equivalent to a U.S. Government Class 5 security container. This container is rated at 30 man minutes against surreptitious entry, 10 man minutes against forced entry, and 20 man hours against radiological techniques. In lieu of a U.S. Government Class 5 safe or steel cabinet, a bank safety deposit box is acceptable for storage of the drug.

4. All authorized registrants handling M99 are required to maintain complete and accurate records to ensure full accountability for the substance. These records must include specific quantities administered and any amount lost when a target is missed. It is important that the records be complete so that at any time the registrant can fully account for all of the products received.

Schedule III drugs are less dangerous narcotics and ultra short acting barbiturates, including methohexitol.

Schedule IV drugs are dangerous but are less subject to potential abuse. Diazepam has recently been added to this schedule.

A summary of the Controlled Substances Law follows:

By law, anyone who deals with controlled substances in any way is required to register with the Drug Enforcement Administration, that is, practitioners, wholesalers, researchers, manufacturers, and importers. The only person who can legally possess these substances without such registration is the ultimate consumer who possesses a legitimate prescription.

Veterinarians who wish to become registered must: (1) be licensed to practice by the state, (2) be free of felony convictions for drug related offenses, (3) have separate registration for each address where practice is conducted, (4) pay a fee, (5) apply for registration on application form BND-224 (available from the Registration Branch, Drug Enforcement Administration, P.O. Box 28083, Central Station, Washington, D.C. 20005), and (6) specify which schedules of drugs one wishes to use. Chemical restraint drugs are included in Schedules II, III, and IV.

All records and inventories must be kept for 2 years. An inventory must be made biannually. Records of receipt of all drugs in Schedule II must be kept separate from all other business records and the actual date of receipt recorded. Records of receipt of drugs in Schedules III, IV, and I must be kept in an "easily retrievable manner." All drugs in Schedule II must be ordered on an official BND-222 order form and this must be signed by the registrant. A practitioner who has controlled substances in his possession must store them in a safe or a substantially constructed, locked cabinet. Any theft of controlled substances must be reported to a regional office of the Drug Enforcement Administration at the time the theft is realized. Local police must also be notified.

All laws controlling drugs are subject to change. Contact regional offices of the Drug Enforcement Administration for current regulations.

DEFINITION OF TERMS

Analgesia: Absence of pain.

Analgesic: A drug abolishing pain without producing unconsciousness or sleep.

Anesthesia: Without sensation or loss of sensation, with an accompanying reversible depression of nervous tissue either locally or generally.

Ataractic: An agent capable of producing ataraxia. (Tranquilizers fit this classification.)

Ataraxia: Impassiveness or calmness; perfect peace or calmness of mind; detached serenity without depression of mental faculties or clouding of consciousness.

Catalepsy: A condition characterized by waxy rigidity of muscles. The extremities tend to remain in any position in which they are placed.

Cataleptic: (1) A drug inducing catalepsy. (2) A person or animal in a state of catalepsy. (Phencyclidine analogs are cataleptics.)

Hypnosis: A condition resembling deep sleep, resulting from moderate depression of the central nervous system.

Narcosis: Has a variable meaning, but is generally similar to the following: analgesia accompanied by deep sleep.

Neuroleptic: (NEURO + LEPSIS; a taking hold) A drug or agent that produces symptoms resembling those of disorders of the nervous system. (Tranquilizers fit into this category.)

Neuroleptoanalgesia: (1) A state produced by the combination of a neuroleptic and an analgesic. (2) A combination of a tranquilizer and a narcotic analgesic. (Innovar-Vet is an example of a drug that induces neuroleptoanalgesia.)

Sedation: A mild degree of central nervous system depression in which an animal is awake but calm, free of nervousness, and incapable of fully responding to external stimulation.

REFERENCES

1. Alford, B. T.; Burkhart, R. L.; and Johnson, W. P. 1974. Etorphine and diprenorphine as immobilizing and reversing agents in captive and free-living ranging mammals. J. Am. Vet. Med. Assoc. 164:702-5.
2. Banditz, R. 1972. Sedation, immobilization and anesthesia with Rompun® in captive and free-living wild animals. Vet. Med. Rev. 3:204-26.
3. Beck, C. C. 1972. Chemical restraint of exotic species. J. Zoo Med. 3:3-66.
4. Crockford, J. A.; Hayes, F. A.; Jenkins, J. H.; and Feurt, S. W. 1958. An automatic projectile type syringe. Vet Med. 53:115-19.
5. Fazekas, J. F.; Shea, J. G.; Ehrmantraut, W. R.; and Alman, R. W. 1957. Convulsant action of phenothiazine derivatives. J. Am. Med. Assoc. 165:1241-45.
6. Fowler, M. E., and Hart, R. 1973. Castration of an Asian elephant using etorphine anesthesia. J. Am. Vet. Med. Assoc. 163:539-43.
7. Hankoff, L. D.; Kaye, H. E.; Engelhardt, D. M.; and Freedman, N. 1957. Convulsions complicating ataractic therapy, their incidence and theoretical implications. N.Y. State J. Med. 57:2967-72.
8. Haarthoorn, A. M. 1965. Application of pharmacological and physiological principles in restraint of wild animals. Wildl. Monog. 14, p. 74.
9. ———. 1966. Restraint of undomesticated animals. J. Am. Vet. Med. Assoc. 149:875-80.
10. ———. 1970. The Flying Syringe. London: Geoffrey Bles.
11. ———. 1973. Review of wildlife capture drugs in common use. In E. Young, ed. The Capture and Care of Wild Animals. Capetown, South Africa: Human and Rousseau.
12. ———. 1976. The Chemical Capture of Animals. London: Baillière, Tindall.
13. Haigh, J. C., and Hopf, H. C. 1976. The blowgun in veterinary practice: Its uses and preparation. J. Am. Vet. Med. Assoc. 169:881-83.
14. Lord, W. 1958. The historical development of an instrument for live capture of animals. Southeast. Vet. 9:147-49.
15. Pistey, W. R., and Wright, J. F. 1961. The immobilization of captive wild animals with succinylcholine II. Can. J. Comp. Med. 25:59-68.
16. Short, R. V., and King, J. M. 1964. The design of a crossbow and dart for immobilization of wild animals. Vet. Rec. 76:628-30.
17. U.S. drug enforcement agency guidelines for handling and using M99 and M50-50. 1975. (Mimeo.) U.S. Dep. of Justice.
18. Young, E., ed. 1973. The Capture and Care of Wild Animals. Capetown, South Africa: Human and Rousseau.

5 STRESS

THE THREE objectives of this chapter are: (1) to introduce and describe the importance of the stress concept in relation to the restraint and handling of animals; (2) to explore methods of preventing the development of the stress syndrome; and (3) to detail some important pathophysiological bases for the occurrence of this phenomenon. This plan provides for a general overview of problems associated with stress, followed by an in-depth discussion of the phenomenon.

Every organism must cope with potentially fatal forces in the environment. Primitive organisms react to cold, heat, dryness, excessive moisture, and lack of nutrients. More complex organisms develop systems to inform the animal of destructive environmental changes and stimulate responses, allowing the animal to adapt to the new conditions (Fig. 5.1). Lacking this reaction, a zebra would not flee from a stalking lion nor a person pull a hand back from the flame.

Claude Bernard first delineated the interaction of animals with the environment [16]. He defined an internal environment characterized by its constancy and an external environment characterized by its variability. Cannon, in 1914, coined the term "homeostasis," defining it as the steady state obtained by the optimum action of counteracting processes (physiological regulation) [8].

Overstimulation of some normally protective mechanisms has been found to bring destructive influences to bear on organisms. In documenting these adverse changes, Selye coined the terms "general adaptation syndrome" (GAS), "noxious stimuli," and "alarm reaction" (Fig. 5.2) [32-35].

Hundreds of scientists, including physiologists, endocrinologists, biochemists, neurologists, and behaviorists, have attempted to unravel the complex biochemical mechanisms involved in the development of the stress syndrome. In spite of intensive research, the stress concept is controversial and difficult to define precisely.

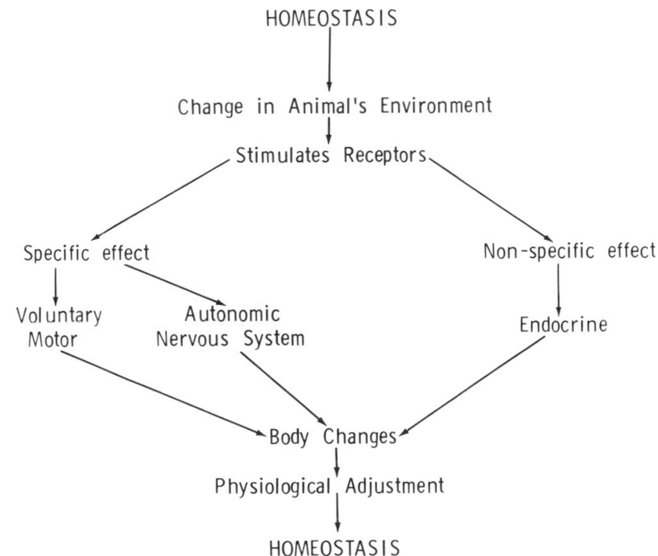

FIG. 5.1. Neuroendocrine pathways for regulation of homeostasis.

It is clear that restraint and handling procedures produce some of the most stressful episodes of an animal's life. Therefore it is incumbent on those manipulating animals to understand the basic physiologic reactions triggered in an animal when normal activity is restricted.

DEFINITION OF TERMS

Stress—The cumulative response of an animal resulting from interaction with its environment via receptors [34].

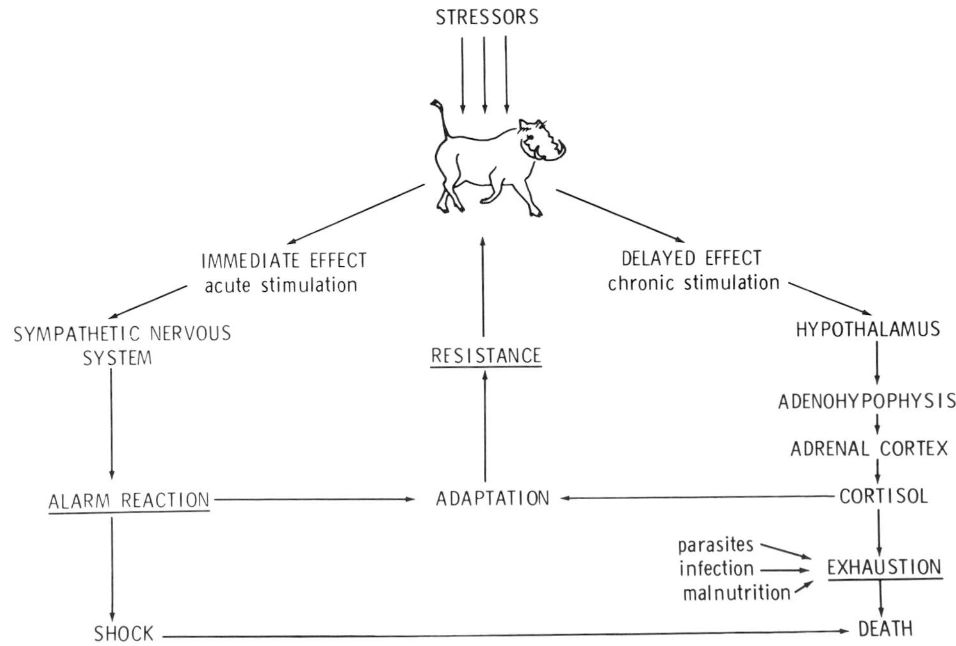

FIG. 5.2. Schema of the "General Adaptation Syndrome" depicting the stages of alarm, resistance, and exhaustion.

Stress is an adaptive phenomenon. All responses are primarily directed at coping with environmental change. Behavioral repertoires may be dependent on the stressful interaction of an animal with its environment. Intense or prolonged stimulation induces detrimental responses in the animal.

Some researchers use the term "stress" only when stimulation is destructive. I prefer the broader definition given above.

Stressor—A stress-producing factor; any stimulus that elicits a nonspecific response when perceived by an organism.

Specific response—The response by an organism that is appropriate for the stimulated receptor. For instance, when cold receptors are stimulated, the body experiences a sensation of coolness. Various somatic and behavioral changes occur that conserve heat and stimulate increased heat production such as shivering, piloerection, or postural changes. The animal is adjusting to a new situation (homeostatic accommodation).

Nonspecific response—In addition to specific responses to an individual stimulus, the body reacts with nonspecific responses to many stimuli and stressors. When a cold receptor is stimulated, an impulse also acts upon the hypothalamic adenohypophyseal adrenal pathway (HAAP). As the animal attempts to adapt to the temperature change, numerous nerve pathways respond, biochemical and endocrine systems react, and subtle changes occur in the body. The same changes may be produced in response to fright, excessive or strange sounds, unexpected touches, or any other type of stimulus. Nonspecific effects may not be immediately identifiable.

Physiological adaptation—The development of qualitatively new protective processes.

Exhaustion—The failure of a physiological adaptive mechanism.

BASIC CONCEPTS

The animal is stimulated by stressors (environmental changes) via receptors. The nervous system analyzes and processes impulses from the receptors and feeds responses back through various components of the nervous system to effector organs, producing either a specific or nonspecific reaction or both. Both reactions are important in restraint procedures. Stressors initiate these reactions, so it is important to understand those stressors that operate during restraint.

Stressors can be classified as somatic, psychological, behavioral, and some miscellaneous stressors.

Somatic stressors (Fig. 5.3) that may act during restraint and handling procedures include strange sounds, sights, odors, and possibly tastes, since animals bite at unusual objects. Additional stressors include unexpected touches, changes in position, abnormal stretching of muscles and tendons, heat and cold changes, pressure changes, effects of chemicals or drugs used in restraint, and insufficient oxygen as a result of struggling or the application of excessive pressure. In prolonged restraint and handling procedures, thirst and/or hunger may also act.

Each reaction to a stressor has adaptive significance.

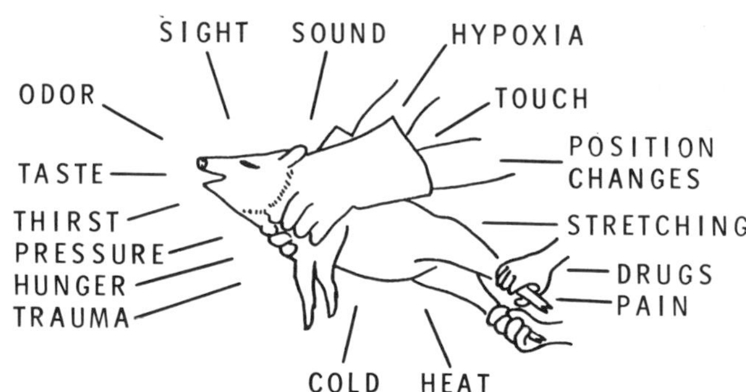

FIG. 5.3. Somatic stressors acting during restraint.

Reactions carried to the extreme can become detrimental and may elicit potentially fatal responses in the restrained animal.

Psychological stressors (Fig. 5.4) act intensely on the higher primates, including man. They also play an important role in the adaptation of other wild species to the captive environment and to restraint practices. Apprehension is a mild psychological stressor that may intensify to become anxiety, fright, or, in its most severe form, terror. The look in the eyes of certain animals when being captured and restrained gives the strong impression that these animals are terror stricken. Some animals become angry and progress toward rage. The emotions of fright or anger are adaptive. Certainly they are involved in the flight or fight reaction. It is when animals are in a constant state of fright or anger that harmful, nonspecific reactions occur.

Anxiety ⟶ Fright ⟶ Terror

Anger ⟶ Rage

Frustration ⟶ No Means of Escape

FIG. 5.4. Psychological stressors acting during restraint.

Frustration is also a psychological stressor for animals under restraint. An animal faced with a strange situation in its normal environment will escape or fight. The animal becomes frustrated when both normal alternatives are prevented by restraint; it can neither escape nor defend itself. Intense frustration experienced over a long period elicits dangerous, nonspecific reactions.

Closely allied to restraint stressors are the many *behavioral stressors* (Fig. 5.5) that act prior to actual restraint or are experienced following restraint, such as unfamiliar surroundings, overcrowding, territorial and hierarchial upsets, upset biological rhythms, lack of social contact or, conversely, lack of isolation, and lack of habitual or imprinted foods. Combined with restraint, these factors are stressors lacking adaptive context.

Miscellaneous stressors (Fig. 5.6) include malnutrition, toxins, parasites, infectious agents, burns, surgery, drugs, chemical and physical immobilization, and confinement. These stressors may act over a long period, contributing to the exhaustion of general adaptation systems (Fig. 5.2).

Overcrowding
Territorial Upset
Hierarchial Upsets
Disruption of Biological Rhythms
Lack of Social Contact - Isolation
Lack of Habitual or Imprinted Foods
Unfamiliar Surroundings

FIG. 5.5. Behavioral stressors that may compound restraint stress.

Malnutrition
Toxins
Parasites
Infectious Agents
Chemical Agents - Drugs
Immobilization
Burns

FIG. 5.6. Additional stressors compounding restraint stress.

Fatal adrenal shock may be triggered if an animal in an imminent exhaustion phase is subjected to a restraint procedure that overtaxes the already depleted body reserves.

Considering all the influences that act upon an animal during restraint, it is not difficult to perceive that such procedures subject an animal to an intensely stressful period. Usually, when the stimulation is of short duration, the animal adjusts without harm. Domestic animals seldom suffer serious adverse effects. However, certain wild species may injure themselves during the alarm phase or pass quickly into fatal shock. They may also be weakened if previous confinement has produced a state of chronic stress. Any restraint procedure may then add the crucial stressful event that triggers an acute or exhaustive response as a result of a lack of body reserves.

BODY RESPONSE TO STRESS STIMULATION

Response to the stimulation of a receptor may follow one of three pathways: voluntary motor, adrenal medulla, or hypothalamic adenohypophyseal adrenal (Fig. 5.7).

Voluntary Motor

The stimulus eliciting a voluntary motor response may be initiated peripherally or internally. The impulse is relayed to the thalamus and thence to the neocortex, where it is categorized and integrated for transmission to the motor areas, which relay the information back through the lower brain centers, through the spinal cord, and thence to the peripheral nerves (Fig. 5.7).

Responses elicited via the voluntary motor system may include avoidance, struggling, escape attempts, running, hiding, defensive or protective postures, vocalization, and aggressive behavior [23].

Each species responds to stimuli with innate and conditioned reflexes unique to its genetic makeup and learned experiences. A squirrel monkey may cower in a corner when approached with a net. A colobus monkey or a langur may attack. The toothless giant anteater defends itself by driving the claws of its forepaw into the unwise restrainer who grabs it by the nose.

Further examples will be discussed in the sections dealing with specific animal groups. In general, animals respond in a manner characteristic for the species when placed in a crisis situation. Restraint practices must be adapted to minimize or counter the injurious effects of such responses on both the animal and the restrainer.

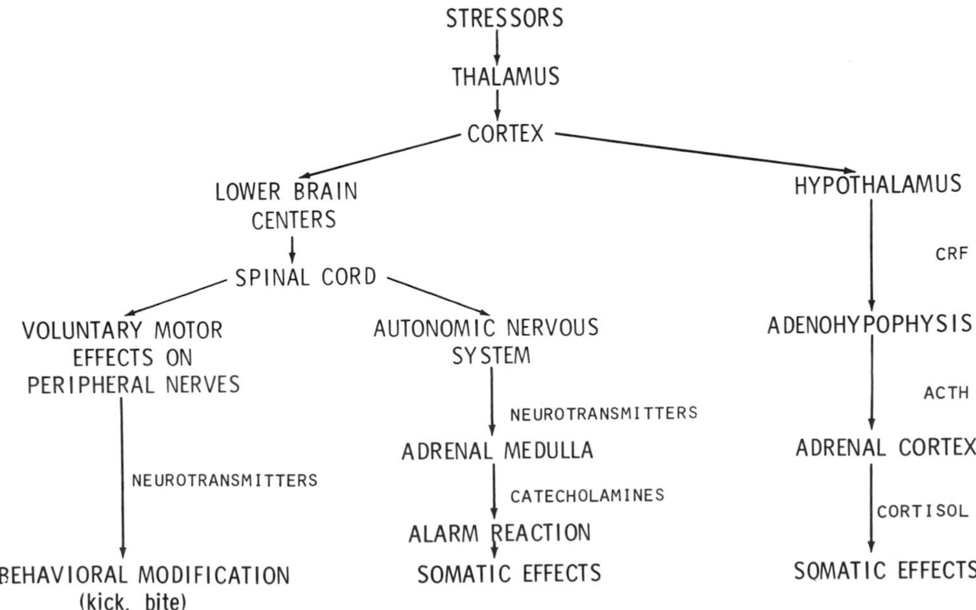

FIG. 5.7. Neuroendocrine pathways by which stressors affect the body.

Sympathetic Nervous System-Adrenal Medulla

Stimulation of the sympathetic nervous system and the adrenal medulla induces the flight or fight reaction [9], or, as described by Selye [33], the alarm reaction (Fig. 5.8).

The alarm response is adaptive. Nonetheless it seriously hampers restraint practices. The syndrome is familiar; we have each experienced the phenomenon and have empathized with frightened fellow human beings, pets, and livestock species. The intensity of the alarm response appears to be greater in those species that are flighty and nervous. The muntjac deer is an excellent example; it is extremely difficult to handle this deer without precipitating fatal shock.

The major medical problem associated with the alarm reaction is trauma inflicted on the animal as it seeks to escape. Contusions, concussions, lacerations, nerve injuries, hematomas, and fractures are common sequelae to injudicious restraint practices. These injuries may directly cause death or result in disability, which further compounds stress. Many individuals have difficulty accepting the idea that the death of a wild animal may truly result from apparently healed injuries received weeks or months earlier.

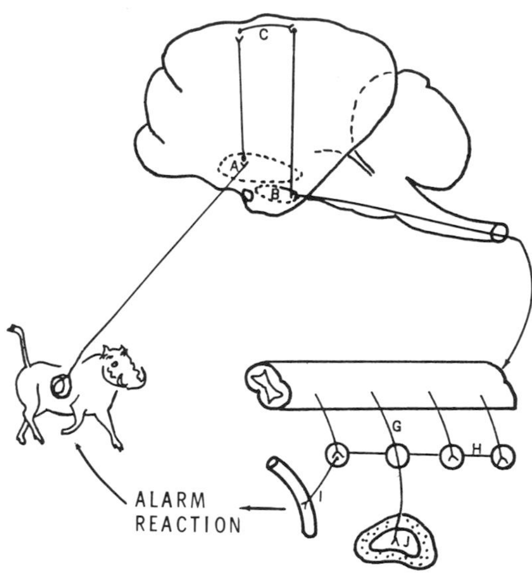

Vasoconstriction Skin & Intestine	Bronchodilatation
	Piloerection
Vasodilatation Muscles & Heart	Mydriasis
Blood Glucose ↑	Coagulation time ↓
Metabolic rate ↑	Pain threshold ↑
Muscle Fasciculation	Blood Pressure ↑
Alertness ↑	Secretions ↓

FIG. 5.8. Neuroendocrine pathways and clinical signs of alarm response: **A.** Thalamus. **B.** Hypothalamus. **C.** Neocortex. **G.** Preganglionic fiber of sympathetic nerves. **H.** Sympathetic trunk. **I.** Postganglionic fiber of sympathetic nerves. **J.** Adrenal medulla.

An injury may necessitate close confinement and therapy which cannot be tolerated by that species. The chronic restraint syndrome may develop. Other consequences may be cardiac arrest and shock. These are described in detail in Chapter 7.

The alarm response changes the body's reaction to many drugs, including some commonly used for chemical restraint. These restraint agents are potentially lethal unless used with wisdom and understanding.

Although thermoregulation is discussed in detail in Chapter 6, it is important to reiterate here that heat is produced by muscle activity associated with the alarm response. To prevent harmful hyperthermia and resultant hypoxia, monitor body temperature and take appropriate cooling action if required.

The alarm response may also produce profound changes in hematocrit and hemoglobin levels. Blood samples collected from excited animals are likely to provide false values.

Although it may be helpful to recognize the signs and results of the alarm response, it is more important to understand and institute techniques to prevent or minimize alarm. Basically this consists of diminishing or eliminating the sensory stimulations that initiate the response. Catch a budgerigar in a darkened room, hood a falcon, avoid chasing an antelope, use squeeze cages, and administer the appropriate chemical restraint agent to unalarmed animals. Select the restraint technique (some are described in later chapters) that allows manipulation of the animal at hand with the least amount of stimulation.

Hypothalamic Adenohypophyseal Adrenocortical Pathway (HAAP)

The third response pathway involves the neuroendocrine system (Figs. 5.7–5.10). Continuous adrenal cortex stimulation and excessive production of cortisol elicit many adverse metabolic responses. Psychological as well as physical changes occur. The following collective clinical signs have been observed in human beings, dogs, and horses with known adrenocortical hyperfunction. Similar signs can be expected in other species. However, the overall clinical syndrome in a given species may vary widely from the expected responses.

Clinical signs may include muscle weakness and trembling, bilaterally symmetrical alopecia, atrophy of temporal muscles, enlarged abdomen, weight loss, increased susceptibility to bacterial infections, impaired antibody response, vaccination failure, high blood pressure, poor wound healing, frequent urination, and high consumption of water.

Chronic stress specifically affects circulating leukocyte numbers. Lymphocyte and eosinophil activity is suppressed. Neutrophil numbers increase. Other clinicopathologic parameters are altered, making it difficult to utilize blood counts and blood chemistries as aids in diagnosing diseases in wild animals.

Behavioral changes include increased aggressive and antisocial tendencies. The chronically stressed animal may refuse to eat or drink. Instead of food refusal, some individuals engorge themselves (hyperphagia). Hypersexuali-

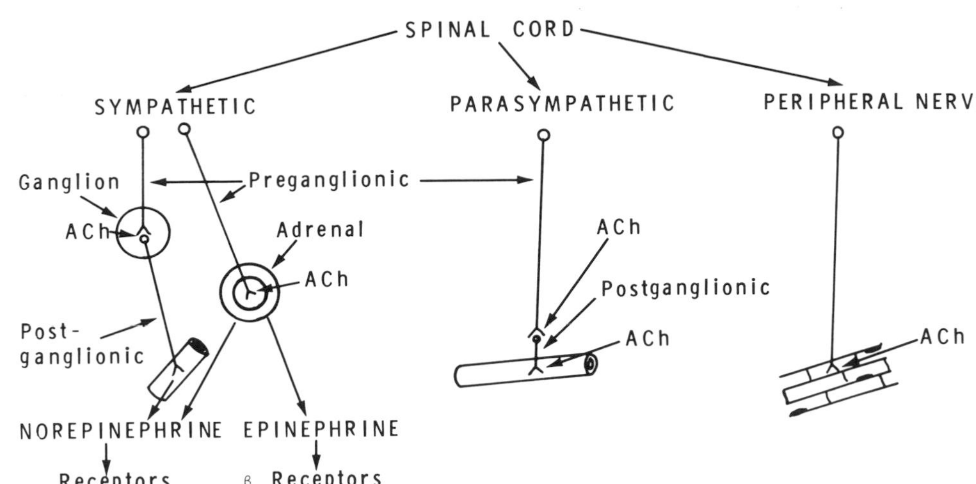

FIG. 5.9. Neurohormonal pathways from spinal cord to autonomic and voluntary nervous systems (ACh = acetylcholine).

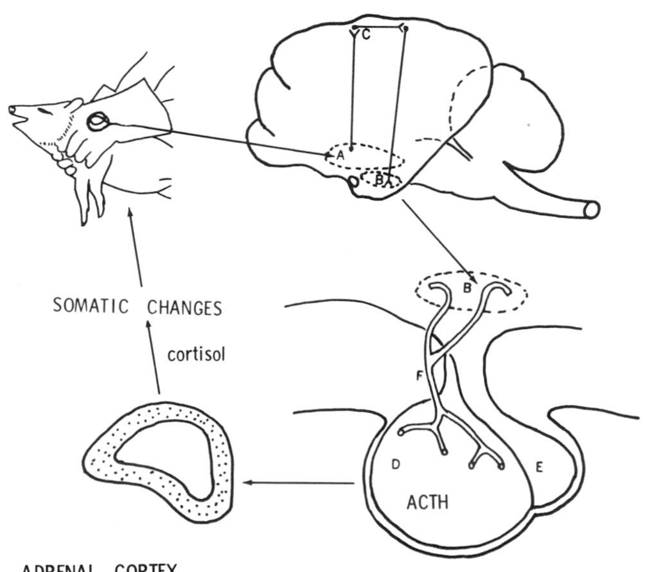

FIG. 5.10. Hypothalamic adenohypophyseal adrenal pathway: **A.** Thalamus. **B.** Hypothalmus. **C.** Neocortex. **D.** Adenohypophysis. **E.** Neurohypophysis. **F.** Hypothalamic adenohypophyseal portal vein.

ty is evidenced by masturbation and excessive copulation. Conversely, hyposexuality may be seen.

To prevent chronic stress, minimize noxious stimuli. Evaluate the whole program of animal care, from pen and cage design and proper feeding to opportunities for social interaction. Succinctly, prevention of stress means providing a physical and social environment that approximates the normal habitat as closely as possible.

PATHOPHYSIOLOGY OF STRESS

Receptors are the communicating link between the animal and its environment (Table 5.1). Species and individual animals within a species vary widely in regard to sensitivity to stimuli. Nocturnal owls utilize acute auditory receptors to capture food and move in darkness. Many rodents and other species have tactile hairs (vibrissae) which relay information important to that species. Psychological stressors are perceived by the body through visual, auditory, tactile, and other receptors.

TABLE 5.1. Receptors

Teleceptors (stimuli received from a distance)
Sight (visual)
Sound (auditory)
Odor (olfactory)
Exteroceptors (cutaneous stimuli)
Heat (thermal)
Cold (thermal)
Touch
Pressure (proprioceptor)
Pain (nociceptor)
Interceptors (visceral and internal stimuli)
Hunger
Thirst
Taste (chemoceptor)
Oxygen and carbon dioxide tension (carotid body)
Deep pressure
Body position (vestibular)

The development of unique receptors has allowed certain species to utilize a particular food or occupy a favorable niche in the environment, gaining an advantage over other species. Receptors reach the most sophisticated level of organization in the higher mammals.

Clinicians handling animals must appreciate the biological variation of sensory perception among species in order to minimize stimulation of highly developed receptors. Overstimulation may cause changes in biological responses so drastic that normally adaptive mechanisms become life destructive.

Afferent nerves carry impulses from the receptors via cranial and peripheral nerves, relaying impulses to appropriate sensory areas of the cortex. After interpretation by the cortex, interconnecting neurons transmit impulses to motor areas of the cortex and thence to lower brain centers such as the thalamus, hypothalamus, and limbic systems for further integration [10,12,13,14,21,22,26,37, 38,39].

The hypothalamus is composed of numerous nuclei, which are collections of neurons. Afferent and efferent nerves pass to and from these nuclei in passing to interconnections with the limbic system, thalamus, basal ganglia, and other midbrain pathways. All these systems act in concert with the cortex. The hypothalamus integrates regulatory functions important to restraint such as thirst, hunger, thermoregulation, and release of catecholamines and other neuroendocrine secretions.

Closely associated with the hypothalamus is the limbic system, which is a complex collection of centers in the lower brain. Behaviorists have given a great deal of attention to the limbic system because it is involved with many functions such as olfaction, acquired and instinctual behavior patterns, and emotions. The limbic system and the hypothalamus, together, are concerned with the control of biological rhythms, sexual behavior, motivation, and emotions such as fear and rage. Higher primates are capable of intense emotional responses and develop some unique behavioral patterns under stress.

Although much of the basic research presently being carried out is directed at elucidating behavioral problems in the human being, animals often serve as models to develop hypotheses. Information gained from research programs involving animals should contribute to the knowledge of those behavioral changes that can be detrimental to the continued existence of such animals [25].

Alarm Response

The hypothalamus contains centers controlling both sympathetic and parasympathetic autonomic pathways. Emotional or physical stimulation common to restraint procedures usually triggers a sympathetic response typical of the alarm reaction.

Preganglionic fibers of the sympathetic system innervate the adrenal gland (Fig. 5.9). Stimulation of the sympathetic system causes release of acetylcholine (ACh) within the adrenal medulla. Depolarization of the chromaffin cells by ACh effects the release of epinephrine and smaller amounts of norepinephrine, both of which are kept in cellular stores.

Stimulation of the sympathetic system results in the production of norepinephrine at the effector organ. Thus in a stimulated animal both norepinephrine and epinephrine are elaborated, producing variable responses determined by the blood concentrations of each substance and their action on α and β adrenergic receptors (Table 5.2).

The clinical signs of the alarm reaction are depicted in Figure 5.8. Harmful or lethal consequences include ventricular fibrillation [24], hypoglycemia, hyperthermia, and shock. Detailed discussions of these conditions are found in Chapters 6 and 7.

Hypothalamic Adenohypophyseal Adrenal Pathway

Immediately adjacent to the hypothalamus is the hypophysis or pituitary gland. Morphologically this organ is composed of a posterior segment under direct neural stimulation and an anterior segment or adenohypophysis which has no neural connections. The adenohypophysis shares a common blood supply with the hypothalamus via portal vessels.

TABLE 5.2. Chart of sympathetic stimulation responses [12]

Effector Organ	Norepinephrine Response (α receptors)	Epinephrine Response (β receptors)
Eye	Mydriasis	Relaxation of ciliary muscle for far vision
Heart	No response	Increased heart rate; increased contractility
Blood vessels		
Coronary	Vasodilatation	Vasodilatation
Skin	Vasoconstriction	Vasodilatation
Skeletal muscle	Vasodilatation	Vasodilatation
Abdominal viscera	Vasoconstriction	No response
Lung		
Bronchial muscles	No response	Relaxation
Intestines		
Motility and tone	Decrease	Decrease
Sphincters	Contraction	No response
Skin		
Pilomotor	Contraction	No response
Sweat glands	Slight localized secretions	No response
Salivary gland	Thick viscous secretion	Amylase secretion

The adenohypophysis is an important regulatory endocrine gland. It secretes at least seven hormones important to body functions: adrenocorticotropic hormone (ACTH), thyroid-stimulating hormone or thyrotropin (TSH), growth hormone or somatotropin (STH), follicle-stimulating hormone (FSH), luteinizing hormone in the female (LH) or interstitial cell-stimulating hormone in the male (ICSH), prolactin or luteotropic hormone (LTH), and intermedin or melanocyte-stimulating hormone (MSH).

Each of these adenohypophyseal hormones are under the direct control of releasing factors (polypeptides) produced by neurostimulation of the hypothalamus. They are carried to the adenohypophysis via portal blood vessels from the hypothalamus. The hormones of the anterior pituitary stimulate the respective target organs to produce hormones affecting somatic tissue. ACTH stimulates the adrenal cortex to produce cortisol, while catecholamines, sex hormones, and thyroid hormones may moderate cortisol production.

In the normal feedback mechanism, cortisol inhibits the secretion of cortisol-releasing factor (CRF), hence decreasing the production of ACTH. However, mechanisms activated by highly stressful situations can override this inhibiting effect. Other hormones stimulated by adenohypophyseal-tropic hormones operate under similar feedback mechanisms.

The mammalian adrenal cortex contains three microscopically distinct layers: the zona glomerulosa (outer), which produces mineralocorticoids such as aldosterone; the zona fasciculata (middle), which produces cortisol and other glucocorticoids; and the zona reticularis (inner), which produces sex steroids. Stress effects are limited primarily to stimulating cortisol production. Prolonged stress is deleterious to the adrenal gland, and specific diseases of the adrenal cortex and adenohypophysis may alter the production of any or all of the adrenocortical steroids.

The clinical syndromes of adrenocortical malfunction

have been identified in some species (man, dog, horse, and laboratory animals). Virtually nothing is known about the effects of hypocorticism or hypercorticism in nondomestic species.

One must be cautious when evaluating the adrenal gland at necropsy. Little data is available describing normal weights and shapes. In mammals, the cortex surrounds the medulla. In amphibians, birds, and reptiles, the cells are intermixed.

Biological Effects of Cortisol

Cortisol stimulates protein catabolism. It is also glyconeogenic, lipolytic, antiinflammatory, and antiimmunologic. Specifically, excessive levels of cortisol increase protein breakdown and nitrogen excretion, with resultant decrease in muscle mass (weight loss). Weakness of abdominal muscles causes a potbellied appearance. Hepatomegaly and a distended bladder contribute to the enlarged abdomen. Atrophy of the hair follicles and thinning of the epidermis are followed by bilateral symmetrical alopecia.

Failure of collagen synthesis causes resorption of the osteoid framework of bone, resulting in osteoporosis. Wound healing and scar tissue formation are impaired through inhibition of fibroblast proliferation and collagen failure. Wound dehiscence following surgery is common.

Protein catabolism and lipolysis contribute to the pool for glyconeogenesis. Slight to moderate hyperglycemia has a diuretic effect, producing polyuria and polydipsia. Prolonged hyperglycemia stimulates the β cells of the pancreas to produce more insulin.

Cortisol reduces the heat, pain, and swelling associated with the inflammatory response. This effect is useful in the treatment of many diseases. The antiinflammatory action of cortisol is brought about by reducing capillary endothelial swelling, thus diminishing capillary permeability. Additionally, capillary blood flow is decreased by the action of cortisol. Both of these actions are helpful in shock therapy.

The integrity of lysosomal membranes is enhanced by cortisol. Under such circumstances, bacteria and other particulate matter are engulfed by phagocytes, but hydrolytic enzymes (which would destroy the organisms) are not released from the lysosomes.

Interference with DNA synthesis causes atrophy of lymphoid tissue throughout the body. Cell-mediated immune responses are also diminished. This could interfere with tuberculin testing programs.

Within a few hours of cortisol stress response, reduction in the number of circulating lymphocytes is 50% or greater. Lymphocyte levels return to normal in 24-48 hours following cessation of stress.

The effect of stress on the leukocyte count varies with the species and depends upon the normal relative leukocyte distribution. Species with normally high percentages of lymphocytes such as mice, rabbits, chickens, and cattle respond with a lymphopenia and neutrophilia and a *decrease* in total leukocytes. Dogs, cats, horses, and human beings, having relatively low lymphocyte counts, respond with an *increase* in leukocytes [6] (Fig. 5.11).

Eosinophil production decreases in response to elevated levels of cortisol. Eosinophil production is directly related to histamine production such as occurs in the event of tissue injury or allergic reactions. Cortisol neutralizes histamines and inhibits regranulation of mast cells, thus reducing further histamine production. The elevated production of cortisol during stress results in eosinopenia. Catecholamines also cause eosinopenia; thus emotional stress may also elicit a stress hemogram [30].

	Mice, Rabbits Chicken, Cattle (High % Lymphocytes)	Dogs, Cats Horse, Man (Low % Lymphocytes)
Neutrophilia	+	+
Lymphopenia	+	+
Eosinopenia	+	+
Total leukocytes	decrease	increase

FIG. 5.11. Stress hemogram variations among species.

In addition, cortisol stimulates increased production of circulating erythrocytes. Serum calcium levels decrease as well, through inhibition of calcium absorption from the gastrointestinal tract.

Prolonged stress results in hypertrophy of the adrenal gland, caused by continual stimulation by ACTH and resulting in the excessive production of cortisol. Ultimately the adrenals become exhausted and atrophic. Manipulation of an animal in a state of incipient adrenal insufficiency is likely to cause fatal shock.

Hyperadrenocorticism and Hypoadrenocorticism

Hyperadrenocorticism and hypoadrenocorticism are clinical syndromes of human beings, dogs, and horses, usually caused by the presence of a tumor of the pituitary or adrenal gland. These syndromes may serve as guides to the types of signs that develop in any animal suffering from chronic stress. It is important to recognize that subtle differences may appear in the syndrome as it develops in each mammalian species. Little is known about these diseases in reptiles, amphibians, and fish, but it is probable that similar reactions occur in these animals.

Hyperadrenocorticism (Cushing's syndrome in human beings) is characterized by malfunctioning protein metabolism, evidenced by the development of reddish striae and thinning of the skin, loss of bone matrix accompanied by demineralization, poor wound healing, muscle wasting and weakness, capillary fragility, bruising, and impaired growth in children.

Abnormal carbohydrate metabolism is demonstrated by overt diabetes mellitus and abnormal glucose tolerance curves. Centripetal fat distribution and "moon faces" also occur, indicating upset lipid metabolism.

Mineralocorticoid (aldosterone) production contributes to electrolyte balance. Malfunction causes sodium retention, potassium loss, hypertension, and hypervolemia. Hematopoietic effects typical of cortisol stimulation include lymphopenia, eosinopenia, neutrophilia, and erythrocytosis.

Persons with hyperadrenocorticism may also have impaired immunological tolerance, gastric ulceration, increased incidence of renal calculi, and psychoses.

In dogs the predominant signs of hyperadrenocorticism consist of polydipsia, polyuria, bilateral alopecia, and an enlarged, pendulous, flaccid abdomen when the rest of the body is normal or thin. Other generalized signs that may be present include lethargy, polyphagia, hyperpigmentation, and testicular atrophy [31].

Hypoadrenocorticism may be acute or chronic. The acute syndrome is characterized by shock, resulting from complete inability of the animal to adapt to stressful situations. Persons restraining captive wild animals should be fully aware of the possibility that an animal may be in the exhaustion phase of the GAS and that the added stress of handling may trigger acute adrenal insufficiency. Should this syndrome develop during restraint, the animal must be treated quickly to avert hypovolemic shock and imbalance of serum electrolytes caused by increased potassium levels as a result of hypoaldosteronism.

Chronic hypoadrenocorticism (Addison's disease) in man is characterized by weakness, fatigue, weight loss, increased pigmentation of the skin, anoxia, and, less commonly, diarrhea.

Hypoadrenocorticism in dogs is characterized by progressive weakness, weight loss, and gastrointestinal upset such as vomiting and anorexia. Dehydration accompanied by polyuria and polydipsia may also develop. The exhaustion phase of the GAS (Fig. 5.2) may be characterized by adrenal insufficiency.

Chronic Restraint Stress Syndrome

Seventy-five years ago, Goldscheider noted physiological changes in physically restrained rabbits [15]. Since then, numerous scientists have recognized and documented the effects of chronic stress due to restraint, usually as an unanticipated and complicating response of control animals [25,36].

Volumes of reports of the effects of prolonged restraint have resulted from experiments conducted by scientists involved in the space program [2,6,7,20]. Confining animals and persons in vehicles used for space travel has necessitated severe restriction of muscular activity (hypokinesia), which is particularly stressful. Counterweighting, which unloads the leg muscles, has a similar effect. Consequently weightlessness, which would likewise unload the legs, should also contribute to a chronic restraint syndrome. Various nonhuman primates, small laboratory animals, and chickens have been subjected to weightlessness in chronic restraint trials.

Hypokinesia produces an interesting phenomenon in chickens. A chicken suspended in a restraint harness develops the chronic restraint syndrome within a few hours. However, if the bird can press its feet against a spring perch, it will tolerate more severe restraint without apparent ill effects. A certain level of leg pressure is necessary for homeostasis in chickens [36].

Dogs and monkeys become disoriented and develop muscle weakness. Lymphopenia and eosinopenia are typical hematological responses to chronic stress.

Mineral metabolism is altered during chronic restraint [18,40,41,42]. Calcium excretion increases. Osteoporosis, causing bone fragility, may develop due to progressive decalcification of bone. Bone decalcification occurs in animals larger than the dog but is not seen in smaller laboratory animals. Conversely, the development of gastric erosion and ulceration is limited to rats, mice, and other small laboratory species weighing less than 8 kg. These lesions have not been produced in rabbits, dogs, or monkeys [3,4].

Hypothermia has been noted during chronic restraint in mammals, likely because of increased heat loss through peripheral vasodilatation [17]. Birds, better insulated by feathers, do not become hypothermic [36].

The signs of chronic restraint stress have been duplicated by the administration of cortisol [1,42].

Not all animals are equally susceptible to restraint stress. In general, the larger the animal, the greater the adverse effect of restraint. Position in a hierarchy bears on the response of an individual animal to restraint. Dominant animals are apt to suffer more from being restrained than subordinates. Wild animals are affected to a higher degree than domestics. Individual variation is apparent within species. Furthermore, conditioning helps animals tolerate restraint stress.

Sexual differences in response to stress have been observed [43]. Unanesthetized chickens, restrained for 1-3 hours by being placed on their backs, experienced changes in cardivascular parameters. In hens, heart rate, cardiac output, and stroke volume increased, while blood pressure and hematocrit decreased. In roosters both blood pressure and heart rate increased. The heart rate increase persisted for many hours following restraint.

Animals kept in severely restrictive shipping crates for long ocean voyages are likely candidates to develop chronic restraint syndrome.

Since simple restraint may profoundly influence physiological and pathological processes, investigators should carefully consider the protocol used in animal experimentation. Furthermore, experimental animals should be maintained in cages large enough to eliminate the development of any restraint/restriction symptoms [11]. At present, minimum cage size recommendations are not based on experimental evidence. Minimum cage size standards need to be accurately determined and instituted.

Chronic restraint stress can be compared with the bedrest syndrome in human beings. Prolonged bed rest induces physiological, psychological, and even pathological complications. Normally a person remains supine only while sleeping. Remaining in this position for prolonged periods upsets certain body functions [5]. During prolonged bed rest, altered arterial pressure, poor venous return, interference with ciliary action of the bronchioles, muscle atrophy, joint stiffness, and disuse osteoporosis may develop. People also frequently lose the ability to adapt to an erect position and may faint when attempting to stand (orthostatic hypotension).

Diseases resulting from such system alterations are deep

vein thrombosis, pulmonary embolism, pneumonia, decubitis, constipation, difficult micturation, muscle weakness, pathological fractures, vasomotor instability, and psychological disorders. Knowledge of these human system disorders has profound implications for those who must continually restrain a wild animal in a squeeze cage for prolonged therapy. Likewise, the animal that must remain recumbent for days should be constantly monitored and every effort made to minimize and alleviate the effects of such confinement.

Lesions Resulting from Stress

The lesions produced by harmful stress are difficult to document. Pathologists often negate a diagnosis of death caused by stress. Many of the effects of stress are functional, leaving no definitive physical lesions to mark their presence. Nonetheless tissues and organs are weakened by prolonged insult [19,27,28,29], lowering resistance to disease. The actual cause of death may be pneumonia, parasitism, or starvation, but stress has paved the way for development of these terminal ailments.

Adrenocortical hypertrophy, lymphoid aplasia, gastrointestinal ulceration, and arterial calcification are lesions that may be produced in experimental animals through prolonged stress. Similar lesions are found in both wild and domestic animals.

CONCLUSION

Although the exact role of stress in the development of lesions is not understood, lesions attributed to stress have been frequently observed in the field. Orphaned California sea lion pups exhibit a high incidence of gastric ulcers. Marsupials suffer from arterial calcification, particularly when crowded and during inclement weather. Lymphoid aplasia and adrenal hypertrophy are common lesions found in wild animals at necropsy.

The manifold manifestations of somatic and psychological stress mimic the signs of other generalized or specific diseases. This compounds differential diagnosis problems for the clinician who is faced with disease, particularly in wild animals. The similarity of the signs of stress to those of other diseases has led some to unwisely minimize the possibility of restraint damage in a wild animal. Any person undertaking the restraint, particularly the repeated restraint, of a wild animal should expend every effort to eliminate or minimize the numerous noxious stimuli accompanying the procedure. Otherwise the animal may suffer from deleterious effects induced by these stimuli, either immediately or as a delayed response.

REFERENCES

1. Bartlett, R. G., Jr., and Miller, M. A. 1956. The adrenal cortex in restraint hypothermia and in adaptation to the stress of restraint. J. Endocrinol. 14:181-87.
2. Besch, E. L.; Smith, A. H.; Burton, R. R.; and Sluka, S. J. 1967. Physiological limitation of animal restraint. Aerosp. Med. 38:1130-34.
3. Brodie, D. A. 1962. Ulceration of the stomach produced by restraint in rats. Gastroenterology 43:107-9.
4. Brodie, D. A., and Hanson, H. M. 1960. A study of the factors involved in the production of gastric ulcers by the restraint technique. Gastroenterology 38:353-60.
5. Browse, N. L. 1965. The Physiology and Pathology of Bed Rest. Springfield, Ill.: Charles C Thomas.
6. Burton, R. R., and Beljan, J. R. 1970. Animal restraint applications in space (weightless) environment. Aerosp. Med. 41:1060-65.
7. Burton, R. R.; Smith, A. H.; and Beljan, J. R. 1971. Effect of altered "weight" upon animal tolerance to restraint. Aerosp. Med. 42:1290-93.
8. Cannon, W. B. 1914. The emergency function of the adrenal medulla in pain and major emotions. Am. J. Physiol. 33:356-72.
9. ______. 1929. Bodily Changes in Pain, Fear, Hunger and Rage, 2nd ed., p. 404. New York: Appleton.
10. Dillon, R. S. 1973. Handbook of Endocrinology. Philadelphia: Lea & Febiger.
11. Draper, W. A., and Bernstein, I. S. 1963. Stereotyped behavior and cage size. Percept. Mot. Skills 16:231-34.
12. Eyzaguirre, C. 1969. Physiology of the Nervous System. Chicago: Yearbook Medical Publishers.
13. Ezrin, C.; Goodden, J. O.; Volpe, R.; and Wilson, R., eds. 1973. Systematic Endocrinology. New York: Harper & Row.
14. Ganong, W. F., and Forsham, P. H. 1960. Adenohypophysis and adrenal cortex. Annu. Rev. Physiol. 22:579.
15. Goldscheider, A., and Jacob, P. 1894. Ueber die Variationen der Leukocytose [Changes in leukocytosis]. Z. Klin. Med. 25:373-448.
16. Grande, F., and Visscher, M. B., eds. 1967. Claude Bernard and Experimental Medicine, pp. 179-89. Cambridge, Mass.: Schenkman.
17. Grant, R. 1950. Emotional hypothermia in rabbits. Am. J. Physiol. 160:285-90.
18. Groover, M. D.; Seljeskog, E. L.; Haglin, J. J.; and Hitchcock, C. R. 1963. Myocardial infarction in the Kenya baboon without demonstrable atherosclerosis. Angiology 14:409-16.
19. Gross, A. L.; Krough, L.; Miesse, J. W.; and Roberson, K. T. 1966. Calcium, phosphorus, and magnesium mobilization resulting from inactivity in monkeys. U.S. Aerosp. Med. Tech. Rep., pp. 66-94.
20. Hoffman, R. A.; Dozin, E. A.; Mack, P. B.; Hood, W. N.; and Parrott, M. W. 1968. Physiologic and metabolic changes in *Macaca nemestrina* on two types of diets during restraint and nonrestraint. I. Body weight changes, food consumption and urinary excretion of nitrogen creatinine-creatinine. Aerosp. Med. 39:693-98.
21. Issacson, R. L. 1974. The Limbic System. New York: Plenum Press.
22. James, V. H. T., ed. 1968. Recent Advances in Endocrinology. Boston: Little, Brown.
23. Jenkins, W. L., and Kruger, J. M. 1973. Modern concepts of the animals' physiological response to stress. In E. Young, ed. The Capture and Care of Wild Animals, pp. 172-83. Capetown, South Africa: Human and Rousseau.
24. Klide, A. M. 1973. Mechanisms of death associated with restraint and handling of wild animals. Proc. Annu. Meet. Am. Assoc. Zoo Vet., 1972, 1973, pp. 102-3.
25. Knize, D. M.; Weatherby-White, R. C. A.; Geisterfer, D. J.; and Paton, B. C. 1969. Restraint of rabbits during prolonged administration of intravenous fluids. Lab. Anim. Care 19:394-99.
26. Kurtsin, I. T. 1968. Physiological mechanism of behavior disturbances and corticovisceral interrelations in animals. In M. W. Fox, ed. Abnormal Behavior in Animals, pp. 107-16. Philadelphia: W. B. Saunders.
27. Lapia, B. A., and Cherkovich, G. M. 1971. Environmental changes causing the development of neuroses and corticovisceral pathology in monkeys. In L. Levi, ed. Society, Stress, and Disease, vol. 1, pp. 266-95. London: Oxford Univ. Press.
28. McKioch, D. 1971. The development of gastrointestinal lesions in monkeys. In L. Levi, ed. Society, Stress and Disease, vol. 1. London: Oxford Univ. Press.
29. Mitchell, G. 1968. Persistent behavior pathology in rhesus monkeys following early social isolation. Folia Primatol. 8:132-47.
30. Schalm, O. W.; Jain, N. C.; and Carroll, E. J. 1975. Veterinary Hematology, 3rd ed., p. 497. Philadelphia: Lea & Febiger.
31. Schechter, R. D. 1974. Hyperadrenocorticism. In R. W. Kirk, ed. Current Veterinary Therapy V, pp. 783-86. Philadelphia: W. B. Saunders.
32. Selye, H. 1936. A syndrome produced by diverse nocuous agents. Nature 138:32.
33. ______. 1950. The Physiology and Pathology of Exposure to Stress, p. 150. Montreal: Acta. (This is the major treatise on the subject, although the first report was in 1936.)
34. ______. 1973. The evolution of the stress concept. Am. Sci. 61:692-99.
35. Selye, H., and Heusner, G. 1956. Fifth Annual Report on Stress, 1955-56, pp. 16-32. New York: M. D. Publications.
36. Smith, A. H., ed. Response of animals to reduced acceleration fields.

Principles of Gravitational Biology. Government document from NASA contract with George Washington Univ., Washington, D.C. (NSR-09-010-027).
37. Sodeman, W. A., and Sodeman, W. A., Jr. 1967. Pathologic Physiology: Mechanisms of Disease, 4th ed. Philadelphia: W. B. Saunders.
38. Swenson, M. J., ed. 1970. Duke's Physiology of Domestic Animals, 8th ed., pp. 1112, 1113, 1231. Ithaca, N.Y.: Cornell Univ. Press.
39. Ungar, G. 1963. Excitation, pp. 313-23. Springfield, Ill.: Charles C Thomas.
40. Urist, M. R. 1960. Cage layer osteoporosis. Endocrinology 67:879-80.
41. Urist, M. R., and Deutsch, N. M. 1960. Osteoporosis in the laying hen. Endocrinology 66:377-91.
42. ______. 1960. Effects of cortisone upon blood, adrenal cortex gonads, and the development of osteoporosis in birds. Endocrinology 60:805-18.
43. Whittow, G. C. 1965. Cardiovascular changes in restrained chickens. Poult. Sci. 44:1452-59.

SUPPLEMENTAL READING

General

Barnett, S. A. 1955. Competition among wild rats. Nature 175:126-27. (On depletion of adrenal cholesterol and/or rise in adrenal weights of subordinates)

Deitrick, J. E.; Whedon, G. D.; and Shorr, E. 1948. Effects of immobilization upon various metabolic and physiologic functions of normal men. Am. J. Med. 4:3-36.

Sedgwick, C. J. 1972. Clinical philosophy for exotic animal practice. Proc. Annu. Meet. Am. Anim. Hosp. Assoc., pp. 30-55.

______. 1973. General adaptation syndrome, a curse or a blessing. Proc. Annu. Meet. Am. Assoc. Zoo Vet., pp. 172-73.

Von Ehler, U. S. 1956. Stress and catechol hormones. In H. Selye and G. Heusner, eds. Fifth Annual Report on Stress, 1955-56, pp. 125-37. New York: M. D. Publications.

Welch, B. L. 1965. Psychophysiological response to the mean level of environmental stimulation: A theory of environmental integration. Symposium of medical aspects of stress in the military climate. USGPO Publication, pp. 39-99, 778-814.

Restraint and Exercise Stress

Berendt, R. F. 1968. The effect of physical and chemical restraint on selected respiratory parameters of *Macaca mulatta*. Lab. Anim. Care 8:391-94.

Chapman, C. B., and Mitchell, J. H. 1965. The physiology of exercise. Sci. Am. 212:88-96.

Falls, H. B. 1968. Exercise Physiology. New York: Academic Press.

Morehouse, L. E., and Miller, A. T. 1971. Physiology of Exercise, 6th ed. St. Louis: C. V. Mosby.

Morris, D. 1964. The response of animals to a restricted environment. Symp. Zoolog. Soc. London, pp. 99-118.

Shephard, R. J. 1969. Endurance Fitness. Toronto: Univ. of Toronto Press.

Behavioral, Social, and Psychological Stress

Berkson, G. 1967. Abnormal stereotyped motor acts. In J. Zubin and H. F. Hunt, eds. Comparative Psychopathology, pp. 79-94. New York: Grune-Stratton.

______. 1968. Development of abnormal stereotyped behaviors. Dev. Psychobiol. 1:118-32.

Cassel, J. 1970. Physical illness in response to stress. In S. Levine and N. S. Scohn, eds. Social Stress, pp. 189-209. Chicago: Aldine.

Fox, M. W., ed. 1968. The influence of domestication upon behavior of animals. In Abnormal Behavior in Animals, pp. 64-76. Philadelphia: W. B. Saunders.

Froberg, J.; Karlsson, C.; Levi, L.; and Lidberg, L. 1971. Physiological and biochemical stress reactions induced by psychosocial stress. In L. Levi, ed. Society, Stress and Disease, vol. 1. London: Oxford Univ. Press.

Hediger, H. 1968. The Psychology and Behavior of Animals in Zoos and Circuses. New York: Dover Publications.

Lazarus, R. 1971. The concepts of stress and disease. In L. Levi, ed. Society, Stress and Disease, vol. 1, pp. 53-58. London: Oxford Univ. Press.

Levi, L., ed. 1967. Endocrine reactions during emotional stress. In Emotional Stress. New York: American Elsevier.

Lissak, K., and Fudroczi, E. 1965. The Neuroendocrine Control of Adaptation. Elmsford, N.Y.: Pergamon Press.

Mason, W. A. 1963. The effects of environmental restriction on the social development of rhesus monkeys. In C. H. Southwick, ed. Primate Social Behavior, pp. 161-73. Princeton, N.J.: D. Van Nostrand.

Sassenrath, E. N. 1970. Increased adrenal responsiveness related to social stress in rhesus monkeys. Horm. Behav. 1:283-98.

Sassenrath, E. N.; Hine, L. J.; and Kita, A. H. 1969. Social behavior and corticoid correlates in *M. mulatta*. Proceedings of the Second International Congress of Primatology, vol. 1, pp. 219-31. Basel and New York: Karger.

Snyder, R. L., ed. 1975. Behavioral stress in captive animals. In Research in Zoos and Aquariums, pp. 41-76. Washington, D.C.: National Academy of Science.

Walls, S., and Fox, M. W. 1973. Wild animals in captivity: Veterinarians' role and responsibility. J. Zoo Anim. Med. 4:7-17.

6 THERMOREGULATION

LIFE IS dependent on energy. Animals obtain energy through the chemical action of ingested nutrients plus radiant energy from the sun. Products of energy use within the body are further chemical action, heat, and work. The animal body is an energy transformer, utilizing energy to produce work, locomotion, growth, maintenance, reproduction, and useful products such as wool and milk. Excess heat must be dissipated from the body by radiation, conduction, convection, or evaporation [5].

For practical purposes, the body temperature of a given animal is understood to be the temperature recorded on a thermometer inserted into the rectum deeply enough to reflect the core or deep temperature of the animal. Other temperatures may be of concern to the animal restrainer (Fig. 6.1). The temperature at the skin surface may be either higher or lower than core temperature. Heat may be lost or gained, depending on the temperatures of the skin and the surface upon which the animal is placed. The coat temperature of well-insulated species will likely be close to ambient air temperature. The effect of insulating layers, both external and internal, on body temperatures is of critical importance during restraint.

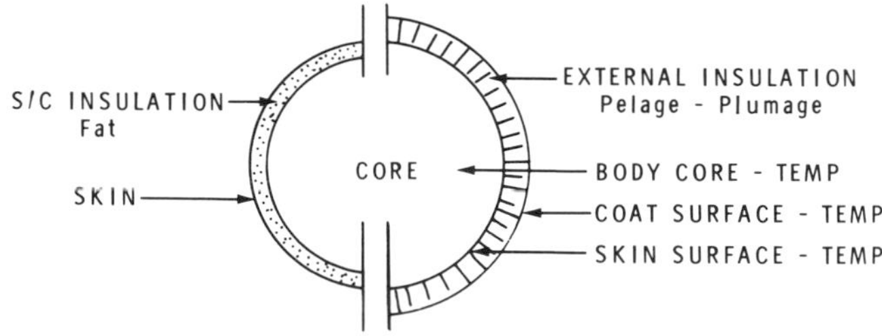

FIG. 6.1. Schematic of insulation methods and their relationship to temperatures.

Homeotherms (endotherms), including most birds and mammals, are capable of physiological responses that initiate heat production or conservation. Body temperatures usually remain within narrow limits. The body temperature of poikilotherms such as reptiles, amphibians, fishes, and invertebrates fluctuates with the ambient temperature. The primary source of heat for these animals must be external. The primary method available to poikilotherms for cooling, if the ambient temperature rises to a point incompatible with life, is to move into a cooler environment. Thus heat regulation in this class of animals is behavioral.

Thermoregulatory problems of wild animals during restraint are likely to be more difficult to prevent than those of domestic species. Domestic animals, bred for docility and accustomed to people, usually accept restraint practices. Wild animals, on the other hand, usually struggle against restraint until completely exhausted. Violent muscular activity generates significant quantities of heat. Unless the animal is physiologically and physically in a position permitting heat dissipation, hyperthermia will result. Unalleviated hyperthermia can cause death in a matter of minutes.

The degree of temperature elevation is directly related to the duration and intensity of muscular effort [16], modulated by inherent heat regulatory mechanisms. Small species overheat more quickly than large species, due to a higher metabolic rate.

Restraint of animals during periods of high ambient air temperatures and/or high relative humidity is fraught with danger. Under such conditions muscular activity will generate heat more rapidly than it can be dissipated.

PHYSIOLOGY

Many physiological and behavioral mechanisms are involved with heat regulation in animals (Table 6.1). Details can be obtained from the references [4,5,6,8,9,10,12, 14,17].

Hot and cold receptors in the skin act as detectors, alerting the body to environmental conditions that may be destructive. When the appropriate receptor is stimulated, the impulse is relayed to special cells in both the anterior and posterior hypothalamus. The temperature of blood flowing through the hypothalamus may directly affect thermosensitive cells.

Information is integrated with specific motor responses to either increase heat production, and/or conserve heat, or increase heat loss (Fig. 6.2). The whole system functions as a thermostat.

Heat Production

Heat is gained by increased production or by absorption from the environment. The body produces heat through basal metabolic activities, muscle tone, shivering, exercise, fever (disease), and by utilization of special energy stores such as brown fat. Heat from the environment is absorbed by radiation, conduction, and convection (Figs. 6.3, 6.4).

Heat and Moisture Conservation

Heat conservation mechanisms are not as important to the animal restrainer as are mechanisms of heat production, except in a negative way. An animal that is an efficient heat conserver may require special attention to provide cooling if it overheats during restraint. Heat is conserved through vascular responses such as peripheral vasoconstriction and by countercurrent heat exchange systems (Fig. 6.5).

Peripheral vasoconstriction is important for heat conservation; it allows the skin temperature to drop without jeopardizing the core temperatures. Arteriolar constriction decreases blood flow to the skin. Venous constriction increases the velocity of blood flow, which decreases the exposure time of the blood to cold. All species are capable of

TABLE 6.1. Mechanisms for coping with hyperthermia

Mammal Group	Evaporation						Behavioral						Other						Prime Cooling Mechanism
	Sweating	Panting	Moisture on skin	Saliva on skin	Spraying	Swimming	Postural adjustment	Decreased activity	Seeking shade	Nocturnal activity	Burrowing	Torpidity	Insulation (detrimental)	Thermal windows	Conduction	Convection	Radiation	Heat storage	
Order Monotremata (echidna, platypus)	0	++	+	+	0	±	0	+	+	±		±	0	0	+	+	+	0	Panting
Order Marsupialia (kangaroo, wallaby)	0	++	+	+	0	0	0	+	+	±		±	0	0	+	+	+	0	Panting
Order Insectivora (shrew, hedgehog)	0	++	±	±	0	0	0	+	+	±	+	±	0	0	+	+	+	0	Panting
Order Primates Suborder Prosimiae (lemur, tarsier)	0	+	0	0	0	0	0	+	+	+	0	±	0	0	+	+	+	0	Behavioral
Suborder Simiae Superfamily Ceboidea (New World)	0	+	0	0	0	0	0	+	+	0	0	0	0	0	+	+	+	0	Panting, behavioral
Superfamily Cercopithecoidea (Old World)	±	+	0	0	0	0	0	+	+	0	0	0	0	0	+	+	+	0	Panting, behavioral
Superfamily Hominoidea (ape)	++	+	0	0	0	0	0	+	+	0	0	0	0	0	+	+	+	0	Sweating
Order Chiroptera (bat)	0	0	0	0	0	0	0	+	0	+	0	+	0	+	0	+	0	0	Behavioral
Order Tubulidentata (anteater)	0	0	0	0	0	0	0	+	+	+	0	0	0	0	0	+	0	0	Behavioral

TABLE 6.1. *(continued)*

Mammal Group	Evaporation						Behavioral							Other						Prime Cooling Mechanism
	Sweating	Panting	Moisture on skin	Saliva on skin	Spraying	Swimming	Huddling, nest building, postural adjustment	Cyclic activity	Decreased activity	Seeking shade	Nocturnal activity	Burrowing	Torpidity	Insulation (detrimental)	Thermal windows	Conduction	Convection	Radiation	Heat storage	
Order Edentata (sloth)	0	0	0	0	0	0	+	+	+	+	+	+	0	0	0	0	+	+	0	Behavioral
Order Pholidota (pangolin)	0	0	0	0	0	0	+	+	+	+	+	+	0	0	0	0	+	+	0	Behavioral
Order Lagomorpha (rabbit, hare)	0	±	+	±	0	0	+	+	+	+	+	+	0	0	±	0	+	+	0	Behavioral
Order Rodentia (rodents) Suborder Sciuroidea	0	0	+	++	0	0	+	+	+	+	+	+	±	±	+	0	+	+	±	Behavioral
Suborder Myomorpha	0	0	+	++	0	0	+	+	+	+	+	+	±	±	+	0	+	+	±	Behavioral
Suborder Hystricomorpha	0	0	+	++	0	0	+	+	+	+	+	+	±	±	+	0	+	+	±	Behavioral
Order Cetacea (whale, dolphin)	0	0	+	0	0	+	0	0	0	0	0	0	0	+	+	+	+	0	0	Conduction

TABLE 6.1. *(continued)*

Mammal Group	Evaporation						Behavioral							Other						Prime Cooling Mechanism
	Sweating	Panting	Moisture on skin	Saliva on skin	Spraying	Swimming	Postural adjustment	Cyclic activity	Decreased activity	Seeking shade	Nocturnal activity	Burrowing	Torpidity	Insulation (detrimental)	Thermal windows	Conduction	Convection	Radiation	Heat Storage	
Order Carnivora																				
Family Canidae (dog, wolf)	0	++	0	+	0	±	0	+	+	+	±	0	0	+	0	0	+	+		Panting
Family Ursidae (bear)	0	+	0	+	0	±	0	+	+	+	0	0	±	+	0	0	+	+		Panting
Family Procyonidae (raccoon)	0	+	0	+	0	±	0	+	+	+	±	0	0	+	0	0	+	+		Panting, behavioral
Family Mustelidae (skunk, otter)	0	+	0	+	0	±	0	+	+	+	±	0	0	+	0	0	+	+		Panting, behavioral
Family Viverridae (mongoose, civet)	0	+	0	+	0	±	0	+	+	+	±	0	0	+	0	0	+	+		Panting, behavioral
Family Hyaenidae (hyaena)	0	+	0	+	0	0	0	+	+	+	±	0	0	+	0	0	+	+		Panting, behavioral
Family Felidae (cat)	0	+	0	+	0	0	0	+	+	+	+	0	0	+	0	0	+	+		Panting, behavioral
Order Pinnipedia (seal, sea lion)	0	0	+	0	0	+	0	0	0	0	0	0	0	+	+	+	+	0	0	Conduction
Order Proboscidea (elephant)	±	0	+	0	+	+	0	0	+	+	0	0	0	0	+	+	+	+	0	Thermal windows, behavioral
Order Perissodactyla																				
Family Equidae (horse, zebra)	+	0	0	0	0	0	0	0	+	+	0	0	0	0	0	0	+	+	0	Sweating
Family Tapiridae (tapir)	0	±	+	0	0	+	0	0	+	+	0	0	0	0	0	±	+	+	0	Conduction, behavioral
Family Rhinocerotidae (rhino)	0	±	0	0	0	+	0	0	+	+	0	0	0	0	0	0	+	+	0	Behavioral
Order Artiodactyla																				
Suborder Suiformes																				
Family Suidae (swine)	0	++	+	0	0	+	0	+	+	+	±	0	0	+	0	0	+	+	0	Panting
Family Tayassuidae (peccary)	0	++	+	0	0	+	0	+	+	+	±	0	0	+	0	0	+	+	0	Panting
Suborder Tylopoda																				
Family Camelidae (cameloid)	+	±	0	+	0	0	0	+	+	+	0	0	0	+	0				+	Behavioral
Family Hippopotamidae (hippo)	±	0	+	0	0	+	0	+	0	0	+	0	0	+	0	+	+	0	±	Conduction, behavioral
Suborder Ruminantia							0													
Family Cervidae (deer)	0	++	0	0	0	0	0	+	+	+	±	0	0	+	±	0	+	+	0	Panting, behavioral
Family Giraffidae (giraffe)	0	++	0	0	0	0	0	+	+	+	0	0	0	0	0	0	+	+	0	Panting, behavioral
Family Bovidae (cattle, sheep, antelope)	±	++	±	0	0	±	0	+	+	+	±	0	0	+	±	0	+	+	±	Panting, behavioral, thermal windows

TABLE 6.1. *(continued)*

Animal Group	Evaporation						Behavioral								Other						Prime Cooling Mechanism
	Sweating	Panting	Moisture on skin	Saliva on skin	Spraying	Swimming	Postural adjustment	Cyclic activity	Decreased activity	Seeking shade	Nocturnal activity	Burrowing	Torpidity	Habitat selection	Insulation (detrimental)	Thermal windows	Conduction	Convection	Radiation	Heat storage	
Turtle, tortoise	0	±	±	±	0	±	+	+	+	+	+	+	+	+	0	±	+	0	+	+	Behavioral
Lizard		±	0	0	0		++	+	+	+	+	+	+	+	0	+	+	0	+	+	Behavioral
Snake			0	0	0	+	++	+	+	+	+	+	+	+	0	0	+	0	+	+	Behavioral
Crocodilians			+	0	0	+	+	+		0	+	0	±	+	0	0	+	0		+	Behavioral
Frog, toad	0	0	+	0	0	+	+	+		+		+	+	+	0	0	+	0	+	0	Behavioral
Salamander	0	0	+	0	0	+	±	±	+	+		+	+	+	0	0	+	0	+	0	Behavioral
Fish	0	0	+	0	0	+	0	0	0	0	0	0	±	+		0	+	0	0	+	Conduction/selection of proper temperature

TABLE 6.1. *(continued)*

	Evaporation						Behavioral							Other	
Bird Group[a]	Sweating	Pantihg	Moisture on skin	Saliva on skin	Spraying	Swimming	Postural adjustment	Cyclic activity	Decreased activity	Seeking shade	Nocturnal activity	Burrowing	Torpidity	Insulation (detrimental)	Thermal windows
Tinamous	0	+	0	0	0	0	+	0	+	+	?	0	0	+	+
Ratites (ostrich, rhea, emu, kiwi)	0	0	0	0	0	0	+	0	0	+	0	0	0	+	0
Grebe	0	0	±	0	0	+	0	0	0	0	0	0	0	+	0
Loon	0	0	±	0	0	+	0	0	0	0	0	0	0	+	+
Penguin	0	+	0	0	0	+	+	+	0	0	0	0	0	+	0
Tubenoses (albatross, petrel, fulmar)	0	0	±	0	0	+	+	0	0	0	0	0	0	+	+
Pelican, cormorant, frigate bird, gannet	0	+[b]	±	0	0	+	+	0	0	0	0	0	0	+	0
Heron, stork, spoonbill	0	+[b]	±	0	0	+	+	0	0	0	0	0	0	+	0
Flamingo	0	0	±	0	0	+	+	0	0	0	0	0	0	+	0
Waterfowl (duck, goose, swan, screamer)	0	0	±	0	0	+	+	0	0	0	0	0	0	+	0
Raptors (hawk, falcon, eagle, vulture)	0	+	0	0	0	0	+	+	+	+	0	0	0	+	0
Pheasant, quail, grouse, chicken, turkey, guinea fowl	0	+[b]	0	0	0	0	+	+	+	+	0	0	0	+	±
Crane	0	+[b]	+	0	0	+	+	0	+	0	0	0	0	+	±
Wader, gull, tern	0		+	0	0	+	+	0	+	0	0	0	0	+	0
Pigeon, dove	0	+[b]	0	0	0	0	+	0	+	+	0	0	0	+	0
Parrot, cockatoo, macaw	0	±	0	0	0	0	+	0	+	+	0	0	0	+	0
Cuckoo	0	0	0	0	0	0	+	+	+	+	0	0	0	+	0
Owl	0	±	0	0	0	0	0	+	+	0	+	0	0	+	0
Nightjar	0	0	0	0	0	0	0	+	+	0	+	0	0	+	0
Swift	0	0	0	0	0	0	0	+	+	0	+	0	0	+	0
Hummingbird	0	0	0	0	0	0	0	+	+	+	0	0	+	+	0
Mousebird	0	0	0	0	0	0	0	+	+	+	0	0	0	+	0
Trogan	0	0	0	0	0	0	0	+	+	+	0	0	0	+	0
Roller	0	0	0	0	0	0	0	+	+	+	0	0	0	+	0
Woodpecker	0	0	0	0	0	0	0	+	+	+	0	+	0	+	0
Perching bird, songbird	0	+	0	0	0	0	0	+	+	+	0	0	0	+	0

[a]Prime cooling mechanism is behavioral response.
[b]Gular flutter.

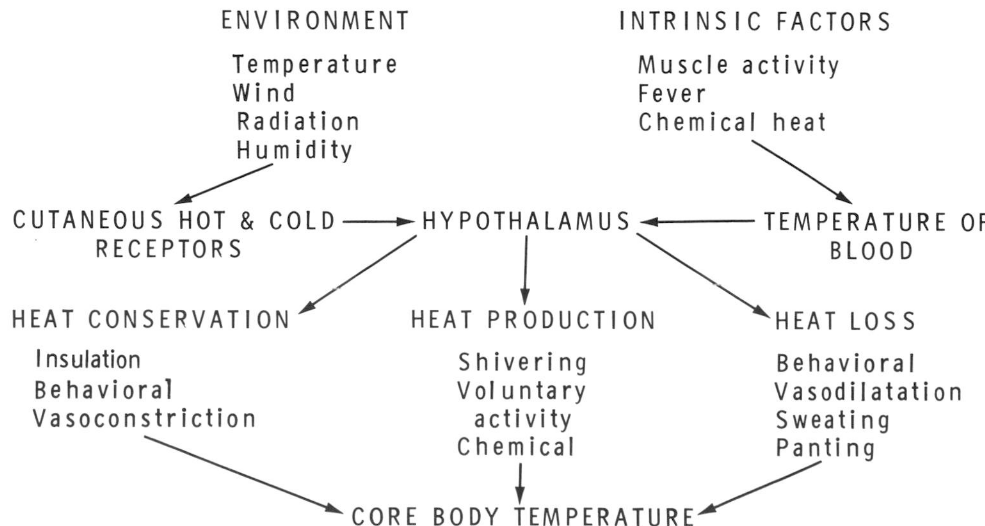

FIG. 6.2. Thermoregulatory mechanisms.

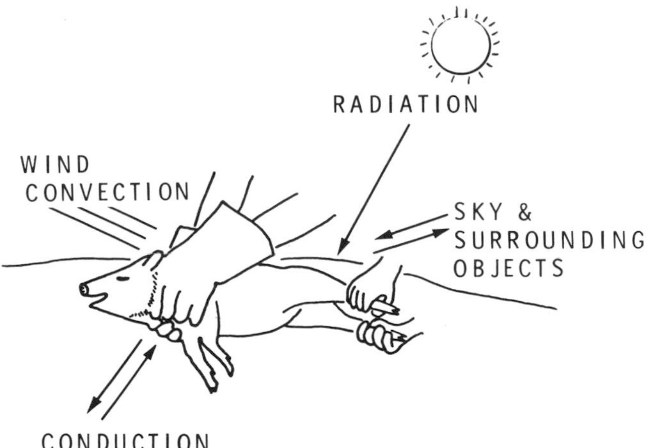

FIG. 6.3. Thermal influences on a restrained animal.

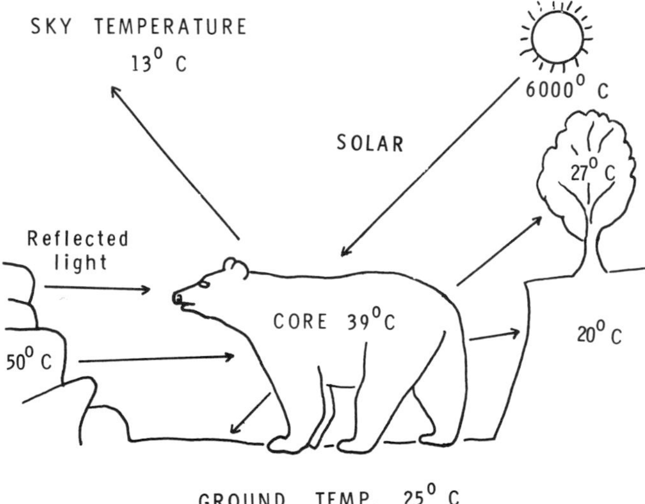

FIG. 6.4. Radiant heat flow to and from an animal.

peripheral vasoconstriction to some degree, but arctic species, highly adapted to cold, exhibit a marked ability to respond in this manner.

Countercurrent heat exchange systems are also highly developed in animals living in extremely cold environments [5]. The animal restrainer must not interfere with these systems but should take advantage of them when handling such species. Captivity in unnatural habitats may adversely alter an animal's ability to acclimate to extremely cold situations. For instance, in winter gulls can walk on ice at −30 C (−22 F) without harm. If gulls are acclimated to warm laboratory conditions and then allowed to walk on ice, their feet freeze [5]. Animals in zoos often live in unnatural environments. Such animals should not be expected to respond in the same manner as free-living wild animals.

Piloerection is also a mechanism that conserves heat by increasing the thickness of the pelage or plumage insulation layer.

Desert species use one or more of the following mechanisms to conserve moisture: voiding of highly concentrated urine, production of dry feces, allowing body temperature to elevate during the heat of the day (heat storage), peripheral vasoconstriction, and insulation or piloerection to prevent evaporation.

Mechanisms for Cooling

Heat is dissipated through conduction, convection, radiation, evaporation, and excretions such as urine and feces.

The conversion of moisture to water vapor (evaporation) requires 0.58 kcal of heat for each gram of water converted [11]. Evaporation takes place from the skin surfaces of all species to a limited degree (insensible heat loss), as long as the relative humidity is not so high that the air is already saturated with moisture [10].

Species with numerous cutaneous sweat glands, such as the horse and human being, maintain effective cooling mechanisms as long as the volume of water intake is sufficient. Surface evaporation can be augmented by smearing the skin with saliva (marsupials) or sponging it with water (elephants).

Evaporation also takes place from the lungs of all breathing animals. Some species such as the dog pant, increasing evaporative cooling from the respiratory tract. Panting is more than rapid breathing. Otherwise the dog would hyperventilate and become alkalotic. The open-mouth breathing of an ungulate may assist in dissipating heat, but other metabolic changes harmful to the animal may also take place.

Morphological Adaptations for Cooling

The large ears of the African elephant increase the surface area for cooling by evaporation, simple conduction, convection, and radiation. The ears are kept moving to assist in heat dissipation. Desert species such as the fennec or bat-eared fox also have large ears to aid in thermoregulation.

It has been demonstrated that the horns of bovid species function in thermoregulation [15]. Do not cover horns for prolonged periods during restraint procedures, since this may inhibit cooling and cause the brain temperature to rise to a dangerous level.

Behavioral changes often allow animals to exist in microclimates within an environment that is generally assumed to be incompatible with the animal's ability to control body temperature. Nocturnal species are active

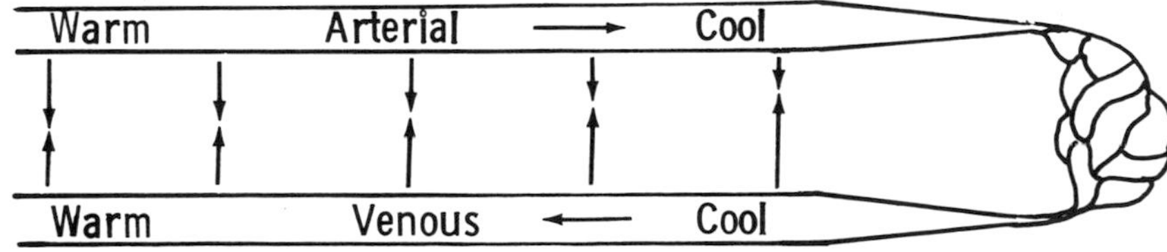

FIG. 6.5. Diagram of countercurrent heat exchange system. Peripheral arteries and veins are adjacent to one another. Warm arterial blood coming from the core gives up heat to cooler venous blood coming from the capillary bed. Central body temperature is not jeopardized.

when it is cool. Rodents burrow during the day when the aboveground ambient temperatue is too high for comfort. The hippopotamus forages at night when it is cool and submerges in water during the hotter part of the day. Desert bovids seek any available shade during the heat of the day to lessen heat gain from the hot desert environment. Many other species rest quietly during the hotter part of the day and forage or move about in the cooler morning and evening.

The crocodile reverses the process. During daylight hours the crocodile basks on the shore, absorbing heat to maintain body temperature during the night when it feeds in the water.

THERMOREGULATORY MEDICAL PROBLEMS

Hyperthermia (heat exhaustion, heat cramps, sun stroke, overheating)

DEFINITION. Hyperthermia is excessive elevation of the body temperature.

ETIOLOGY. Predisposing factors to the development of hyperthermia include prolonged high ambient temperatures, high humidity, and excessive muscular exertion or metabolic activity. Muscular exertion is a particularly important source of heat production during restraint and is more dangerous in fat or heavily insulated animals.

Placing animals in poorly ventilated shipping crates exposed to high ambient temperatures is sure to cause hyperthermia. A brachycephalic dog is less able to cool by panting because it cannot move sufficient air through its narrowed upper airway. Temperatures in enclosed automobiles parked in the sun quickly reach 49–54 C (120–154 F). Pets left in cars may die from hyperthermia.

Dehydration, lack of salt, adrenal insufficiency, and the use of vasodilatory drugs (such as alcohol) contribute to overheating. Additional contributing factors include reduced cardiac efficiency or cardiac failure. Reduced cardiac efficiency may be the result of malnutrition, lack of exercise, infection, or intoxication.

Trauma (extensive contusions, fractures, or lacerations) causes the release of pyrogens as products of tissue destruction. This reaction also takes place following surgery, so a slight elevation in body temperature may be expected. Although usually not dangerous, this slight elevation may tip the balance if body temperature is already at a precariously high level because of heat produced during restraint for the surgery.

The temperature elevation seen in animals suffering from infectious diseases results from increased metabolic activity and enhances phagocytosis and immune body production; it also decreases the viability of disease organisms. Prolonged elevation of temperature for days causes the development of certain physiological conditions that affect restraint practices. Stores of liver glycogen are depleted, with resultant decreased energy stores and potential hypoglycemia. The animal may be forced to call upon body protein for energy, resulting in weight loss, weakness, and increased nonprotein nitrogen in the blood. Elevated body temperature increases the need for fluid intake. If the need is not met, the animal dehydrates.

Restraint techniques may inhibit heat-dissipating mechanisms. Canids use panting for evaporative cooling. As the body temperature elevates, the normal resting respiratory rate of approximately thirty changes to 300–400, increasing the rate of evaporative cooling from the respiratory mucous membranes. It is a common practice to muzzle dogs and other canids to prevent biting. With the muzzle in place, the animal cannot pant. With high ambient temperatures and a struggling animal, hyperthermia is inevitable.

Another example is the use of stockinette to restrain the wings of hawks. Hawks dissipate heat by extending their wings, exposing lightly feathered areas beneath the wings to the air for convection cooling. By clasping the wings close to the body and adding a layer of insulation in the form of the stockinette or nylon hose, heat dissipation is effectively prevented.

Hyperthermia increases metabolic activity and cellular oxygen consumption (10% for each degree C rise in the human being) [7]. In mammals at body temperatures above 41 C (105.8 F), oxygen utilization exceeds the oxygen supplied by normal respiration, initiating hypoxic cellular damage. The brain, liver, and kidneys are most likely to manifest such damage. Protein begins denaturing at approximately 45–47 C (113–116.8 F) in all species [6,11]. Normal body temperatures of birds are 40–42 C (104–107.6 F); thus a struggling bird has a narrow temperature safety margin.

CLINICAL SIGNS. Clinical signs of hyperthermia include increased heart and respiratory rates, accompanied by open-mouth breathing. Species capable of sweating sweat and salivate profusely in the early stages. As the temperature continues to elevate, the animal dehydrates, and both sweating and salivation decline and may cease. As hyperthermia accelerates, the pulse becomes weak and the animal shows signs of restlessness, dullness, and incoordination. Convulsions (cerebral anoxia) and collapse, rapidly followed by death, result if temperatures rise and remain for long above 42–43 C (107.6–109.4 F).

Other metabolic and pathological changes associated with hyperthermia include hypoxemia, metabolic acidosis, hypercalcemia, myoglobinuria, hemoglobinuria, disseminated intravascular coagulation, hemolytic anemia, and renal shut down.

Hyperthermia is characterized by peripheral vasodilatation in an attempt to cool the blood. This results in relative hypovolemic shock. Hyperthermia produces signs similar to those of septicemia, high fevers, and other convulsive syndromes. These should be considered in differential diagnosis.

THERAPY. Cool the animal as quickly as possible. Techniques for cooling include spraying the body surface with cold water or immersing small animals in cold water. In those species having a dense coat, ruffle the hair to allow the water to penetrate to the skin. Cold water enemas and alcohol baths may be beneficial. The animal can be packed in crushed ice. Provide adequate ventilation—circulating air with a fan if necessary—to assist in convection heat removal.

Hypovolemic shock should be treated with rapid administration of cold lactated Ringers and corticosteriods. Sodium bicarbonate should be given to counteract metabolic acidosis. Supplemental oxygen is required to combat hypoxemia.

Hyperthermia may devitalize tissues, resulting in delayed illness such as pneumonia or nephritis. Monitor or observe the animal for several days following a known hyperthermic episode.

Hypothermia (cold stress, freezing, exposure)

DEFINITION. Hypothermia is a decreased body temperature caused when heat loss exceeds heat gain. Hypothermia is normally less damaging to animals than hyperthermia, but if the body temperature of homeotherms falls below 34 C (93.2 F), thermoregulation is impaired, requiring artificial rewarming. Below 30 C (86 F) thermoregulation is completely eliminated [11, p. 1068].

ETIOLOGY. Predisposing factors include exposure to wind (convection cooling) (Table 6.2), a soiled or moistened coat, restraint on a cold surface, and restricted exercise. Another important cause is impairment of central thermoregulatory controls by anesthetics or chemical restraint drugs.

Two roloway monkeys were chemically restrained with ketamine hydrochloride for routine tuberculin testing. Upon completion of the test, each monkey was returned to a large outdoor enclosure and laid on a cold concrete slab. The environmental temperature was 15-18 C (60-65 F). Normally, after ketamine immobilization, an animal revives in 20-40 minutes and is then capable of increased muscle activity to assist internal heat production. One and a half hours later, the attendant found both animals still comatose. The rectal temperature of each animal had dropped below 32 C (90 F). One animal died, the other recovered following intensive treatment.

A chemically restrained animal may be incapable of shivering to generate heat in a cold environment. Hypothermia may quickly ensue.

Newborns are particularly susceptible to hypothermia because of undeveloped thermoregulatory powers. The unattended or neglected newborn quickly becomes hypothermic and may die. Adults of small species become hypothermic more rapidly than those of large species because of the relatively larger surface area exposed by a small animal.

Animals subjected to surgery in addition to restraint may become hypothermic due to heat loss from cold tables, exposure of large surgically prepared areas, large open incisions, the use of vasodilatory drugs (acepromazine), and general anesthesia (halothane).

Excessive application of cleansing solutions and alcoholic skin disinfectants is detrimental to the patient's thermal status, particularly in small species.

Animals in shock become hypothermic quickly. Any hypoxic condition predisposes to hypothermia [17, vol. 2].

Just as excessive muscular activity associated with physical restraint may produce hyperthermia, so prolonged immobility may induce hypothermia, particularly in poorly insulated species [1,2,3,13].

CLINICAL SIGNS. Shivering is a standard sign in animals not sedated or anesthetized. Sometimes it is difficult to differentiate between the shivering that occurs with fright or anger and that of hypothermia. Hypothermic animals are usually dull in reaction and slow to respond to stimuli. The primary sign of hypothermia is an excessive drop in body temperature. Temperatures may decrease below the limits of an ordinary clinical thermometer. Critical evaluation of body temperature with a broader calibrated thermometer may reveal that the temperature is as low as 29.5 C (85 F). When the body temperature drops below 32 C (89.6 F), the animal is likely to become comatose and unable to respond to any stimulation. Artificially induced hypothermia is sometimes used as a technique to replace general anesthesia. Accidental hypothermia may decrease the amount of anesthetic agent required and lead to the assumption that a patient is anesthetized when in reality it is simply hypothermic.

A decrease in body temperature is accompanied by a decrease in cardiac output, heart rate, blood pressure, and glomerular filtration rate. Blood viscosity and hematocrit levels increase. Signs noted with temperatures below 30 C (86 F) may include slow and shallow breathing, metabolic acidosis, "sludging" in the microcirculation, ventricular fibrillation, and coagulation disorders.

THERAPY. Rapid warming of the whole body is essential to maintain life. Local heat applications to the limbs are insufficient for warming the whole body. The most effective way is to immerse the animal in a warm water bath. Water temperature should be maintained between 40.5 C (105 F) and 45.5 C (114 F). Hold the animal's head out of the water and ruffle the hair coat to make sure that heat ex-

TABLE 6.2. Wind chill chart

Estimated Wind Speed		Actual Thermometer Reading Degrees Celcius					Actual Thermometer Reading Degrees Fahrenheit				
		10	−1	−12	−18	−23	50	30	10	0	−10
(km/hr)	*(mi/hr)*	*(relative temperature)*					*(relative temperature)*				
Calm	Calm	10	−1	−12	−18	−23	50	30	10	0	−10
8	5	9	−3	−14	−21	−26	48	27	6	−5	−15
16	10	4	−9	−23	−29	−36	40	16	−9	−21	−33
24	15	2	−13	−28	−38	−43	36	9	−18	−36	−45
40	25	−1	−18	−34	−42	−51	30	0	−29	−44	−59

change is taking place at the skin surface. Large animals that cannot be immersed in water baths can be sprayed with warm water or wrapped in warm blankets and the body surface massaged. Warm water enemas are helpful. Warm broth or other liquids given via stomach tube are indicated. Intravenous infusions of warm saline are effective in raising the body temperature. Surgical exposure of a suitable vein may be necessary to effect intravenous administrations because of the decreased blood pressure.

Small animals can be rubbed dry and warmed by being placed next to the skin of the attendant until more effective warming techniques can be applied.

If a surgical patient becomes hypothermic, the incision site can be flushed with warm (not over 42 C) saline. Circulating water type heating pads are effective in preventing hypothermia in surgical patients and in treating accidental hypothermia. Electric heating pads or blankets are not recommended because they can easily become too hot. Hypothermic and shock patients normally suffer from skin vasoconstriction and are thus incapable of carrying intense heat away from the skin. Electric heating pads have caused skin burns and sloughs. Be cautious when applying heat directly to the skin.

Hot water bottles can be used to raise the ambient air temperature in a small enclosed area (heat tent). A hot water bottle should be wrapped in a towel if it is to be used near the skin. Plastic milk cartons or plastic bags can be used in lieu of a standard hot water bottle.

The air surrounding the patient can be warmed with infrared heat lamps, forced-air driers, electric floor heaters, surgical lamps, or commercial radiant heat infant warmers.

When an animal has been warmed with water, be sure the coat is thoroughly dried as soon as the temperature equilibrates. Damp hair chills the skin. Monitor the temperature frequently during warming to prevent overheating. Observe the animal for a sufficient time following equilibration to be certain it is thermoregulating on its own.

DEFINITION OF TERMS

Basal metabolism (BMR): The metabolic rate necessary to produce the energy required by an animal (in a thermoneutral and postabsorptive state) to carry out basic maintenance functions while at rest. These include blood circulation, respiration, kidney function, and specific dynamic action. BMR is related to body size and surface area. Animals with high BMRs represent a greater risk for hyperthermia during restraint.

Conduction: Direct transfer of heat between an animal and contiguous objects. The rate of transmission of heat is proportional to the temperature gradient. The direction of flow is from higher to lower temperature. An animal lying on hot soil absorbs heat. Contrarily, an animal lying on a cold concrete surface loses heat.

Convection: The transmission of heat by movement of a medium surrounding or within an object. An animal may either absorb or dissipate heat because of air or water movement over the surface of the body. Blood carries heat to and from organs by convection.

Critical temperature: Those ambient temperature levels above and below which life is threatened. High and low critical temperatures vary with the species of animal; one cannot extrapolate critical levels for other species from the information found in the human being.

Countercurrent heat exchange system: An anatomical arrangement of adjacent veins and arteries that warms the blood at the periphery, which assists in stabilization of core body temperature (Fig. 6.5).

Dehydration: A decrease of tissue and cellular body fluids.

Evaporation: The conversion of liquid to vapor. Water absorbs 0.58 kcal for every gram of water evaporated. This is an extremely important cooling mechanism for many animals.

Frostbite: A condition resulting when living tissue is frozen. Gangrene usually develops in the affected tissue. Frostbite commonly affects the extremities of the limbs, the tail, and the tips of the ears.

Heat cramps: Spasms of muscles following reduction of sodium chloride levels in the plasma. Spasms may be part of the syndrome of heat exhaustion or may occur independently.

Heat stroke or sun stroke: Inability of the heat regulatory mechanisms to maintain body temperature. It is characterized by acute onset and extremely high body temperatures up to 42–45 C (108–113 F).

Heat exhaustion: A state of collapse brought on by insufficient blood supply to the cerebral cortex as a result of dilatation of the blood vessels in response to heat. The disease is not as acute nor so rapidly fatal as heat stroke.

Homeostasis: (1) That group of mechanisms functioning to produce stability of the internal environment of an organism. (2) The maintenance of body functions within ranges compatible with life, reached by actions or reactions initiated in response to environmental change.

Homeotherm or endotherm: An animal capable of thermoregulation by intrinsic mechanisms. The term "warm blooded" is sometimes used but is neither descriptive nor accurate.

Hyperthermia: Any disorder resulting in an elevated body temperature. Hyperthermia is not necessarily a fever. A fever is hyperthermia plus toxemia, in many cases associated with an infectious process.

Hypothermia: A state characterized by subnormal body temperatures.

Poikilotherm: An animal that must rely primarily on external sources for heat or coolness to maintain a suitable body temperature. Behavioral adaptations provide primary thermoregulatory mechanisms. These animals are popularly referred to as "cold blooded," an inaccurate term.

Radiation: All rays within the electromagnetic wave spectrum transfer energy through space without heating the intervening air. The sun is the most important source of radiant heat. However, all warm objects, including animals, emit radiant energy. The direction of flow depends on the temperature gradient.

Thermoneutral zone: The ambient temperature range within which an animal can carry out normal body functions while at rest without resorting to special heating or cooling mechanisms.

REFERENCES

1. Bartlett, R. G., and Miller, M. A. 1956. The adrenal cortex in restraint hypothermia. Endocrinology 14:181-87.
2. Bartlett, R. G., and Quimby, F. H. 1958. Heat balance in restraint (emotionally) induced hypothermia. Am. J. Physiol. 193:557-59.
3. Bartlett, R. G.; Bohr, V. C.; and Helmendoch, R. H. 1956. Comparative effect of restraint (emotional) hypothermia on common laboratory animals. Physiol. Zool. 29:256-59.
4. Calder, W. A., and King, J. R. 1974. Thermal and caloric relations of birds. In D. S. Farner and J. R. King, eds. Avian Biology, vol. 4. New York: Academic Press.
5. Dill, D. B., ed. 1964. Adaptation to the environment. Handbook of Physiology, sec. 4. Washington, D.C.: American Physiological Society. (This is an extremely valuable general reference.)
6. Gordon, M. S. 1968. Animal Function: Principles and Adaptations. New York: Macmillan.
7. Guyton, A. C. 1971. Body temperature, temperature regulation, and fever. In Textbook of Medical Physiology, 4th ed., p. 842. Philadelphia: W. B. Saunders.
8. King, J. R., and Farner, D. S. 1961. Energy metabolism, thermoregulation, and body temperature. In A. J. Marshall, ed. Biology and Comparative Physiology of Birds, vol. 2. New York: Academic Press.
9. Prosser, G. L., ed. 1973. Temperature. In Comparative Animal Physiology, 3rd ed. Philadelphia: W. B. Saunders.
10. Richards, S. A. 1973. Temperature Regulation. London: Wykenham Publications.
11. Ruck, T. C., and Patton, H. D. 1965. Physiology and Biophysics, 19th ed. Philadelphia: W. B. Saunders.
12. Schmidt-Nielsen, K. 1964. Desert Animals. London: Oxford Univ. Press.
13. Squires, R. D.; Jacobson, F. H.; and Bergey, G. E. 1971. Hypothermia in cats during physical restraint. Naval Air Development Center, Crew Systems Dep. NADC-CS-7117.
14. Swan, H. 1974. Thermoregulation and Bioenergetics. New York: American Elsevier.
15. Taylor, C. R. 1963. The thermoregulatory function of the horns of the family bovidae. Ph.D. diss., Harvard University.
16. Taylor, C. R., and Rountree, V. J. 1973. Temperature regulation and heat balance in running cheetahs: A strategy for sprinters? Am. J. Physiol. 224:848-51.
17. Whittlow, G. C., ed. 1970, 1971, 1973. Comparative Physiology of Thermoregulation. Vol. 1, Invertebrates, Fish, Amphibians, Reptiles, Birds; Vol. 2, Mammals; Vol. 3, Special Aspects of Thermoregulation. New York: Academic Press.

SUPPLEMENTAL READING

Best, C. H., and Taylor, N. B. 1966. The Physiological Basis of Medical Practice, 8th ed., pp. 1413-27. Baltimore: Williams & Wilkins.

Blaxter, K. L. 1967. The Energy Metabolism of Ruminants, rev. ed., vol. 1. London: Hutchinson.

Brody, S. 1945. Bioenergetics and Growth. New York: Reinhold.

Davson, H. 1970. Textbook of General Physiology, 14th ed., vol. 1, pp. 298-367. London: J & A Churchill.

Folk, G. E., Jr. 1966. Introduction to Environmental Physiology, pp. 77-182. Philadelphia: Lea & Febiger.

Hutchison, J. C. D. 1954. Heat regulation in birds. In J. Hammond. Progress in the Physiology of Farm Animals, 1st ed. London: Butterworth's.

Kleiber, M. 1961. The Fire of Life. New York: Wiley.

7 MEDICAL PROBLEMS DURING RESTRAINT

PERSONS who are responsible for restraint procedures must be continually alert to prevent or deal with medical problems or emergencies arising during restraint. Emergencies can arise even under ideal conditions. The behavior of any animal is unpredictable when it is excited as a result of a restraint procedure. Injuries may occur or metabolic changes, inapparent to the eye, may take place. Either or both can result in incapacitation or death.

It is not my intent to discourage the use of animal restraint techniques by dwelling on the myriad adverse conditions that may arise therefrom. Rather, it is to emphasize some severe potential problems in order to encourage restrainers to take precautions to prevent or alleviate them. *Prevention* must be the byword. To prevent medical problems, recognize and minimize the animal's exposure to potential etiological factors. When emergencies arise, be prepared to take immediate remedial action.

The objective of this chapter is to provide for the person untrained in veterinary medicine an overview of medical problems and to review basic medical techniques for the veterinary clinician. Further details can be obtained from standard veterinary medical texts and references listed at the conclusion of the chapter.

PREPARATION FOR RESTRAINT PROCEDURES

Plan carefully and anticipate problems. Think through in detail each section of the procedure. A written plan may be necessary for the novice. In any case, plan possible counteractions for every conceivable contingency. Think safety—first, for people involved in the procedure; second, for the animal. Consider whether or not the designed procedure will permit completion of the necessary task.

Make certain that all tools are on hand and in proper repair. It is tragic to snare an animal only to find that the release mechanism is malfunctioning. The animal may strangle before a tight snare can be released.

Severe hemorrhage and respiratory arrest are two conditions requiring immediate attention. Digital pressure and pressure bandaging will usually control hemorrhage until ligation and/or suturing can be done. The restrainer should always be prepared to clear airways, assist respiration, and provide supplemental oxygen.

Animals do die during restraint. Figures 7.1-7.3 list the commonest causes of death, based on rapidity of mortality. Each will be discussed in some detail. Keep these killers uppermost in mind when preparing for a restraint procedure.

Of only slightly less concern to the animal restrainer are animals that become unconscious during the manipulative period. Figure 7.4 lists conditions likely to result in unconsciousness. Note that many of these conditions can end in death if proper therapeutic measures are not quickly instituted.

Hypnosis is not a medical problem but rather a tool that may be used by the restrainer (see Chapter 2). However,

PERACUTE DEATH FROM RESTRAINT (Minutes)

- Ventricular fibrillation
- Cholinergic bradycardia (fatal syncopy)
- Anoxia—strangulation, pulmonary edema
- Hemorrhage
- Hypoglycemia
- Brain concussion or contusion

FIG. 7.1.

ACUTE DEATH FROM RESTRAINT (Minutes-Hours)

- Adrenal insufficiency
- Gastric dilatation—bloat
- Hyperthermia
- Hypothermia
- Acidosis
- Hypocalcemia
- Hypoglycemia
- Fracture of cervical vertebrae

FIG. 7.2.

DELAYED DEATH FROM RESTRAINT (Hours-Days)

- Capture myopathy—cardiac necrosis
- Gastric dilatation—bloat
- Gangrenous pneumonia—regurgitation
- Hypothermia
- Shock—trauma

FIG. 7.3.

CAUSES OF UNCONSCIOUSNESS IN A RESTRAINED ANIMAL

- Adrenal insufficiency
- Anoxia
- Brain concussion, contusion
- Cholinergic bradycardia
- Hemorrhage
- Hypnosis
- Hypoglycemia
- Hypothermia
- Shock
- Ventricular fibrillation

FIG. 7.4.

hypnosis complicates differential diagnosis in an unconscious animal.

TRAUMA

Hemorrhage (bleeding)

DEFINITION. Hemorrhage is loss of blood from the vascular system. Hemorrhage may be internal, taking place within tissues (hematoma), organs (as into the intestine), or into body cavities; or external, in which vessels are opened to the surface, allowing blood to escape.

ETIOLOGY. Lacerations are the most common cause of hemorrhage during restraint. Disruption of small vessels and capillaries is of little consequence to larger animals, but transection of large arteries or veins can be life threatening to any animal. Contusions may likewise result in extensive blood loss, because many animals have pliable skin that stretches to accommodate large quantities of subcutaneous fluid. Hemorrhage leading to the development of a hematoma may originate from capillary oozing or rupture of a single large vessel.

Hemorrhage accompanies fractures. If sharp bone ends are not quickly immobilized, they may sever arteries and veins coursing near the free ends of the bone. Fractures of the femur are especially dangerous because of potential laceration of the large femoral artery, which would result in rapid exsanguination.

CLINICAL SIGNS. Surface hemorrhage is obvious, but it is difficult for the inexperienced person to assess the extent of blood loss. Blood dispersed over the floor or walls of an enclosure appears to be more voluminous than it really is. Animal owners often become highly excited at seeing such blood and assume the animal is dying. Except when major arteries and veins are transected, larger animals rarely die because of lacerations. Any blood loss is much more dangerous for tiny animals.

Blood volume in vertebrates, as a percentage of body weight, varies from species to species, from 5 to 16% [1,19]. The average blood volume for human beings is reported as 7.7% body weight [22]. The blood volume of birds varies from 5 to 13%, depending on species, age, sex, and functional status [12]. Loss of 15-20% of the blood volume produces no clinical signs in human beings. Losses of 30-35% can be life threatening, and loss of 40-50% is usually fatal [2]. The blood volume of a 4,500 kg (9,900 lb) elephant is approximately 360 L (95 U.S. gal). A 50 g parakeet circulates 4 ml of blood. A 20% loss amounts to 0.8 ml. As can be seen, a few drops of blood lost from a parakeet is a matter of serious concern, whereas a loss of 2 L is inconsequential to a large animal.

Internal hemorrhage is not visible to the eye. Nonetheless, the consequences are equally as dangerous as those of external hemorrhage. Clinical signs associated with internal hemorrhage are pale mucous membranes and a rapid, shallow pulse. The most serious complication of hemorrhage is shock from vascular hypovolemia. Clinical signs are the same as above. Large hematomas that can be seen or palpated externally may develop from internal hemorrhage. With limb fractures, the swelling may be evident. Hematomas over the stifle or on the head are also apparent and can be easily identified by tapping with a large bore needle, following customary surgical preparation at the site.

THERAPY. Identify the source of hemorrhage and stop the bleeding. Standard first aid techniques, taught for use in human beings and domestic animals, may be impossible to apply to a wild animal. Pressure bandages, digital pressure, or tourniquets—to constrict the blood vessels—are important techniques for stopping hemorrhage. When major vessels have been severed, it is necessary to isolate the vessel and ligate it. All these techniques require the animal to be in hand and quiet. With wild animals, this is the first problem to solve.

It is especially important that blood vessels be securely ligated and hemorrhage control is complete before releasing wild animals, since a second restraint to control subsequent bleeding magnifies stress. Wild animals also suffer from greater catecholamine response, with its accompanying higher blood pressures. When the animal is released, elevated blood pressure may destroy the clot, reinstituting hemorrhage.

If a developing hematoma is noticed, a pressure wrap or cold therapy may prevent further hemorrhage. When hemorrhage into a hematoma ceases, the serum gradually separates from the formed clot and the lesion becomes a seroma. This process may take 1-2 days. Small seromas may be resorbed. Seromas over 4 cm require incision for drainage. Little drainage can be accomplished with a needle and syringe. Usually it is prudent to wait 3-5 days before incising to allow healing of the ruptured vessel and adequate separation of the serum from the clot. If hemorrhage into the hematoma continues, consider the possibility of a coagulation defect. If the blood coagulates normally, it may be necessary to make a large incision to search for the ruptured vessel and ligate it to stop the bleeding. This is drastic therapy, instituted only as a last resort.

Institute replacement therapy if sufficient blood has been lost to threaten life. Although replacing a loss of blood volume with saline, dextrose, or other electrolyte solutions may help alleviate shock, whole blood is required when massive hemorrhage has occurred. Whole blood may be extremely difficult to obtain in the case of wild animals. Nearby zoos and research facilities have heretofore responded quickly to an urgent call for blood to save the life of a valuable and rare animal.

Cross matching should be carried out whenever possible, though little is known about the blood groups of most wild animals. Even if simple cross matching shows extreme incompatibility, at least one blood transfusion may be indicated in a life-threatening situation. The clinician must make a value judgment in this instance. It is unlikely that one transfusion will induce fatal anaphylactic shock, a common result of mismatched transfusions in human beings. It is highly improbable that a wild animal has been exposed to many types of blood proteins. A second blood transfusion may be much more hazardous because the animal may have developed antibodies against such blood.

A hemogram is of little value in determining the extent of acute blood loss. The ratio of blood constituents remains essentially the same even following massive hemorrhage and a pronounced drop in total volume.

Laceration (wound, cut, bite, puncture, goring)

DEFINITION. A laceration is a wound resulting from disruption of the integrity of the skin, exposing underlying muscles, blood vessels, nerves, bones, and other tissue. Lacerations are a common result of restraint procedures.

ETIOLOGY. Many objects tear or incise the skin. Objects protruding into a cage or pen are extremely hazardous to animals, particularly during restraint when the excited animal exercises little caution to avoid them. Pieces of wire used to mend fencing, bolts on doors, boards with rough edges or exposed ends are all potential sources of lacerations. It is important to recognize that objects need not be sharp to inflict such wounds. Blunt objects can severely lacerate when struck with the terrific force exerted by a frightened animal attempting to escape by throwing itself against a fence.

Other types of lacerations include those inflicted by the bites of cagemates during the excitement of capture. This is a particular problem when dealing with primates in groups.

Hierarchial status may be upset as an animal attempts to elude capture. Free-fighting may occur if one animal tries to hide behind another. An animal may even bite itself during these stressful periods. This behavior is known as "displacement activity," that is, the animal is unable to escape so it resorts to other types of objectionable behavior such as biting itself or cagemates.

Goring wounds can be caused by animals with horns or antlers. Serious or fatal wounds can be inflicted when animals become impaled on objects in or around enclosures. Animals may jump onto the tops of fences and become impaled on posts or other objects attached to the top of the fence. One horse was fatally injured when it jammed the center divider pole of a trailer into the thoracic cavity, rupturing the anterior vena cava.

THERAPY. If only superficial structures have been exposed, standard wound treatment with debridement, suturing, or leaving the wound open may suffice. Give the wound more intensive treatment if vital structures such as joint capsules, tendons, nerves, or arteries are exposed.

Antibiotics are of questionable value in the treatment of most lacerations. However, certain wounds seem prone to become highly septic, requiring antibiotic treatment. Bites from primates and snakes are particularly septic and should be treated with debridement and by establishing drainage—in addition to treatment with antibiotics. Many wounds resulting from bites and tears do not heal with first intention. Establish satisfactory drainage and leave the wound open to heal by granulation.

The severity of goring or other puncture wounds is difficult to assess. Sometimes, though not always, it is possible to establish the extent of the lesion with a blunt probe. Clean the wound as thoroughly as possible and continue to monitor the animal to determine whether vital structures have been penetrated. Punctures into thoracic and abdominal cavities are particularly dangerous and life threatening. Standard technniques of therapy are detailed in surgical textbooks.

Abrasion (scrape)

DEFINITION. An abrasion is a wound caused by erosion.

ETIOLOGY. Abrasions are frequently caused during restraint and may be self-inflicted through escape attempts. Abrasions can occur as an animal bolts against a fence or wall. A common mistake in the construction of animal enclosures is to place the posts on the inside of the fence, providing objects that are potential weapons for producing abrasions, contusions, and lacerations.

Abrasions may be inflicted by the restrainer. Animals immobilized with drugs are frequently grasped and dragged from one spot to another, abrading the undersurface. Large animals that are impossible to lift should be moved on a large, heavy canvas or a piece of plywood used as a stoneboat, so that hair and skin are not eroded by dragging.

An animal fleeing capture may slip, fall, and slide along a rough concrete surface, seriously damaging the skin and underlying structures.

A horse fell through a trailer floor while being transported and ground off the soles of its feet. Escaping animals frequently run without regard for the pain caused by erosion of their feet. Some panicked animals have worn hoofs or nails off to the bone. It is impossible to repair such injuries. In one instance a group of foals, frightened by handling for medication, escaped through an open gate and ran down a highway. Before the animals could be stopped and rounded up, two foals had worn the hoofs away, exposing the third phalanx. Both animals had to be destroyed.

CLINICAL SIGNS. Abrasions vary from the minimal damage of hair scraped off the surface to severe trauma occurring as a result of total erosion of the skin, exposing or even wearing away such vital structures as nerves, arteries, and bones.

THERAPY. Treat each abrasion according to its severity. Obviously, in the case of simple hair loss, no therapy is needed. Many abrasions can be treated by simply cleaning the wound and applying soothing ointments. Abrasions involving vital structures must be treated surgically as well as medically.

Contusion (bruise)

DEFINITION. A contusion is injury to a tissue without laceration of the skin.

ETIOLOGY. Contusion results from a blow to the body surface, either self-inflicted (as when an animal collides against an object) or by an object striking the animal. Contusions often result when animals attempt to elude capture. They also may be inflicted by the handler hitting the animal with a stick or the metal hoop of a net. A contusion caused by a light blow may be limited to the skin and subcutaneous tissue. Heavier blows can seriously damage the periosteum of the bone or crush blood vessels and nerves, resulting in permanent disability.

CLINICAL SIGNS. Bruising may not be evident immediately following injury. Sometimes no surface injury is visible, even when deep structures are severely damaged. Contusion of bone can result in a periostitis that may not be apparent for several days after occurrence of the injury.

Usually within moments following a blow, extravasation of blood or serum into the damaged area causes swelling, pain, and sometimes discoloration, depending on the thickness of the skin and the degree of pigmentation. Heat is another sign of contusion, as the inflammatory response develops quickly.

The color of a contusion varies with the length of time since the injury. Color is usually associated with hemorrhage. As hemoglobin progresses through various degenerative stages and is converted to other metabolites, various shades of red, blue, and green may be seen. Discoloration is not necessarily indicative of tissue necrosis.

PATHOGENESIS OF THE LESION. A contusion traumatizes cells, rupturing capillaries and other blood vessels and destroying certain cells through simple mechanical damage. Within moments the damaged area fills with extravasations of either plasma or whole blood escaping from the damaged vascular system. Such extravasation will continue until clotting mechanisms function or sufficient pressure develops to inhibit the further escape of fluids. How quickly this occurs depends on the location of the contusion.

In enclosed spaces such as the calvarium, the hoof, or next to the bone, pressure from extravasation may inhibit the actions of vital nerves, interfere with circulation, or interfere with brain function. When clotting takes place, serum in the area incites inflammation.

THERAPY. Immediately apply cold compresses and/or ice packs. Cold inhibits the extravasation of fluids into the area, reducing swelling and pain. It allows time for the normal protective mechanisms of the body to close off damaged vessels and may prevent the massive swelling that causes pressure damage. Ice can be applied directly or in a plastic bag. Do not, under any circumstances, put salt into ice and water, because salt lowers the temperature to the freezing point, damaging tissue. An animal may resent the temporary pain of the initial application of cold, but as the coolness anesthetizes the area, the pain subsides.

Simple contusions, not secondarily involving broken skin, usually do not require antibiotics as adjunct therapy.

Rope Burn

DEFINITION. Rope burn is an abrasive injury caused by the friction and heat generated as a rope moves rapidly over the body.

ETIOLOGY. The etiology of rope burn involves the entangling of an animal in lariats or nets used for restraint. Rope burns are usually inflicted during operations requiring the roping of animals or when hobbles are applied. Improper placement of hobbles or ropes used for casting may cause rope burns if the animal has sufficient freedom to rapidly kick against the rope, sawing it back and forth against the skin.

PATHOGENESIS OF THE LESION. The damage caused by rope burn is a combination of abrasion and the excessive heat generated by friction. The resulting injury is as serious as a burn produced by cautery or open flame. Deeper tissues are injured as well as surface structures.

CLINICAL SIGNS. Abrasion follows the path of the rope over the body or limb. Hair is removed and the skin discolored. In cases of severe rope burn, the skin may be penetrated, exposing subcutaneous structures. Within minutes after the injury is inflicted, fluid exudes from the surface. Swelling and inflammation quickly ensue.

THERAPY. Apply cold water or ice packs immediately to the surface as major first aid. Maintain the packs for at least an hour. Once the initial effect has been overcome, treat the burn with ointment to soften the skin surface and prevent extravasation of fluids. Dress limbs to keep dirt and other debris from contaminating open wounds.

HEAD AND NECK INJURIES

Brain Concussion and Contusion

DEFINITION. Concussion is the transient loss of brain function as the result of a blow to the head. It is a functional condition with no gross structural changes. A contusion is an extension of the concussion, wherein the brain sustains physical damage (hemorrhage, edema).

ETIOLOGY. Any blow to the head can result in concussion. Blows may be self-inflicted when an animal flings itself against a wall or ceiling. Birds frequently injure themselves by flying into glass walls or windows unperceived as barriers.

The use of sight barriers is extremely important when dealing with wild animals. Many wild animals do not perceive a chain link fence to be a barrier and can severely injure themselves if suddenly moved into an enclosure surrounded by this type of fence. Animals in flight may also attempt to escape through openings too small to accommodate them.

CLINICAL SIGNS. Immediate unconsciousness brought about by a blow to the head is the primary clinical sign of concussion. Unconsciousness may be transitory or prolonged. If it persists longer than 24 hours, it is likely that organic damage has been done. Other signs are transitory vasomotor and respiratory malfunction, immediate loss of reaction to stimuli, loss of the corneal reflex, dilated pupils, and flaccid muscles. Vomiting occurs in some cases.

The clinical signs of contusion of the brain include either immediate or delayed unconsciousness. If the animal is not rendered unconscious initially, hemorrhage into the cranial cavity may produce delayed unconsciousness. Additional signs of brain contusion are seizures, ophisthotonos, absence of pupillary response, signs of cranial nerve damage, and failure of vital respiratory and cardiac functions.

THERAPY. Protect the animal from further injury during recovery periods. Shade it from sun and light. Such an animal usually has impaired thermoregulatory ability and may become either hyperthermic or hypothermic if not protected from exposure. The animal may be disoriented upon regaining consciousness. Protect it from hazards such as ponds, lakes, low fences, moats, and so on. Keep the animal in a darkened, quiet stall or cage to reduce stimulation. Cold compresses applied to the head have initial value but are of little benefit once clinical signs have developed.

Usually little can be done to aid a wild animal suffering from a serious head injury. While principles of neural surgery used in small animal practice can be applied, the prognosis is grave when dealing with wild species.

Fractures of the Skull

DEFINITION. A fracture is a break in the continuity of bone.

ETIOLOGY. Skull fractures can result from any blow to the head. Blows on the side of the skull (temporal region) or at the back of the head (occipital or nuchal region) are particularly dangerous. Animals that throw themselves over backward may strike the back of the head on surrounding objects or on the ground. Basilar skull fractures are a usual consequence.

CLINICAL SIGNS. Depression on the surface of the skull is the clinical sign of a fracture involving the calvarium. Rarely will crepitation be present. One should not palpate vigorously because of the danger of further depressing the fracture. Radiographs are the basis for a definitive diagnosis of fracture.

The animal with a fractured skull will exhibit varying degrees of central nervous involvement, such as paralysis or, if certain basal functions are disrupted, respiratory or cardiovascular failure. The animal will probably be unconscious. A superficial wound exposing the bone may or may not be seen.

THERAPY. The treatment of skull fractures is exceedingly difficult and must be conducted by a veterinary orthopedic surgeon. General treatment consists of protecting the animal from extremes of environment and preventing further injury from thrashing or convulsions. Monitor vital functions and take appropriate steps to keep respiratory and cardiovascular systems operating.

Horn and Antler Damage

Appendages of the head are frequently liable to injury as a result of restraint practices. Some of these structures are designed for active combat and will withstand considerable pressure. The bighorn sheep has massive horns capable of withstanding heavy blows. On the other hand, the fine horns of some antelope species are easily fractured or traumatized.

Antlers are bony extensions of the frontal bones exhibited by members of the family Cervidae. These structures are shed and replaced by new antlers annually. As the antler begins to develop, it is covered with a highly vascularized epithelium known as "velvet." When the antler is mature the velvet is scraped off, leaving a highly polished, hard, branched structure on the head.

The antler is soft and easily broken during the developmental stage. Grasping immature antlers as handles during restraint or banging them against a wall or chute may easily fracture them. When developing antlers are fractured, extensive hemorrhage may result from laceration of the velvet. Less severe injuries during the velvet period may produce disfigured and asymmetrical antler growth.

When the velvet has been scraped off, the antler is hard and may serve as a handhold for the head, but if too much pressure is applied, the mature antler will also fracture. Since mature antlers are avascular, the only damage is deformity until the animal sheds again. Mature antlers of males can be sawed off a few inches from the skull to reduce danger to persons or other animals.

A horn is the combination of a specialized cornified epithelium overlying a bony core that is an extension of the frontal bone of members of the family Bovidae.

Two types of horn injuries may occur during restraint. The first is contusion of the horn, resulting in separation of the outer covering from the bony core. The second type of injury is fracture of the bony core, separating it from the frontal bone.

ETIOLOGY. The horns of species such as the American bison are prone to injuries that pull the horn off the cornual process of the frontal bone. When these animals are confined in a chute for tuberculin testing or collecting blood samples, they may tear off the horn by raking the head up and down the side of a chute. Bleeding usually stops without treatment. The outer core grows from the base of the horn at the corium and regenerates in time.

Antlers and horns can be damaged if an animal strikes its head against fences, walls, cages, or shipping crates. An animal can fracture a horn caught in a fence or in a crack in a door or wall. Flimsy horns and antlers can be broken by grasping them as handholds. If antlers must be used as handholds, grasp the base of the horn or antler to minimize the danger of injury.

CLINICAL SIGNS. Deformity of the structure is the most prominent sign of a fracture. Increased mobility can be either observed or palpated. Hemorrhage may be present, depending on the stage of development of antlers or whether the horny covering has been torn off the bony core of a horn.

When making a clinical examination of a fractured horn or antler, it is important to determine whether or not the skull has been fractured. Usually only the bony core of the horn fractures, but occasionally the frontal bones also fracture. Frontal bone fractures are serious because the frontal sinus is involved. Furthermore, such a fracture may open into the brain cavity.

THERAPY. It is impractical to attempt to repair fractures of antlers. Amputate the distal segment. Incise and remove the distal segment of an antler in velvet and apply a pressure bandage to the proximal segment to control hemorrhage.

A horn shell torn off the bony core cannot be reattached since the blood supply has been destroyed. If the bony core is intact, apply a soothing ointment such as furacin ointment. In time a new horny shell will develop. If the horny shell is still in place but the bony core is fractured, attempt to stabilize the horn by clamping it to the opposite horn. A good clamping technique is to insert wooden blocks between the two horns and bind them together. Another technique is to wrap plumber's tape or perforated metal tape around the horns. Attach one to the other with a turnbuckle. Splinting the horn and wrapping the head with bandage and cotton may sufficiently stabilize the horn of a young animal.

Fractures of the frontal bone must be carefully reduced and the horn set into its proper place. Fix it with an appropriate splint or fixation device. It is important to recognize that injuries to the horn often result in hemorrhage from the nose because of the respiratory connection via the cornual process of the horn into the frontal sinus, thence to the maxillary sinus, and into the nose.

Facial Paralysis

DEFINITION. Facial paralysis is a clinical syndrome produced by the disruption of facial nerve function. The facial nerve innervates muscles of the eyelid, cheek, and upper lips.

ETIOLOGY. During restraint, a blow to the jaw or to the side of the head may injure the facial nerve, especially if the blow strikes at a point where the nerve courses near a bone. A blow to that nerve can result in either temporary or permanent cessation of function.

CLINICAL SIGNS. Signs noted are dependent on the location of the damage to the nerve. If it occurs at the base of the ear, all the classical signs may be present. These are closure of the upper eyelid, inability to clear food from the cheek pouches, and pulling of the upper lip to the side of the face opposite the injury. Additionally, the facial nerve supplies some ear muscles. If the injury occurs near the ear, the ear may droop or fail to prick in response to auditory stimulation. Swelling over the lateral aspect of the face or jaw may or may not appear.

A second potential cause of facial paralysis is infection or abscess in or near the point at which the facial nerve emerges from the skull, near the base of the ear. An abscess may be noticed before restraint is begun, or it may be discovered as a result of the restraint practice.

A temporary or permanent rope halter left on an animal in prolonged lateral recumbency may produce point pressure on the facial nerve.

THERAPY. Most facial paralyses are transient, self-correcting in a few hours to a few days. In persistent paralysis, hot compresses may mobilize the constituents causing pressure on the nerve. Otherwise little can be done. Some clinicians recommend the use of large doses of steroids to slow the inflammatory process, but others feel such therapy is of little benefit.

Paralysis from Stretching the Neck

DEFINITION. Impaired neural or muscular function as a result of stretching the head and neck into abnormal positions.

ETIOLOGY. It is not unusual for the restrainer to grasp the head or neck when attempting to control an animal. Nerves and muscles of the head can be damaged by excessive stretching or twisting.

CLINICAL SIGNS. The head may be carried in a peculiar position. Onset may be noted immediately or several hours after restraint has been carried out. The animal may be unable to elevate the head, or the head may be deviated to one side as the result of unilateral damage to a nerve or muscle.

The most serious manifestation of neck injury is "wobbler syndrome" (idiopathic ataxia) of foals. Many affected foals have been injured by mishandling during restraint procedures for medication or examination. The cervical vertebrae have been twisted, resulting in a subluxation. Pressure is exerted on the spinal cord, causing disruption of motor function in the hindquarters.

THERAPY. No specific therapy has been found to be effective for conditions produced by damage to the neck. Symptomatic and/or supportive therapy is indicated.

LIMB INJURIES

Fracture (broken bones, a break)

DEFINITION. A fracture is disruption in the continuity of a bone. A complete discussion of various types of fractures, with clinical signs and recommendations for treatment, is not appropriate for this book, but it is important to know that the limbs of many species are easily fractured during restraint procedures. Extreme care must be exercised to minimize the pressures exerted on fragile bones by heavy-handed restrainers.

Fractures occur when limbs are caught in chutes, stanchions, walls, cargo doors, screen mesh, or wires. Bones can also be broken by injudiciously slamming the hoop of a capture net into an animal or by an animal jumping against trees, walls, or cages. Some fractures are a result of the stress inherent in capturing or moving groups of animals, which frequently leads to increased aggression, manifested by kicking or butting.

A startled animal usually initiates flight with a sudden jump. If the footing is poor the animal may fall. A fall as an animal attempts to whirl away on one leg may result in fracture of the femur or tibia in both wild and domestic species, particularly equine species.

Fractures may occur through unusual accidents. In one instance, when removing a Siberian ibex from a crate, a handler reached into the front and grabbed the horns. The animal bolted out of the crate and two or three people immediately jumped on it. In the scuffle the animal's tibia was broken.

In another instance, a horse was being cast for surgery. The animal was pulled into lateral recumbency, but the hind limbs were not flexed and pulled up tightly enough against the body. The animal was able to extend its hind leg and exert sufficient pressure to fracture its femur.

If animals become twisted and pinned into peculiar positions, fractures are likely to occur.

CLINICAL SIGNS. Fractures of the distal extremities are usually easy to recognize because of the extreme abnormal mobility and abnormal positioning of the limb. In many cases crepitation or grating of the bone ends can be discerned. Fractures of the upper limbs, particularly of heavy-bodied animals, are extremely difficult to diagnose. In many instances radiographs of these bones cannot be made; diagnosis is based on the continued inability of the animal to support weight on the limb.

THERAPY. Treatment of fractures is a complicated procedure requiring the special talents of the veterinary surgeon.

Fractures are emergencies, requiring quick treatment to prevent trauma to contiguous vessels and nerves from the jagged bone ends. In the case of domestic animals, immediately apply a splint to immobilize the leg. In wild species, it is usually necessary to sedate or immobilize the animal to apply a splint.

When dealing with wild animals, any suspected fracture should be fully evaluated before the animal is released from restraint. Once it has been released, efforts to recapture it may cause further damage of sufficient magnitude to result in complete disuse of the limb and perhaps the death of the animal.

Sprain

DEFINITION. A sprain is damage of the ligaments and tendons surrounding a joint, caused by twisting and pulling. A sprain may be more serious than a fracture and result in prolonged incapacitation.

ETIOLOGY. Joint injuries are caused in the same manner as bone fractures. A sprain results when a joint is twisted in such a manner as to tear the collateral ligaments, joint capsules, or other tendonous structures supporting the integrity of the joint.

CLINICAL SIGNS. Sprains are usually indicated by disuse of the limb and therefore are seldom observed until after a restrained animal has been released. The joint may swell because of the stretching of collateral ligaments, tendons, joint capsules, and other tissues around the area. Additional clinical signs include heat, pain, and immobility.

DIFFERENTIAL DIAGNOSIS. Most sprains are noticed a few hours to days after a restraint procedure. Other conditions may simulate a sprain. An abscessed joint, arthritis, contusion of a joint, and circulatory conditions resulting in edema of the structures are typical ailments, with symptoms similar to those of sprain.

THERAPY. Therapy is similar to that for fracture.

Laminitis (founder)

DEFINITION. Laminitis is inflammation of the highly vascular laminae attaching the bone of the digit to the horny covering of the hoof. Laminitis is associated with severe pain and disfunction of the limb involved. It is a common disease of the horse, but any species of hoofed animal may develop laminitis.

ETIOLOGY. Restraint laminitis is caused by contusion of the hoof. Trauma to the hoof may occur through an animal's attempt to escape by banging the hoofs against a solid surface such as a concrete or wooden wall or against the side of a truck. During manipulative procedures in chutes, animals often thrash, striking any object in the way. Contusions of the hoof wall are a possible consequence. Animals chased for some distance on a hard surface may develop laminitis.

PATHOGENESIS OF THE LESION. Contusion of the hoof wall results in congestion as a result of the increased blood supply to the area. The hoof is a finite space, and engorgement exerts pressure on sensitive tissues, causing severe pain. Severe trauma at a single point can result in a hematoma beneath the hoof wall similar to that suffered when a thumbnail is struck with a hammer. Usually, following the initial congestion, the body resorbs the excess blood and structures are undamaged. If pressure from congestion continues, the tissues of the lamina may atrophy, causing the hoof to separate from the digit of the foot. Chronic laminitis is a common ailment of the horse.

CLINICAL SIGNS. Laminitis is characterized by severe pain, heat over the affected hoof, and disuse. Unfortunately laminitis caused by restraint is seldom evident until after the animal has been freed. Adequate subsequent observation and/or examination can usually define the injury of a domestic animal, but in wild species a subsequent examination requires resubjecting the animal to restraint. Furthermore, it is difficult to make a correct diagnosis because there is no evidence of swelling or discoloration of the hoof. Since complete disuse of the limb is the only symptom, laminitis must be differentiated from sprains, fractures, and nerve injuries.

THERAPY. The most valuable immediate treatment for acute traumatic laminitis is the application of cold water or ice for at least an hour. The difficulty lies in making the original diagnosis quickly. Usually by the time the diagnosis is definite, it is too late for cold water to have much value. In fact, at this stage, alternate hot and cold applications are indicated to relieve the congestion of the hoof.

Traumatic laminitis does not usually respond to medication. Local application of cold and heat is the only recommended therapy. Analgesics (e.g., phenylbutazone) may relieve pain but are detrimental because they allow the animal to walk on the damaged feet.

Nerve Injury (paralysis, radial paralysis, perineal paralysis, brachial paralysis)

DEFINITION. Nerve injury is diminished function of either or both sensory and motor nerves.

ETIOLOGY. Nerve injury can result from the same types of accidents as those that cause fractures, sprains, or contusions. Nerves are injured during the restraint process by excessive stretching of the limbs or head or by a blow to the nerve as it crosses a bony prominence. In attempting to control or place limbs in less active positions, they may be unduly stretched, disrupting normal nerve function.

Delayed paralysis may result when heavy-bodied animals are immobilized and kept in lateral recumbency for long periods. Two mechanisms may be involved in this phenomenon. The first operates through anoxia, caused by excessive pressure on the arterial blood supply—familiar to us as the foot "going to sleep." This occurs when a heavy-bodied animal lies against a lower leg, causing ischemia. Usually the ischemia is transient and is alleviated as soon as the animal gets up. Temporary paralysis may be apparent, but function of the limb returns within moments.

The second mechanism operating in cases of delayed paralysis is also associated with ischemia. This type often occurs during equine surgery, when the horse is restrained in lateral recumbency. If an animal under general anesthesia or restrained by immobilizing agents is totally relaxed, pressure on blood vessels in the brachial plexus results in ischemia. If ischemia lasts for 1-2 hours, the endothelium of the capillaries may be slightly damaged. As the animal recovers and arises, blood rushes into the ischemic areas. Because of the damaged endothelium, capillary permeability is increased, allowing extravasation of plasma into the area, which exerts pressure on nerves.

Thus an animal may experience paralysis immediately after arising from anesthesia or immobilization. The use of the leg quickly returns, only to become disfunctional again in an hour or so as a result of swelling in or around the nerve site (brachial plexus).

Radial paralysis may result from a blow to the forearm over the radius and ulna at the point where the radial nerve lies in close proximation to the bone.

Peroneal paralysis is commonly noted in animals trussed up in a casting harness or rope in such a manner that the hind limb is tightly flexed for a long time. The problem may also be caused by excessive stretching or by continued pressure on the nerve.

Paralysis associated with pressure on the brachial nerve seldom occurs in a partially conscious animal. The struggling of such an animal involves sufficient muscular activity to support normal blood circulation in the limb and maintain capillary integrity. The problem develops only in a recumbent animal that is kept entirely relaxed for a long time.

CLINICAL SIGNS. The most prominent signs are those of motor malfunction. Animals with brachial paralysis cannot use the forelimb. They stagger and may fall when they try to walk. Most wild animals are able to support themselves on three legs, but some refuse to do so. Radial paralysis is indicated by the inability of the animal to extend the front leg forward without dragging the hoof along the ground. Peroneal paralysis is characterized by the inability of the animal to support weight on the hind limb. The joint knuckles at the fetlock, and the animal may walk or attempt to walk on the anterior surface of the fetlock. Sensory perception is usually disrupted as well. Sensory disruption can easily be identified if the animal is in hand but is less easily evaluated with a free wild animal.

THERAPY. Most nerve injuries are temporary in nature, but the animal may injure itself further, and, more seriously, before nerve function returns. In the case of peroneal paralysis, it is important to support the leg and protect the anterior aspect of the fetlock to prevent joint trauma. Wild animals are particularly difficult to treat in these cases because they will not tolerate slings or other supports. Generally therapy is aimed at protecting soft tissue structures and bones from trauma until nerve function returns. Unfortunately little can be done to assist the animal afflicted with permanent nerve injury. The process of nerve regeneration is long and drawn out; it is usually not possible to provide a wild animal with adequate nursing care for a sufficient length of time to allow regeneration to take place.

Differential diagnosis of nerve injuries must include fractures, sprains, and capture myopathy (discussed later in this chapter).

Injury to Toenails, Claws, and Hoofs

Toenails, claws, and hoofs can be contused, or torn from the digit.

ETIOLOGY. A hoof can be torn from the foot if the foot is entangled in fences or in cracks or crevices in walls. Claws or toenails are frequently torn as animals scratch or claw while attempting to escape.

CLINICAL SIGNS. Signs of claw injury include hemorrhage and loss of the claw from the surface of the digit or the hoof from the foot. In some instances the nail remains attached by the corium. In other cases it is torn completely away from the foot. Abrasions of the nails occur from prolonged scratching at hard surfaces in attempts to escape. In these cases the shape of the claw is changed and bleeding from the tip of the nail may be noted.

THERAPY. If the nail or claw has been entirely torn away, control the hemorrhage and protect the bed of the nail by applying bandages as long as is necessary for regrowth of the nail. In some instances a damaged coronary band prevents regeneration, and amputation of the digit becomes necessary.

Damage to Feathers and Scales

The skin appendages of birds, reptiles, and fish are as liable to damage as is the hair of mammals. All these structures are protective in nature. Abrasions that remove such structures expose more sensitive tissues and may allow infections to penetrate the body.

Mature feathers are avascular, except at the tip of the follicle. Removal of certain feathers or cutting them at the tip to inhibit flight is a routine clinical procedure. A newly developing feather (pin feather), however, is intensely vascular. Exercise special caution when handling birds known to be developing new feathers. Manipulation of an animal at the time feathers are erupting may tear the feathers and cause serious hemorrhage. If a pin feather is torn, the feather shaft remaining in the follicle must be removed completely to stop hemorrhage. Remove it by grasping the shaft with forceps and pulling straight out.

METABOLIC CONDITIONS

Stress and Thermoregulatory Problems

Stress is the ever-present cloud that plagues both animal and restrainer. The concept is difficult to understand, yet of great significance. Chapter 5 is devoted to this topic.

Hyperthermia and hypothermia are both significant medical problems occurring during restraint procedures. The medical importance of these conditions warrants a detailed separate discussion (see Chapter 6).

Acidosis

DEFINITION. Homeostasis necessitates maintenance of a delicate acid-base balance in the blood. Normal mammalian blood pH varies between 7.35 and 7.45. Minor changes in either direction trigger serious metabolic consequences: pH less than 7.35 = acidosis; pH greater than 7.45 = alkalosis [1,22].

ETIOLOGY. The prime cause of acidosis in restrained animals is excessive muscular activity associated with excitement, chase, and resistance to handling [28]. Acidosis is the result of lactic acid buildup during anaerobic oxidation in the muscle cells. One minute of exhaustive exercise may cause a drop to a pH of 6.8 [11,17]. Other respiratory and metabolic activities of the body contribute to the acid-base balance. A malnourished animal, for example, is utilizing its own protein reserves for energy and has usually developed ketosis with excessive hydrogen ion production. Metabolic acidosis can also be induced by starvation, chronic interstitial nephritis, acute renal insufficiency, diarrhea, and dehydration [31]. Thus evaluation of the disease state of any animal to be restrained is important to permit the clinician to guard against aggravation of a preexisting acidosis.

Respiratory acidosis may develop whenever there is interference with normal respiratory function. Airways are frequently obstructed during restraint. Pneumonia, emphysema, and anesthesia without forced or assisted respiration may contribute to the development of acidosis.

Acidosis may contribute to other serious metabolic and electrolyte upsets. Acidosis associated with exercise persists for several minutes after running or struggling has ceased, even if animals are trying to accommodate by hyperventilation. Thus animals are commonly manipulated or anesthetized while in an acidotic state. In the acidotic state, serum calcium is elevated, which combined with hypoxia sensitizes the cardiac muscle to the effects of catecholamines. Ventricular fibrillation and death may be the sequel to such a double metabolic insult.

CLINICAL SIGNS. Neurological signs are primary indicators of acidosis. The animal is listless and exhibits mental confusion. It may lapse into coma and/or suffer from seizures progressing to convulsions. The skin is frequently characterized by lack of turgor since dehydration is commonly associated with acidosis. The animal breathes rapidly to exhale excessive carbon dioxide.

THERAPY. It is important to establish open airways and assist respiration to eliminate carbon dioxide. Direct therapy for correcting acidosis is intravenous infusion of sodium bicarbonate solution (4-6 mEq/kg), usually in conjunction with other parenteral fluids such as saline or dextrose.

Alkalosis

Alkalosis is of minimal importance as a restraint problem. Respiratory alkalosis may be produced by hyperventilation, but usually this is possible only by forced breathing while an animal is on an anesthetic machine. Metabolic alkalosis is common in certain digestive tract diseases associated with pyloric obstruction, gastritis, gastric foreign bodies, vomiting, and abomasal disease in cattle. Additional disease conditions characterized by alkalosis include salt poisoning, hyperadrenalism (which may be a factor in stress), and certain brain stem diseases.

Hypoxia/Anoxia

DEFINITION. Hypoxia is decreased availability of oxygen in the tissues. Anoxia is total absence of oxygen. Hypoxia may be general, in which all tissues lack sufficient oxygen, or it may occur only in localized organs such as the brain and cardiac muscles, which are particularly susceptible to insufficient oxygen.

ETIOLOGY. Some species can breathe through both mouth and nostrils; others breathe primarily through the nose (horses, elephants). Airways may be obstructed by tight ropes or snares around the neck, too tight a grip by a handler, or by twisting the neck. Strangulation can occur easily if an animal sticks its head through a net or a webbing containing spaces too large for that species. Do not obstruct the nostrils of any animal during restraint. Gloves diminish tactile discrimination and may mask excessive pressure applied to the thoracic cavity while gripping the animal (Fig. 7.5).

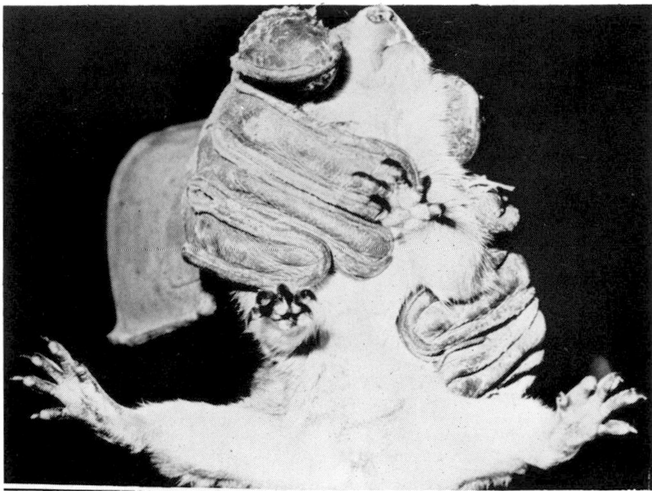

FIG. 7.5. Gloves decrease tactile sense and may result in an animal being squeezed to the point of suffocation.

Birds breathe with a bellows type of respiration that necessitates movement of the keel or sternum forward and down for inspiration and backward and up for expiration (see Fig. 24.1). Any restraint procedure that interferes with such movement will quickly produce suffocation.

Other causes of hypoxia include regurgitation with inhalation of ingesta, bloat, and a concurrent respiratory disease such as pneumonia or emphysema. Wild animals are capable of masking signs of severe respiratory disease until the condition is almost terminal. In a number of instances animals have quickly suffocated and died during restraint. Necropsy revealed a functional lung capacity of approximately 10% of normal.

Pulmonary edema is a special problem seen when unconditioned domestic or captive wild animals are forced to exercise. A female bison dropped after five trips around a large pen to escape roping. She died of pulmonary edema. Her free-living counterpart could probably have run for miles without harm.

CLINICAL SIGNS. Minimal hypoxia causes dyspnea, cyanotic mucous membranes, and accelerated pulse. As hypoxia deepens, cerebral anoxia produces unconsciouness. An animal will likely begin to struggle vigorously at this point or even convulse. To a casual observer, this may convey the impression that the animal is struggling against restraint.

If cerebral and cardiac anoxia are prolonged for more than 4-5 minutes, damage is irreparable and death ensues.

THERAPY. Correct the mechanism causing hypoxia and provide supplemental oxygen. When dealing with wild animals, it is wise to have emergency oxygen available at all times.

If the animal is still breathing, insert a tube into the nostril for oxygen insufflation. A color change in the mucous membranes should be noted quickly. If respiration has ceased, the trachea must be intubated. If suitable tubes and specula are available, oral intubation is sufficient. If not, an emergency tracheotomy must be performed and oxygen supplied under pressure.

Hypocalcemia (eclampsia, puerperal tetany, milk fever)

DEFINITION. Hypocalcemia is decreased concentration of calcium ions in the blood.

ETIOLOGY. Calcium is required for numerous chemical reactions in the body, including those of normal nerve and muscle function. Decreased circulating calcium produces many systemic manifestations.

Critical serum calcium levels are normally maintained through calcium ingestion. If the calcium level in the diet is inadequate, the deficiency is made up by drawing on bone stores.

Malnutrition predisposes restrained animals to hypocalcemia. Wild animals in captivity, particularly those in private ownership, are frequently not supplied with a diet providing sufficient levels of calcium, vitamin D, and phosphorus in the required ratio. Bone decalcification is the initial response. Eventually serum calcium drops to such a low level that the clinical syndrome of hypocalcemia appears.

Hypocalcemic tetany may be induced during restraint by either hypoxia or respiratory alkalosis brought on by hyperventilation. Alkalosis increases the quantity of calcium bound to protein, thus reducing ionized calcium [28]. Forced hyperventilation during anesthesia may also produce alkalosis.

CLINICAL SIGNS. Hypocalcemia results in hyperirritability of nerves and muscles, causing muscle cramping, muscle twitching, laryngeal spasm, carpopedal spasm, stridor, and generalized convulsions.

Caged birds are frequent victims of hypocalcemia. Clinical signs include wing fluttering, falling from perches, and tetanic convulsive seizures on the floor of the cage. A lizard exhibiting signs of hypocalcemia falls into extensor rigidity with the limbs extending backward along the body. In mammals, the nictitating membrane may extend over the eye because the ocular muscles contract, pulling the bulb back into the orbital socket. Respiration may be inhibited by spasm of respiratory muscles.

Milk fever in the cow and puerperal tetany in the bitch and the mare are also manifestations of hypocalcemia. Cattle are unique: instead of tetany, paresis results from hypocalcemia. In the mare and bitch, classical signs of tetany are more prominent.

In the bitch, signs noted are nervousness and anxiety, frequent whining, and difficult locomotion. The animal may stagger and walk stiffly. Usually, increased muscular activity results in an elevated body temperature. Ultimately the animal collapses and falls to the floor. The legs and neck are rigidly extended. Twitching may develop, followed by periods of relaxation. The convulsive seizures may progress to severe tetanic spasms and ultimate death.

THERAPY. Treat acute hypocalcemia with an intravenous solution of calcium gluconate (Table 7.1). In small species, or if no vein can be raised, the intramuscular route can be used.

Intravenous solutions of calcium salts must be ad-

TABLE 7.1. Parenteral dosage of calcium gluconate for acute hypocalcemia

	Percent Calcium Gluconate Solution	Dose Actual Calcium Gluconate	Dose Actual Calcium Gluconate	Ml Solution per kg B.W.	Ml Solution per lb B.W.
Large animals (cow, antelope, sheep)	20 (200 mg/ml)	100-200 mg/kg	50-100 mg/lb	0.5-1	0.25-0.5
Medium-sized animals (dog, cat)	10* (100 mg/ml)	100-200 mg/kg	50-100 mg/lb	1-2	0.5-1
Small birds and reptiles	1† (10 mg/ml)	0.1-0.2 mg/g	. . .	0.01-0.02 ml/g	. . .

*1 ml 20% calcium gluconate in 1 ml water = 10% solution calcium gluconate.
†0.5 ml 20% calcium gluconate in 9.5 ml water = 1% solution calcium gluconate.

ministered carefully. Monitor the heart rate. If tachycardia develops, slow the infusion.

Hypoglycemia (hypoglycemic shock, insulin shock)

DEFINITION. Hypoglycemia is abnormal decrease of glucose levels in the blood.

ETIOLOGY. Captive wild animals, particularly those kept as pets, may be malnourished, with depleted glycogen reserves. Energy needs increase at the time of restraint. With no reserves for the body to call upon, glucose levels drop. Hypoglycemic shock may ensue.

Many animals enter a state of torpidity, characterized by decreased metabolic activity, when exposed to extremes of environmental temperature or when food becomes unavailable. Such animals are capable of maintaining themselves for an extended period in the torpid state; however, they are particularly prone to develop hypoglycemic shock if they are disturbed and called upon to quickly mobilize energy reserves. Hibernators must be handled carefully when torpid. Alligators and other crocodilians become seasonally torpid. Handling them during this time is hazardous.

CLINICAL SIGNS. Hypoglycemia deprives the brain of the nutritive substrate upon which it is dependent for oxidative metabolism. Thus hypoglycemia results in anoxia of the neurons in the central nervous system.

Hypoglycemia is characterized by tetany varying from slight, transient tremors to incoordination, twitching, and convulsions. Signs of autonomic nervous system malfunction include copious salivation, tachycardia, and profuse sweating.

Continued cerebral anoxia causes irreversible brain damage [9]. Hypoglycemic convulsions can be controlled by sedation, but sedation does not prevent brain damage from anoxia. When conducting a differential diagnosis, hypoglycemia must be considered, and ruled out, before sedation is administered to a convulsing animal. Failure to correctly diagnose and treat a hypoglycemic animal may result in mental retardation, partial paralysis, ataxia, epilepsy, or death.

THERAPY. Administer a 10-50% dextrose solution intravenously. Response should be immediate. If intravenous injection is impossible, inject the solution intramuscularly.

Table 7.2 lists the daily basal metabolic caloric requirements of animals from 0.05 to 2,000 kg, using the formula: $70.5 \times (\text{B.W. in kg})^{.73}$ = kcal/24 hr.

One ml of 50% dextrose yields 2 kcal. One ml of 5% dextrose yields 0.2 kcal. A 10 kg animal requires 379 kcal/24 hr or 190 ml of 50% dextrose to supply the needed calories. In a 10 kg animal, 50 ml of 5% dextrose is not likely to alleviate hypoglycemia for more than a few minutes.

Epinephrine, 1:1000, injected subcutaneously, produces an immediate gluconeogenic effect and is helpful in relieving acute hypoglycemia. Give a large animal 0.5-1 ml/50 kg of a 1:1000 solution. Give a small animal 1 ml/10 kg. Dilute the 1:1000 solution 1:10 before administering.

Long-term therapy consists of providing adequate nutrition and making certain the animal consumes it. Comprehensive and long-term therapy should be instituted following emergency treatment.

TABLE 7.2. Basal daily caloric requirements for mammals and birds

Body Weight (kg)	Kcal/24 hr	Body Weight (kg)	Kcal/24 hr
0.05	8	95	1,958.49
0.1	13	100	2,033.20
0.2	21.50	110	2,179.86
0.3	29	120	2,322.98
0.4	35.50	125	2,392.77
0.5	42	130	2,462.57
0.6	48	140	2,599.34
0.7	54	150	2,735.40
0.8	59.50	160	2,865.83
0.9	65	170	2,995.55
1	70.50	180	3,122.45
2	116.89	190	3,248.64
4	193.95	200	3,372.02
6	260.78	225	3,675.17
8	321.69	250	3,969.15
10	378.52	275	4,254.68
12	432.52	300	4,533.86
14	484.06	350	5,073.89
15	509.01	400	5,593.47
20	627.95	450	6,095.43
30	843.89	500	6,583.29
35	944.70	600	7,522.37
40	1,041.99	700	8,417.70
50	1,225.96	900	10,109.70
60	1,400.13	1,000	10,920.45
70	1,567.22	1,500	14,678.10
75	1,648.29	1,800	16,771.95
80	1,727.25	2,000	18,111.45
85	1,805.51		
90	1,883.06		

Note: Basal energy requirements for mammals and birds are based on the formula $70.5 \times W^{.73}$ = kcals required for 24 hours (W = body weight in kg). Actual daily requirements for an animal may be two to three times this figure depending on energy required to keep warm, degree of activity, lactation, and growth.

Dehydration

DEFINITION. Dehydration is excessive loss of body fluids.

The mammalian body is composed of 60-80% water. Higher amounts of body fat decrease the percentage of water [19]. Except for desert animals, adult mammals require water consumption in amounts of approximately 40 ml/kg of body weight daily to maintain normal water balance in a basal metabolic state. This is equivalent to the fluid lost in urine and feces and through insensible evaporation via skin and lungs.

Desert-adapted animals have developed specialized methods of water conservation, such as concentrating urine, voiding dry hard feces, and storing heat [24], and require lower volumes of water intake or none at all.

ETIOLOGY. Dehydration may be caused by water deprivation prior to restraint, failure of a newly acquired animal to recognize the water source, a frozen water source, failure to use automatic waterers, failure to provide sufficient water during hot weather, overheating, prolonged chase for capture, severe diarrhea, persistent vomiting, hemorrhage, or loss of fluid as a result of burns.

CLINICAL SIGNS. Early signs of mild dehydration (3% B.W. loss) in nondesert-adapted species are low urine output, dryness of the mouth, and some loss of skin elasticity. Feces become dry and hard. Fluid is lost first from interstitial fluid compartments. Homeostatic mechanisms function to keep plasma volume constant as long as possible [15,16].

Moderate dehydration (5% B.W. loss) is accompanied by marked loss of skin elasticity. The eyes are sunken. Blood pressure may fall as a result of decreased plasma volume. Weakness, fever, and weak pulse may be observed.

Marked dehydration (10% B.W. loss) involves circulatory failure from decreased plasma volume. Signs of shock and coma are evident [15].

Severe dehydration (12-15% B.W. loss) results in renal failure, marked by uremia and acidosis. At this point severe kidney damage may preclude recovery even though fluid is supplied.

Death usually follows a fluid loss of 20-25% body weight. Laboratory examination shows elevation of packed cell volume (PCV), hemoglobin, and plasma proteins. Blood test results of dehydrated anemic animals may show normal values [23]. Desert-adapted species can cope with dehydration levels incompatible with life in other species.

THERAPY. Provide fluids orally. If the animal cannot drink, gastric intubation is indicated. Fluid is absorbed rapidly if circulation is functioning. As fluid is also readily absorbed from the colon, enemas are effective in rehydration, especially if vomiting animals are unable to retain ingested liquids.

In acute dehydration stress, intravenous fluids must be given. Veins are often collapsed, making it necessary to expose the vein (via a skin incision) to effect administration.

Physiological saline solutions or 5% dextrose solutions are satisfactory for intravenous treatment of dehydration. Do not use hypertonic solution. Tap water or saline-dextrose solution can be given orally or rectally. A customary mistake is to underestimate the volume of fluid required. A resting animal's basal fluid requirement is 40 ml/kg body weight daily. As much as three to five times that amount may be required to make up the deficit of marked dehydration. A 240 kg animal may require as much as 48 L of water to restore fluid balance.

Adrenocortical Insufficiency (Addison's disease, adrenal failure)

DEFINITION. Adrenal insufficiency is failure of the adrenal cortex to produce sufficient corticosteroids to maintain homeostasis.

ETIOLOGY. Responses to prolonged, intense stress may exhaust the adrenal cortex, resulting in adrenocortical atrophy. Prolonged glucocorticoid (cortisone) therapy causes iatrogenic atrophy of the adrenal cortex. Sudden withdrawal of cortisone causes acute insufficiency.

CLINICAL SIGNS. Acute adrenocortical insufficiency is a rapidly fatal shock syndrome. Increased serum potassium levels lead to bradycardia and heart block. Other electrolytic changes cause hypotension, vasomotor collapse, renal failure, and uremia. Hypoglycemic shock may also be involved.

Chronic adrenal insufficiency in the dog is characterized by progressive weakness, weight loss, lethargy, chronic intermittent gastrointestinal upsets (vomiting, diarrhea), dehydration, polyuria, and polydipsia.

Some signs of insufficiency in the dog reflect a deficiency of mineralocorticoids (aldosterone) and glucocorticoids (cortisol). Adrenal exhaustion from stress may or may not affect aldosterone production. The clinical picture of chronic adrenal exhaustion in most animals has not yet been delineated.

Laboratory procedures may assist in diagnosis. Lymphocytosis, eosinophilia, slight to moderate hypoglycemia, hyperkalemia, hyponatremia and hypochloremia, increases in blood urea nitrogen, and hemoconcentration may be defined in laboratory tests [23].

Animals with chronic adrenocortical insufficiency cannot tolerate exercise or any other stress. Restraint is likely to precipitate acute collapse.

THERAPY. Begin intensive shock therapy at once. However, if adrenal insufficiency is suspected, do not use solutions containing potassium (lactated Ringers). Start therapy by administering 5% dextrose in physiological saline. Prednisolone (solu-delta-cortef) in doses of 50 mg/kg intravenously are used in the dog. This can be added to the intravenous fluids [21].

If bradycardia is pronounced—suggesting aldosterone deficiency—give desoxycorticosterone acetate (DOCA) at a rate of 0.1 mg/kg intramuscularly [30].

It is imperative to carefully monitor the patient and institute appropriate remedial measures if complications arise.

Shock

DEFINITION. Shock is a clinical condition characterized by signs and symptoms arising when cardiac output fails to fill the arterial tree with blood under sufficient pressure to provide organs and tissues with an adequate blood flow [6,8,27,29].

With reduced tissue perfusion, oxygen available to the tissues diminishes. Oxygen deficit causes deterioration of the heart and circulatory system, compounding the problem. Irreversible shock occurs when deterioration proceeds to the point that tissue cannot be rejuvenated.

ETIOLOGY. Shock results from severe physical and psychological insult. Shock is often the terminal manifestation of traumatic or metabolic disorders that develop during restraint (Table 7.3).

TABLE 7.3. Shock classifications important in restraint

Cardiogenic—failure of the heart as a pump
Ventricular fibrillation—catecholamine response
Cardiac standstill—cholinergic response
Cardiac tamponade—pressure on the heart
Hypovolemic (actual)—decrease in blood volume
Hemorrhage—whole blood loss
Plasma extravasation—contusions, burns
Dehydration—exercise, hyperthermia
Hypovolemic (relative)—change in vascular bed, increasing its relative capacity
Neurogenic response—pain, fear, anger
Endotoxins—concomitant enteric infections
Toxins—drugs

CLINICAL SIGNS. Typical signs of shock include decreased blood pressure, pale mucous membranes, depression, coolness of the skin, muscular weakness, coma, rapid breathing, rapid and weak pulse, dilated pupils, and decreased body temperature.

Laboratory tests may be helpful in establishing a diagnosis. Shock is characterized by hemoconcentration; increased levels of nonprotein nitrogen, glucose, and potassium; decreased levels of alkali reserves, chloride, and P_{O_2}; and inhibited blood coagulation.

The lesions of shock are nonspecific. A definitive diagnosis cannot be made at necropsy because of species variation of the lesions of affected organs. Necropsy may, however, show suggestive lesions which include hyperemia or petechial hemorrhage of the liver, kidney, and mucosa of the gastrointestinal tract, lungs, and serous membranes; empty and bloodless spleen; effusions into the serous cavities; ischemia of peripheral muscles; and intravascular coagulation of blood.

Psychogenic or neurogenic shock is produced when an animal is subjected to pain or experiences intense emotions such as fear or anger. The pathogenesis of neurogenic shock is not known; however, it is mediated through interference with the balance of vasodilators and vasoconstrictors of the arterioles and venules. It may be closely related to the cholinergic bradycardia syndrome but usually produces less serious consequences.

In neurogenic shock, the blood volume is sufficient but vascular muscle tone is lessened, allowing increased reservoir capacity in both arterioles and venules. Pooling in these vessels effects a decrease in venous return to the right side of the heart, subsequently reducing cardiac output. This syndrome may be seen after or accompanying acute gastric dilatation.

The signs of neurogenic shock differ slightly from those of typical hypovolemic shock in that the pulse rate is usually slow, accompanied by decreased blood pressure. The skin is characteristically warm and may be flushed.

THERAPY. The crucial triad of therapeutic measures to treat shock consists of (1) eliminating the cause of the shock, (2) providing supplemental oxygen, and (3) restoring circulating blood volumes to normal levels.

Establish a patent airway and provide supplemental oxygen. If possible, intubate with a cuffed endotracheal tube to allow assisted or controlled ventilation. Intubation may be impossible when dealing with certain nondomestic species because of the extremely small size of the trachea and inaccessibility to the glottis through the oral cavity of these species. If endotracheal intubation is not possible, use a face mask, nasal catheter, transtracheal catheter, pediatric incubator, oxygen tent (enclose a cage with a plastic bag), or a special oxygen cage to achieve ventilation. Tracheotomy and insertion of an endotracheal tube may be indicated if airways of the nasal or oral cavities are obstructed.

Use the most expedient method to supply oxygen to an unconscious animal. If the animal is conscious, select the least stressful method available. Applying severe physical restraint to place a nasal catheter may do more harm than good. It is unwise to utilize chemical restraint or sedation to supply supplemental oxygen.

Begin intravenous fluid therapy simultaneously with providing a patent airway. Lactated Ringers solution or physiological saline are suitable solutions. A minimum volume which can be given in a few hours is 88 ml/kg. Implement careful monitoring of cardiovascular function to determine if additional fluids are required [21].

Glucocorticosteroid therapy is somewhat controversial but is generally recognized as beneficial. Dosages have been established for dogs. Hydrocortisone sodium succinate is given intravenously at the rate of 50 mg/kg [21]. Dexamethasone (Azium) is given concurrently (also intravenously). The dosage is 4 mg/kg, repeated every 4 hours.

The development of metabolic acidosis is inevitable in cases of shock. If blood-gas analysis is available, the precise amount of sodium bicarbonate needed to counteract acidosis can be calculated, using the formula: $NaHCO_3$ required = base deficit in mEq/L × 0.03 × B.W. in kg.

Without blood-gas analysis, give an initial dose of 4.5-5.6 mEq of sodium bicarbonate per kg of body weight slowly, intravenously, and add a similar amount to intravenous fluid administered over the next several hours [21].

Ischemia of the liver parenchyma and the intestinal epithelium predispose these tissues to necrosis [7]. Bacterial invasion of the body is common during and following shock. Broad-spectrum antibiotic therapy is essential to prevent septicemia. Use a high dosage and continue therapy for 5 days.

In a hospital situation, with laboratory backup and availability of monitoring equipment, additional therapeutic measures may be instituted. The use of diuretics, ionotropic agents, sedatives, and anticoagulants may be indicated in specific situations. To alleviate disrupted thermoregulation, keep the patient warm. Monitor the body temperature.

Cholinergic Bradycardia (syncope, fatal syncope, fainting, vagal reflex, vagal bradycardia)

DEFINITION. Cholinergic bradycardia is the slowing of the heart rate produced when vagal stimulation overrides the usual adrenergic response of alarm.

ETIOLOGY. Usually during restraint the typical adrenergic alarm response is initiated, characterized by vasoconstriction and hypertension. Centers in the hypothalamus normally stimulate the sympathetic system. However, similar centers in the hypothalamus can also stimulate the parasympathetic system.

Under intense stimulation—in some animals—the cholinergic response overpowers the adrenergic, resulting in a precipitous fall in blood pressure and slowing of the heart rate. Unconsciousness may result.

Syncope (fainting) is an ordinary phenomenon in the human being. Cerebral ischemia brought about by rapid carotid hypotension causes unconsciousness. In human beings, this effect is usually transitory. As soon as the person lies flat, normal pressure is restored and consciousness returns.

Why do animals die? The answer is not known. It is known that animals can slip from syncope into irreversible hypovolemic shock. If syncope occurs in an animal in hand, death may be prevented by permitting it to lie out flat, giving it a chance to recover.

Fatal syncope can also occur in human beings if the upright position is maintained. Crucifixion causes death through a form of recurrent syncope, brought about in this manner. Some fainting individuals wedged in an upright position in a panic-stricken crowd have suffered similar mortality [25]. The diving reflex of marine mammals is a normal cholinergic bradycardia. The response may be initiated in a seal by grasping the muzzle and clamping the nostrils shut [20].

Other reflexes that may initiate cholinergic bradycardia are triggered by ocular pressure, carotid sinus pressure, and increased abdominal pressure (valsalva maneuver—forced expiration against a closed glottis) [25]. It is important for the animal restrainer to be aware of these reflexes because any of the pressures mentioned can be applied during restraint.

In the human being, actions initiating a vagal response are cold water on the face, tilting the head downward, elevating the feet, or standing in water. Convulsive seizures may result in syncope because of carotid sinus pressure [26]. Individuals vary in their susceptibility to this phenomenon.

Aspirated vomitus in the trachea or acute pleural irritation such as might occur from trauma to the thoracic wall may initiate reflex vagal stimulation and bradycardia [13].

When an animal senses the futility of struggling to extricate itself from a hopeless situation, it may die. The precise mechanism of such death is not known, but bradycardia is a prominent clinical sign. The zebra or gnu that is finally dragged down by a lion usually gives up without additional fighting although physically capable of further struggle. Perhaps at this point, bradycardia and cerebral anoxia deaden the pain of the inevitable outcome.

Physicians have recorded many unexplained deaths of persons faced with either a real or imaginary hopeless situation. Imprisonment as a prisoner of war, loss of a loved one, confinement in a nursing home, physical incapacitation, and voodoo curses have led to death. Such persons simply give up and resign themselves to death. This type of fatality is called submissive death or the helplessness syndrome [25].

Animal experimentation has documented that some rats, monkeys, chickens, dogs, and even cockroaches die when faced with circumstances over which they have no control. Instead of responding with the usual adrenergic stimulation, a cholinergic response takes place.

When wild Norway rats were physically restrained until they had ceased struggling and were then placed in a tank of warm water, some immediately sank to the bottom and drowned. Bradycardia was documented via electrocardiographic monitoring. Cagemates placed in the water without prior physical restraint began to swim [25]. Researchers could condition animals to tolerate more restraint by subjecting them to intermittent periods of physical restraint. The animals learned that the situation was not hopeless.

Submissive death may explain the high mortality of newly captured wild animals. Placing such an animal in a confining crate and shipping it to a strange environment is ample cause for an animal to give up. Successful wild animal capture and shipping can be accomplished by slow conditioning.

In Africa, professional collectors go to great lengths to capture animals quickly and move them without delay to a conditioning center, where they are released into a large cage or pen. The pen is constructed to be as compatible as possible with the animals' previous environment. Human interference is minimized.

Related to submissive death, but of less intensity (nonfatal), is the response of an animal that feigns death when grasped. This catatonic response (frozen posture) is known by many names: animal hypnosis, tonic immobility, death feint, playing possum, catalepsy, and mesmerism.

A small caiman (< 1 m) placed on its back and held for 15 seconds will struggle briefly and relax. Although released, it will remain immobile in that position for several minutes.

Chickens are easily mesmerized, and some rabbits become catatonic each time they are handled. This reaction can be beneficial if restraint for therapy is desired. However, diagnostic examination of a catatonic animal yields little of value.

Professional collectors capitalize on mesmerism to handle newly captured dangerous animals. A zebra or giraffe can be manually handled and crated immediately after a chase with little danger. Approaching the same animal in a catch pen during the conditioning period would be foolhardy.

Cholinergic bradycardia is more likely to affect an animal weakened from malnutrition, parasitism, or a variety of other subclinical illnesses, but an otherwise healthy animal that has resigned itself to helplessness will rapidly deteriorate. Anorexia is common. Such animals are more susceptible to infection from viral and bacterial agents.

CLINICAL SIGNS. Cholinergic bradycardia is characterized by slowing of pulse and heart rate, heart stoppage, unconsciousness, and death. Lesions of cholinergic heart stoppage are minimal. The heart is usually engorged with blood, indicative of a gradual but steady decrease in the heart rate.

THERAPY AND PREVENTION. It is not likely that cholinergic bradycardia will be detected clinically in time to allow treatment. However, atropine sulfate, 0.04 mg/kg, given intravenously will block the cholinergic (vagal) response. No experimental work has been done to determine the prophylactic efficacy of administering atropine to wild animals prior to restraint.

Prevention primarily consists of using the least amount of restraint possible to minimize the feeling of helplessness. Carry out restraint procedures quickly. Diminish external stimuli to decrease stressor effects.

Cardiac Tamponade

DEFINITION. Cardiac tamponade is cardiac failure as a result of the inability of the heart to expand sufficiently to fill with blood.

ETIOLOGY. External pressure on the heart from pericardial or pleural effusion causes cardiac tamponade. Heart-based tumors and excessive external thoracic pressure may simulate such conditions. Gloves reduce tactile discrimination and frequently result in the handler squeezing tighter than necessary to restrain an animal. Tiny species can be easily injured in this manner.

CLINICAL SIGNS. Cardiac failure results in shock and rapid death. If the syndrome is recognized and the etiology can be quickly reversed, the animal may live.

THERAPY. Correct the cause and supply oxygen.

Ventricular Fibrillation (heart flutter)

DEFINITION. Ventricular fibrillation is uncoordinated rapid contraction of the ventricular cardiac muscles.

ETIOLOGY. This phenomenon is caused by many conditions but the primary cause during restraint is elevated levels of catecholamines (epinephrine and norepinephrine).

During the alarm response, a normal tachycardia develops under catecholamine stimulation. If, however, the cardiac muscle has been previously sensitized to catecholamines by acidosis or hypoxia, or both, such stimulation may lead to fibrillation.

CLINICAL SIGNS. In ventricular fibrillation, agonal struggling may simulate normal resistance to restraint. The ventricles can neither fill nor pump blood normally. Circulatory failure is followed quickly by unconsciousness and death. No pulse can be palpated nor can the heart beat be picked up by auscultation. An electrocardiogram would provide a definitive diagnosis, but the animal will die before such diagnostic techniques can be employed.

THERAPY. The prognosis for ventricular fibrillation is grave. Electrical stimulation with a defibrillator is the only effective therapy. This hospital procedure is not likely to be available to the animal restrainer in the field. Preventive medicine involves minimizing conditions causing acidosis or hypoxia.

Capture Myopathy (stress myopathy, white muscle stress syndrome, overstraining disease, "Vangspier sindroom" [Afrikaans]) [3,4,18]

DEFINITION. Capture myopathy is a muscle disease associated with the stress of capture and restraint. The disease is characterized by degeneration and necrosis of striated and cardiac muscles. Numerous species of birds and mammals are affected.

ETIOLOGY. Capture myopathy has been reported from diverse places on the globe. It is most likely to occur in operations for capture of free-living wild animals. Wild sheep in western Canada and numerous mammals and birds in Africa have been affected.

The disease usually develops within 7-14 days after capture or transport. However, it has been observed in as short a time as 6 days and after as long a period as 30 days following handling [3]. It can be seen both in animals that exert themselves maximally and in those that are relatively quiet. It occurs with either physical or chemical restraint. Predisposing factors include fear, anxiety, overexertion, repeated handling, failure to allow an exhausted animal to rest before transportation, prolonged transportation, and constant muscle tension such as may occur in protracted alarm reaction. Muscles cramped into strange positions in crates or sacks may develop local muscular anoxia and necrosis. A variety of stressors may function in concert or individually, precipitating development of the classical syndrome.

PATHOGENESIS OF THE LESION. South African workers have presented good evidence that an acidemia is one cause of the lesion [10]. Profound muscle exertion results in a metabolic conversion from aerobic to anaerobic oxidation. Lactic acid builds up more rapidly than it can be metabolized, producing a marked local and systemic acidosis. Necrosis of the muscle cells may result. This is thought to be the cause of equine myoglobinuria azoturia (tying-up syndrome).

In human beings, muscular necrosis (rhabdomyolysis, paroxysmal paralytic myoglobinuria) has been related to hypokalemia. Potassium is involved in normal vascular function. Muscle stimulation drives potassium out of the cells, causing local vasodilatation. Vasodilatation provides additional oxygen to support muscle activity. This response cannot occur in a hypokalemic muscle. Prolonged, intense muscle stimulation results in muscular ischemia and potential necrosis. Hypokalemia also results from high cortisol levels—common in restrained animals.

Nutritional myopathy (white muscle disease) is a common disease of domestic cattle, sheep, and horses. A close correlation exists between the occurrence of the disease and nutritional deficiency of vitamin E and selenium. No such correlation has been shown in capture myopathy.

There may be more mechanisms for production of the lesion than are presently known. Much remains to be learned about the pathogenesis of this disease.

CLINICAL SIGNS. Cardiac failure with resultant peracute death follows cardiac necrosis. Skeletal muscle necrosis causes signs similar to those of the tying-up syndrome of horses. Painful, stiff movement of the hind legs is seen. The muscles of the back, rump, and upper leg may be swollen, hard, and hot. Paresis progressing toward paralysis and prostration occurs. Myoglobinuria, dyspnea, and tachycardia may be present in acute stages. Secondarily, trauma of the exposed surfaces of the limbs may be brought about through the animal's struggles to stand.

NECROPSY. Light, grayish streaks are observed in affected muscles; hemorrhages are evident. Histologically, cellular degeneration—with or without hemorrhage—is seen. Fibrosis is present in protracted cases. Adrenocortical atrophy may also be seen.

THERAPY. Prevention must be paramount. Minimize all the factors listed in the etiology. If it is felt that severe stress has been inflicted, treat intensively for acidosis with intravenous sodium bicarbonate. When blood pH determinations are not available, start with a dose of 4–6 mEq sodium bicarbonate per kg body weight. This dose may be repeated in the next few hours. Keep the animal well oxygenated.

Once muscle necrosis has occurred, general nursing care, plus hot packs to the affected muscles, may give some relief, but the prognosis is unfavorable.

Regurgitation

DEFINITION. Regurgitation is the forceful expulsion of the contents of stomach or rumen backward through the esophagus and the mouth or nose. Usually the ingesta is cast out of the mouth with no untoward effects.

Ruminant species normally cast a bolus of ingesta (the cud) back up the esophagus into the mouth for rechewing. This is not regurgitation.

Camelids (camels, llamas, alpacas) are prone to spew foul-smelling ingesta on persons attempting to restrain them. Although unpleasant for the person involved, this type of regurgitation is harmless to the camel.

ETIOLOGY. Regurgitation is a symptom of many diseases and is a serious consequence of both physical and chemical restraint. A recently fed excited animal may regurgitate the meal. Physical restraint may inhibit normal evacuation of vomitus through the mouth, causing fatal inhalation of food particles into the trachea and lungs.

Reflex response to stimulation of the tracheal mucosa by inhalation of food particles may initiate cholinergic bradycardia and fatal syncope. If sufficient food is inhaled, airways are obstructed and strangulation occurs. Gangrenous pneumonia is the universally fatal consequence of inhaling ingesta.

The cardiac sphincter of the stomach may relax in a chemically restrained animal. Improper positioning of the body may place too much pressure on the abdomen of such animals and force ingesta up through the esophagus.

Regurgitation is more common in ruminants that must remain recumbent for prolonged procedures. To prevent regurgitation, maintain a ruminant species in the sternal position with the shoulders higher than the hindquarters. If lateral recumbency is necessary, it is even more important to keep the shoulders and head elevated.

Regurgitation in the horse and other equids is a grave sign because the elongated palate directs regurgitated ingesta through the nose instead of the mouth. Inhalation of ingesta is facilitated under these circumstances.

CLINICAL SIGNS. Stomach contents coming from the nostrils or mouth is diagnostic. Gagging or retching may momentarily precede evacuation. Regurgitation of an animal under chemical restraint is usually passive, with ingesta flowing out of the mouth or nose. The odor of the stomach or rumen contents is fetid and may be a clue to the presence of ruminal ingesta in the mouth.

Inhalation of ingesta may produce coughing spasms, dyspnea, strangulation, and immediate death. Gangrenous pneumonia and death in 3–7 days is the fate of an animal that survives the initial inhalation episode.

PREVENTION AND THERAPY. Once inhalation of ingesta has occurred, death is certain. Inhalation must be prevented. Avoid prolonged immobilization in lateral recumbency. The sternal position is safest in all species. The cardiac sphincter of the stomach is in the most natural position in sternal recumbency. If possible, keep head, neck, and shoulders slightly higher than the rest of the body. This will aid in preventing bloat and regurgitation. Avoid pressure on the abdomen of a restrained animal. Do not restrain animals that have recently fed except as a last resort.

If retching or regurgitation begins, lower the head and neck quickly to allow regurgitated material to exit through the mouth easily and completely. The hazard of regurgitation can be eliminated in anesthetized animals by intubation of the trachea with an inflatable cuff.

Bloat (tympany, gastric dilatation, hoven)

DEFINITION. Bloat is overdistension of the stomach or intestines, usually of ruminants. Bloat is a frequent cause of mortality in species such as the impala and other antelopes, giraffe, and African buffalo kept restrained in lateral recumbency [3].

ETIOLOGY. Restraint procedures are seldom carried out on animals subjected to prior fasting. Thus the rumen and stomach may contain ingesta undergoing fermentation and digestion. Bloat may be caused by excessive gas formation, but it more likely is a result of the inability of the animal to eructate normally.

It is dangerous to keep ruminants in lateral recumbency for extended periods. In this position, ruminal fluid covers the esophageal opening, preventing the escape of gases.

Some drugs such as succinyl chloride and curarelike drugs may contribute to bloat production by relaxing the voluntary muscles of the esophagus. The voluntary control of regurgitation and eructation is subject to wide species variation.

A starved animal given unlimited access to food may engorge itself. This is particularly dangerous when dry food is provided. Water taken into a stomach already full of dry food produces marked swelling, with classic signs of bloat, and such pressure can kill.

CLINICAL SIGNS. If given the opportunity, the animal shows evidence of colic by frequently lying down and getting up, kicking at the abdomen, and rolling.

Rumen motility is increased initially, but as bloat progresses the rumen becomes atonic. Tympanitic sounds are heard on percussion. Dyspnea is marked, accompanied by cyanosis and rapid pulse. Pressure on abdominal vessels prevents adequate circulation. Electrolyte changes and shock ensue. Regurgitation is a common sequel, with potential for inhalation of ingesta and, consequently, the development of gangrenous pneumonia.

THERAPY. In ruminal or stomach distension, slowly reduce intraluminal pressure by means of a stomach tube. Internal pressure on the cardia may impede the passage of a stomach tube, but gentle pressure will usually overcome the obstruction. It may be necessary to manipulate the tube in order to locate and release all the gas pockets within the rumen. Alternatively, a large needle (12 gauge) or a trocar and cannula may be used to deflate the rumen. Penetrate the rumen through the skin in the left paralumbar fossa. If time permits, prepare the area surgically, but if the patient is in critical condition, immediate penetration is imperative.

If gaseous intestinal distension is detected early, rapid evacuation is permissible. If distension has been prolonged, a gradual release of gas is essential. All the veins will have collapsed [14]. Capillary permeability will have increased. Too rapid reduction of intraabdominal pressure will cause venous distension, splanchnic pooling, and possibly shock.

Intestinal distension is less prevalent than bloat during restraint. If it does occur, it is likely to be as a result of bowel obstruction caused by a twisted loop of bowel or by drug paralysis.

Therapy for bowel obstruction is more complicated than for gaseous distention. More definitive diagnostic techniques must be used to determine the etiology. Medical and/or surgical intervention may be required to relieve the condition.

NECROPSY. Bloat associated with restraint is usually acute and readily diagnosed antemortem. In some cases the animal may bloat and die after release from restraint. Differentiation between bloat—which killed the animal—and postmortem gaseous distention is not easy. Pressure ischemia of the abdominal viscera—especially of the liver—present in bloat, may be a key to such differentiation.

Death from bloat is caused by anoxia. The blood is dark and does not clot. Dyspnea prior to death is indicated by petechial hemorrhage of the tracheal mucosa and congestion or hemorrhage of lymph nodes draining the upper respiratory tract.

POSTRESTRAINT COMPLICATIONS

Reduced resistance to disease is an ill-defined consequence of restraint stress (see Chapter 5). Disruption of the integrity of skin and mucous membranes may allow opportunistic microorganisms to gain entrance and cause infection. Pneumonia, general sepsis, wound infection, and enteric infection are frequent sequelae to restraint procedures.

When working with wild species, one cannot discount the accumulative effects of the chronic stress syndrome. Animals suffering from chronic stress may develop adrenal exhaustion and hypoadrenal shock during any subsequent stressful period such as inclement weather, introduction of new animals, or reduced food intake.

Handling may initiate a period of anorexia. Snakes commonly refuse to feed for weeks to months after being pinned. Mammals and birds may be similarly affected, especially if roughly handled. Agalactia may be induced in a lactating female, causing the infant to become hypoglycemic.

Capture myopathy has been known to develop as long as 30 days following a capture operation. Continued tonic muscle activity or metabolic changes can precipitate ischemia and necrosis.

SPECIAL PROBLEMS

Infant animals are subject to special problems during restraint procedures. Youngsters are much more liable to trauma because of their size and inexperience. Infants are easily crushed, gored, trampled on, and bitten by penmates or cagemates or even by the dam, who may be excited by capture operations.

Wild species under stress may exhibit drastic behavioral changes. The mother's protective behavior may become so aggressive that she kills the infant. It is wise to separate young animals from larger specimens before attempting to capture either.

Young animals are more susceptible to heat and cold stress than are adults. Intense excitement places added burdens on the young. An infant prevented from nursing the mother for an extended period may develop hypoglycemia. Furthermore, a drugged or excited mother may suffer temporary or even permanent agalactia, causing the infant to suffer from malnutrition subsequent to the restraint procedure.

If at all possible, postpone shipping until after weaning so that young and their larger dams can be crated separately.

HUMAN INJURY DURING RESTRAINT

Trauma to restrainer or assistant is common during restraint procedures. Be prepared to administer first aid.

Lacerations, contusions, abrasions, fractures, concussions, bite wounds, and kicks are all consequences of failure to adequately prepare for a restraint procedure [5]. In almost every instance, a human mistake precedes an injury.

As with the animal, the most important first aid measures for a person include stopping hemorrhage and maintaining adequate air exchange. All individuals working around animals should know rescue breathing techniques (mouth-to-mouth, etc.).

Traumatic shock occurs more frequently in persons than in animals. Be prepared to alleviate shock by having the patient lie down, lowering the head, and providing warmth.

The most effective therapy for many traumatic cases is to apply ice packs. Crush ice and put it into a plastic bag. Do not add salt or any other foreign substance to the ice so the bag can be placed directly against the skin without freezing the tissue. The patient may experience transient pain while the skin is cooling. A thin towel can be placed between the skin and the ice for sensitive individuals.

Coolness minimizes extravasation of plasma and induces vascular constriction with resultant decreased hemorrhage, lessening swelling and pain. Function of the injured tissue returns sooner than if the inflammatory process is allowed to run its course. Standard first aid should be carried out until a physician can be consulted.

The person in charge of a restraint procedure is legally obligated to protect those participating, including the animal's owner, who may be assisting. It is incumbent on the supervisor to select an appropriate restraint procedure and to instruct all concerned how the technique should be safely carried out.

Veterinarians who deal with restraint procedures daily must be certain that assistants are capable of carrying out a given procedure. Do not assume that a person's general knowledge extends to handling animals in unusual situations or rely on someone's "common sense." There is no such thing as common sense. Common sense is only the result of direct or vicarious experience. One person may be experienced in the handling of cattle but ill equipped to deal with dogs or wild animals. Do not assume that

TABLE 7.4. Possible etiology of signs observed during restraint

Convulsions (seizures)
- Anoxia
- Hypocalcemia
- Hypoglycemia
- Hyperthermia

Hypoxia, pneumonia, struggling, cardiac failure
- Brain contusion
- Catatonia (drug reaction)
- Epilepsy
- Fracture of cervical vertebrae
- Acidosis

Tetany
- Hypocalcemia
- Hypoglycemia
- Hypothermia (shivering)

Rapid breathing
- Hyperthermia
- Hypoxia
- Acidosis

Elevated body temperature
- Latent infection
- Increased muscle activity
- Drug effects on central thermoregulation
- Restraint practices that prevent heat dissipation
- Catatonia
- Convulsive disorders

Decreased body temperature
- Drug effects on central thermoregulation
- Failure to provide proper environment
- Prolonged anesthesia

Pale mucous membranes
- Anemia
- Shock
- Hemorrhage

Dark mucous membranes
- Normal pigmentation of mucosa in that species
- Hypoxia (strangulation, pneumonia, pulmonary edema)

Bloat (stomach or intestines)
- Improper positioning of body during restraint
- Twisted intestine
- Ileus
- Drug-induced ileus

Loose stool
- Fright
- Previous enteric disease
- Response to drugs

Regurgitation
- Pressure on thorax or abdomen
- Improper positioning of body during restraint
- Relaxation of cardia by drug action
- Excitement

Carrying a limb
- Fracture of a bone
- Severe sprain
- Contusion to hoof or claw

Unable to stand on leg
- Nerve damage
- Ruptured tendon or ligaments
- Capture myopathy
- Fracture

Unable to use hindquarters
- Fracture of thoracic or lumbar vertebrae
- Fracture of pelvis
- Thrombus in iliac vessel

Inability to use all four limbs
- Fracture of cervical vertebrae
- Brain contusion
- Hypoglycemia

Urination and frequent defecation
- Fright
- Response to drugs

everyone knows that anteaters have strong recurved claws, which are dangerous weapons.

Warn assistants of the tremendous strength possessed by animals, particularly by wild species. Prescribe safe distances from hoofs, claws, horns, and antlers. Emphasize the incredible quickness with which even large species such as the rhinoceros and elephant can move.

Assistants, particularly owners of animals, are reluctant to "hurt" the animal and frequently fail to grasp it firmly. The struggling animal escapes and in the process injures the handler. Even customarily affectionate pets may injure their owners when under restraint tension.

Animal protein allergies may cause problems for some individuals. A person may be sensitive either to general animal dust or be species specific. One of my colleagues is extremely sensitive to black buck and bush buck antelope (cutaneous and respiratory signs develop moments after contact with these species) but not to other antelope species.

The owner of the animal or other assistants often faint during restraint procedures. Critical problems may be created if the person who is holding the animal faints. The suddenly freed, excited animal may injure anyone nearby. The fainting person usually recovers quickly, following momentary recumbency.

If an individual who fears an animal is pressed into assisting with animal restraint, there is danger for all concerned. Such a person cannot be relied on to properly respond to directions. It is extremely dangerous to permit a person with ophidiophobia (fear of snakes) to handle poisonous species.

CONCLUSIONS

Numerous medical problems can develop during restraint procedures. Table 7.4 tabulates signs that can be observed and lists a group of potential etiologies for that sign. Differential diagnosis can be made utilizing information presented in this chapter.

REFERENCES

1. Altman, P. L., and Dittmer, D. S. 1974. Acid base balance. In Biology Data Book, 2nd ed., vol. 3, pp. 1830-49. Washington, D.C.: Federation of the American Society for Experimental Biology.
2. Ballinger, W. F.; Rutherford, R. B.; and Zuidema, G. D., eds. 1973. The Management of Trauma, 2nd ed., p. 74. Philadelphia: W. B. Saunders.
3. Basson, P. A., and Hofmeyr, J. M. 1973. Mortalities associated with wildlife capture operations. In E. Young, ed. Capture and Care of Wild Animals, pp. 151-71. Capetown, South Africa: Human and Rousseau.
4. Donaldson, L. E. 1970. Muscular dystrophy in cattle suffering heavy mortalities during transport by sea. Aust. Vet. J. 46:405-8.
5. Douglas, L. G. 1975. Bite wounds. Am. Fam. Physician 11:93-99.
6. Fine, J. 1965. Shock and peripheral circulatory insufficiency. In W. F. Hamilton and P. Dow, eds. Handbook of Physiology. Sec. 2, vol. 3, Circulation, pp. 2037-69. Washington, D.C.: American Physiological Society.
7. ______. 1967. Mechanisms of defense in massive injury. Proc. U.S. 34th Annu. Meet. Am. Anim. Hosp. Assoc., pp. 6-18.
8. Glenn, T. M. 1974. Steroids and Shock. Baltimore: University Park Press.
9. Goodman, L. S., and Gilman, A. 1960. The Pharmacological Basis of Therapeutics, 2nd ed., p. 1627. New York: Macmillan.
10. Harthoorn, A. 1974. A relationship between acid base balance and capture myopathy in zebra *(Equus burchelli)* and an apparent therapy. Vet. Rec. 94:337-41.
11. Hermansen, L., and Osnes, J. 1972. Blood and muscle pH after maximal exercise in man. J. Appl. Physiol. 32:304-8.
12. Jones, D. R., and Johansen, K. 1972. The blood vascular system of birds. In D. S. Farner and J. R. King, eds. Avian Biology, vol. 2, p. 159. New York: Academic Press.
13. Jubb, K. V. F., and Kennedy, P. C. 1970. Pathology of Domestic Animals, 2nd ed., vol. 1, p. 124. New York: Academic Press.
14. ______. 1970. Pathology of Domestic Animals, 2nd ed., vol. 2, p. 52. New York: Academic Press.
15. Kirk, R. W., and Bistner, S. I. 1975. Handbook of Veterinary Procedures and Emergency Treatment, 2nd ed. Philadelphia: W. B. Saunders.
16. MacBryde, C. M. 1970. Dehydration, fluid and electrolyte imbalances. In C. M. MacBryde and R. S. Blacklow, eds. Signs and Symptoms, 5th ed., pp. 746-803. Philadelphia: J. B. Lippincott.
17. Mattson, J. L. 1973. Understanding respiratory physiology as a foundation for restraint of wild animals. Proc. Annu. Meet. Am. Assoc. Zoo Vet.
18. Munday, B. L. 1972. Myonecrosis in free-living and recently-captured macropods. J. Wildl. Dis. 8:191-92.
19. Prosser, C. L., ed. 1973. Circulation of body fluids. In Comparative Animal Physiology, 3rd ed., p. 824. Philadelphia: W. B. Saunders.
20. Ridgway, S., ed. 1972. Homeostasis in the aquatic environment. In Mammals of the Sea, pp. 597. Springfield: Charles C Thomas.
21. Ross, J. N., Jr. 1975. Heart failure and shock. In S. J. Ettinger, ed. Textbook of Veterinary Medicine, pp. 825-64. Philadelphia: W. B. Saunders.
22. Ruch, T. C., and Patton, H. D., eds. 1965. Physiology and Biophysics, 19th ed., pp. 1054-68. Philadelphia: W. B. Saunders.
23. Schalm, O. W. 1965. Veterinary Hematology, 2nd ed. Philadelphia: Lea & Febiger.
24. Schmidt-Neilson, K. 1975. Animal Physiology, Adaptation and Environment, p. 255. London: Cambridge Univ. Press.
25. Seligman, M. 1974. Giving up on life. Psychol. Today (May):80-85.
26. Sharpey-Schafer, E. P. 1956. Emergencies in general practice-syncope. Br. Med. J. 1:506-9.
27. Shires, G. T.; Carrico, C. J.; and Canizaro, P. C. 1973. Shock, pp. 3-11. Philadelphia: W. B. Saunders.
28. Sodeman, W. A., and Sodeman, W. A., Jr. 1974. Pathologic Physiology, Mechanisms of Diseases, 5th ed. Philadelphia: W. B. Saunders.
29. Thal, A. P.; Brown, E. B., Jr.; Hernreck, A. S.; and Bell, H. H. 1971. Shock: A Physiologic Basis for Treatment. Chicago: Year Book Medical Publishers.
30. Wolland, M. 1974. Primary adrenocortical insufficiency. In R. W. Kirk, ed. Veterinary Therapy V, pp. 783-86. Philadelphia: W. B. Saunders.
31. Yagil, R.; Etzion, Z.; and Berlyne, G. M. 1975. Acid base parameters in the dehydrated camel. Tijdschr. Diergeneeskd. 100:1105-7.

SUPPLEMENTAL READING

Blood, D. C., and Henderson, J. A. 1974. Veterinary Medicine, 4th ed. Baltimore: Williams & Wilkins.

Catcott, E. J., and Smithcors, J. F., eds. 1972. Equine Medicine and Surgery, 2nd ed. Santa Barbara, Calif.: American Veterinary Publishers.

Dunne, H. W., and Leman, A. D., eds. 1975. Diseases of Swine, 4th ed. Ames: Iowa State Univ. Press.

Ettinger, S. J. 1975. Textbook of Veterinary Internal Medicine: Diseases of the Dog and Cat, vols. 1, 2. Philadelphia: W. B. Saunders.

Fox, M. W. 1965. Canine Behavior. Springfield, Ill.: Charles C Thomas.

______. 1971. Behavior of Wolves, Dogs and Related Canids. New York: Harper & Row.

______. 1974. Understanding Your Cat. New York: Coward, McCann and Geoghegan.

Fox, M. W., ed. 1968. Abnormal Behavior in Animals. Philadelphia: W. B. Saunders.

Hofstad, M. S., ed. 1972. Diseases of Poultry, 6th ed. Ames: Iowa State Univ. Press.

Hungerford, T. G. 1969. Diseases of Poultry Including Cage Birds and Pigeons, 4th ed. Sydney, Australia: Angus and Robertson.

Jones, L. M. 1977. Veterinary Pharmacology and Therapeutics, 4th ed. Ames: Iowa State Univ. Press.

Kirk, R. W., ed. 1977. Current Veterinary Therapy VI. Philadelphia: W. B. Saunders.

Marsh, H. 1965. Newsom's Sheep Diseases, 3rd ed. Baltimore: Williams & Wilkins.

2 Domestic Animals

8 HORSES, DONKEYS, MULES

CLASSIFICATION

Order Perissodactyla
Family Equidae: horse, pony, ass, burro, donkey, mule, hinny

ALTHOUGH horses are no longer used extensively as work animals, there are more horses in the United States at the present time than existed at the zenith of the horse's use for power. Racing and general recreation account for the phenomenal increase in numbers.

Horses are seen in all shapes and sizes, from tiny ponies less than 1 m (39 in.) tall and weighing less than 45 kg (100 lb) to large draft horses, 160 cm (68 in.) tall and weighing 1,364 kg (3,000 lb).

Donkeys and mules are intermediate in size and share many characteristics with the horse. Restraint practices are essentially the same.

See Table 8.1 for the names of gender used in the equine.

DANGER POTENTIAL

Horses can kick, strike, bite, and press persons against walls. Even a foal or a pony can injure a handler's foot by stepping on it. The bite of a horse is serious. A large stallion is capable of grasping a person by an arm or shoulder and lifting him/her off the ground. Once the horse has grasped a person, the jaw remains closed and the tissue is pulled through the teeth. Serious contusion results (Fig. 8.1).

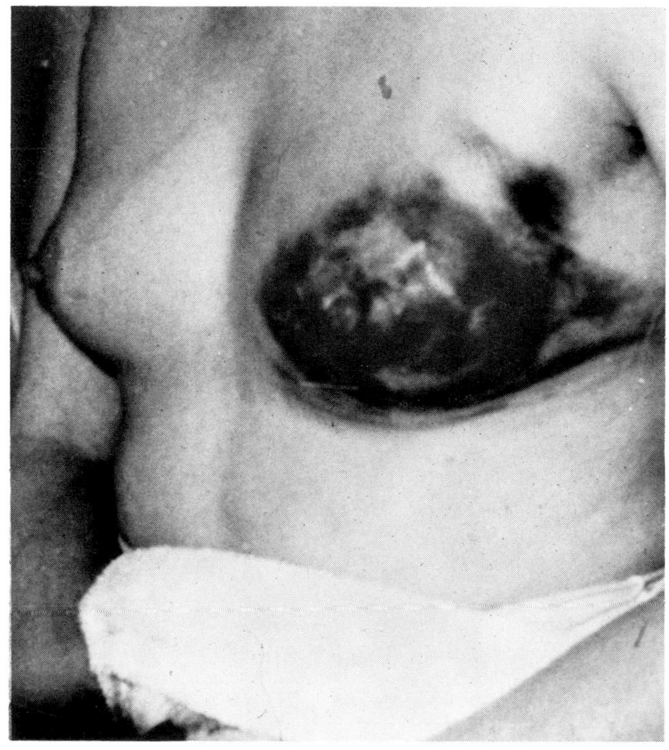

FIG. 8.1. Woman's breast 2 days after being bitten by a pony stallion.

Horses usually kick directly backward with either or both hind feet. A few horses are adept at kicking forward and outward (cow-kicking) and can reach up to their front legs. Mules and donkeys are experts at this. Handlers can be kicked in the head and killed as they bend over to work on the legs of such a horse. It is wise when working around a horse's legs to keep your head above the knee or the hock at all times.

Horses can strike with one foreleg from a standing position, or they can rear and strike with both legs. Striking often occurs when an attempt is made to apply a twitch. The safest place to stand is next to the left foreleg, close to the body, to prevent kicking, striking, and biting (see Fig. 8.5).

PHYSICAL RESTRAINT

Horses are startled by quick movements and loud noises but respond to voice commands. Speak firmly, but do not shout. Voice and other mannerisms betray fear, quickly detected by a horse. If you lack confidence, remain silent.

Horses enjoy being petted, and this is soothing during handling procedures. The wise handler continually speaks

TABLE 8.1. Names of gender of equine

	Adult				Immature	
Species	Male	Female	Neutered	Newborn	Male	Female
Horse, pony	Stallion, stud	Mare	Gelding	Foal	Colt	Filly to 3 years
Donkey, ass, burro	Jack	Jenny	Gelding	Foal		
Mule (male donkey + female horse = hybrid)	Mule	Mare	Gelding	Foal		

to the horse and maintains physical contact by petting. Since horses may be ticklish, petting must be firm, avoiding sensitive areas like the flanks or around the ears and eyes.

Mules and donkeys are stubborn and require more patience and perseverance than the horse.

Horses are usually handled with a halter and lead shank. Halters can be temporarily constructed of rope (see Fig. 3.37). Permanent halters are constructed of cotton and nylon rope (Fig. 8.2), leather (Fig. 8.3), or webbing (see Fig. 8.5). Leather halters are usually the weakest unless triple-stitched and made of heavy straps. Haltering is accomplished as illustrated in Figures 8.4 and 8.5. The horse will stand more quietly for haltering if a rope is placed around its neck first. The lead shank can be a simple rope tied to the halter ring. Most shanks use a snap for easier attachment to the ring. Frequently the snap is the weakest link and may break at an inopportune time.

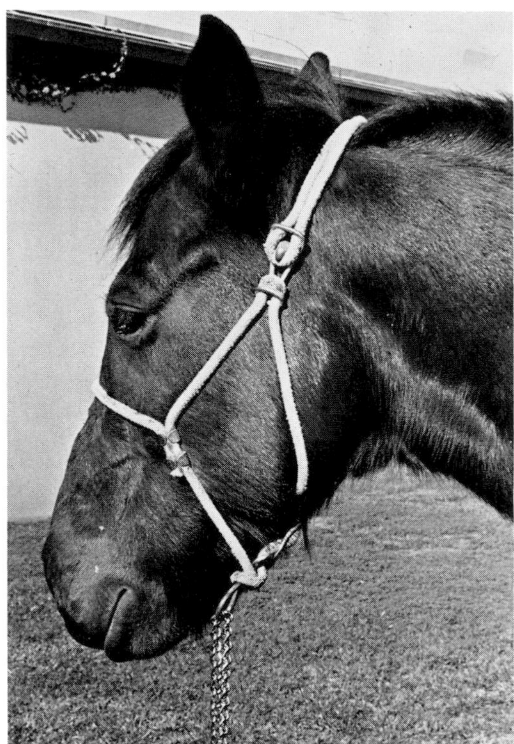

FIG. 8.2. Rope halter.

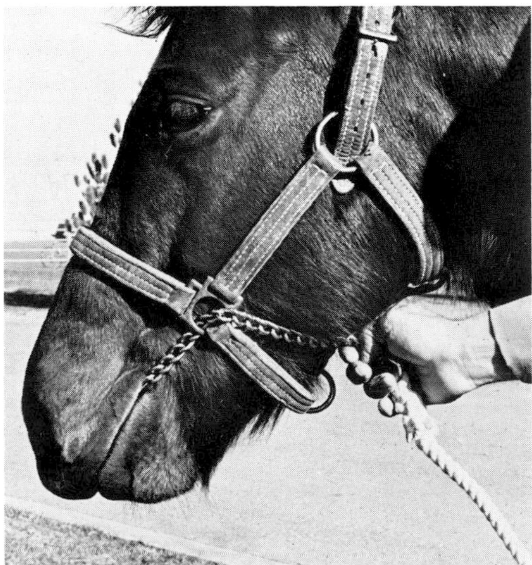

FIG. 8.3. Triple-stitched leather halter used with a chain shank in the mouth.

The snap-chain shank is popular (Fig. 8.5). The chain can be attached to a rope or leather strap. The snap can attach directly to the halter ring or be threaded through the ring and snapped to itself, adding strength to the snap and making it possible to grasp the shank closer to the head of the horse without having to grab the chain.

The chain shank is frequently used for further restraint. The chain can be placed through the mouth and attached to the cheek ring on the opposite side, serving as a bit or bridle (Fig. 8.3). If the mouth is jerked or otherwise injured by the chain, the animal will resist any bit. The chain can also be placed over the bridge of the nose (Fig. 8.6 left). By gently tugging the shank, one can divert the animal's attention. The pressure exerted tends to pull the nose down.

Frequently the chain is placed under the chin (Fig. 8.6 right) instead of over the bridge of the nose. Mechanically this is a mistake. When the chain is pulled, the horse naturally throws its head up to relieve pressure underneath the chin, making the horse unmanageable instead of controlling it.

Keep the lead rope high and short when tying a horse (Fig. 8.7) to prevent it from entangling its feet (Fig. 8.8). A horse will fight to extricate itself from such a predicament, and serious rope burns and injury to the cervical vertebrae may occur in the ensuing struggle.

The twitch is the most important manual tool used in equine restraint. The principle is based on reaction to pressure applied to the sensitive lip. This diverts the horse's attention from procedures taking place elsewhere on the legs or body. Used injudiciously, serious damage can be done to the lip. For example, a judo expert assisting me nearly tore the lip off his horse because he twisted the twitch too tightly.

The most satisfactory twitch consists of a short length of chain attached to a hardwood handle approximately 2 ft long. A rope loop can also be used. A pick handle with two holes drilled at the sides to admit the rope to pass through is a satisfactory base.

The rope twitch is gentler but is slower to twist up. Another disadvantage is that if the horse pulls loose from the handler, the loop is slow to untwist. In such a situation the horse usually swings its head and flails the twitch handle around, endangering both horse and handler. A chain twitch in the same situation drops off quickly.

To use a twitch, grasp the twitch and the cheek piece of the halter with the right hand. Place the fingers of the left hand partially through the loop of the twitch (Fig. 8.9). Do not insert the whole hand to the wrist, as this complicates placement of the twitch. Bring the left hand over the bridge of the nose and gently move it to the upper lip (Fig. 8.10 left). Once the operator is prepared to grab the lip, it should be done firmly to prevent the horse from pulling

FIG. 8.4. Sequence in haltering a horse (Part 1).

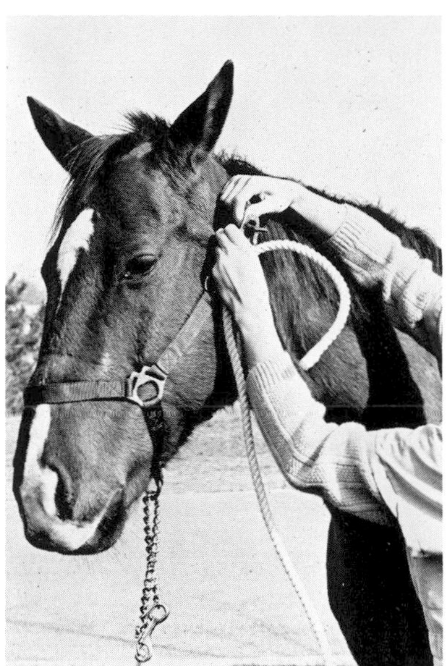

FIG. 8.5. Sequence in haltering a horse (Part 2). Proper position for holding or leading a horse.

FIG. 8.6. Use of the chain shank. Over the nose provides excellent mild restraint. Under the chin causes head tossing and is not recommended.

FIG. 8.7. Horse tied properly, high and short.

FIG. 8.8. Horse tied improperly, allowing a leg to be entangled in the rope.

away. The more times the horse successfully escapes the operator, the more difficult it will be to place the twitch. Once the fingers have a firm grasp of the nose, the rope or chain is brought over the lip and the right hand begins to twist the loop (Fig. 8.10 right). Twist firmly to maintain a grip, but not so tightly that severe pain is felt, or the horse will resist by pulling away or even striking. It is important for the operator to maintain a grip on the handle with *both* hands or the handle may be pulled away. The halter shank should be used to pull the head toward the left side. Do not use the twitch as a lever. The pressure on the lip should be a twist, not a pull.

FIG. 8.9. Applying twitch: First grasp cheek piece of halter and twitch in the same hand.

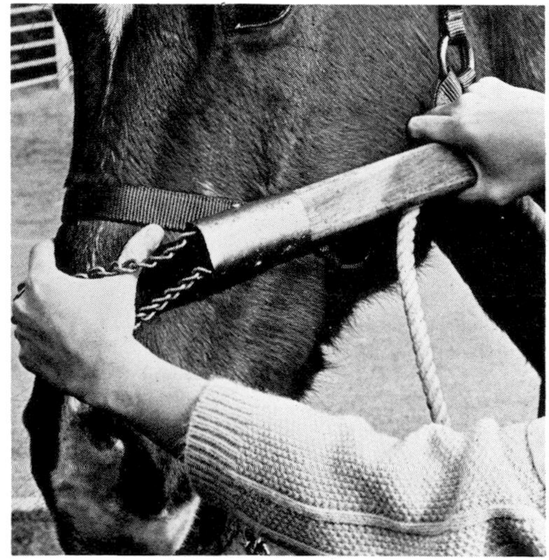

FIG. 8.10. Sequence in twitching a horse.

When twitching, pull the head to the left so that the operator is not exposed to the front legs if the horse should strike. The handler should stand close to the shoulder (Fig. 8.11). *Never stand in front of the horse.*

Once the twitch is in place, do not maintain a constant pressure or the lip will become numb and fail to provide the necessary restraint. It is more desirable to carry out a rocking motion with the handle so the twist is released and tensed periodically. If the animal shows signs of fidgeting or failing to respond to the twitch, shake the nose more vigorously. Jerking the nose or twisting too hard will cause permanent damage to the lip. It is not likely that a twitch left in place for more than 15 minutes will remain effective. Periodic rest is necessary for prolonged procedures.

Removing the twitch requires as much care as placing it on the nose. The moment of release seems to be a stimulus for the horse to pull away and perhaps even to strike. Remove the twitch quickly to prevent the horse from jerk-

FIG. 8.11. Proper *(left)* and improper positions for holding a twitch.

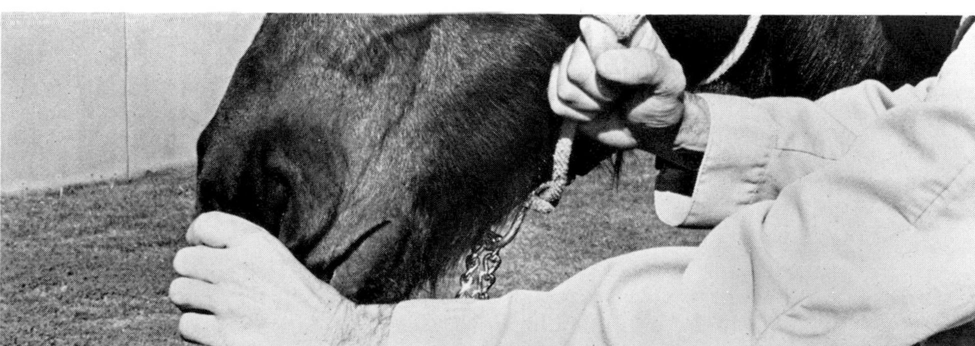

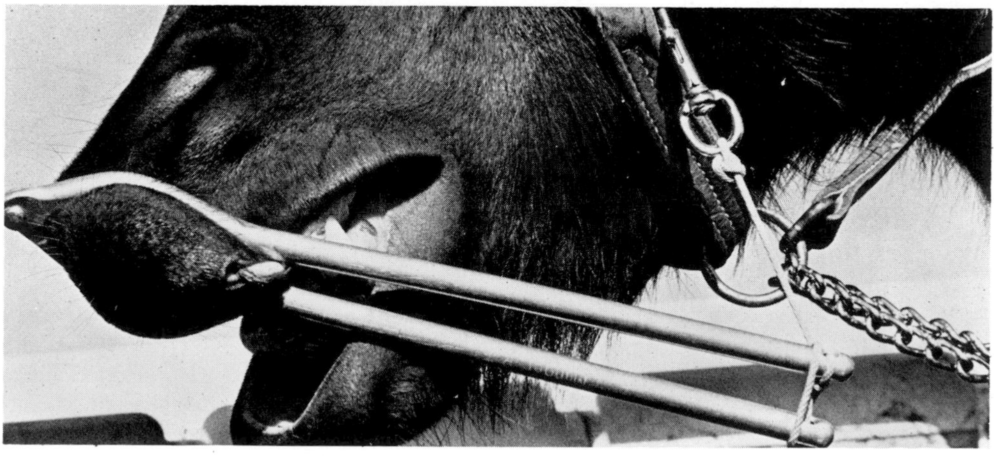

FIG. 8.12. Using the hand as a twitch *(upper)*. Self-retaining twitch *(lower)*.

ing it out of your hand and swinging it around. After twitching, it may be desirable to massage the lip to restore circulation. The twitch can be used repeatedly if the horse is not injured by rough handling during the procedure.

The hand can be used as a mild twitch (Fig. 8.12 upper). The lone operator can use a self-retaining twitch (Figs. 8.12 lower, 8.13). Placing a twitch on the ear is hazardous, as the ear may be permanently damaged.

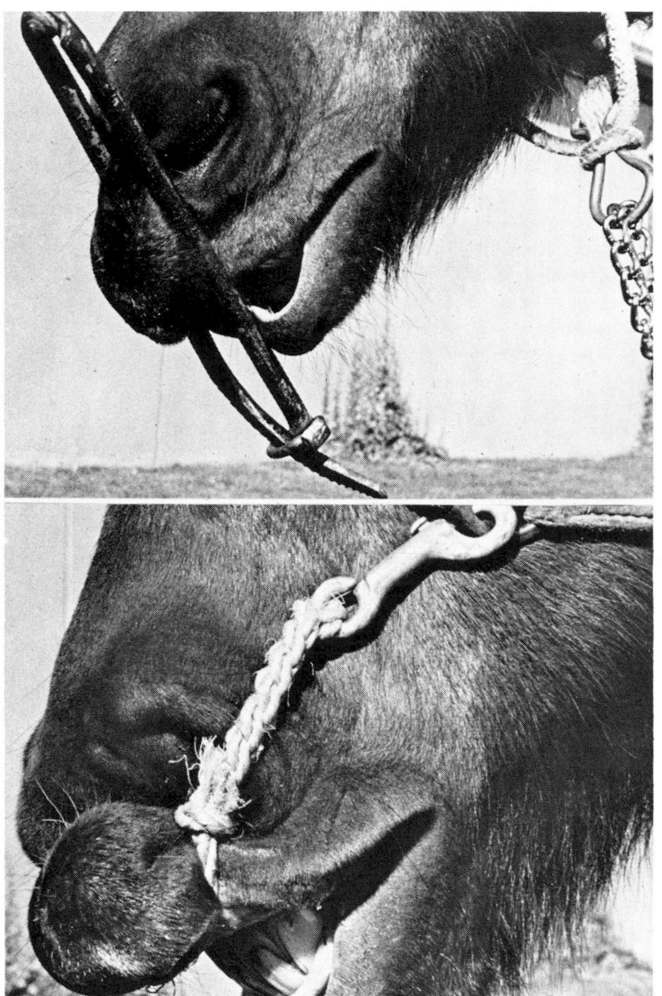

FIG. 8.13. Self-retaining twitches.

A horse's attention may be diverted by pressing on the eyelid or grasping a fold of skin at the shoulder (Fig. 8.14). The lip chain utilizes the chain shank and is placed as illustrated in Figures 8.15 and 8.16. Tension must be constantly applied or it will slip off. Use caution, because the rough chain can severely traumatize the mucous membrane of the lip.

Some horses will fight a twitch but can be restrained by applying pressure to one or both ears. To do this, stand in front of the shoulder alongside the neck. Grasp the left ear with the right hand. The left hand grasps the halter or is placed over the bridge of the horse's nose. Pull the horse's head toward its left side to keep the animal slightly off balance. As soon as the animal feels pressure on the ear, it will pull away toward the right. To ear a horse, lay the

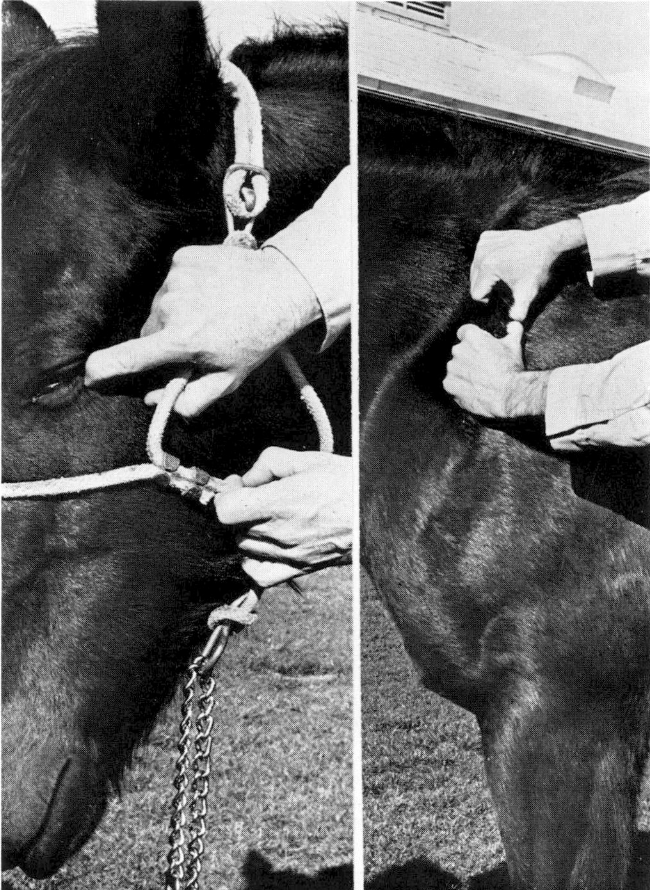

FIG. 8.14. Forms of mild restraint: Pressing a finger on the eyelid; lifting a fold of skin at the shoulder.

right hand on the top of the poll with the palm down, the fingers together, and the thumb extended (Fig. 8.17 upper). Push the hand forward to the base of the ear so the web between the thumb and the forefinger is tight against the base. Then bring the fingers and thumb together and squeeze the ear. The closure of the hand tends to lift the ear from the top of the head (Fig. 8.17 lower).

Even if the horse pulls the operator toward the right, the grip should be maintained. The horse is not apt to strike in this position, being more concerned with trying to pull away from the pressure on the ear. The tension on the ear can be relaxed and increased by simply opening and closing the hand. At no time should the ear be twisted or the pinna bent and cramped since these actions are likely to injure the pinna, causing the horse to become head shy.

An alternate method of earing the horse can be carried out using both hands. One person must hold the halter. Another approaches the horse in the same manner as previously described, on the left side, only in this instance the right arm is draped over the neck and both hands brought up as before. The right hand is placed on the right ear and the left hand on the left ear. In this position the handler can control the head and neck. The animal is less able to rear because of the added weight of the handler on the neck.

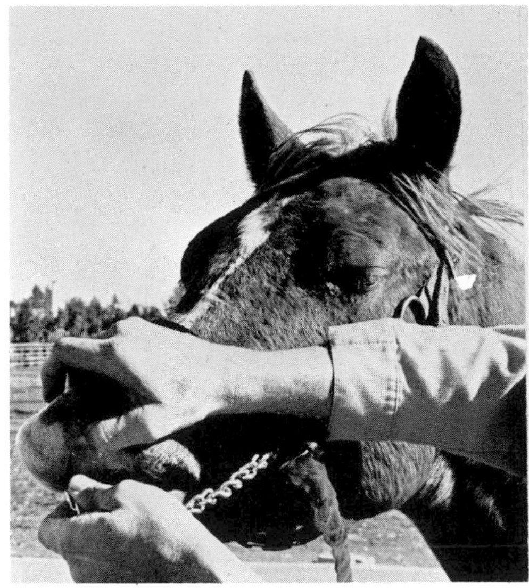

FIG. 8.15. Applying lip chain.

FIG. 8.16. Lip chain properly held.

War bridles are rarely indicated now that tranquilizers are available. War bridles are severe forms of restraint and should be used only as a last resort on obstreperous horses. Extreme caution must be used in applying to avoid damage to commissures of the mouth. One of the many types of war bridles is illustrated in Figure 8.18.

Manipulating Feet and Legs

The foreleg may be lifted by a rear or front approach. With the rear approach on a left foreleg, place the right shoulder against the horse's left chest (Fig. 8.19 left). Press the right forearm forward against the back of the knee and run the hand down the leg to grasp the pastern or cannon bone (Fig. 8.19 right). When approaching from the front, pull the knee forward to make the horse lift its leg (Fig. 8.20).

Examine the foot while holding it with one hand, or by straddling the leg and supporting the foot with the knees (Fig. 8.21). The most comfortable position is with the knee slightly flexed. The feet should be about 12 in. apart with the toes turned in, driving the knees together. Once the limb is lifted, it is important that the handler maintain the

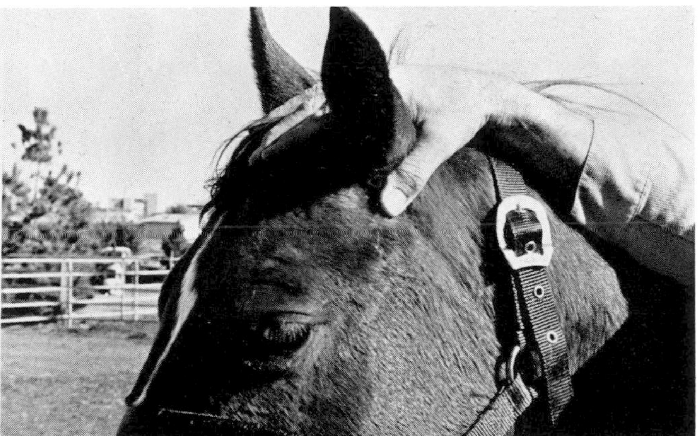
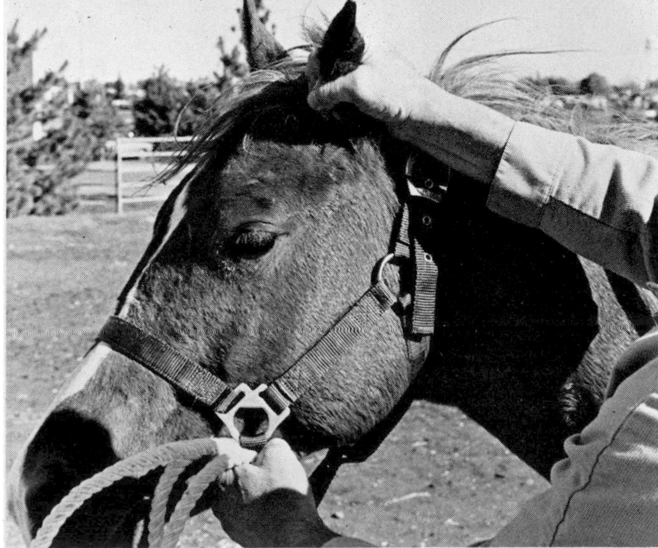

FIG. 8.17. Earing a horse.

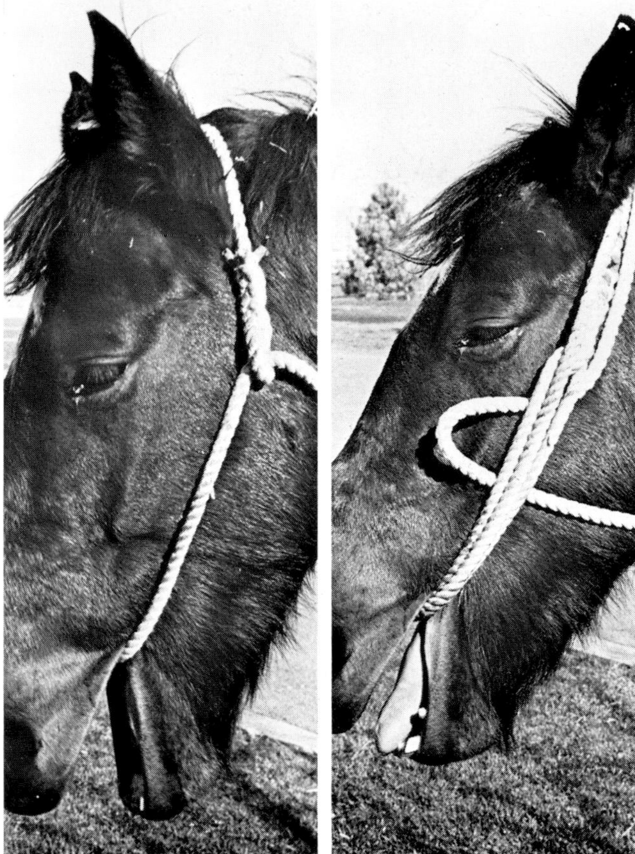

FIG. 8.18. War bridle.

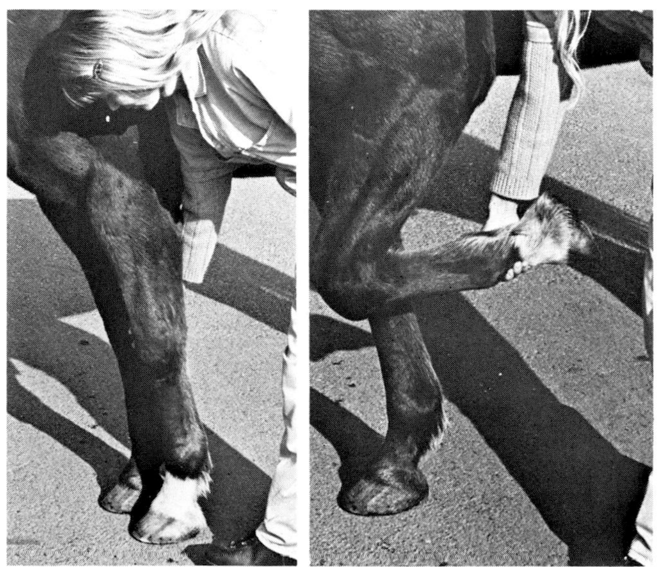

FIG. 8.19. Lifting a foreleg from rear approach.

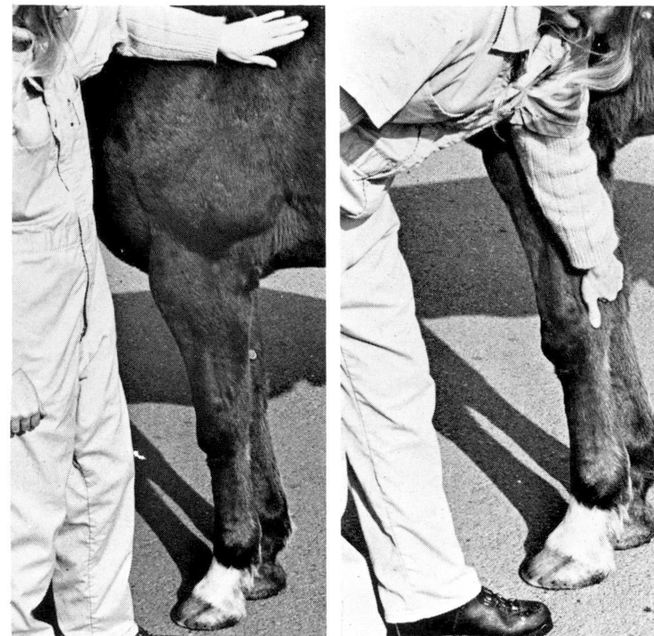

FIG. 8.20. Lifting a foreleg from front approach.

grip and not allow the horse to set its foot down again until desired. If the horse is permitted to pull its foot away and place it on the ground, it will attempt to do this again and again. Fortunately most horses that have experienced some handling are content to lift the leg, providing it is not pulled into an abnormal position and the animal is allowed to balance on the other three legs. Holding one leg up is an excellent way to minimize kicking while someone works elsewhere on the horse. Do not underestimate a horse's ability to balance, however. Some can kick even while balancing on two legs.

The left hind leg is lifted as follows: Face the back of the horse and move the right hand over the rump of the horse

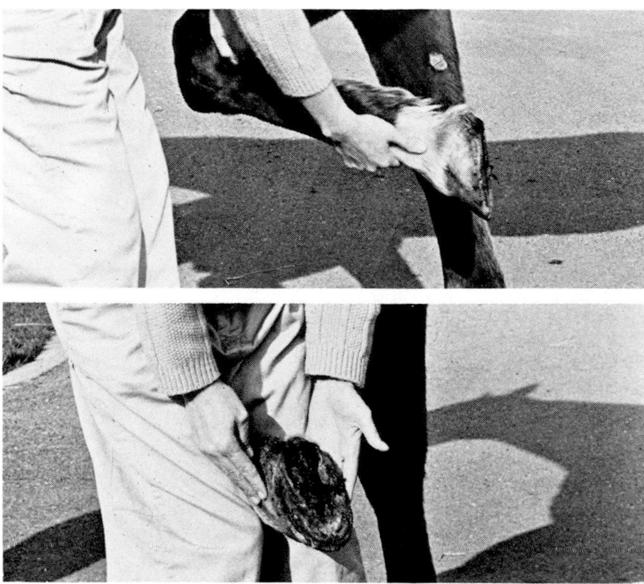

FIG. 8.21. Lifting and holding a foreleg

and down its hind leg (Fig. 8.22). This manipulation alerts the horse that you are going to be working on that hind leg. If it is inclined to kick, it will probably indicate this as well. Do not tickle the flank area. The right hand is brought down over the thigh and hock, ultimately resting on the midtarsal region. Pull the leg forward and upward (Fig. 8.23). Some horses may kick, but a person in the proper position will not be injured. Next place the left hand over the hock and step back under the leg (Fig. 8.24). Finally, the leg is supported by the legs and body of the handler (Fig. 8.25), leaving both hands free to examine and manipulate the foot.

A foal does not usually respond well to a halter or to being led. If a detailed examination must be given, grasp the foal by the tail and hold under the neck (Fig. 8.26). Older foals and small yearlings can also be held in this manner by individuals with adequate strength. Rarely, a foal will reach around and bite when grabbed like this. A word of caution: If the grip on the tail is held tight for too long, the foal will attempt to sit down.

Sometimes a foal is difficult to catch. The mare can be placed against a wall so that the foal will wedge between her and the wall (Fig. 8.27). If the mare is not a kicker, the handler can step in behind and tail the foal. The mare handler must prevent the foal from escaping forward.

Never stand directly facing the hind legs. Stand sideways, to prevent receiving a kick in the abdomen or groin.

A foal can be taught to be led by placing a rope over the

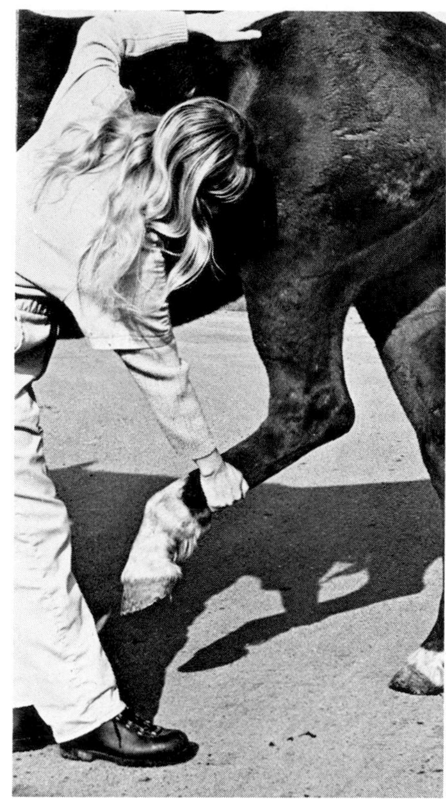

FIG. 8.23. Lifting hind leg.

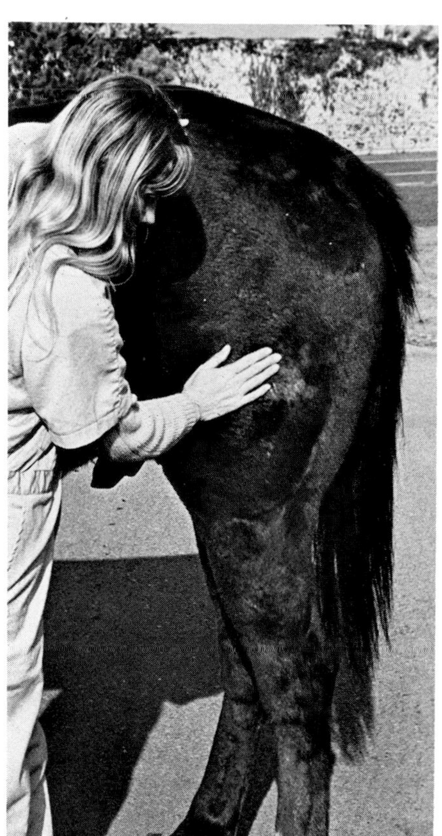

FIG. 8.22. Approach to lifting a hind leg.

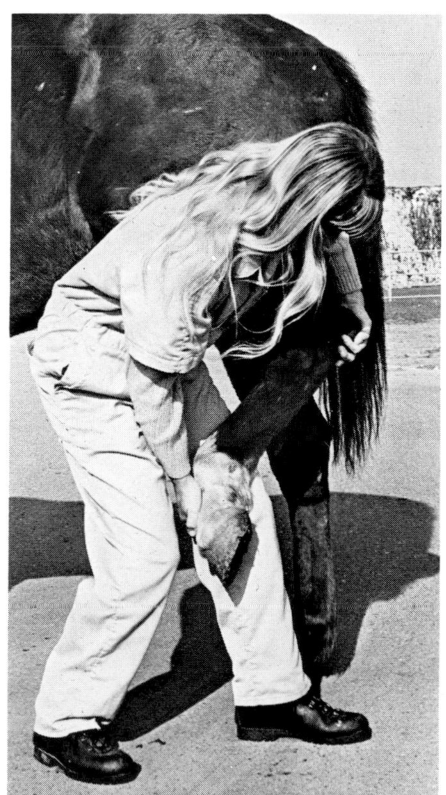

FIG. 8.24. Moving hind leg backward.

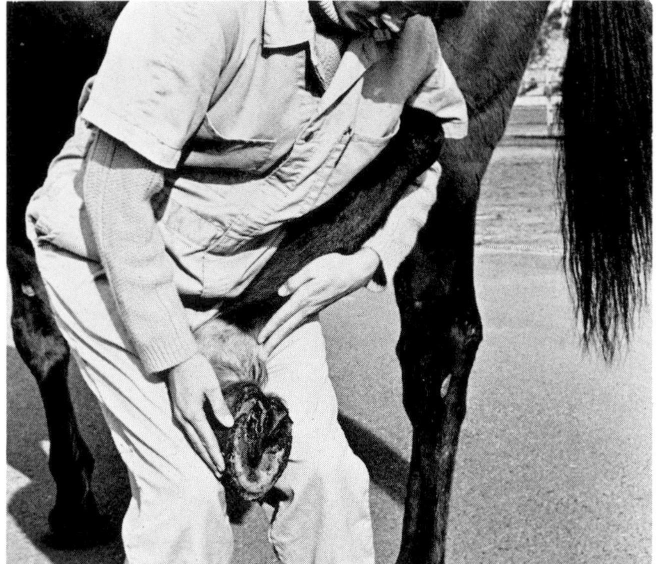

FIG. 8.25. Hind leg held against body and legs.

FIG. 8.26. Proper way to hold a young foal.

rump and through a halter (Fig. 8.28). A foal can be cast quite simply by pulling the tail between the hind legs and maintaining pressure. Soon the foal will begin to relax and slump to the ground (Figs. 8.29, 8.30). Maintain the cast position by pulling the tail up in front of the stifle and placing pressure on the neck (Fig. 8.31).

A foal is extremely curious. In a pasture it is sometimes possible to catch a foal by simply squatting down and allowing it to come up and investigate. As it approaches the handler, it can be grasped in the manner previously described. This procedure sometimes works with wild animals, too. A partially tranquilized elk calf walked up to me when I was squatted down.

FIG. 8.27. Capturing a foal by holding the mare against a wall.

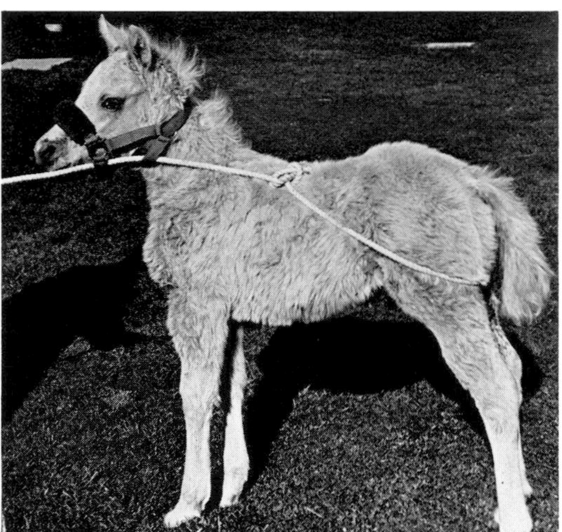

FIG. 8.28. Use of rump rope to lead a foal.

FIG. 8.29. Casting a foal by pulling tail between the hind legs.

FIG. 8.30. Foal relaxing and slumping to the ground.

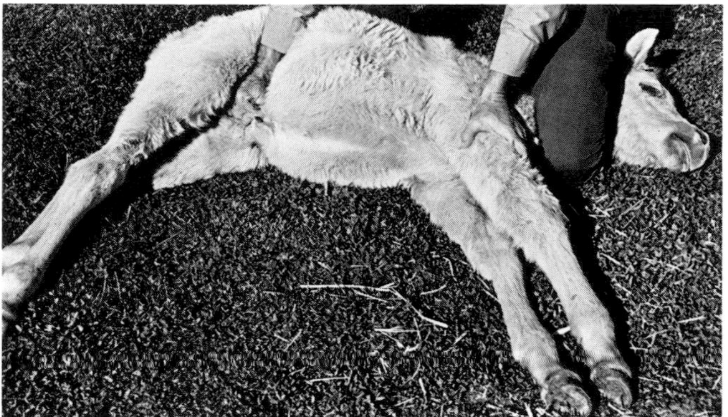

FIG. 8.31. Holding foal down by maintaining grip on tail and pressure on stifle.

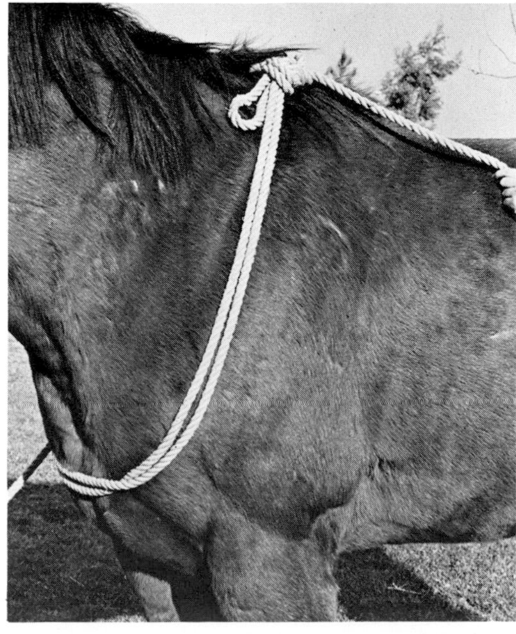

FIG. 8.32. Rope casting harness—neck loop.

Casting Harness (Hobble)

It is frequently necessary to restrain the horse on the ground in a recumbent position. This can be carried out by using general anesthesia or immobilizing agents, but casting harnesses are more often used. Casting hobbles have numerous designs. The simplest consists of a 30 m (70 ft), 12–15 mm ($\frac{1}{2}$–$\frac{5}{8}$ in.) diameter nylon rope plus two web or leather hobbles. The hobbles can be improvised (see Fig. 8.48); two ropes can be combined to provide the necessary length.

A neck rope is constructed by tying a bowline in the doubled rope (Fig. 8.32). Place the knot over the withers. The loop should be large enough to slip over the point of the shoulder. Thread the rope through one side of the hobble and then place it around the leg above the hock (Fig. 8.33). Thread the rope through the other side of the hob-

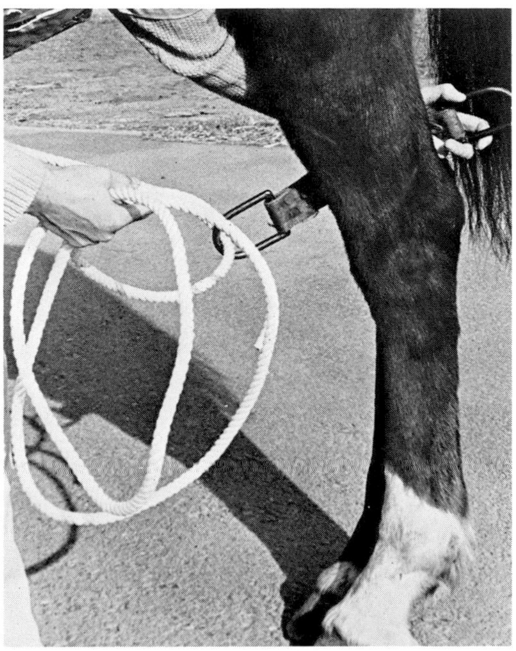

FIG. 8.33. Rope casting harness—placing hobble above hock.

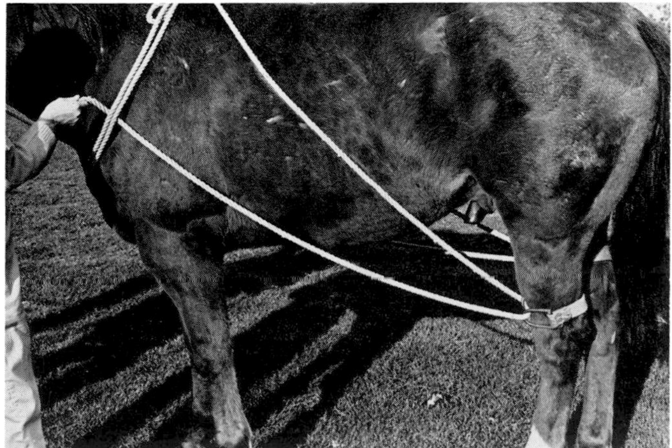

FIG. 8.34. Rope casting harness—ropes in place above hocks. Hobbles will be moved to the pasterns when horse is ready to go down.

ble and run the end back through the neck rope (Fig. 8.34).

At this point administer a sedative or anesthetic. Drop the hobbles to the pasterns as soon as the horse quiets down and the handlers are in the proper position to pull the horse down (Fig. 8.35). Draw the right leg forward. Pressure on the left hip tends to push the horse so that it will fall on its right side. The head is also pulled to the right. Once the horse is down, hold the head to prevent concussion caused by the horse attempting to rise and slamming down again (Fig. 8.36). A blow to the lateral side of the head is most likely to cause injury. If the horse attempts to lift its head, pull the nose upward and toward the back of the horse. Then if the head is slammed down, the blow will be on the back of the neck and cause less damage.

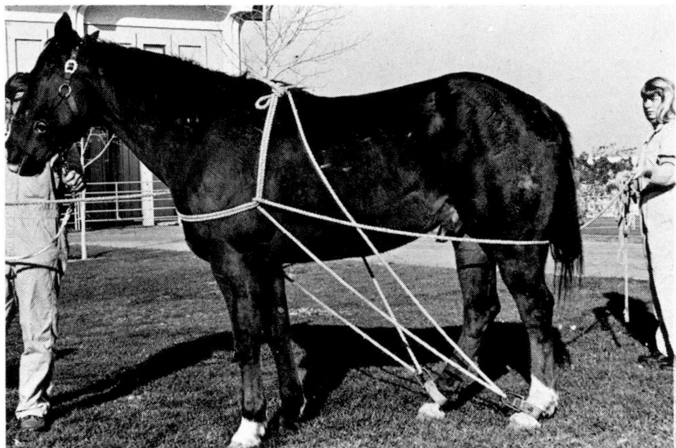

FIG. 8.35. Horse in position to be cast.

FIG. 8.36. Holding head of cast horse.

Legs should be flexed against the body. Otherwise, a powerful horse may extend and fracture a femur. Make certain the neck loop is pulled over the point of the shoulder (Fig. 8.36). Immobilize the upper hind leg by wrapping a half hitch around the pastern and beginning a figure eight around the hock (Fig. 8.37). When the first

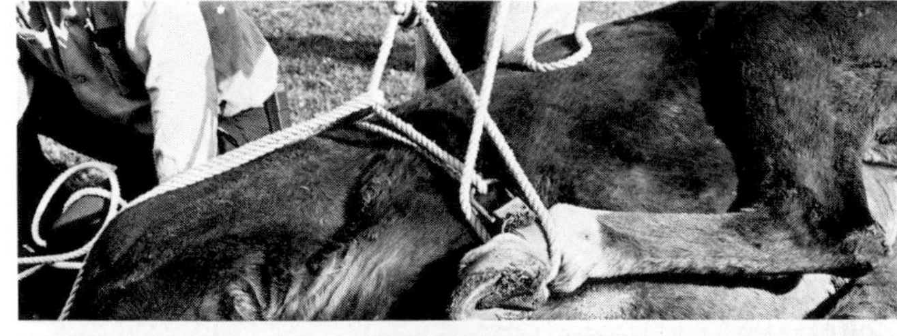

FIG. 8.37. Casting harness. Securing hind leg.

figure eight is completed, stand behind the horse and cinch this hitch tight to flex the hock to a maximum. Place another figure eight over the first one (Fig. 8.38). Maintain tension on the rope with one hand while pushing a loop in the running end between the hock and the ropes (Fig.

FIG. 8.38. Completed figure eight.

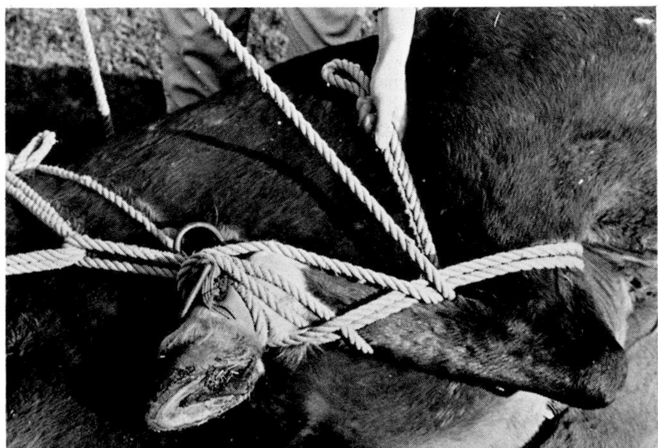

FIG. 8.39. Securing hitch: Loop is inserted through both legs of figure eight to prevent slippage over the hock.

8.39). Secure this segment by running another loop through the first loop and throwing it over the foot (Fig. 8.40) or by threading the remaining rope through the loop and cinching it down tight (Fig. 8.41). This prevents the ropes from slipping over the hock if the horse struggles.

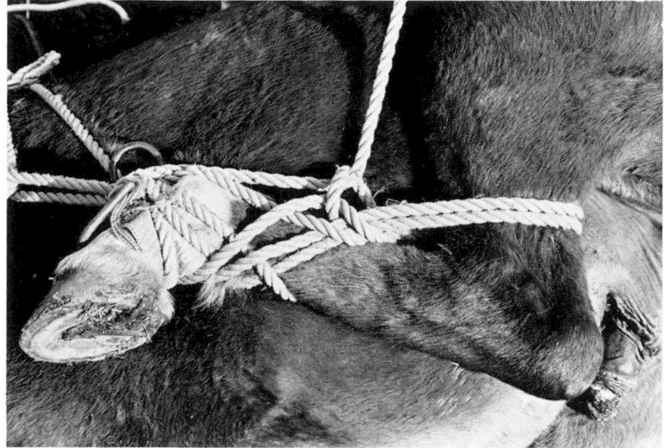

FIG. 8.40. Anchoring the securing hitch by throwing loop over the foot.

FIG. 8.41. Alternate way to anchor securing hitch: Thread standing end of the rope through the loop.

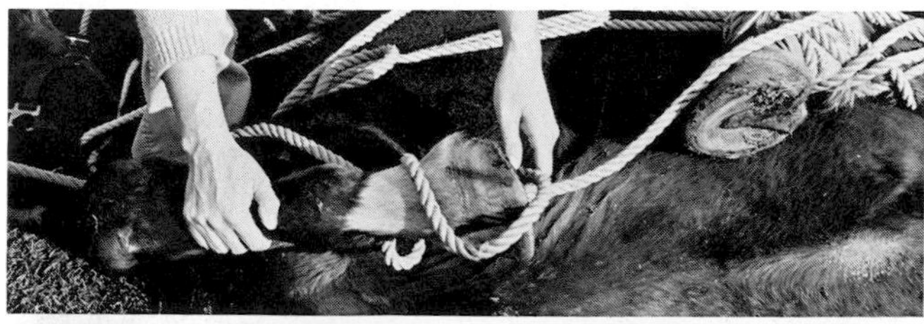

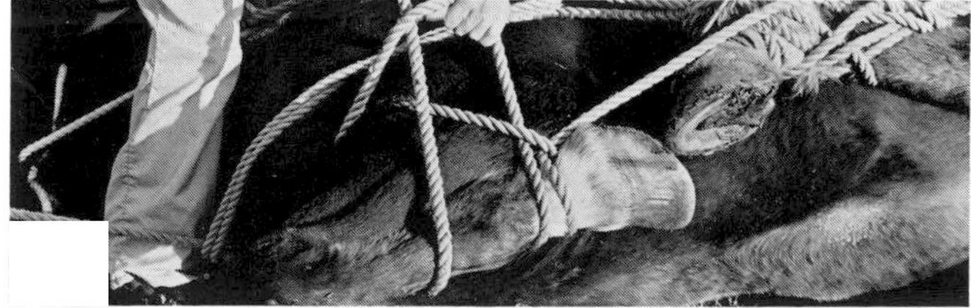

FIG. 8.42. Casting harness. Securing foreleg.

Secure the front leg by flexing the leg and forming a half hitch around the pastern. Then throw a double half hitch over the knee and anchor as high up on the leg as possible (Fig. 8.42). Thread the remaining rope around other ropes to take up any slack (Fig. 8.43). Turn the horse over by hooking the rope on the right leg over the right foot. Stand at the back of the horse and rock it up on its back and on over. At the same time, rotate the head. Tie the right side in the same manner as the left; the casting harness is completed (Fig. 8.44).

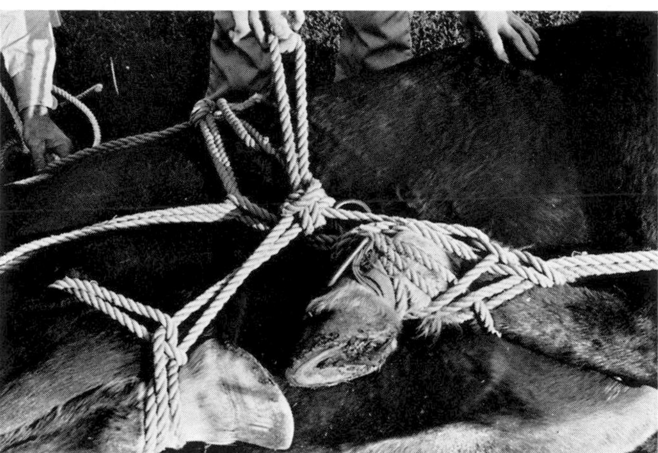

FIG. 8.43. Taking up slack in all ropes.

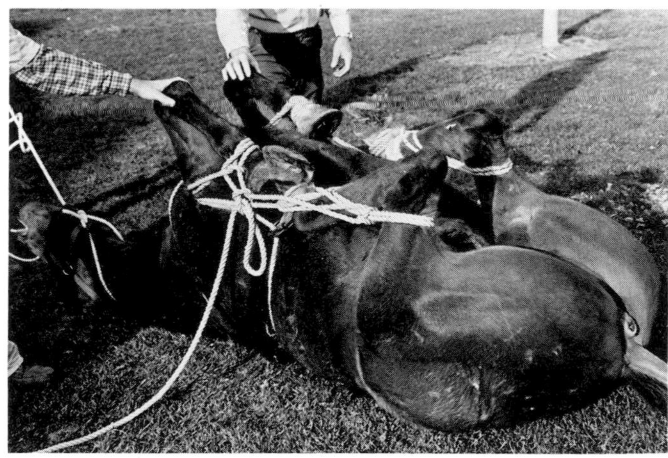

FIG. 8.44. Completed casting harness.

The head can be controlled by threading the shank through the neck rope and back through the nosepiece of the halter (Fig. 8.45). A cradle to hold the horse in dorsal recumbency can be made by placing a bale of straw or hay on either side. A rope is run beneath the horse and the bales, then brought up around the bales, and tied to the rope on the legs (Fig. 8.45). This keeps the legs spread if abdominal surgery is contemplated. A word of caution: It is unwise to cast a horse without sedating it first. Continual struggling against restraint may result in myositis, similar to azoturia.

A sideline is used to prevent a horse from kicking backward when an operator has to work at the rear, for ex-

FIG. 8.45. Head secured to chest rope on a cast horse to prevent trauma to the head. Bales of straw keep a cast horse in dorsal recumbency.

ample, when conducting pregnancy examinations or genital examinations for infertility. The sideline is placed by tying a large bowline loop around the neck. The running end is brought over the side of the shoulder and around the hind pastern. The rope is brought back up to the neck rope and secured by a halter tie (Fig. 8.46).

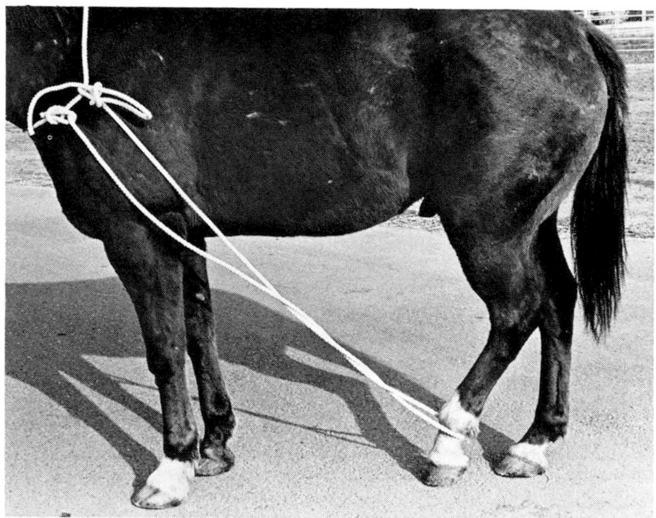

FIG. 8.46. Sideline without using hobble.

The danger of rope burns can be obviated by wrapping the pastern with cotton first (Fig. 8.47). However, if the horse is a kicker, wrapping may be hazardous. Hobbles are safer and can be constructed of webbing, leather, loops of rope, or a roll of burlap (Figs. 8.48–8.52).

Breeding hobbles are routinely used to prevent injury to valuable stallions (Fig. 8.53). The simplest, safest, and most effective hobble consists of a system of straps and

FIG. 8.47. Pastern is wrapped to prevent rope burn with rope sideline.

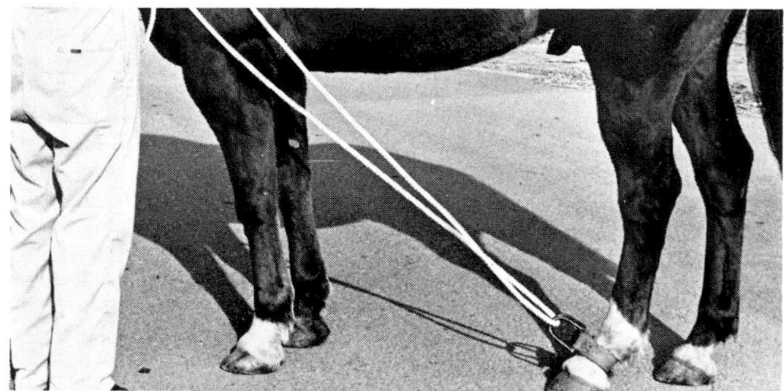

FIG. 8.48. Web hobble used on sideline.

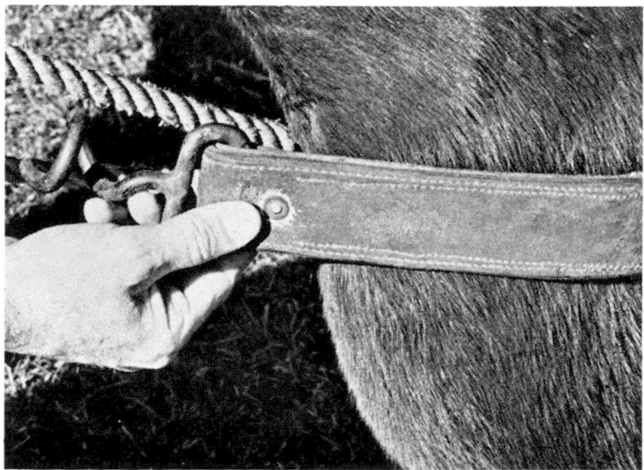

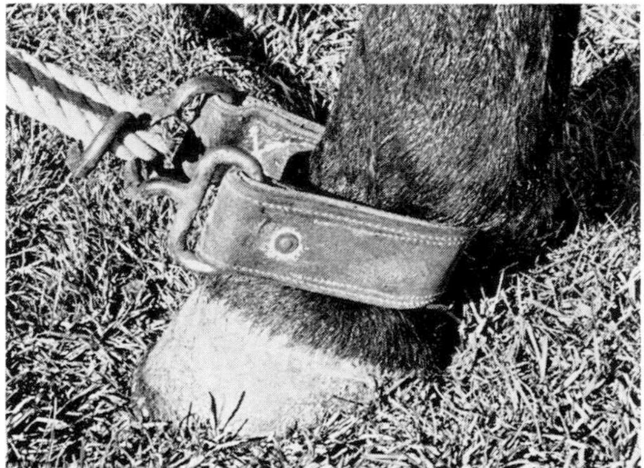

FIG. 8.49. Leather hobble with corkscrew ring for rapid attachment.

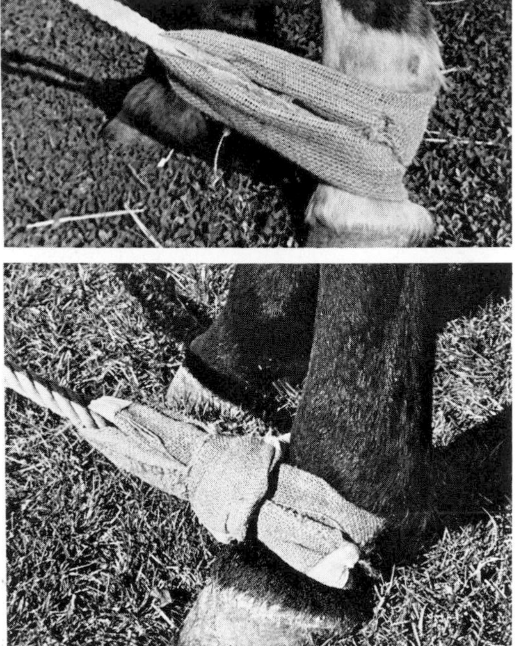

FIG. 8.50. Burlap loop used as pastern hobble.

ropes. One strap encircles the neck. Another strap is attached to a ring on the neck strap and goes between the front legs. This strap ends in the ring, which is the focal point of the hobble. Figure eight hobbles are applied to the hocks. A length of nylon rope with a strong snap on one end is attached to one hobble. The rope is threaded through the ring and tied at the other hock. A burlap roll can be used in lieu of web or leather hock hobbles (Fig. 8.54).

FIG. 8.51. Making a hobble from burlap or jute bag. Cutting a strip *(upper)*. Rolling to form donut *(lower)*.

FIG. 8.52. Burlap roll taped to prevent unrolling.

Slings are used to provide partial and temporary support for a weak or injured horse. A horse cannot hang free in a sling for more than a few minutes, so it must be conscious and capable of some degree of self-support. An excellent heavy-duty sling is illustrated in Figure 8.55. Chest and butt straps prevent the horse from slipping forward or back. The belly band should be spread with a single-tree or a similar device. The horse must be lifted with a block and tackle. Refer to Chapter 3 for a consideration of the size of rope and type of blocks necessary. The head is tied to prevent the horse from swiveling around the block and tackle.

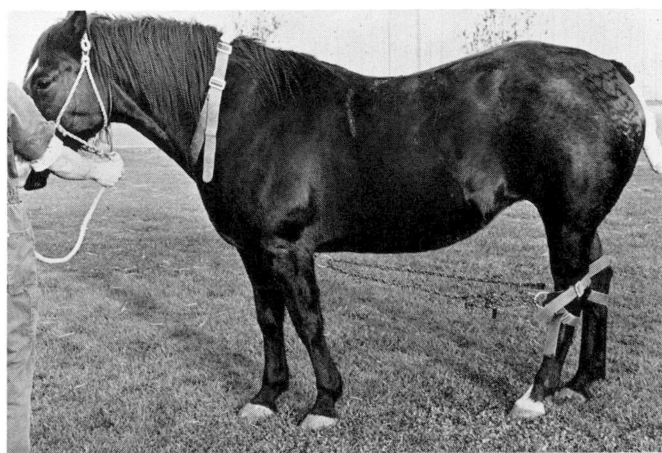

FIG. 8.53. Breeding hobble. Figure eight hock hobbles can be constructed of leather or webbing.

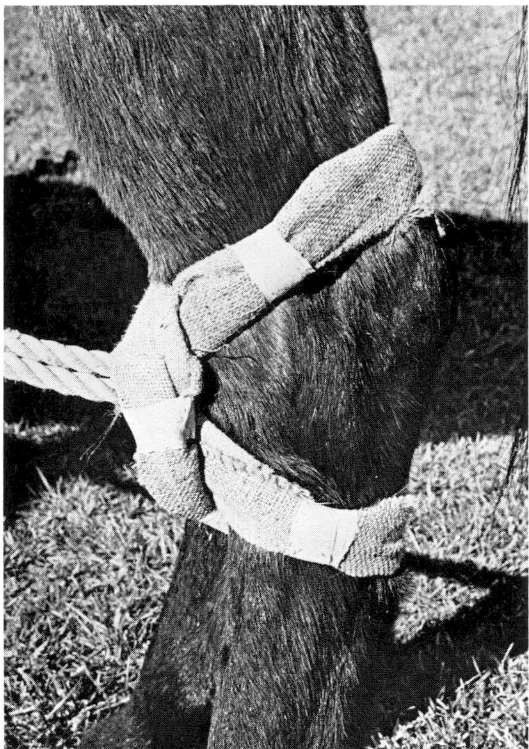

FIG. 8.54. Burlap roll used as hock hobble.

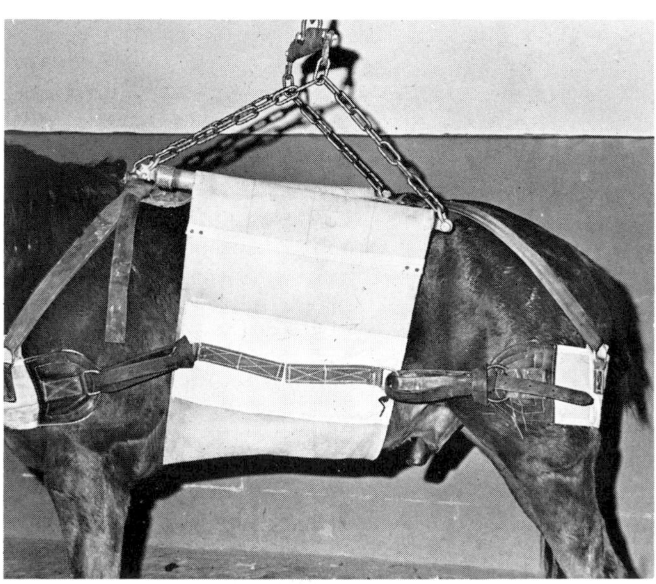

FIG. 8.55. Canvas sling.

Rope slings (see Figs. 3.47–3.49) can be used to lift a horse to its feet or to extricate it from a predicament such as falling into a ditch. Continual support with a rope sling will compromise circulation.

Blindfolding usually has a calming effect on most horses. Special blinders are available for trailering horses. Towels can also be used (Fig. 8.56). Blindfolded horses must be watched to prevent them from stepping off tailgates or getting into other predicaments. Occasionally a horse becomes terrified when blindfolded and is unmanageable. These horses cannot be blindfolded, and other techniques must be used to control them.

FIG. 8.56. Blindfolded horse.

Cradles are used to prevent a horse from chewing on bandages or surgical sites. A simple one constructed of rope and doweling is illustrated in Figure 8.57.

A horse in a trailer or in a tie stall must be approached with caution. Such horses are easily excited; unless you know the horse you cannot determine whether or not it may kick. Some horses are prone to squeeze, as well, if you walk alongside them. The most serious complication occurs if the horse shies and pulls back as its head is approached. Then it finds itself restrained by the rope and may jump forward. Unless there is adequate room to move away, you may be pawed or crushed by the horse. If possible, approach the horse from the front.

A trying problem arises when it is necessary to catch a horse that seems bent on keeping its head in a corner or

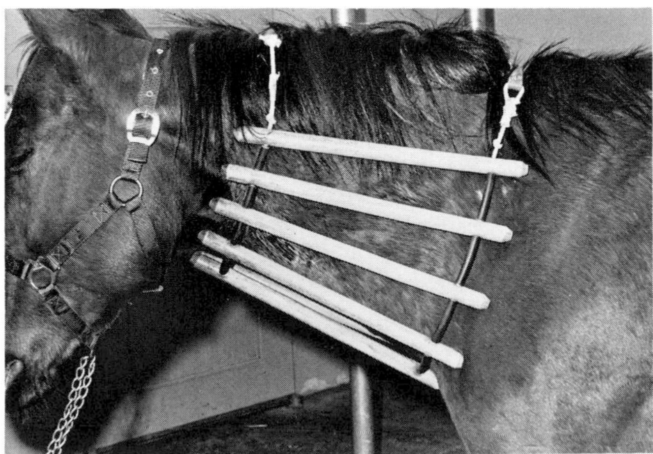

FIG. 8.57. Cradle.

kicks at you with its hind legs each time you attempt to approach. Although the training of horses is not the intent of this book, it is sometimes necessary to use training procedures to catch the animal. Several alternatives are available to catch such a horse.

The handler can throw a rope and catch it. A horse becomes excited if the roper twirls the rope around its head, and it is usually fast enough to duck a tossed rope unless the roper is highly skilled. Therefore this technique is usually not an option for the average handler.

An alternate approach—if the animal is in a confined space out of which it cannot jump or climb—is to punish the horse when it misbehaves or swings its head away from you and to encourage and reward it when it faces toward you. Approach the animal. If it whirls and exposes its hind legs or makes an attempt to kick, use either a whip or a segment of rope to lash the hind legs. Administer only a sharp sting, not a vicious beating. Timing is important. When the horse responds correctly by turning toward you, speak to it gently. Eventually the horse will come to you and allow handling.

A cross tie is useful to control the head and prevent whirling. Place ropes between any contiguous structures such as doorways, corners of stalls, alleyways, or stocks (Fig. 8.58). Tie from the check rings and anchor the ropes or chains high enough to prevent the horse from rearing and entangling its foot in the rope.

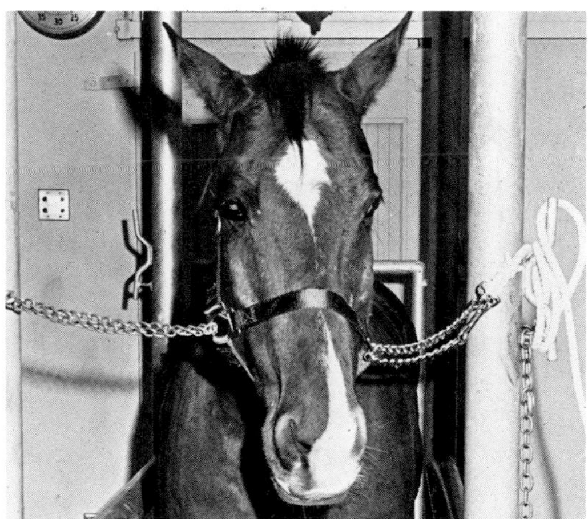

FIG. 8.58. Cross tie.

TRANSPORT

Horses are moved extensively by truck and trailer. They can and should be trained to load easily and ride without stepping on themselves or getting excited. Many hazards are associated with trailering. These can be minimized by making certain the trailer is properly constructed and repairs made promptly.

Protruding objects of any kind are dangerous. Even blunt objects become dangerous when a 1,000 lb horse jams against them. Nuts should always be secured away from areas in which the horse will stand or walk. On dividers, all nuts and bolt heads should be countersunk.

The floor must be given special attention. Most trailers have wood surfaces. Some have no subflooring. Constant wearing by the standing horse plus urine and feces may weaken the wood. I had to destroy a horse that had broken through a floor while being trailered.

Few trailers are now constructed without a cover. Some rental trailers still have just a high front end and no cover. Horses' eyes can suffer severely from wind, insects, and dirt. If you must trailer in an open unit, drive less than 40 MPH or put goggles on the horse. High-speed travel in open trailers also makes breathing difficult.

Be particularly mindful of obstructions on the roof of the trailer. Joints or seams with downward projecting flanges are dangerous. Head tossers are particularly liable to cut their scalps. Do not trailer tall horses in low trailers. Simple balancing when stopping and starting can result in head injuries in a trailer with no head room.

Wire manger dividers in two-horse trailers can become dangerous. If a strand breaks, the eyes and face of the horse can be lacerated.

Some horses are poor travelers. In most cases it is the result of bad experiences. Careful schooling and patience will overcome a few of these. Boots or wraps must be put on the feet of some horses to protect against injury if the horse walks all over itself. If a horse comes out of a trailer lame, examine the pastern and fetlock carefully to see if the horse has stepped on itself while trying to regain its balance. Application of cold water or ice for an hour or so will reduce pain and swelling. If the skin is cut, further first aid is required. Bruises can occur even though boots are put on the horse.

It is wise to put wraps on the lower legs of even the best trailer-mannered horse. A sudden swerve can throw any horse off balance. Be certain that wraps are snug, but not too tight, and that the bandage is applied with uniform tension up and down the leg. On long trips, wraps have the added advantage of minimizing "stocking" of the legs.

"Tail rubbers" are a special problem. A tail wrap should be used; however, some horses will work these off. Most trailers have a chain that is supposed to prevent rubbing, but in some cases a special tail board must be constructed. Train the horse to allow a rope to sit under the tail. I have known horses that have gotten their tails over a chain and then clamped down. Some start kicking. Some react like a foal that is "tailed up"; they relax and sit down. One horse broke its tail by sitting on it in a trailer.

The point of the hock can be bruised easily in a trailer. If the driver starts or stops too quickly, capped hocks will occur sooner or later. The kicker presents a more difficult problem. Hock pads are useful but difficult to keep in place. A pad can be placed behind the horse once it is loaded.

A great hazard occurs at the moment the hind feet step into or out of a step-up trailer. These trailers are low-slung, but a horse can always slip a foot under the floor of the trailer. If it moves forward without pulling the foot backward, a severe laceration or even a fracture of the cannon bone may result. Be cautious about shoving a horse into this type of trailer.

The first experience of trailering is the most critical for a horse. It may occur soon after birth or at any time through adulthood. Most foals will follow their mothers into a trailer or a truck. It helps to have a quiet mare to show the way. Keep the foal close to the mare as you approach the trailer. It is better if the foal is broken to lead, although not essential. Crowd the foal into the trailer as soon as the mare is settled.

Dividers present special problems for a mare and foal. High pipe dividers are dangerous, as small foals will duck under them and even crawl under the mare. The divider can be removed with a quiet mare and a careful driver but, generally speaking, it is dangerous to do so. Even a well-behaved mare may be forced to sidestep for balance and inadvertently injure the much smaller foal. A nervous mare would be almost certain to do so. A solid plywood or metal partition is best. This allows a mare and foal to nuzzle each other and at the same time protects them from accidents.

A young foal should not be tied in the trailer. Injuries to the neck vertebrae and subsequent development of "wobbler syndrome" are all too common in the foal that hauls back on a tie rope.

Be sure the foal cannot jump out over the tailgate. Usually it is best to string ropes over the top of the tailgate to dissuade jumping in the event the foal turns around.

Teach the horse to stand in the trailer until given a command to back. Never open a trailer door or, more especially, unsnap the chain or drop the butt board without first being certain the horse is untied. Any horse may spook, and it is natural for them to pull back. When nothing pushes from behind, they pull harder and then may suddenly jump forward into the manger.

When a horse jumps forward, anyone in the front of the trailer can be severely injured. Do not stand in front of the manger with either the tailgate closed or the horse tied. If you can walk through, or if there is an escape door, or you can walk on the other side of the divider, the horse can be led in. To walk into a closed trailer and squeeze out alongside the horse is foolhardy. Teach the horse to enter by itself or take a longer lead rope up through the front.

Walk alongside the horse, using a short shank, when entering a walk-through trailer. Move slightly ahead of the horse, and *keep looking forward.* Do not try to drag the horse into the trailer; lead it in. Watch out for the horse that jumps into the trailer, since it may step on your heels.

Loading is only part of the experience. Driving style is critical to successful trailering. Quick starts and stops throw a foal around much more than an adult horse. It will remember the trailer as an unpleasant place and will be harder to load the next time.

Yearlings that have never been in a trailer are a special problem, especially if they haven't been handled constantly. They are big enough to do a lot of damage but have not yet learned to accept discipline readily. If time permits, one of the best methods of accustoming the uninitiated horse to a trailer is to park the trailer in the corral. Solidly block it so that it cannot move, and feed the horse in the trailer. When the horse is in the trailer, close the door or snap a chain across it quietly. Continue to talk to the horse as you pet it. Then unsnap the chain or open the door, walk away, and let the horse come out at will.

How does one load a spoiled horse into a trailer? No one trick will work on every horse. People push, pull, curse, and whip, and in the end sometimes sit down and cry when the horse is victorious. Study the horse to figure out the best approach. Once the horse is overexcited, wait until both you and the horse cool off. Then, take a different tack.

Never tranquilize a horse that is spooked and excited by exhausting attempts to load it into a trailer. The drug may have an adverse reaction and excite the horse even further. If a horse is a known bad trailerer, give the tranquilizer to the calm horse an hour before loading. Overtranquilization is undesirable since it depresses the horse and may upset its equilibrium in the trailer.

If a horse does not go into the trailer after a few attempts, try one of the following:

1. If at least three people are available, put one on the shank and the other two alongside the horse, facing the tail. They can reach around the rump and lock hands or arms, then turn and push forward (Fig. 8.59). They must stay in close to the horse to control sidestepping. The person on the lead shank only directs the horse but does not pull. The two in the rear begin pushing the horse slowly forward, giving it a chance to smell the trailer and get acquainted with it. The pushers keep up steady pressure until the butt chain is snapped.

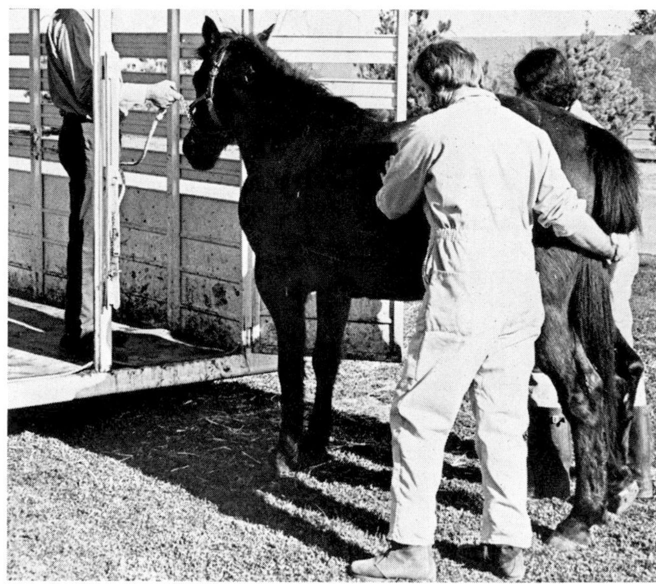

FIG. 8.59. Loading horse into trailer by clasped arms around hindquarters.

Be cautious in using this technique with step-up trailers, since the hind legs may slip under the floorboards. This technique is also somewhat hazardous with a kicker, but unless the horse is a bad "cow-kicker," it will work. It requires two people who are physically capable of staying with the horse and pushing it in. Surprisingly, most horses will respond to light pressure. This is *the most effective* way of putting a horse into a trailer.

2. If alone, place a rope over the rump and through the halter and apply gentle, steady pressure (Fig. 8.60).

3. Another technique with ropes can be used, but it must be done with caution since rope burns frequently

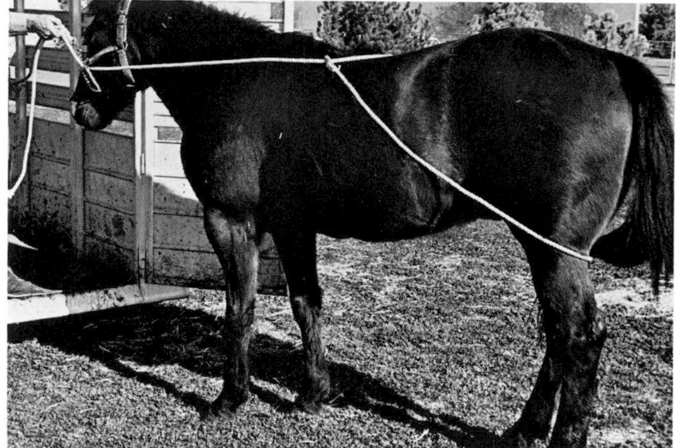

FIG. 8.60. Loading horse into trailer with a rump rope.

result. Tie a long rope to one side of the trailer. Bring the horse up to the tailgate and carefully place the rope around the rump of the horse, keeping it high on the butt. If it drops down to the hock, the horse will usually respond by kicking. Then take the rope through a ring or roof support on the opposite side of the trailer and take up the slack as the horse moves forward. Be ready to release the rope if the horse gets entangled.

4. A variation of this technique is to tie a lariat on each side of the trailer. Two people cross them behind the horse and apply pressure (Fig. 8.61). Also, two people can use one rope as shown in Figure 8.62. These procedures are safer for the horse because the assistants can drop the ropes on command.

5. Preventing a horse from jumping to the side is both desirable and difficult. A narrow alley or loading chute may help. Alternatively, pull the trailer up to the side of a building and eliminate one side of the problem.

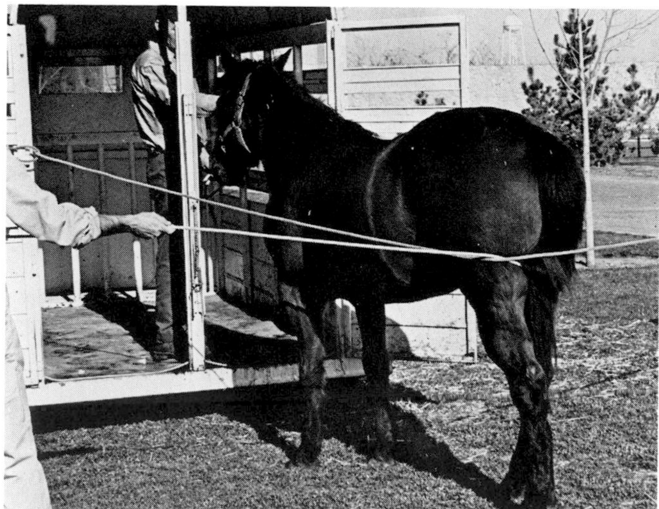

FIG. 8.61. Trailer loading with two ropes crisscrossed behind the horse. Handlers must be prepared to drop the ropes if the horse becomes entangled.

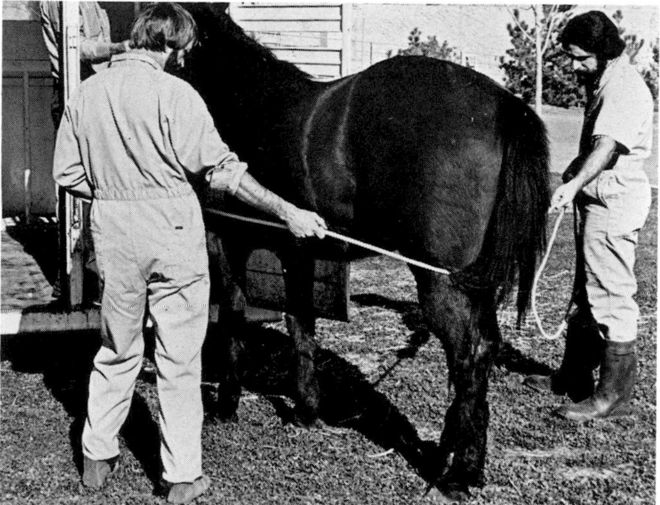

FIG. 8.62. Trailer loading with single rope held by two people.

6. A whip is usually ineffective and tends to aggravate rather than help a difficult situation. A swat with a house broom or the flat of a scoop shovel works well in encouraging a horse to move into a trailer. The noise is more effective than inflicting pain.

A horse should be tied in the trailer with a quick-release knot or a snap that will release while under pressure. Every horseman or horsewoman should be able to tie the halter tie blindfolded. It is the only acceptable knot for tying a horse in a trailer. If a neck rope is used, it should be secured with a bowline knot.

CHEMICAL RESTRAINT

The principles of equine anesthesia are well known and documented [1,2,3,5]. No attempt will be made to list or describe all the techniques or agents used routinely to sedate, tranquilize, or anesthetize horses and donkeys. Rather, those techniques used to sedate or immobilize fractious horses that cannot be approached or caught without danger to the operator will be described here. See Table 8.2 for drugs used in anesthesia and restraint.

Trained horses can usually be given intravenous medication, which will facilitate the speed of the restraint procedure. Some horses or mules can be tied or snubbed to a ring, post, or beam for administration of intravenous medication. Horses that are vicious or fractious to the point of preventing haltering or tying must be given oral medication or intramuscular injections.

Chloral hydrate has been extensively used as a component of anesthetic mixtures for the horse. It may also be used as a sedative, independent of other agents. It is bitter and normally a horse will not consume water containing the drug. If a horse cannot be approached otherwise, remove the water source for 24 hours. Then offer 3-4 L of water containing the appropriate dose of chloral hydrate. They usually will drink it. Within 30-60 minutes the horse will be sufficiently tractable to allow handling.

Intramuscular, subcutaneous, or perivascular infusion

TABLE 8.2. Anesthetic or restraint drug dosages

Drug	Oral		Intramuscular		Intravenous		Comments
	(mg/kg)	*(mg/lb)*	*(mg/kg)*	*(mg/lb)*	*(mg/kg)*	*(mg/lb)*	
Chloral hydrate	110	50	No	No	55	25	Bitter in water; horse must be thirsty; oral or I.V. administration only
Acepromazine maleate	No	No	0.05-0.1	0.02-0.04	0.05-0.1	0.02-0.04	
Promazine hydrochloride	1.65-5	0.75-1.25	0.7-1.1	0.3-0.5	0.5-1	0.2-0.4	
Xylazine hydrochloride	No	No	2.2	1	1.1	0.5	Horse may appear very tranquil, but explode with minimal stimulation
Glycerol guaiacolate	No	No	No	No	110	50	Requires large volume; given rapidly I.V.; muscle relaxant only
Succinylcholine chloride	No	No	0.88-1.1	0.4-0.5	0.09-0.11	0.04-0.05	Destroyed by heat when in solution; muscle relaxant only

of mixtures containing chloral hydrate are extremely irritating to the tissue and may produce a slough.

A horse free in a pasture, and which defies capture, presents a difficult problem. Chloral hydrate in the drinking water can be used if it is possible to remove all other water sources from the pasture. Alternatively, the projectile syringe can be used. Agents used in such capture operations have included succinylcholine chloride (5 mg/kg), xylazine (1-2 mg/kg), acepromazine maleate (0.1-0.2 mg/kg), and etorphine (0.04-0.08 mg/kg).

Succinylcholine chloride is a muscle relaxant and has no anesthetic or tranquilizing effects. It is an efficient immobilizing agent in the horse, rendering the animal incapable of muscular activity. Because the horse is fully conscious, it can feel pain and become excited; therefore the drug should not be used in place of anesthetics or sedatives. Fortunately succinylcholine chloride is compatible with other agents and can be used in combination with them, once the horse is immobilized.

Since intramuscular injection of succinylcholine chloride has not been utilized in equine practice, the dose has not been precisely determined. I have immobilized wild horses in a pasture by pursuing them in a vehicle and using a projectile syringe to inject a dose ten times that of the intravenous dose. In 2-5 minutes the horses lost muscle coordination and fell.

Succinylcholine chloride has a broad therapeutic dose range in horses. Nonetheless, fatalities have occurred following its use, possibly from complications of hypertension. It should not be used on animals that have been wormed or treated in any way with an organic phosphate anthelmintic or insecticide within the previous 2 weeks.

If tranquilizers can be injected into a horse before it becomes excited, calming may result, which will allow handling. When tranquilizers are injected intramuscularly, allow 30-60 minutes before attempting to handle the horse. Early stimulation may negate the tranquilizing effect.

REFERENCES

1. Gable, A. A.; Jones, E. W.; and Vaughn, J. T. 1972. Anesthesiology, chemical and physical restraint. In E. J. Catcott and J. F. Smithcors, eds. Equine Medicine and Surgery, 2nd ed. Wheaton, Ill.: American Veterinary Publications.
2. Hall, L. W. 1971. Equine anesthesia. In L. R. Soma, ed. Textbook of Veterinary Anesthesia. Baltimore: Williams & Wilkins.
3. Jones, L. M. 1957. Veterinary Pharmacology and Therapeutics, 2nd ed. Ames: Iowa State Univ. Press.
4. Leahy, J. R., and Barrow, P. 1953. Restraint of Animals, 2nd ed. Ithaca, N.Y.: Cornell Campus Store.
5. Travernor, W. D. 1971. Muscle relaxants. In L. R. Soma, ed. Textbook of Veterinary Anesthesia. Baltimore: Williams & Wilkins.

9 CATTLE AND OTHER DOMESTIC BOVIDS

CLASSIFICATION

Order Artiodactyla
Suborder Ruminantia
Family Bovidae
Subfamily: Bovinae
Tribe Bovini: European cattle, zebu, yak, kouprey, water buffalo, banteng, gayal

THE DOMESTICATION of wild bovids was a major step forward in the process of civilization. A number of species have wild counterpart populations spread throughout tropical, subtropical, and temperate regions of the world (Table 9.1). Domestic bovids are usually gentle and all can be handled and restrained in much the same manner. Differences in handling are dictated by culture.

Hundreds of millions of cattle of numerous breeds provide milk, meat, and leather (Table 9.2). In some cultures, cattle carry the loads, plow the fields, and pull the wagons. Names of gender are given in Table 9.3.

Asiatic (water) buffalo are second only to cattle in worldwide numbers and economic importance. An estimated 75 million tame water buffalo exist [5]: 50 million in India and Pakistan, 20 million in East and Southeast Asia, and the remainder spread throughout numerous countries, including Japan, Hawaii, and Central and South America.

Just as there are many breeds of domestic cattle, there are many breeds of water buffalo. Some are adapted for a semiaquatic habitat; others thrive without an intimate water relationship. In general, buffalo are handled in the same manner as domestic cattle. Nose rings, nasal septum thongs, and halters are used to control work animals. Stocks are used to restrict movement while giving injections or carrying out minor surgery [3]. Water buffalo bulls are usually more docile than cattle bulls. Some can be handled without nose rings.

The yak is adapted to the cold and bleak existence of the Tibetan steppes. It replaces cattle and buffalo in Nepal, Tibet, and parts of Mongolia at elevations over 2,000 m. Other domestic bovids are of lesser economic importance.

Breeds of cattle differ markedly in their reactions to manipulation. Dairy cows are usually accustomed to being handled by milkers and handlers and, as a result, are likely to be more docile than other breeds. However, the dairy cow can become extremely nervous and may vigorously resist handling if she is not soothed and treated gently.

TABLE 9.2. Weights of domestic bovids

Animal	Bull		Cow	
	(kg)	*(lb)*	*(kg)*	*(lb)*
Cattle				
Holstein-Friesian	998-1,089	2,200-2,400	681	1,500
Jersey	545-817	1,200-1,800	363-545	800-1,200
Brown Swiss	817-1,180	1,800-2,600	590-817	1,300-1,800
Guernsey	568-1,022	1,250-2,250	363-726	800-1,600
Ayrshire	863	1,900	545-681	1,200-1,500
Hereford	817-908	1,800-2,000	545	1,200
Angus	908	2,000	636	1,400
Charolaise	908-1,135	2,000-2,500+	565-908	1,250-2,000
Shorthorn	908	2,000	681	1,500
Santa Gertrudis	908-999	2,000-2,200	636	1,400
Brahman (Zebu)	817	1,800	545	1,200
Water Buffalo	<1,350	<2,970	>150	>330
Yak	<900	1,980	350	770

TABLE 9.3. Names of gender of cattle

Gender	Adult	Newborn	Immature
Male	Bull	Calf	Bullock, bull calf
Female	Cow	Calf	Heifer
Castrated male	Steer, ox, bullock	. . .	Steer, bullock

Cattle having little association with people are easily frightened; techniques used to handle them must involve chutes and stocks where movement can be restricted before they are approached. Beef cattle are usually grazed in pastures, thus handled less, and frequently exhibit flighty reactions.

Although the dairy cow is often easily handled, the same is not true of the adult dairy bull. Dairy bulls are extremely dangerous, and special restraint practices must be observed when working with them. Contrarily, the beef bull is generally as easily handled as the female. Nevertheless, all cattle are capable of injuring a careless handler.

DANGER POTENTIAL

Cattle resist restraint by various actions. The horned animal is capable of quick thrusts sideways and forward with the horns and may fatally gore the unwary individual. A handler working around the head of a horned animal

TABLE 9.1. Domestic bovids [5]

Name	Wild Counterpart	Countries Where Used as Domestic Animal	Countries Where Wild Populations Exist or had Existed in Recent Times
European cattle	Auroch	Worldwide	Europe
Zebu	Auroch	Worldwide	Asia, Africa
Yak	Wild yak	Bhutan, Nepal, Tibet, Mongolia	North Tibet
Kouprey	Banteng X Zebu?	Cambodia, Vietnam, Laos	Cambodia, Vietnam, Laos
Asiatic or water buffalo	Water buffalo (6 subspecies)	India, Pakistan, Southeast Asia, East Asia, Japan, Central and South America	India, China, Nepal, North Africa, Europe, Assam, Mesopotamia
Bali cattle (Banteng)	Banteng	Bali	Java, Borneo, Burma
Gayal	Gaur	India, Burma (little importance)	India, Burma, Malaysia

must be continually conscious of the swinging arc of the head and the extent of reach of the horns.

Both polled and horned animals may butt. They may rush at people and knock them down or crush them against fences or walls. Cattle may also push against people with their bodies, squeezing persons against walls, fences, or other animals.

Cattle seldom use the front feet as weapons, though they may paw the ground to display anger and threaten. However, being stepped on is a minor hazard of working with cattle; even small calves can inflict pain if they step on a toe, and heavier animals may severely bruise or fracture the toes and feet.

Cattle are adept at kicking with the hind feet. The kick is usually forward and out to the side in an arc reaching some distance. Cattle are less likely to kick directly backward, though able to do so. Usually only one leg kicks at a time, as contrasted with the equine species where both hind feet habitually kick simultaneously. Probably the safest place to stand is right at the shoulder, but remember that a cow can kick forward past her shoulder with the hind leg.

The tail is used to swat flies and switches in response to any touch on the skin. The tail may be a source of annoyance during restraint procedures and may also contaminate a prepared surgical field. Furthermore, it may inflict personal injury if the hair of the tail flicks the eyes. The tail becomes an awesome weapon when it is filled with foreign bodies such as burrs or grass awns. These can be removed most easily by immersing the tail in mineral oil and slipping the burrs out. If this proves unsuccessful, a matted entanglement of burrs in the hair of the tail necessitates clipping off the switch and allowing the hair to regrow. Bovine tails are fragile and therefore must be tied or attached only to the animal's own body when restriction of tail switching is required.

Cattle rarely bite. They lack upper incisors.

PHYSICAL RESTRAINT

The temperament of each cow must be considered before approaching her for examination or to apply severely restrictive restraint devices. With beef cattle, it is likely that one must either rope the animal or put it into a chute or stock to halter it or approach closely enough to conduct an examination. Dairy animals can usually be readily approached if confined in a stanchion or tied to a fence (Fig. 9.1).

When working with any species, the handler must alert the animal to prospective movements. Quick motions usually startle animals, so firm, slow, deliberate actions should be the rule. Speaking to animals lets them locate your position and avoids startling them with an unexpected touch. Do not approach any animal directly from the front unless it is secured in a stock. It is natural for an animal to charge forward and butt anyone who makes such an approach. It is most desirable to approach the animal from either the left or right shoulder area. Placing a firm hand on the shoulder lets it know that you are there and that you are confident. Then, if necessary, the approach to the head can take place.

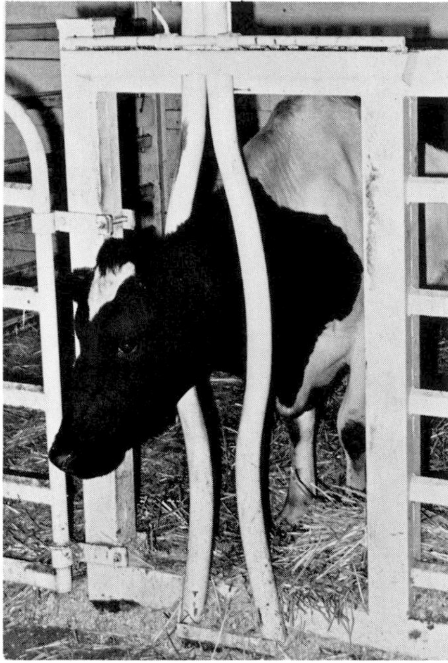

FIG. 9.1. A solid adjustable stanchion.

A rope halter is the basic tool of restraint for working with cattle and many other species (Fig. 9.2). Commercial halters are also available and the temporary rope halters described in Chapter 3 are satisfactory. It is important to place the halter correctly. Frequently a halter is put on upside down (Fig. 9.3) or improperly placed with the rope behind the horns but not behind the ears. This results in the rope crossing over or near the eye, endangering the eye.

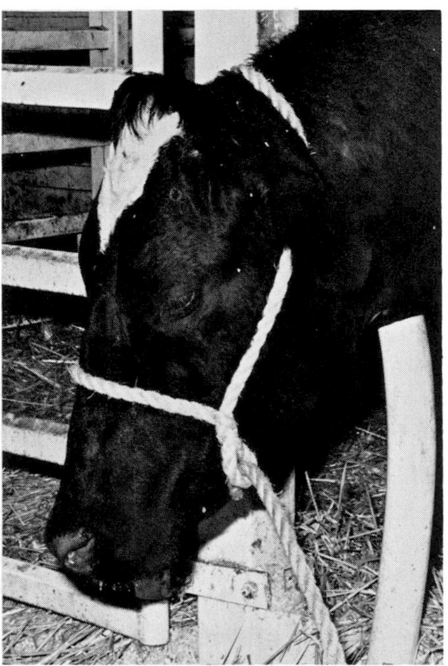

FIG. 9.2. Rope halter properly placed, with the free end exiting beneath the mandible.

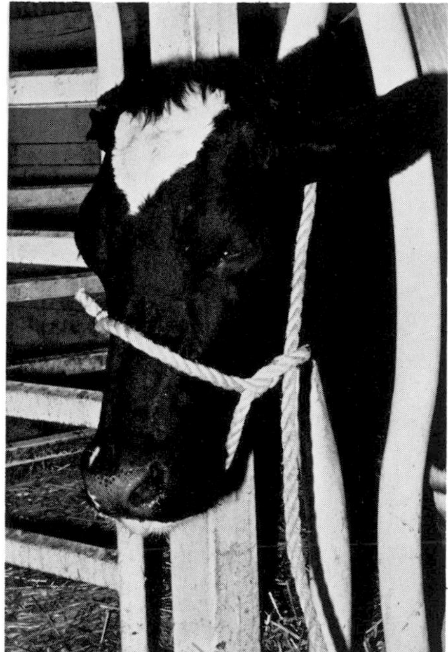

FIG. 9.3. Rope halter improperly placed, with the free end exiting over the poll.

Placing the halter on an animal in a chute or stanchion offers no particular challenge. An animal loose in a box stall may present some difficulty. However, if the nose loop is made slightly larger than the poll loop, one can often flip it over the nose and then over the poll and behind the ears very easily. If it is impossible to approach the animal in this manner, it may be necessary to first place a rope around its neck. Use the shank of the halter or rope the animal first with a honda loop and place the halter on after the animal is subdued.

Once haltered, the animal can be tied to a post, a ring, or any other secure object to carry out additional procedures. It is usually necessary to fix the head by pulling it tightly to the side or upward, or both, and snubbing it to the post with the halter tie (Fig. 9.4). It is somewhat difficult to remove all the slack from the rope when completing the halter tie. Practice is necessary to form the loops closely around the object. Many procedures such as withdrawing blood, giving injections, or examining the teeth and various other body areas can be carried out by controlling the head in this manner. If the halter is to be left on an unattended animal, be certain that it is the type that will not slip and become a noose around the neck.

Cattle have an unusually sensitive nasal septum. A routine restraint practice is to grasp the nasal septum between thumb and finger via the nostrils, forming a nose tong (Fig. 9.5). A large animal is difficult to hold by hand because one cannot maintain sufficient pressure to restrain the animal for more than a few seconds.

For more permanent, more secure restraint, mechanical devices acting on the septum are available. By applying a clamp in the form of a nose tong, one can severely restrict activity. When fully closed, a space of approximately 3.5 mm (⅛ in.) should remain between the two metal balls of the nose lead (Fig. 9.6). This is necessary to prevent

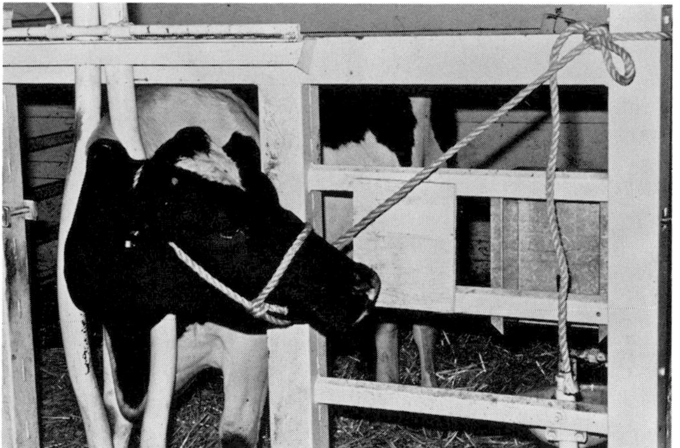

FIG. 9.4. The head is properly secured with rope halter tied for quick release.

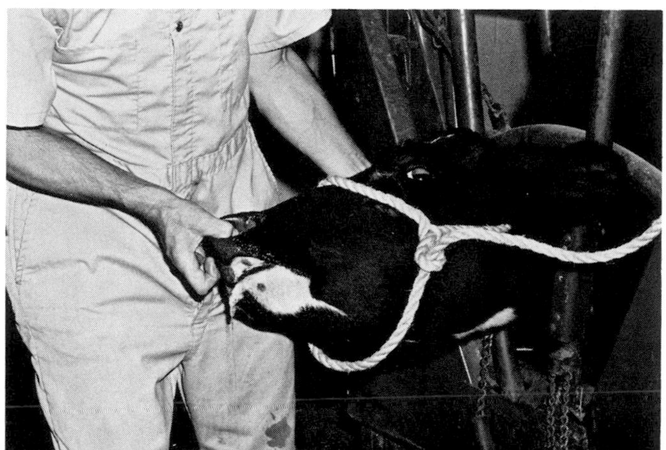

FIG. 9.5. Using thumb and finger as a temporary nose tong.

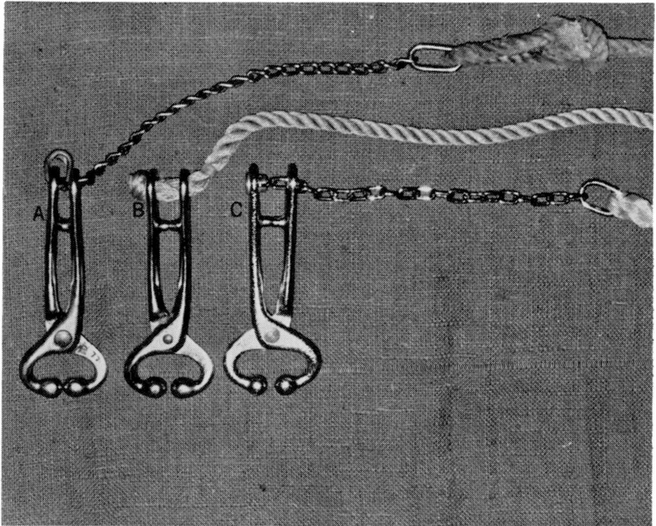

FIG. 9.6. Cattle nose tongs: **A.** Undesirable tong with no space between the clamps. **B.** Tong with rope lead. **C.** Chain and rope lead.

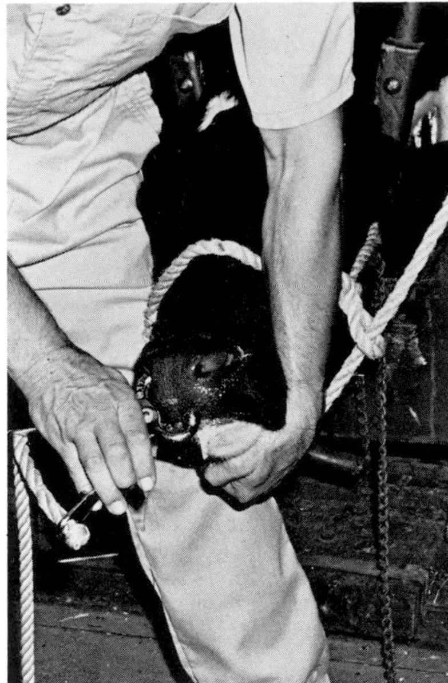

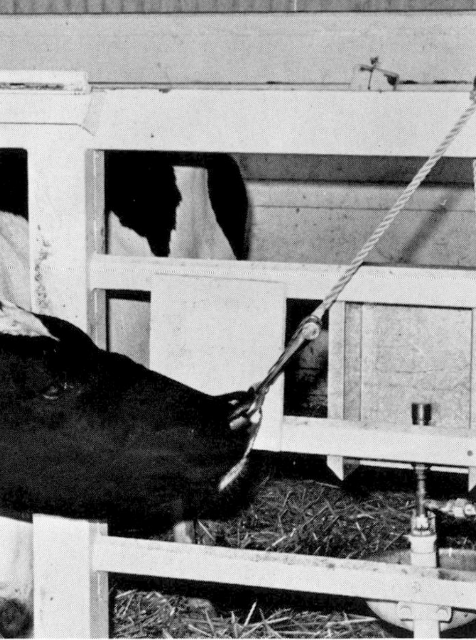

FIG. 9.7. Use of a nose tong: To apply nose tong, first insert one prong, then with a quick rotation insert the other *(left)*. Cow secured with a nose tong *(right)*.

necrosis of the nasal cartilage. Furthermore, one should be certain that the surface of the balls is smooth to avoid scrapes or lacerations. Poor quality nose leads may be die cast. The break in the cast usually leaves a rough edge in the middle of the ball. A file or emery cloth should be used to smooth out such edges or any other roughness on the ball surface.

Placement of the nose tong is not always easy, particularly if an animal has experienced the device previously. The animal frequently darts its head about in an attempt to prevent placement. If the animal is in a squeeze chute or stock, grasp the head or nose of the animal in the manner shown in Figure 9.7. Do not try to push a tong straight into the nose. The nose tong should be placed in the nostrils with a rotating motion. Insert one side of the tong, rotating across the nasal septum to apply the other, as shown. Then quickly close the tong and move away.

To keep the tong in place, tension must be maintained. An assistant must hold the tong, or the rope can be tied above and to the side of the stanchion or chute. Do not leave the animal unattended. Use the halter tie for quick release in case the animal should fall or otherwise get into a predicament. It is not desirable, nor is it humane, to proceed with significantly painful procedures on an animal restrained by a nose tong. Painful procedures require anesthesia.

Water buffalo and oxen in countries other than the United States are usually handled with some variation of the rope thong (Fig. 9.8).

FIG. 9.8. Halter consisting of a rope thong tied around the head and through a hole in the nasal septum.

Adult dairy bulls can never be trusted. Do not approach such animals closely unless they are confined in special chutes or stocks. A dairy bull usually has a ring in the nose and can be controlled by a bull lead attached to the nose ring (Fig. 9.9, 9.10).

The eyelids of bovine species are firm and difficult to evert for proper examination. Rotation of the head exposes much of the scleral surface and some of the conjunctival surface. To do this, with the head in a chute or stanchion, approach the animal from the right side, grasp the nose with the right hand, and press on the horns or grab an ear and press down (Fig. 9.11). This rotates the poll toward the right and pulls the muzzle up. The eye rotates accordingly. Reverse the rotation to expose the lower sclera and conjunctiva. It may be necessary for one person to

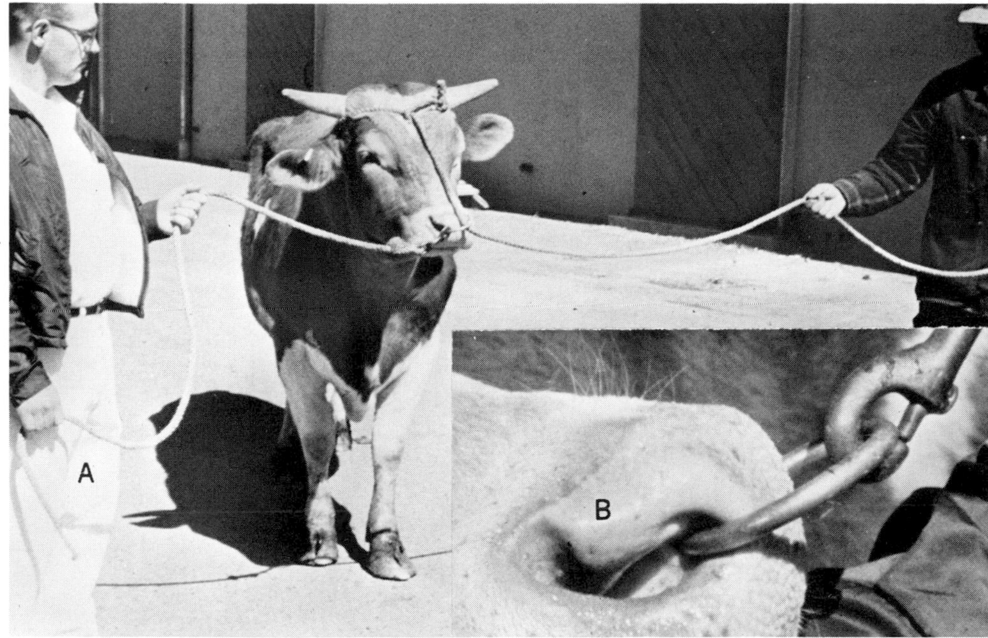

FIG. 9.9. Use of a bullring: **A.** Controlling a bull with two ropes. **B.** Detail of ring through the nasal septum.

manipulate the animal while another person examines the eye. Repeat both maneuvers from the left side to examine the opposite eye.

Complete physical examination of a bovid usually involves an oral and/or pharyngeal examination. The techniques and special devices used for these examinations are illustrated in Figures 9.12–9.18. Although not illustrated, an excellent technique for passage of a stomach tube in calves is to insert a small tube through a nostril.

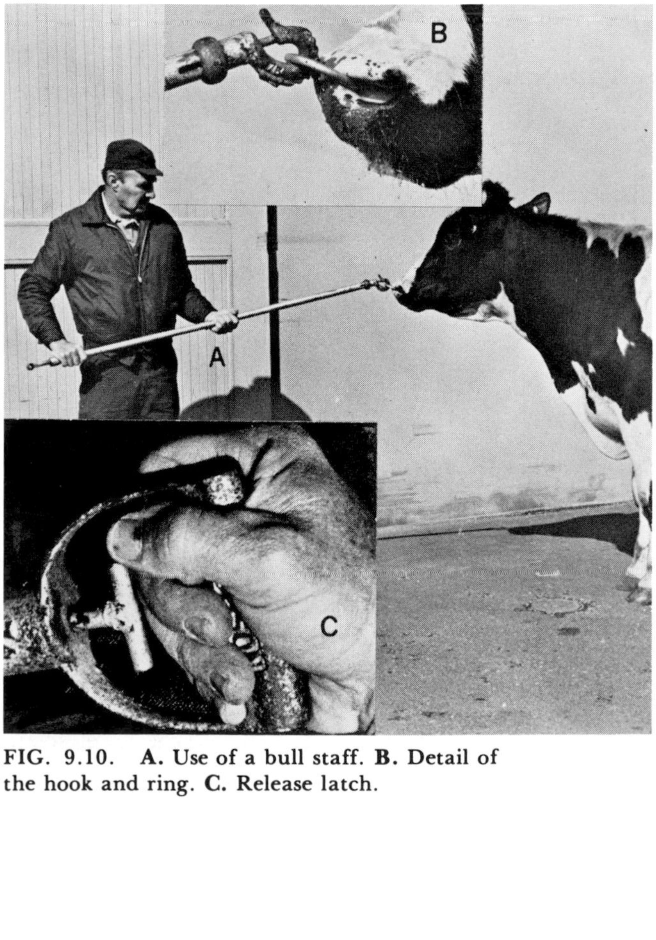

FIG. 9.10. **A.** Use of a bull staff. **B.** Detail of the hook and ring. **C.** Release latch.

Dairy cattle may require hobbles to prevent them from kicking; they usually tolerate the hobbles quite well. Chain hobbles (Fig. 9.19) or rope hobbles (Fig. 9.20) are suitable. The tension is adjustable. Be certain the animal retains the ability to separate its legs widely enough to maintain stability.

Carefully observe an animal being hobbled for the first time to avoid serious and possibly permanent injury. I have seen a frightened cow cast itself and severely lacerate the muscles and tendons on the anterior aspect of the hock

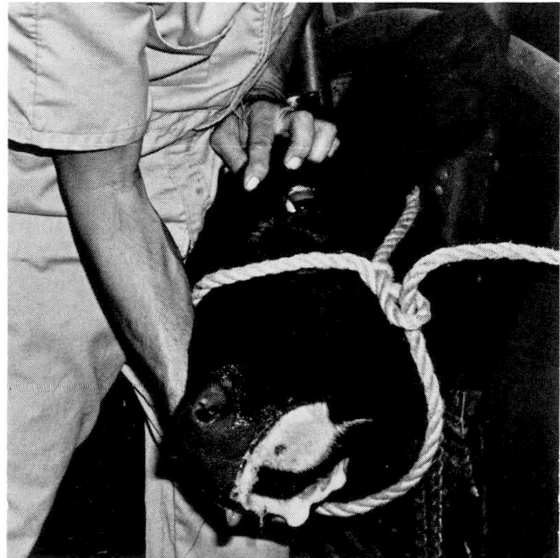

FIG. 9.11. Examining conjunctiva and sclera of the eye by rotating the head.

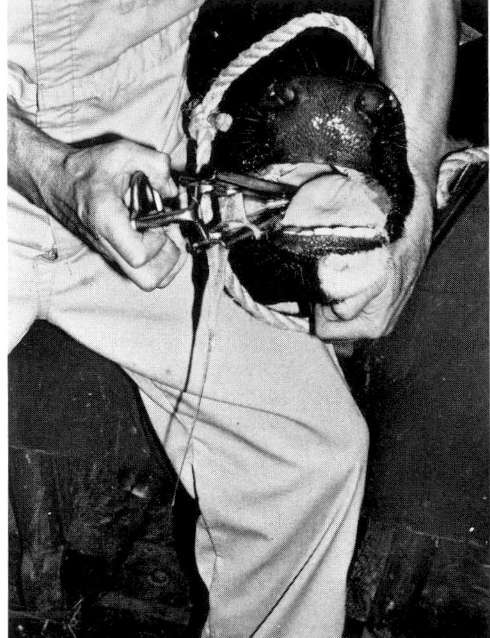

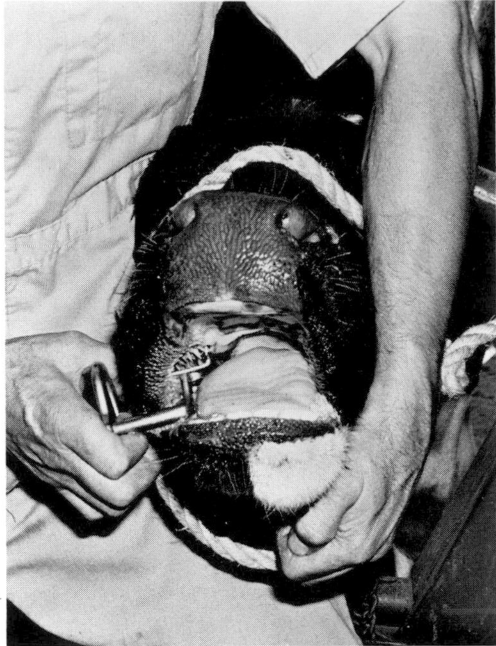

FIG. 9.12. Hauptner dental wedge holds cow's mouth open.

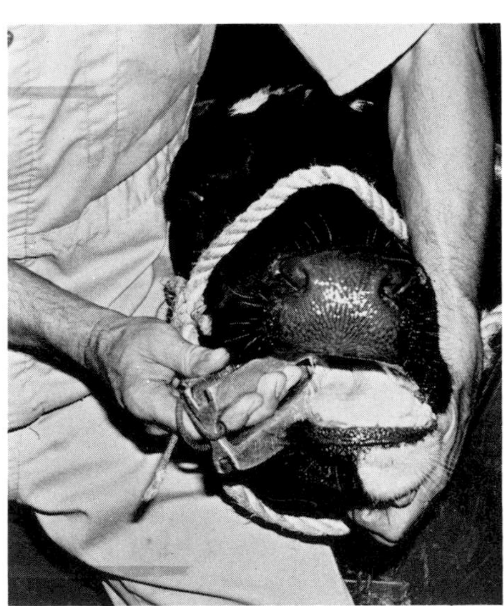

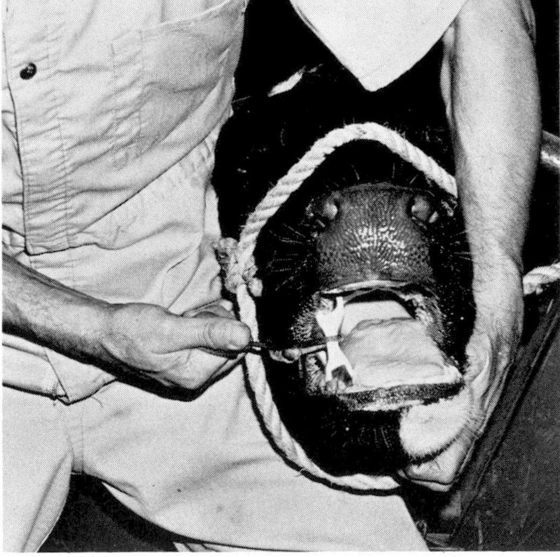

FIG. 9.13. Bovine mouth speculum.

while struggling against hobbles. Chain hobbles should not be left on an unattended animal.

Placement of a rope hobble may be hazardous if the animal is a kicker. To avoid being kicked, a longer rope and two persons are required to apply the hobble. The rope must be long enough for a person to stand out of reach of the kick on either side of the animal.

Another device to minimize kicking is a short length of rope tied around the flank of the animal (Fig. 9.21 left). Pressure in the flank area inhibits kicking but does not provide absolute control; the animal still may kick, but the intensity of the kick is usually diminished. Use caution when applying this type of restraint on a milking cow or a breeding bull, since the pressure of the rope is exerted directly over the udder of the cow or the prepuce of the bull. A special "can't kick" clamp can be placed over the top of the loin and cranked tight in the flank area (Fig. 9.21 right). Temporary flank restraint can be applied by lifting the flank on the side in which protection is desired (Fig. 9.22 left). The knee can also be used to apply

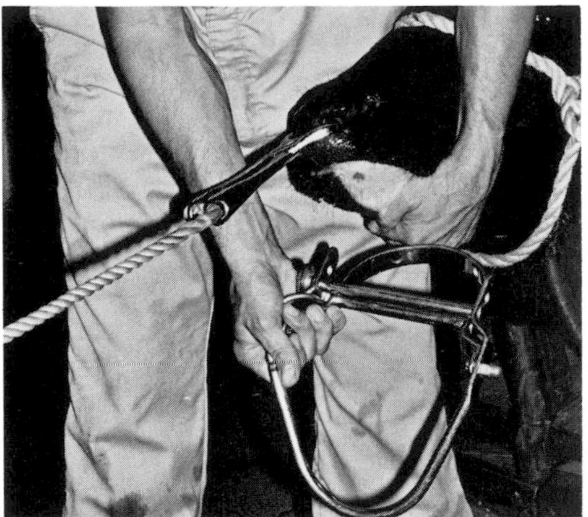

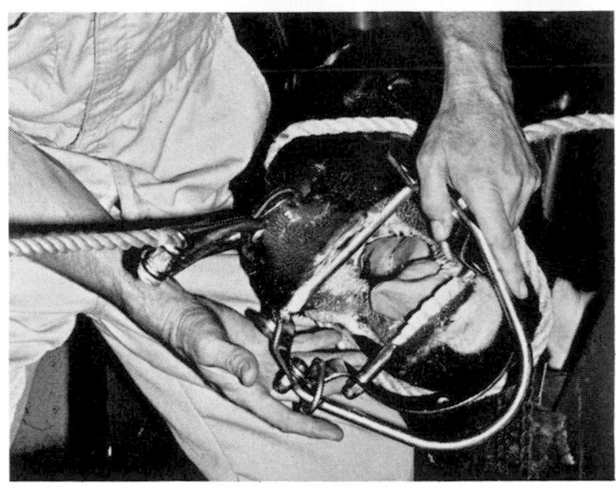

FIG. 9.14. Applying a Hauptner cattle dental speculum.

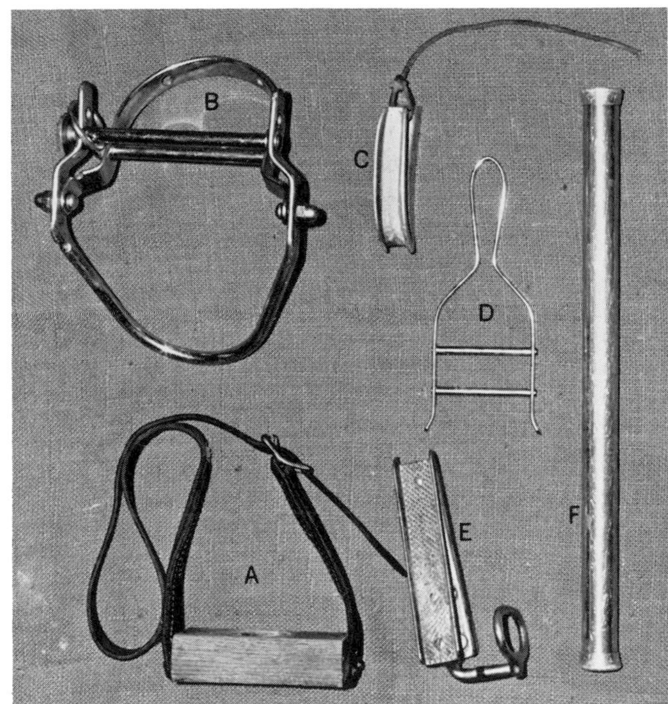

FIG. 9.15. Specula for cattle: **A.** Wooden mouth gag. **B.** Hauptner dental speculum. **C.** Dental wedge. **D.** Calf and swine speculum. **E.** Hauptner dental gag. **F.** Frick speculum.

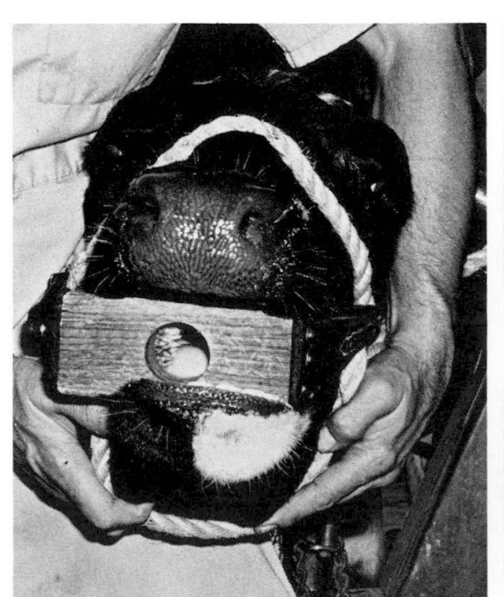

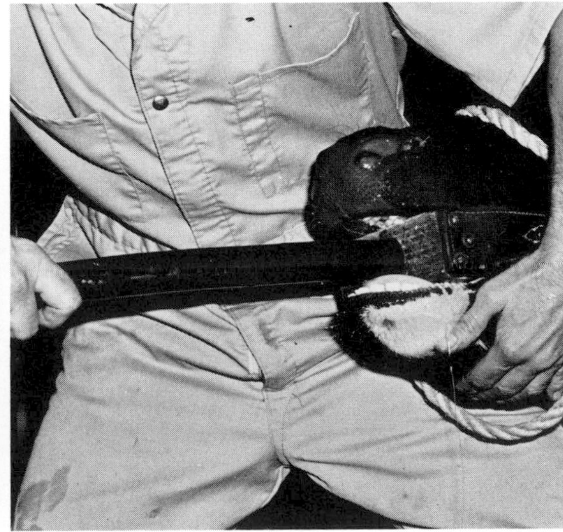

FIG. 9.16. Using a mouth gag to insert large (Kingman) stomach tube.

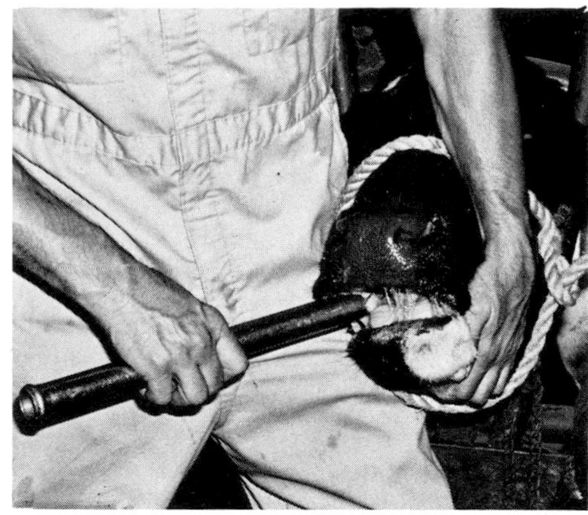
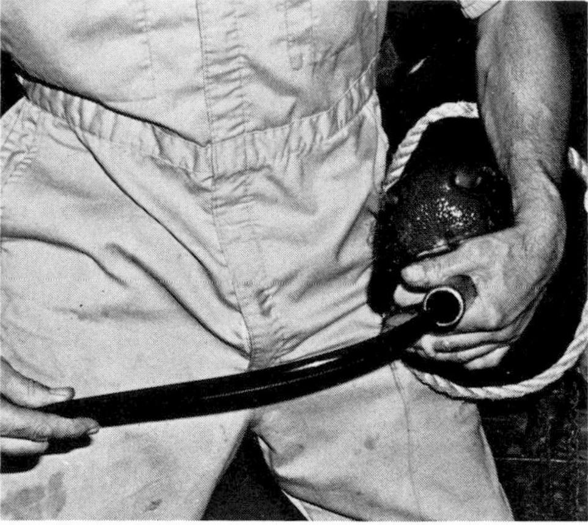

FIG. 9.17. Using a Frick speculum: Inserting speculum through the dental space and over the tongue *(left)*. Holding speculum and passing a stomach tube *(right)*.

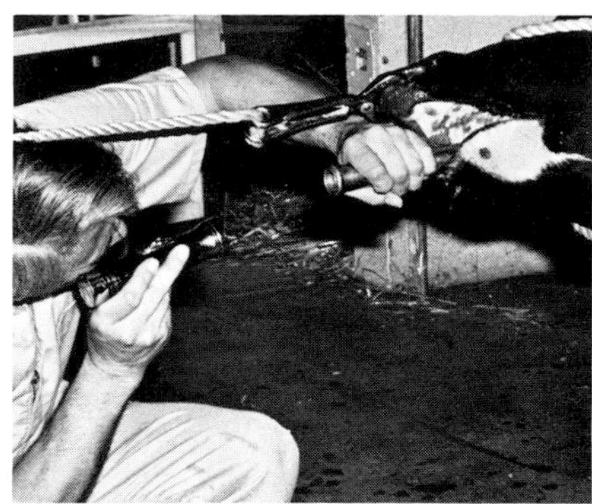
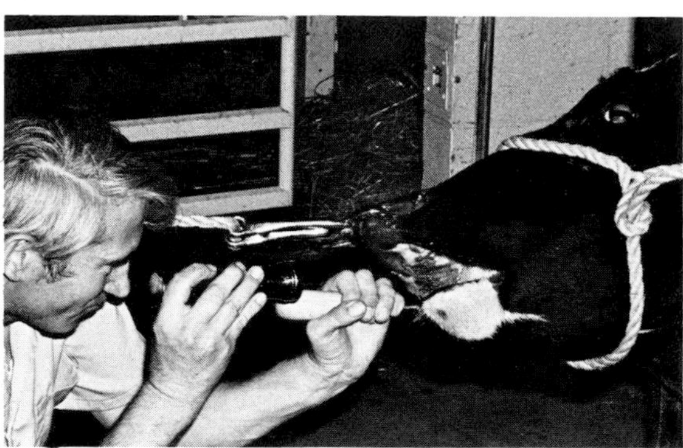

FIG. 9.18. Oral and pharyngeal examination: Frick speculum *(left)*. Depressing base of the tongue with equine dental float *(right)*. (This may be used in conjunction with the Hauptner speculum.)

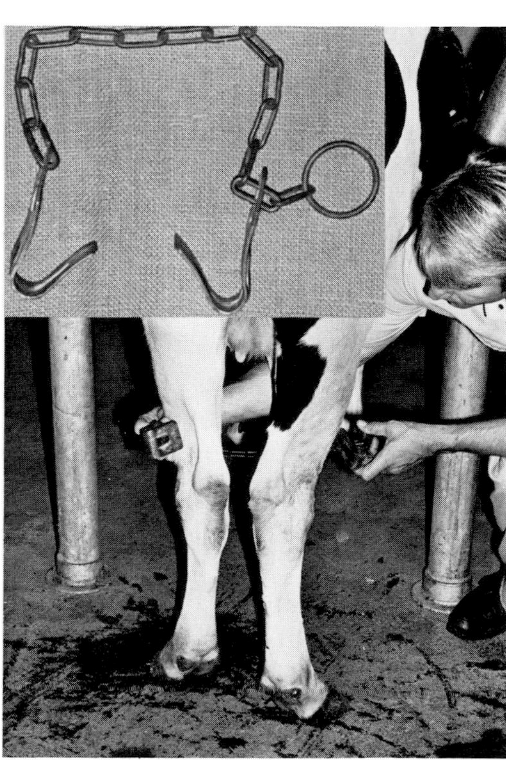

FIG. 9.19. Chain hobbles for dairy cows.

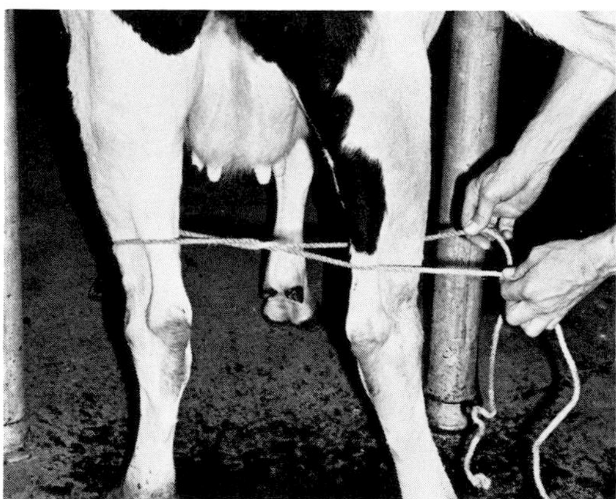
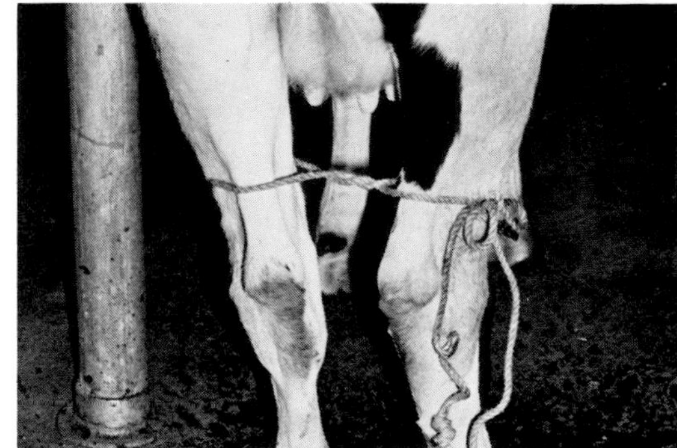

FIG. 9.20. Hock hobbles made from a small rope. Secure with a bowknot for quick release.

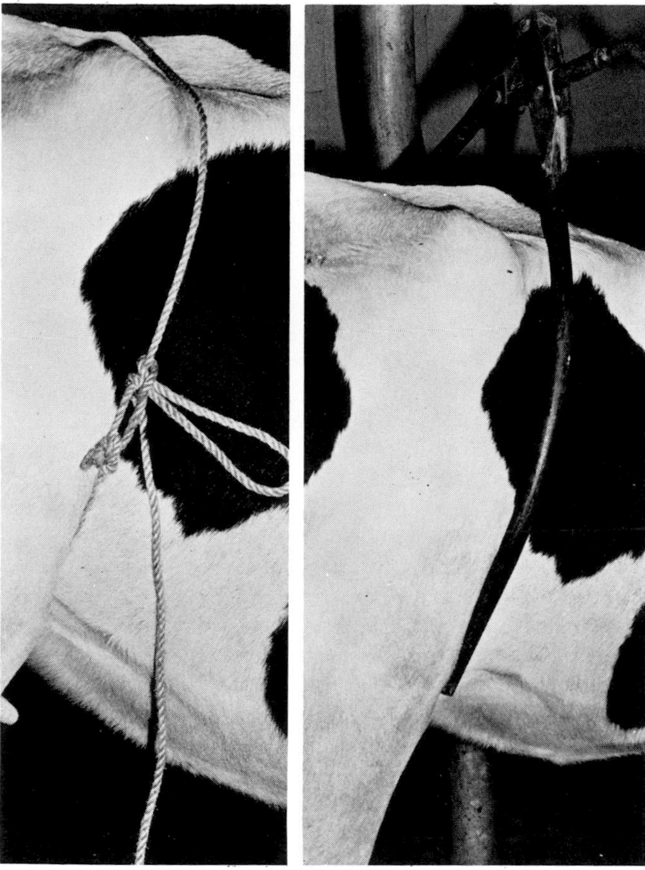

FIG. 9.21. Flank restraint: Use of flank rope *(left)*. Commercial flank clamp *(right)*.

pressure in front of the stifle (Fig. 9.22 right). The animal then cannot reach forward. The animal may, however, graze the leg of the individual carrying out the restraint.

Manipulating the feet and legs of domestic cattle is not as simple as handling those of horses. Dairy cattle that are handled continually may allow a manipulator to pick up the foot and examine the bottom for evidence of foot rot or foreign bodies (Fig. 9.23), but other cattle will probably not permit this.

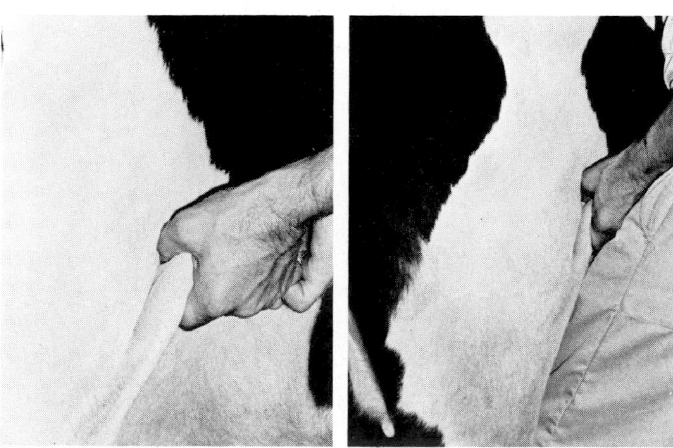

FIG. 9.22. Flank restraint: Grasping flank with hands *(left)*. Using hands and inserting knee in flank *(right)*.

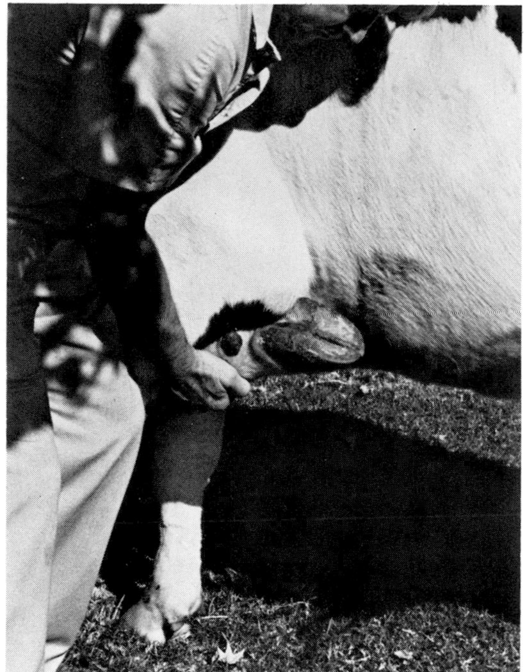

FIG. 9.23. Lifting foreleg of a cow.

Most techniques for examining the hind feet or trimming the hoofs require casting the animal in lateral recumbency or holding the limb up with ropes. A strong manila or nylon rope with a honda (preferably a quick-release honda) is used to lift the hind leg (Fig. 9.24). Leave approximately 1 ft of the rope free at the honda end and tie a clove hitch around the leg above the hock. Slip the running end of the rope through the ring on a beam clamp and back through the honda end of the rope. It is easier to thread the rope through a quick-release honda. The leg is then lifted from the ground by pulling upward on the running end. The mechanical advantage gained by running the rope through a clamp allows exertion of greater pressure than is achieved by manually lifting the foot. Once the limb is elevated to the desired position, a loop hitch prevents the rope from loosening. With the foot in this position, the limb is extremely mobile. Caution should be exercised when approaching the foot because the animal can kick either from side to side or forward and back. The device only prevents the animal from placing the foot back on the ground.

A technique for immobilizing the limb further utilizes either the beam clamp or any other site within the shed or barn where a hitch can be taken. It can also be used with an animal confined in a chute (Fig. 9.25). A loop is placed around the pastern. The rope is extended upward and backward around a ring, a pole, a pipe, or any other sturdy object in front of the animal to pull the limb upward and backward. At the same time, slack is removed from all the ropes. The limb is held extended to the rear and anchored to the front. The rope around the tendon above the hock also serves to partially paralyze or immobilize the animal, permitting any desired manipulation.

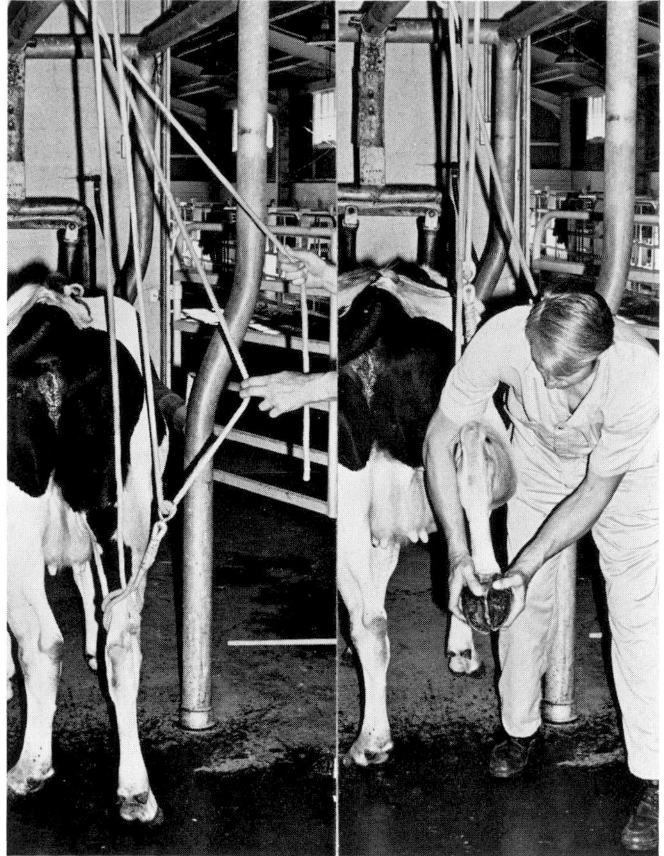

FIG. 9.24. Raising a hind leg. The pull is directly upward, thus there is little fixation of the leg in a forward and backward direction.

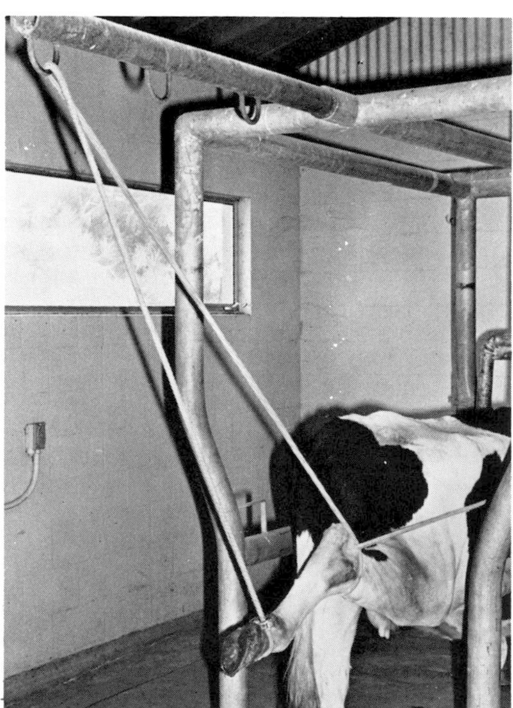

FIG. 9.25. Alternate method for raising and fixing a hind leg. Ropes are anchored forward, upward, and backward.

In barns or sheds built with exposed overhead beams, special clamps may be used to provide a fulcrum from which to lift a leg. The clamp should be attached to a beam directly above the limb (Fig. 9.26).

Special stocks have been made to assist in trimming the feet of large bulls. A heavy beam is attached to upright braces about a foot above the ground. Holes are drilled through the beam at suitable locations. The beam should extend past the upright braces. The animal is led into the stock and stanchioned. A small rope with a loop is placed on the foot, the foot is lifted manually, and the running end of the rope is put through the hole in the beam to anchor the foot.

None of these techniques are entirely suitable for use on extremely wild range cattle. Many range animals struggle sufficiently against restraint devices to seriously injure themselves. To examine or treat the feet or trim the hoofs of such animals, use chemical immobilization or manually restrain the animal into lateral recumbency.

Manipulation of the tail can be an excellent method of restraint. It is primarily applied to the dairy cow that is accustomed to being handled. However, it can be used for beef animals and other species of tailed bovids if they are restricted in lateral movement. Although an excellent restraint, one must be cautious that pressures are properly applied, otherwise the tail can be fractured and permanently disfigured.

When tailing, tie the animal up or put it in a stanchion and stand directly behind it. Lift the tail with one hand, reaching under and grasping the tail at the base with the other hand. Then grasp it close to the base with both hands, pressing the tail upward, straight over the back (Fig. 9.27). When this technique is carried out properly,

FIG. 9.26. Use of a beam clamp for temporary attachment of ropes and pulleys. Ice tongs can also be used.

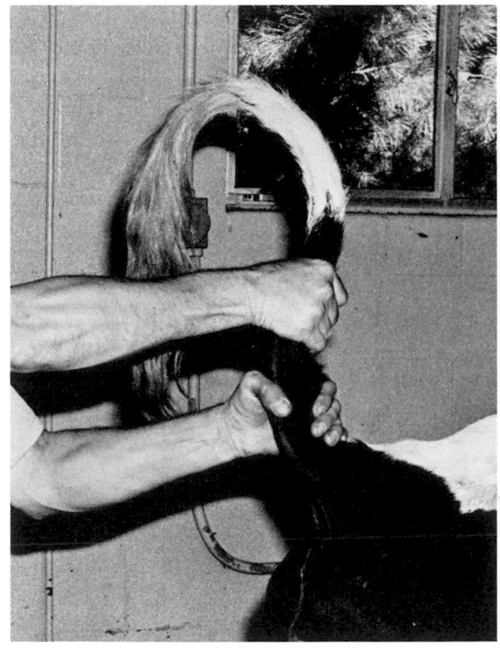
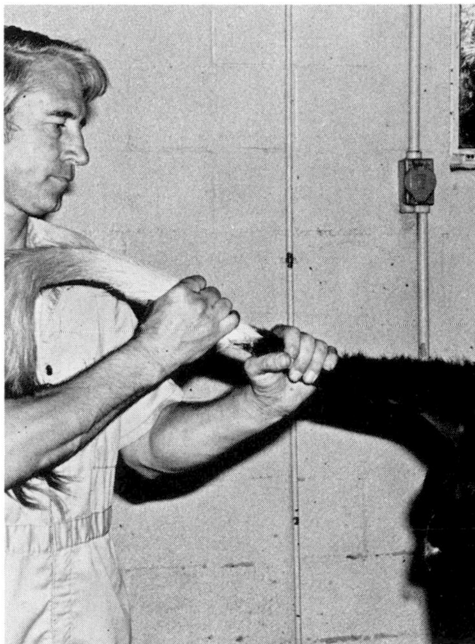

FIG. 9.27. Tailing a cow: Proper position of hands near the base of the tail *(left)*. Improper positioning of hands (too far from base) *(right)*.

pressure will not break the tail, yet will pinch the vertebrae and the caudal nerves sufficiently to make the animal relax and ignore manipulation elsewhere. Once the animal has settled down the pressure can be released, to be reapplied only when a particular procedure requires the animal to stand quietly. It is important for the pressure to be exerted at the base of the tail, not further along it. The tail of the bovine is not as strong as that of the equine, and improper manipulations may fracture the coccygeal vertebrae.

If restriction of the tail's activity is required, tie the tail to some part of the animal's body (Fig. 9.28), not to a stock or a chute. Unlike equines, bovids cannot support the body weight hanging from the tail.

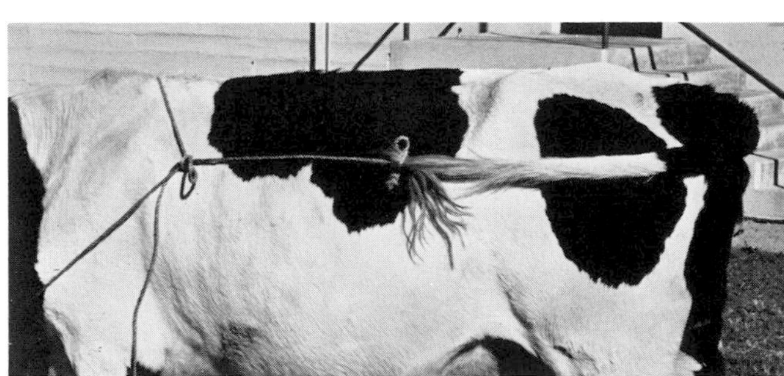
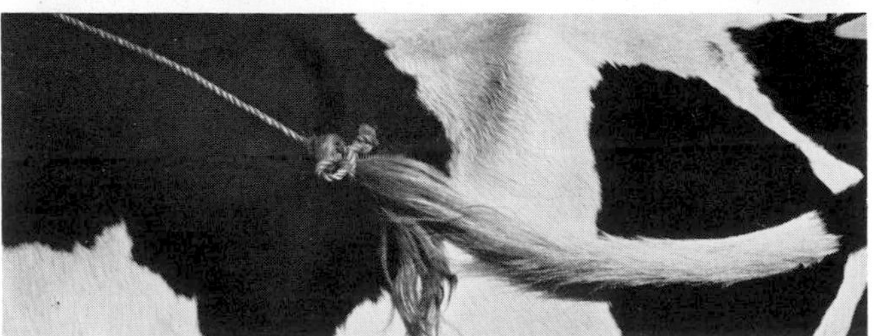

FIG. 9.28. Tail tie on cow: To the neck on the same side *(upper)*. Over the back to the opposite front leg *(lower)*.

FIG. 9.29. Controlling a steer by a foot rope.

A technique used to control tame cattle in countries other than the United States is to place a rope on either a foreleg or hind leg (Fig. 9.29).

Calf Restraint

The newborn calf is easily held by placing one arm underneath and around the neck to the opposite shoulder while holding the other hand over the tail or around the hindquarters (Fig. 9.30).

Calves up to 90 kg (200 lb) can be placed in lateral recumbency either by flanking or by lifting a foreleg. To flank a calf, hold it by the head either with a halter or a honda loop around the neck. Flanking can be carried out on either side. Place the left hand over the neck and almost immediately grasp the animal over the back with the right hand in the right flank. The right knee is in the left flank of the animal (Fig. 9.31). As the calf struggles or jumps, take advantage of the movement by quickly lifting it slightly off its feet, bending your knees to push the left side underneath the animal and quickly pressing the animal onto its left side.

FIG. 9.30. Handling a calf: Restricting *(left)*. Lifting *(right)*.

FIG. 9.31. Flanking a calf: Grasp calf by the opposite flank and over the neck. Press knee against the near flank *(left)*. Lift with both hands, simultaneously pushing knee into the flank *(right)*.

The novice may feel this is extremely laborious and requires lifting too heavy an animal. Obviously this technique is impractical for an animal over 90 kg but is effective with smaller animals. The secret of successful manipulation is to take advantage of the animal's jumps to push and pull it off balance.

An alternate technique for casting both small and large calves is legging. A skilled person can throw a 160 kg (350 lb) calf to the ground using this technique. With the head secured by a lariat or halter, approach the animal on the right side. Going down the neck, grab the right front leg. Grasp the cannon bone and pull the leg out, forward, and upward with one motion (Fig. 9.32 left). Keep the leg straight. Use the leg as a battering ram, driving it against the rib cage (Fig. 9.32 right). This pushes the animal off balance so that it will fall on its left side. All these techniques require practice to achieve proficiency in utilizing the movements of the animal to create the subtle imbalances necessary to tip the animal over.

As the animal goes down, quickly step across the thorax of the animal, grasp the right leg, and apply a loop of a short length of 64 mm (¼ in.) rope (called a piggin string) over the upper metacarpal area (Fig. 9.33 left). Then toss the rope across the animal. While holding the right foreleg with the left hand, reach back with the right hand and pick up both hind legs above the point of the hock (Fig. 9.33 right). Then sit down over the buttocks of the animal. Your right knee is behind the hocks of the calf (Fig. 9.34 left). In this position the calf cannot kick you in the groin. The knee is used to press the hind legs forward. Place the hind legs on top of the right foreleg in a crisscross position. Then grasp the rope and wrap it around all three legs in the metacarpal and metatarsal region (Fig. 9.34 right). The first wrap must be tight or the animal will be able to struggle free. Two or three wraps should be used to anchor the leg firmly (Fig. 9.35). Draw the end of the rope through the last loop to secure the tie (Fig. 9.36).

If the rope has been properly placed, the animal cannot struggle loose for some time. In the event that any one limb is not available to be incorporated into the rope, any three limbs can be tied or all four limbs can be included in the wrap. Tying four legs of a thick-bodied beef animal is not suitable because bringing both front legs close together is not only difficult but also may interfere with respiration.

If a calf is to be held in lateral recumbency without tying the legs, have one person hold the upper foreleg flexed at

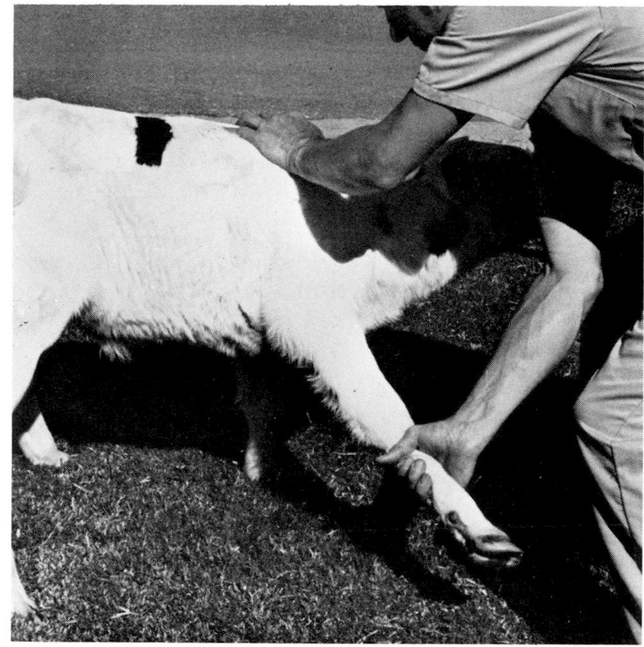

FIG. 9.32. Legging a calf: Grasp front leg and begin to lift *(left)*. Lift the leg higher and drive it against the body, pushing the calf over *(right)*.

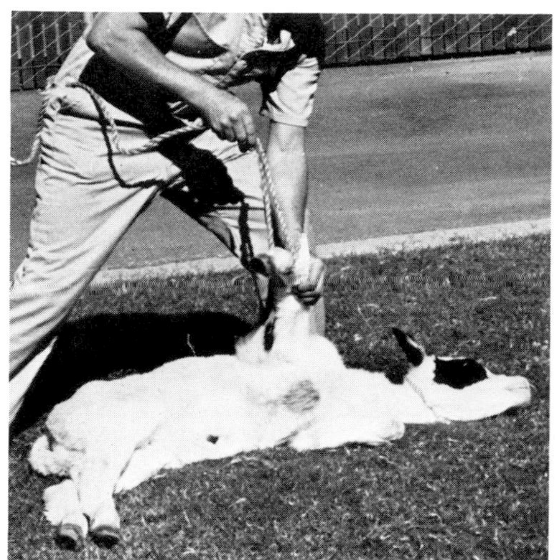

FIG. 9.33. Three-leg tie: Step over the calf and place rope loop over the foot *(left)*. Maintain grip on front leg and pick up hind legs *(right)*.

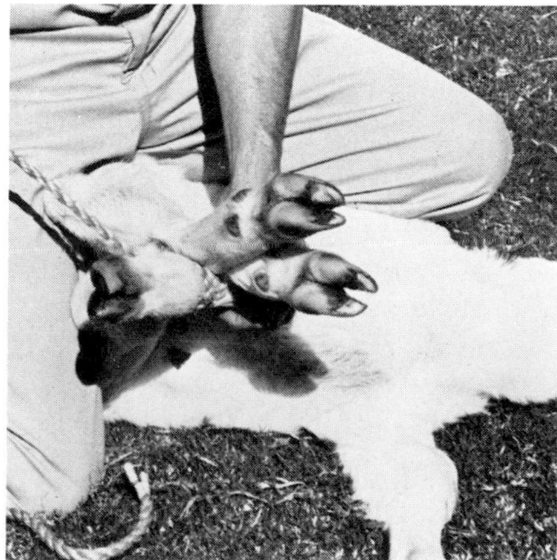

FIG. 9.34. Three-leg tie *(cont.)*: Place hind legs over front leg and hold them in place with the knee *(left)*. Wrap rope around all three legs in midcannon region *(right)*.

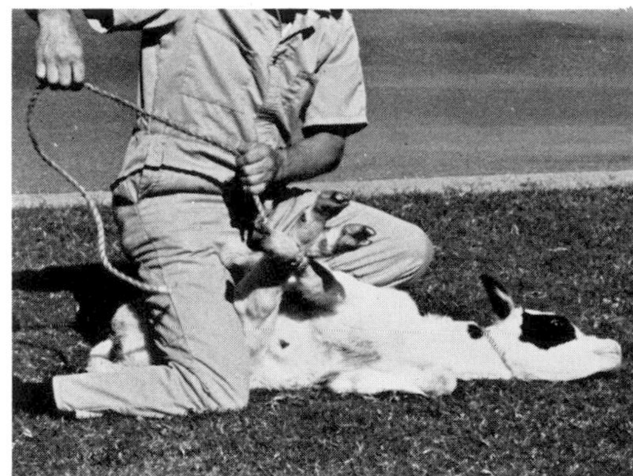

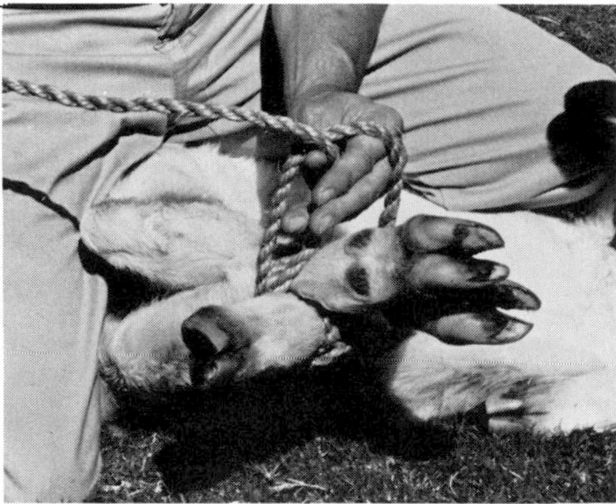

FIG. 9.35. Three-leg tie *(cont.)*: Wrap tightly three times *(left)*. On the third wrap hold a loop open, reach through the loop, and grab the free rope *(right)*.

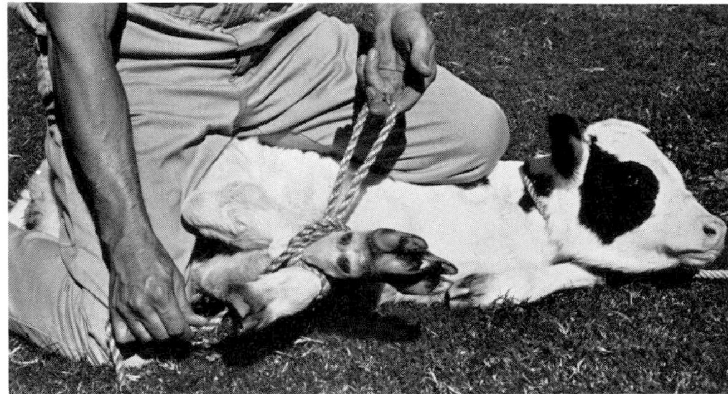

FIG. 9.36. Completed three-leg tie.

the knee. The calf may reach forward and lash out with a hind leg, but this reaction is unusual; in contrast, a wild animal, such as a deer, under similar circumstances would kick viciously to free itself. To control the hind legs more securely, grasp the upper leg and stretch it back while pushing the lower leg forward with the heel over the point of the hock (Fig. 9.37).

Casting Cattle

Small animals can be cast by placing a lark's head hitch around the thorax and abdomen (Fig. 9.38). Pressure is applied by pulling up on the rope.

Large cattle, including bulls, can be cast with the rope casting harness and hobble arrangement used for horses.

FIG. 9.37. Stretching a calf.

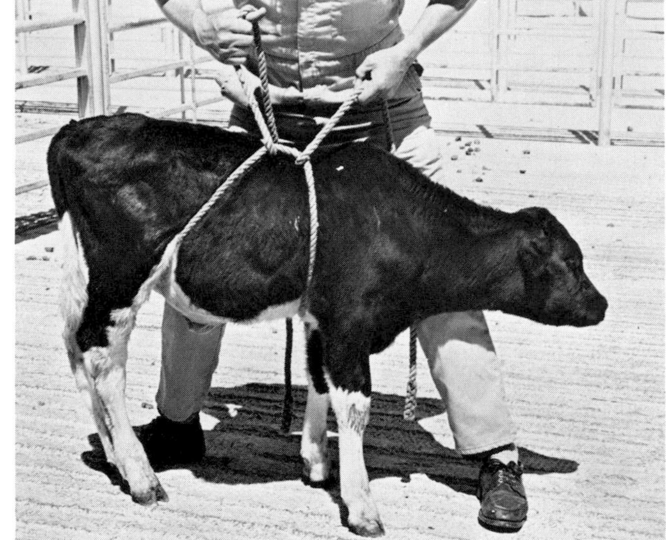

FIG. 9.38. Using a lark's head hitch to cast a calf.

Other suitable methods for placing these animals in lateral or dorsal recumbency include various techniques for applying pressure to the thorax and abdomen. The physiologic mechanism by which this pressure produces weakness and paresis is unknown. Nonetheless it is an effective method of persuading the animal to lie down. With patience and suitable strength, the technique works on large bulls as well as on cows.

The half-hitch method is accomplished as follows (Fig. 9.39): Tie a loose bowline around the neck of the animal.

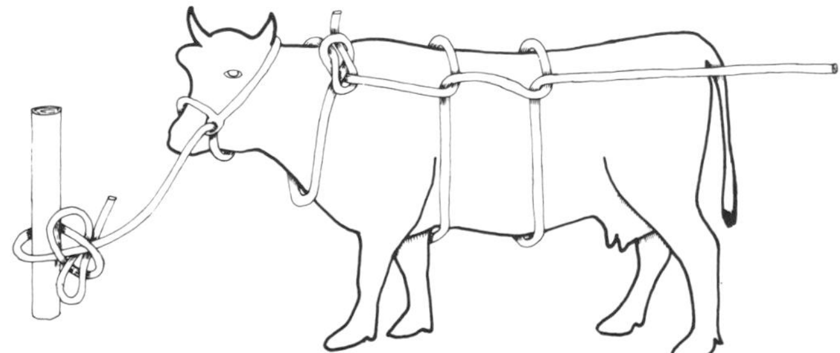

FIG. 9.39. Half-hitch method for casting cattle.

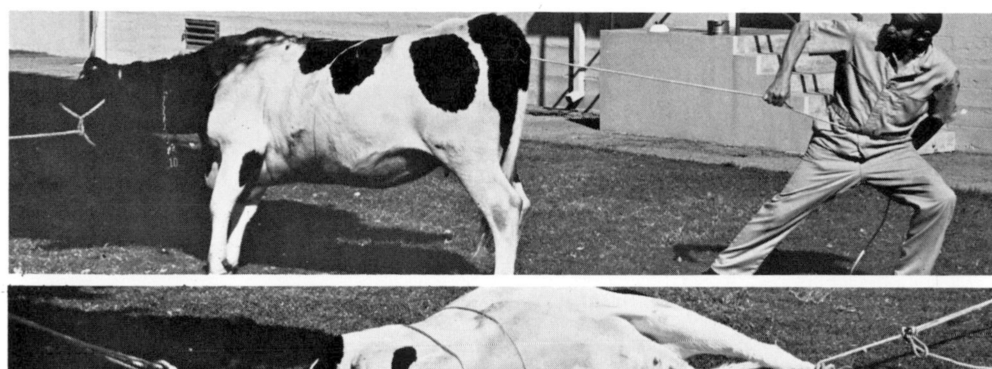

FIG. 9.40. Casting a cow: Half-hitch technique *(upper)*. Forelegs and hind legs stretched to control a cast cow *(lower)*.

Some prefer to tie the rope around the neck in between the front legs, but this is unnecessary in most cases. A half hitch is then placed behind the shoulders over the thorax. Another loop is placed just over the caudal portion of the rib cage. In large animals, a third loop may be placed in the flank area, but when handling milking dairy cows and bulls, one must be aware of potential injury to the subcutaneous abdominal milk vein or to the penis when pulling on the rope.

The animal's head must be securely tied—preferably low to the ground—so that when the animal falls, it will not hang from the head. Remove all slack from the hitches. Steady pressure is placed on the rope by pulling backward (Fig. 9.40 upper). Some animals fall immediately; others resist and even jump forward or sideways to rid themselves of the inconvenience, but if the pressure is consistently maintained, the animal will ultimately sink to its knees and lie over on its side.

I have seen heavy bulls, particularly beef breeds, set their legs in a sawhorse stance and refuse to go down. In these instances, a rope hobble which keeps the limbs together, particularly the hind limbs, forces the animal to fall down when pressure is applied. Once the animal is in lateral recumbency, maintaining the pressure may assist in keeping the animal down, but some will continue to struggle until they can get up. Usually, to maintain recumbency, the animal must be secured by stretching fore and aft with ropes tied to the front and hind legs (Fig. 9.40 lower). For details of the foot wrap, see Figure 3.35.

The crisscross is an alternative technique (Fig. 9.41). Two people are required. A rope approximately 12 m (40 ft) long is divided in half with each half coiled to the center. Place the center of the rope over the neck and pass each coil between the front legs and across to the opposite side. A person must stand on each side to manipulate the ropes. This technique is safest with an animal that does not

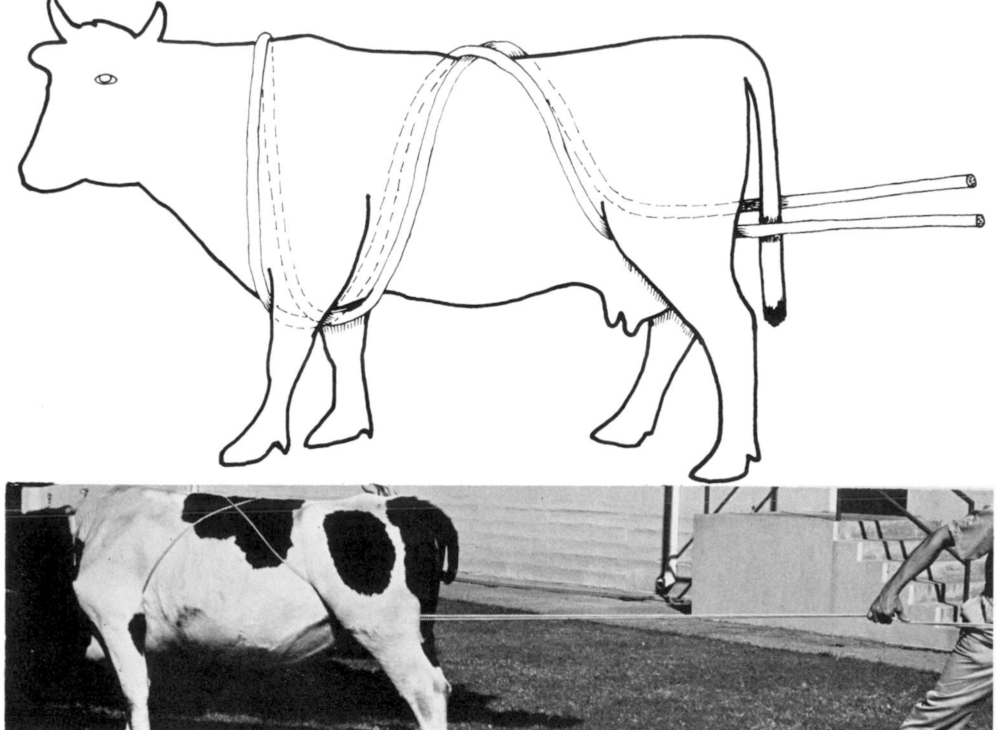

FIG. 9.41. Crisscross method for casting cattle.

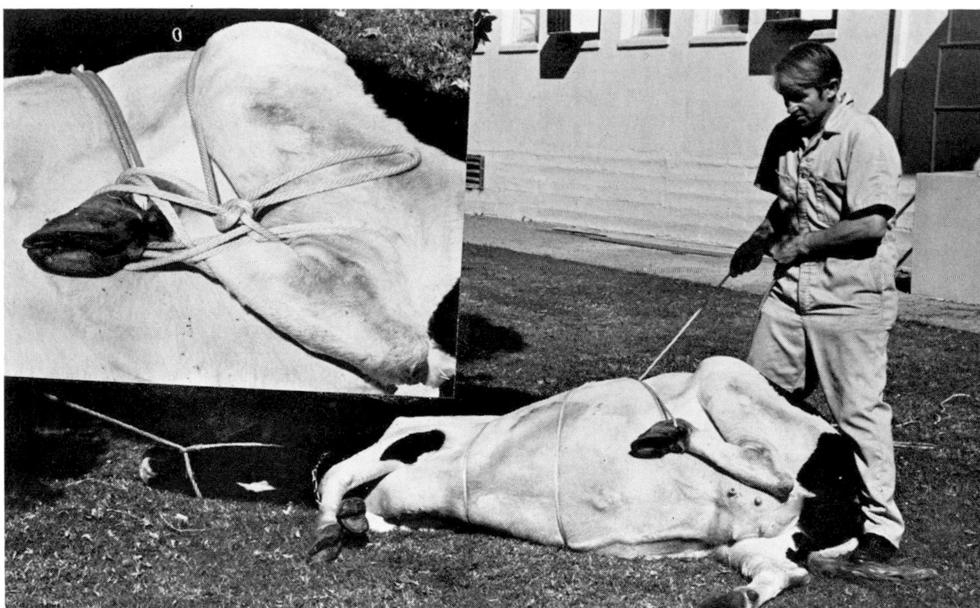

FIG. 9.42. Securing the upper hind leg of a cast cow.

kick, but it can also be used on a kicker by throwing (instead of passing) the rope between the legs. The ropes are then crossed over the top of the thorax to the side. The two persons then exchange ropes and pass them between the hind legs. All slack is removed from the ropes and pressure applied by pulling the ropes from the rear. Either one or both persons can exert the pressure necessary to pull the animal down. The advantage of using this method is that since no ropes cross the abdomen, there is no danger of injury to the milk vein or the penis.

FIG. 9.43. Hip sling for temporarily hoisting a cow.

The crisscross can also be used to pull one hind leg up into a flexed position and tie it (Fig. 9.42), but the metatarsi of most cattle are shorter than those of the horse and this technique is not as suitable for cattle as it is for equine species.

Slinging

Adapted commercial or custom-made slings, similar to those used for horses, are used for cattle. A rope sling can be used to lift an animal to its feet or extract it from a predicament (see Fig. 3.48). A special sling has been designed to lift a dairy animal with prominent tuber coxae (Fig. 9.43). The sling is adjustable to fit various sizes of cattle. This sling must not be left in place for more than a few minutes at a time lest necrosis occur as a result of pressure on the muscles below the tuber coxae. Additional padding such as sponge rubber can be inserted, but no amount of padding will prevent pressure on the bones, cutting off the blood supply to the muscles. For a weak animal, reluctant to get up, this help in arising may be all that is necessary. No sling should be used to support a cow for extended periods. A cow, capable of standing on her own, may refuse to do so while hanging in a sling. If the sling is lowered abruptly, the cow may be startled into bracing herself and stand.

Chutes

Many stocks or chutes are designed to restrain cattle. These vary from crudely constructed pole chutes (Fig. 9.44) to sophisticated commercial models (Figs. 9.45, 9.46). The use of a chute is not without risk to the animal. The design of the chute determines the overall safety as well as the efficiency and ease of manipulation. Items of concern are:

1. The danger of securing the head in such a manner that respiratory passages are obstructed. Suffocation may result if the animal twists, if it falls onto a bar across the base of the neck, or if the head is improperly held in the chute.

FIG. 9.44. Homemade stocks for examination and treatment of cattle.

FIG. 9.45. Commercial cattle chute.

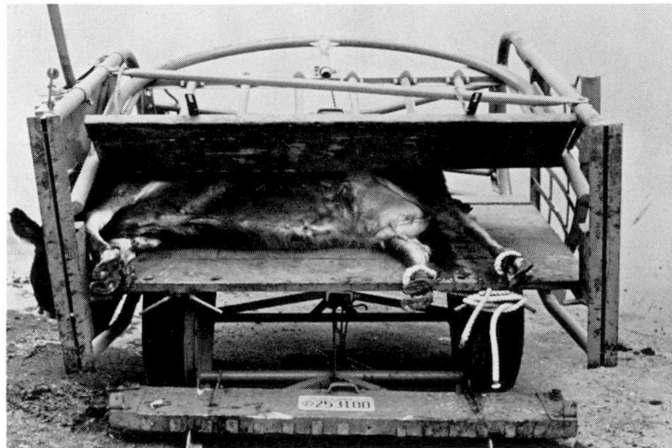

FIG. 9.46. Portable cattle tip chute.

2. Openings in the sides and front of the chute, large enough to allow an animal to catch a foot or put a foot through, provide leverage by which an animal may easily fracture a limb. Access ports are necessary in order to utilize the chute to examine or work on the body, but the best chutes have access ports that remain closed until the animal is in the chute and fully restrained.

3. The ease with which the animal can be released following conclusion of the procedure, or if the animal becomes distressed, is important. A chute is dangerous if it does not permit an animal to arise after it falls down in the chute. Some chutes come apart at the sides so that a downed animal can be pulled out. With others, the head must be released back into the chute before the front gate can be opened. These are more dangerous and thus less satisfactory than those opening to the side.

4. Sharp protrusions on the inside or outside of the chute provide sources of injury to the animal and to those working the chute. Eliminate any projections.

Elaborate or simple arrangements can be designed to funnel cattle into a chute. Special circular lanes or cutting gates can be constructed to sort animals or to remove calves from their dams.

TRANSPORT

Cattle are easily transported. They can be herded into trucks, trains, or trailers. Intercontinental air or sea craft routinely carry bovids in crates designed for one or more individuals.

Group cattle, placed in close quarters for shipping, according to size. To protect calves from injury by the cows, separate them by partitions within crates or ship in different crates.

CHEMICAL RESTRAINT

Tranquilizers are sometimes used to quieten domestic cattle, but complete immobilization is rarely necessary.

Xylazine (Rompun) is now the most widely used of the tranquilizing agents [1,4]. Although it has not been cleared by the Food and Drug Administration for use in cattle, the drug has been given extensive field trials. Xylazine is the current drug of choice for tranquilization and immobilization of intractable cattle. The dosage for domestic cattle is lower than that needed for wild bovids (Table 9.4). Analgesia is variable. Local or epidural anesthesia may be necessary for painful procedures, particularly when limb surgery is performed. The intramuscular route is usually required for intractable cattle.

TABLE 9.4. Xylazine (Rompun) dosage (IM) for cattle [1]

	Sedation	Immobilization
	(mg/kg)	*(mg/kg)*
European breeds	0.05-0.2	0.3-0.6
Feral cattle	0.1-0.3	0.3-1
Zebu	0.2	0.5
Yak	0.3	0.6-1
Water buffalo (domestic)	0.05-0.15	0.2-0.5

Xylazine can be given intravenously at approximately one-half the intramuscular dose to animals already in hand.

Succinylcholine chloride is contraindicated for cattle. Apnea will occur and persist for 30 minutes, even with minimum doses. Assisted respiration may be required to keep a distressed animal alive.

All the standard intravenous and inhalation anesthetic agents are used in cattle.

REFERENCES

1. Bauditz, R. 1972. Sedation, immobilization and anesthesia with Rompun in captive and free-living wild animals. Vet. Med. Rev. 3:204-26.
2. Briggs, H. M. 1969. Modern Breeds of Livestock, 3rd ed. New York: Macmillan.
3. Cockrill, W. R. 1974. Husbandry and Health of the Domestic Buffalo, p. 281. Rome: Food and Agriculture Organization.
4. DeMoor, A., and Desmet, P. 1971. Effect of Rompun on acid-base equilibrium and arterial O_2 pressure in cattle. Vet Med. Rev. 213: 163-71.
5. Wunschmann, A. 1972. The wild and domestic oxen. In B. Grzimek, ed. Grzimek's Animal Life Encyclopedia, vol. 13, pp. 331-98. New York: Van Nostrand Reinhold.

10 SHEEP AND GOATS

CLASSIFICATION

Order Artiodactyla
Family Bovidae
Subfamily Caprinae
Tribe Caprini: domestic sheep, domestic goat

SHEEP and goats provide meat, milk, and fiber to vast numbers of people throughout the world. Many different breeds have been developed to suit the needs of a given culture. Sheep and goats are classified by genders in Table 10.1. The size range of various breeds is listed in Table 10.2.

TABLE 10.1. Names of gender of sheep and goats

Name	Mature Male	Mature Female	Newborn and Young
Sheep	Ram, buck	Ewe	Lamb
Goat	Buck, billy	Doe, nanny	Kid

TABLE 10.2. Weights of sheep and goats

Breed	Male		Female	
	(kg)	*(lb)*	*(kg)*	*(lb)*
Sheep				
Rambouillet	115-135	250-300+	65-100	150-225
Southdown	85-90	185-200	60-70	135-155
Hampshire	115	275	80-90	180-200
Dorset	100	225	80	175
Suffolk	125	275	90	200
Columbia	100-135	225-300	55-90	125-200
Karakul	75-90	170-200	60-70	130-160
Goats				
Angora	80-102	180-225	30-100	70-110
French alpine	75	170	60	135
Saanen	85	185	60	135
Nubian	80	175	60	135
Toggenburg	75	160	55	120

SHEEP

Danger Potential

Sheep are one of the easiest of large domestic animals to handle. Sheep do not bite, strike, or kick. The only danger of injury they offer is from the use of the head as a battering ram. A large horned ram may weigh over 136 kg (300 lb) and can seriously injure a careless handler.

Behavior and Physiology

Sheep have strong flocking instincts and normally move in a group. It is difficult to separate one individual from the group. If one animal can be enticed to pass through a gate, the rest usually follow.

Sheep can be easily guided with panels, either wire or wood. Most sheep will not jump over a 1.2 m barrier. Range-raised sheep may be exceptions and may jump or scramble over a 2 m fence, particularly if separated from a group.

Be cautious when handling sheep, especially if heavily fleeced, during hot weather. The normal body temperature of sheep is high, 39.5 C (103 F). Because of this characteristic, plus the insulating layer of wool, any struggle in hot weather may result in the rapid development of hyperthermia.

Physical Restraint

The most valuable aid in handling sheep is a well-trained dog (Fig. 10.1) directed by an able shepherd. Many breeds of dogs have been specifically developed for this task.

FIG. 10.1. Sheep dog.

Whenever possible, sheep should be crowded into alleyways or narrow chutes for mass medication, examination, and vaccinations (Fig. 10.2). A single sheep is easier to capture if left with the flock. Alone in a large enclosure, a sheep may panic and attempt to escape, but if allowed to stay with the flock it can be approached and grasped quite easily.

Approach the animal quietly with deliberate movements; reach down and place one hand under the chin or breast, stopping its forward motion (Fig. 10.3). With one hand around its chest and the other holding its dock, even the largest sheep can be guided into any desired position. The sheep usually remains upright and able to walk.

Never grab the wool of a sheep when attempting to catch the animal because this damages delicate wool fibers and causes the wool to decrease in quality, or it pulls out. Furthermore, pulling the wool causes subcutaneous hemorrhages that would downgrade a market lamb carcass. If the wool requires examination, one individual should hold the animal as described and another should carefully part the wool with the fingers or hands (Fig. 10.4).

FIG. 10.2. Sheep squeezed into a narrow chute.

The largest ram can be set up on its haunches by application of the proper mechanical principles. The approach described is from the left side; however, one can cast the animal just as easily in the opposite direction. Start from the basic holding position. Place the right knee in the left flank and move the right hand from the dock to the right flank. Change the left hand from encircling the chest to grasp the animal by the lower jaw. Twist the head to the right with the left hand (Fig. 10.5). At the same time, press in on the right flank and whirl the animal. The quick coordinated movement forces the animal to sit down on its left hip and the twirling motion sets it up as illustrated in Figure 10.6

FIG. 10.3. Holding a sheep.

FIG. 10.4. Parting wool.

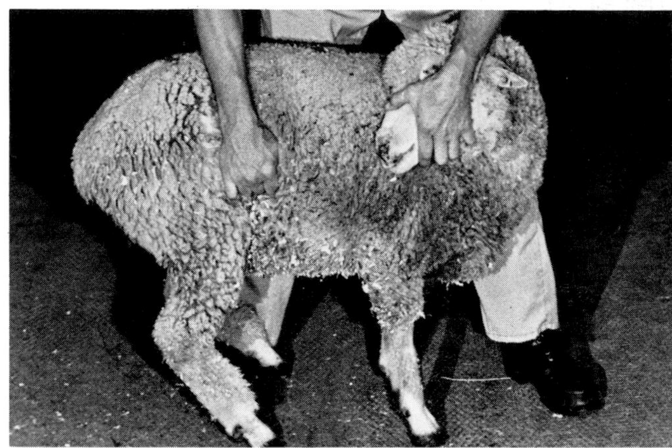

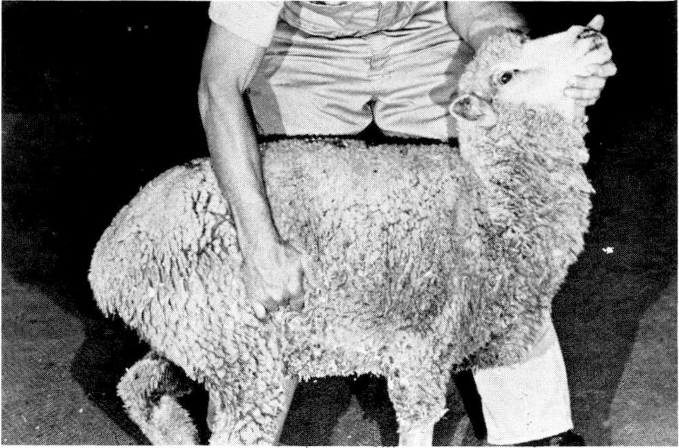

FIG. 10.5. Setting up a sheep: Proper position *(upper)*. Grasp flank and pull head around to the side. Push knee into the flank. Improper position *(lower)*. Do not lift the head up over the back.

The handler's legs should be spread slightly to cradle the sheep's back. The animal should be sitting at approximately 60 degrees to the vertical. If the animal is too perpendicular it will struggle to free itself and perhaps will be able to gain enough balance to throw itself forward and escape. If the animal is too horizontal, too much pressure

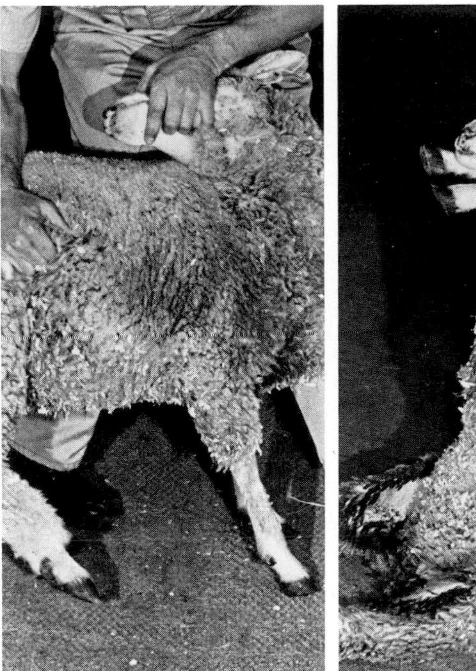

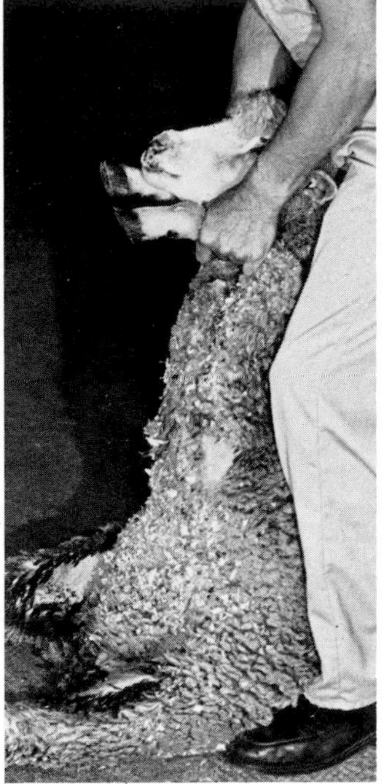

FIG. 10.6. Setting up a sheep *(cont.)*: Rotate animal on the knee by pulling on the flank while pushing the head around *(left)*. Sheep held properly, slightly off vertical *(right)*.

is exerted on the handler's legs and the animal is less accessible for examination and treatment.

If the sheep is properly balanced, both arms of the handler will be free to examine the feet, trim hoofs, examine the mouth, mammary glands, prepuce, or testicles (Figs. 10.7, 10.8). Occasionally, an animal may flail its front legs; care should be taken to protect one's face from such action. Be aware of the relative positions of the head and feet of the sheep at all times.

Rams weighing over 136 kg (300 lb) may seem too large to set up in this manner; some animals set their chins down on the ground, seeming almost to defy a handler to lift and twist the head to the side. Nonetheless, grasping the animal's chin firmly and lifting and twisting it up to the side will enable the handler to set up the largest animal. It is important to twist the head to the side, not just pull it up dorsally (over the back) (Fig. 10.5). It is unnecessary and undesirable to lift the animal off the ground. Simply twist it, using mechanical advantage to position the animal.

The crisscross rope casting technique used with cattle is used on sheep in lieu of a net or snare.

The crook minimizes exciting an animal when the handler steps up close to it (Fig. 10.9).

Lambs are easily handled. Support the body underneath the chest (Fig. 10.10). The lamb can be held for castration or docking in the manner illustrated in Figure 10.11. The forelegs and hind legs are held together.

Halters can be used on sheep, but the nose is short so take care that the halter does not slide off or pull down over the nostrils, restricting air movement.

Various drenching techniques are used to medicate sheep. Large flocks are easily handled by crowding the

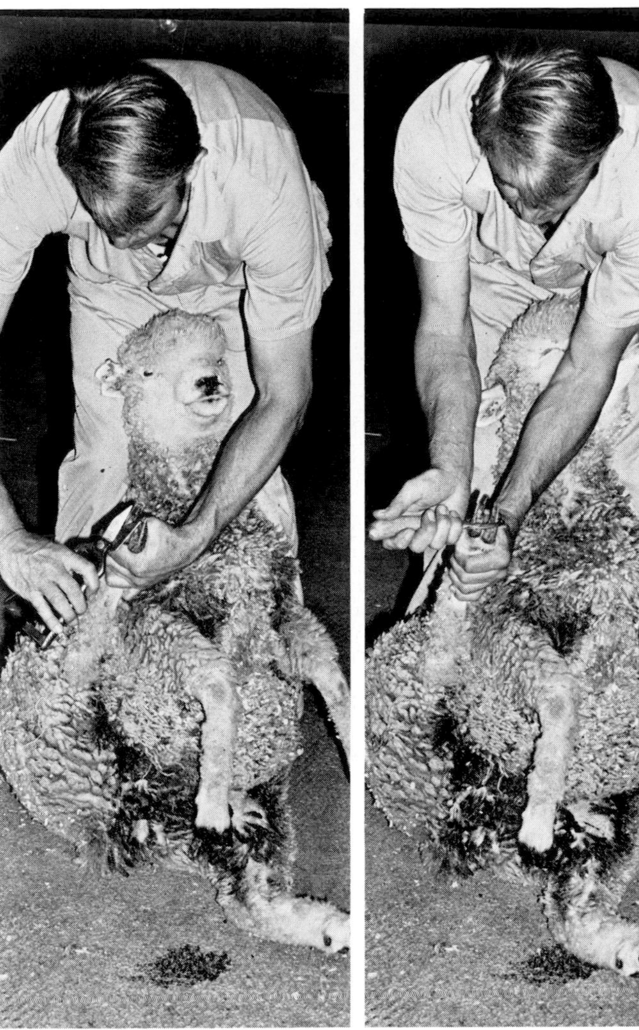

FIG. 10.7. Hoof trimming from set-up position.

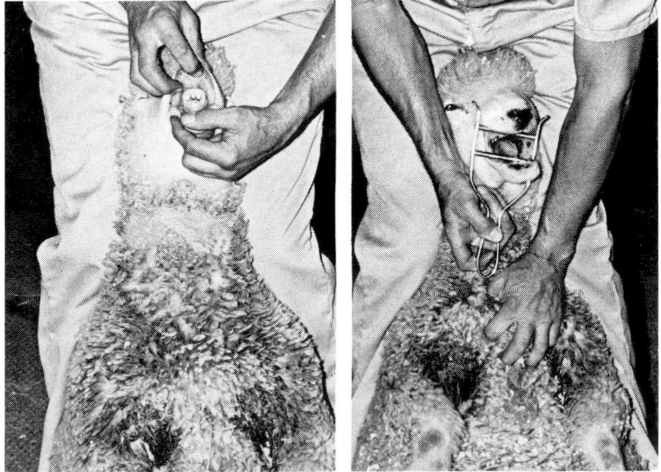

FIG. 10.8. Oral examination from set-up position.

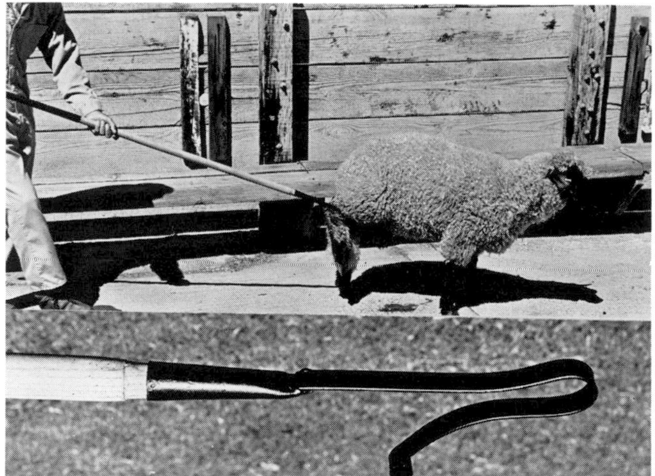

FIG. 10.9. Shepherd's crook.

FIG. 10.10. Holding a lamb.

animals into a narrow chute, three or four abreast. The person conducting the drenching can then walk through the animals and insert the dose syringe nozzle into each mouth without resorting to additional restraint. The presence of the other animals keeps each patient secure for the treatment.

Either straddle the sheep, walk up alongside it, or reach over and lift its head slightly by placing a hand under the chin (Fig. 10.12). Insert the dose syringe in the commissure of the mouth, going over the base of the tongue, and immediately clasp the mouth and nostrils closed while quickly injecting the medication into the pharynx (Fig. 10.13). Automatic drenching guns have been designed for mass medication.

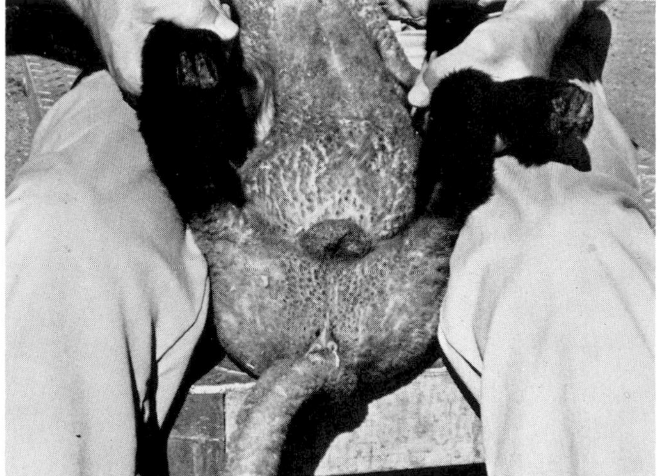

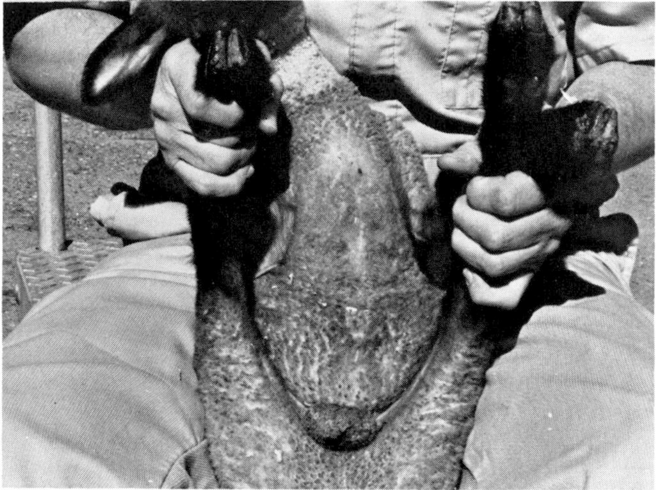

FIG. 10.11. Alternate methods of holding a lamb for castration or docking.

FIG. 10.12. Squeezing sheep into a corner with a panel for medication or examination.

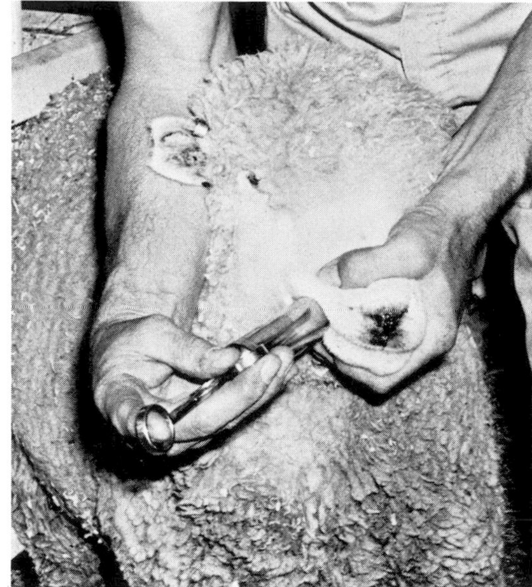

FIG. 10.13. Medicating a sheep with a drench syringe.

Solid medication in bolus or tablet form is given with a balling gun (Fig. 10.14). Insert the tip of the gun through the interdental space.

GOATS

Although goats have the reputation of being able to withstand heavy stresses, in reality they are quite delicate. Their bones are small and easily broken. Usually, rough handling is not necessary. When accustomed to being handled, goats are docile and easily managed (Fig. 10.15). Most respond to gentle treatment.

FIG. 10.15. Female goats are easily handled, even by a child.

Uncastrated males have scent glands which produce a secretion with a very disagreeable odor. The secretion of the odoriferous material is under the control of androgens; hence the glands are not active in castrated males. It is difficult to prevent the odor from impregnating the clothing of anyone handling a buck. Furthermore, the mature male

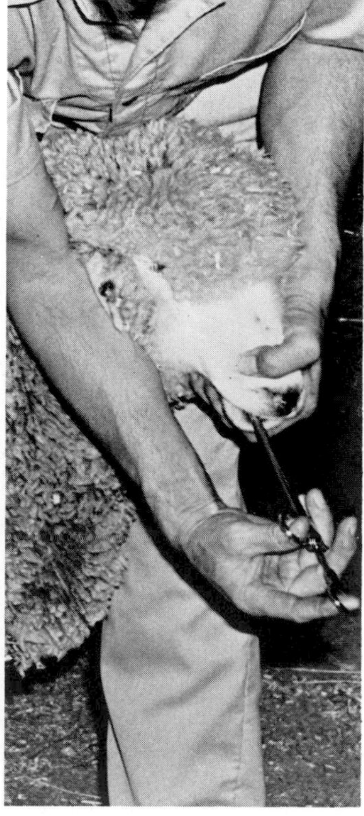

FIG. 10.14. Medicating a sheep with a balling gun.

FIG. 10.16. Goat restraint with a neck chain: Hand held *(left)*. Restrained by a snap anchored to a post *(right)*.

is prone to urinate on its legs, neck, and body. The resulting pungent scent is of significance in breeding behavior.

The scent glands of the male are diffused in the area around the base of the horn. When a young buck is dehorned, if 13 mm (½ in.) of skin is taken around the base of the horn, the gland is excised also. It is important that all males in a breeding group be descented if any are so treated. Otherwise, the females will prefer the scented males over the descented males.

Danger Potential

Goats do not bite, strike, or kick, but usually fuss more than sheep. They vocalize and they may stamp their feet in obvious threat, but once they are grabbed they do not strike. They do, however, use their heads for butting. Species having horns may use them as battering rams. The buck (male) is frequently adorned with heavy horns that can inflict serious injury on the unwary. A male goat is much more likely to initiate an attack than is a male sheep. Most dairy goats in the United States are dehorned as kids or are naturally polled, which lessens the hazard of butting. Nevertheless, even a hornless animal can cause injury.

Physical Restraint

The initial approach to handling a goat is similar to that for sheep in that one arm is placed around the goat's chest as the other hand grasps the dock (tail) area. The similarity ceases here, because goats cannot be "set up" like sheep. Goats are far more agile and less prone to accept such

FIG. 10.17. Goat restraint: Holding a buck by the beard and chain *(left)*. Holding small goat by one leg *(right)*.

restraint. If placed in the set-up position, a goat will lash out with both forefeet and hind feet in a purposeful attack on the face and hands of the handler.

A group of goats can be herded into a chute or corner to single out or capture an individual. Approach a cornered buck with caution. He may attack. Threats are characterized by vocalization and stamping of the feet.

Angora goats kept in range flocks must be confined to sheds or chutes to capture them. Goats are much better jumpers than sheep, so chutes or pen fences must be 2 m (6 ft) or higher. Highly excited goats have been known to attempt to jump over a man. Sometimes all four feet will be planted on the chest or head of the person who gets in the way.

Goats can be roped, but the roper must be skilled. Goats, because of their speed and dodging abilities, are frequently used to sharpen the skills of rodeo calf ropers. Ropers must be prepared to jerk the loop tight quickly if fortunate enough to catch a goat.

Most dairy goats wear a neck chain or collar. Chains are preferred because penmates cannot chew them off. The size of most adult goats makes the chain convenient to hold the goat or lead it (Fig. 10.16). If present, the horns can be used to grasp the goat in lieu of a chain. Although the horns can be used for initial capture, goats dislike being held by the horns or ears. To hold a goat without a collar or a halter, place the open hands on each side of the lower jaw beneath the ears. The beard can be grasped to assist in immobilizing the head (Fig. 10.17 left). A small goat can be restrained by holding one leg (Fig. 10.17 right).

Small halters can be put on temporarily since many goats have been taught to stand quietly when haltered. Halters that are left on will be chewed up. If haltered goats are left unattended, they may either chew the rope or loosen the knot. Some goats become very proficient at escaping from any restriction. Some learn how to unlatch gates, others are prone to climbing over walls, fences, and other restricting devices.

FIG. 10.18. Goat on milking platform.

Dairy goats are likely to be accustomed to being snapped to a post or wall ring for milking, hoof trimming, or examination (Fig. 10.16). Some goats can be milked from a small platform (Figs. 10.18, 10.19). The head is locked in a stanchion. Hoof trimming, examination or treatment of mastitis, and other procedures can be carried out while goats are restrained in this position. Eye examination and balling are illustrated in Figure 10.20.

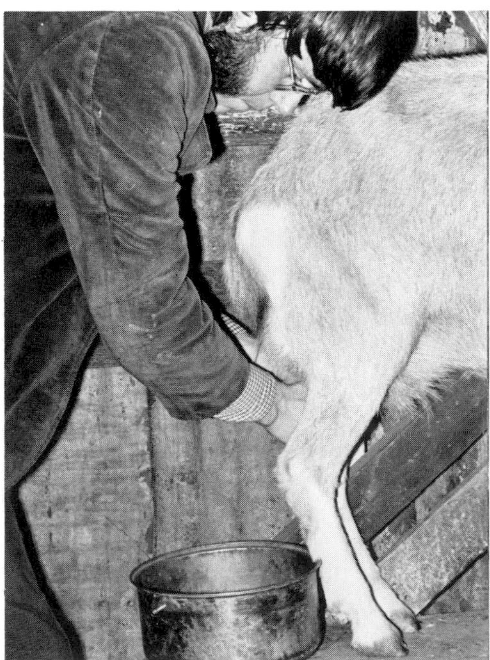

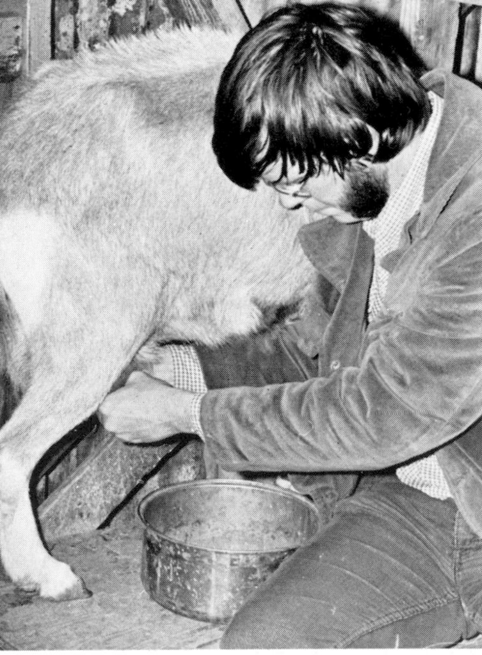

FIG. 10.19. Milking positions with goat on platform.

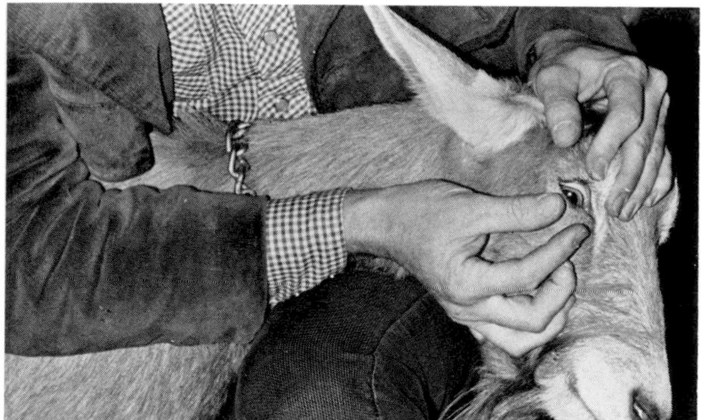

FIG. 10.20. Goat restraint: Examination of the eye *(upper)*. Straddling against a wall to medicate *(lower)*.

The feet and legs of a goat can be picked up like those of a horse. Usually one person holds the hind leg while another manipulates the feet. One person can hold the goat if necessary (Figs. 10.21, 10.22), but the goat is more likely to struggle with one handler than with two.

An adult goat can be placed in lateral recumbency by flanking it (method is similar to that used for casting a calf). Kids are as easily handled as lambs.

Transport

Sheep and goats are easily transported in trucks, trailers, trains, and ships. They tolerate being closely confined.

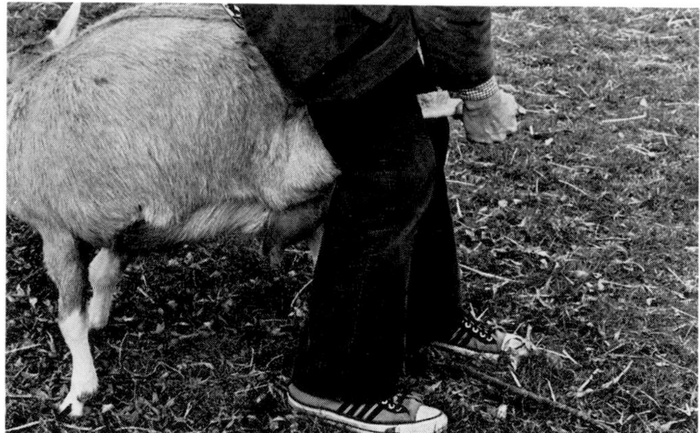

FIG. 10.21. Trimming a hind foot.

FIG. 10.22. Trimming a front foot.

The wool insulation layer of sheep predisposes them to hyperthermia, so arrangements must be made to keep sheep cool.

Chemical Restraint

Domestic sheep and goats are so easily handled that chemical immobilization is rarely practiced. If sedation for minor surgery is required, xylazine (0.3-0.6 mg/kg), intramuscularly, and Tilazol (2.5 mg/kg), intramuscularly, are suitable agents. Monitor sedated animals closely for evidence of hyperthermia.

11 SWINE

CLASSIFICATION

Order Artiodactyla
  Suborder Suiformes
    Family Suidae: swine

SWINE are short necked and short legged, with heavy bodies. They are the descendants of the wild European boar. They are used for meat in tropical, subtropical, and temperate areas of the world. Names of gender of swine are listed in Table 11.1. Weights of different breeds are listed in Table 11.2.

TABLE 11.1. Names of gender of swine

		Male				
Mature Male	Mature Female	Castrated before maturity	Castrated after maturity	Young Female	Young of Either Sex	Newborn
Boar	Sow	Barrow	Stag	Gilt	Shoat	Piglet

TABLE 11.2. Weights of swine

	Male		Female	
	(kg)	*(lb)*	*(kg)*	*(lb)*
Berkshire	410	900	365	800
Hampshire	410	900	320	700
Poland China	445	975	385	850
Duroc	430	950	340	750
Chester white	420	925	330	725
Yorkshire	320	700	370	600

DANGER POTENTIAL

The principal weapon of the pig is teeth. Baby pigs have sharp, needle-like deciduous teeth which inflict nasty wounds that are without exception septic and can be serious. Adult swine tear flesh easily. They have extremely strong jaws capable of crushing bones. Boars also develop elongated canine teeth called tusks, which are fearsome weapons capable of disemboweling a horse and certainly a person. The sow with a litter is a formidable, menacing animal and should be approached with caution.

Pigs do not usually behave as a "herd," but when handling an individual pig in the company of other pigs, be watchful for the development of a mob reaction. On one occasion I attempted to catch a small pig of approximately 18 kg (40 lb) in a large enclosure. The pig was roped around the body and picked up. As soon as the rope was in place, the animal began to squeal. Immediately all the other pigs in the pen crowded around, grunting and threatening to attack. I was forced to drop the pig and move out of the area as quickly as possible.

ANATOMY, PHYSIOLOGY, AND BEHAVIOR

Swine behavior sets them apart from other domestic animals. They have little banding or herding instinct and are stubborn and contrary, resisting all efforts to drive them in a given direction or move them from one place to another. Never enter an enclosure with adult swine without taking safety precautions. Always make note of a quick escape route.

Pigs are unpredictable and frequently become aggressive with little warning. This is particularly true of sows with litters or adult boars. Rely on the husbandman for information about the general behavior of an individual animal.

Pigs naturally pull back when pressure is applied around the upper jaw. This peculiarity makes it possible to manipulate many swine that otherwise could not be handled.

Swine conformation allows a pig to run through underbrush easily. The body is streamlined, the head usually pointed, and separation between the head and body is not well demarcated. The lack of a definite separation at the head and neck prevents the use of halters or ropes around the neck to restrain pigs.

Because of their slick, smooth bodies, it is impossible to handle wet pigs. To successfully work with a pig, it should be clean and confined in a dry enclosure with absorbent bedding.

The strong neck muscles developed by rooting enable swine to lift with considerable force. Any panels on fences used to contain swine must be firmly attached to the ground, otherwise the animal may put its snout underneath and throw the panel over its head to escape or attack.

PHYSICAL RESTRAINT

It is virtually impossible to capture or handle an adult pig in a large enclosure. Move the pig into a small pen by driving, or entice it into the small enclosure with feed. Pigs can be driven into a smaller pen either individually or as a group.

If a group of swine of mixed sizes must be handled, sort them out by size first to prevent large animals from trampling smaller individuals.

Unfortunately swine frequently have the habit of moving in the opposite direction to that desired by the handler. Thus an individual pig may be difficult to move. A snout rope utilizes the natural propensity to pull back against pressure on the upper jaw to enable a handler to point the animal's rump in the desired direction and back the pig into an enclosure.

A bucket or blindfold over the head of the animal triggers the same idiosyncrasy. The pig will move in a negative direction to escape the bucket, and by continuing to hold the bucket over the pig's head, it can be directed as is a ship by its rudder (Fig. 11.1 left).

A cane or a narrow flat stick is excellent for directing a pig (Fig. 11.1 right). The stick or cane is not used to inflict

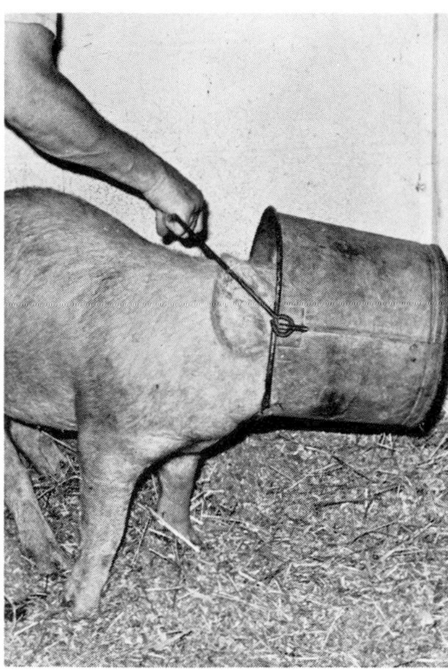

FIG. 11.1. Directing pig backward *(left)*. Cane is used to indicate direction *(right)*.

pain, but is merely tapped on the side of the head to indicate to the animal the desired direction.

A broad leather or canvas strap attached to a short wooden handle is effective in moving swine because it makes a loud noise when slapped against the body.

When directing pigs into a pen or through a gate, use either wire paneling or solid plywood shields. A shield is also a safe structure to work behind when dealing with a dangerous animal. Recognize that a large boar or an extremely aggressive sow could manage to work its snout under the shield and attack, but usually this can be prevented by tilting the top of the board back toward the handler or by slapping at the snout.

The snout, although used for rooting in the ground, is not calloused but is an extremely sensitive organ. One can slow a pig down or even change its direction by tapping on the snout. Do not hit the snout hard, since inflicting severe pain may result in a negative response. The animal then becomes more aggressive rather than subdued, and the handler must then deal with an awesome opponent. A slap on the snout with a board or a cane may stop the rush of the animal, giving the handler time to escape. Kicking the snout may also work, but it is an extremely dangerous practice, because a pig can move its head swiftly and can bite through the shoe, injuring the foot severely.

Swine chased excessively, particularly during hot weather, become exhausted. The deep layer of insulating fat does not allow efficient conduction of heat, and these animals overheat rapidly if harassed too long during manipulative procedures. Any pig that must be restrained for any length of time and which struggles during that period should be examined frequently with a rectal thermometer to determine whether the body temperature is elevated. Temperatures above 40-40.5 C (104-105 F) require that immediate steps be taken to cool the animal with cold water enemas or sprays. Proceed with caution when the animal is under heat stress.

Although more stringent restraint practices may be required, it is often possible to work on adult sows and even boars by speaking to them in a kind, gentle manner, particularly if this is done by the usual husbandman. A sow lying down comfortably can be talked to, scratched on the body, and calmed sufficiently to take her temperature without causing her to jump up (Fig. 11.2). This method should always be attempted first, especially when performing simple procedures. Keep in mind, however, that these animals can turn and bite swiftly. Take precautions to assure that the animal does not twirl and bite the hand.

Ropes are rarely used to capture swine. It is virtually impossible to rope a pig and prevent it from pulling its head back out of the loop because it has no neck. However, the animal can be put into a harness. Place a loop over the head and tighten it gently (Fig. 11.3 left). Form a half hitch, throw it over the head, and allow the animal to walk through it until the loop is past the front legs (Fig. 11.3 right). Then pull it up tight (Fig. 11.4 left). It can then be tied to a post to restrict movement (Fig. 11.4 right).

The snout rope or cable snare is the prime restraint tool for swine. It is used in a variety of procedures requiring the moving or holding of swine. The size of rope used to form a snout loop varies with the size of the animal. To pull down

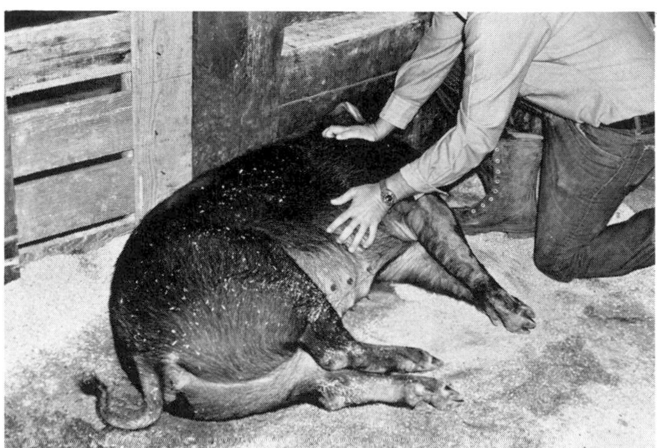

FIG. 11.2. Calming recumbent sow by gentle stroking and a soft voice.

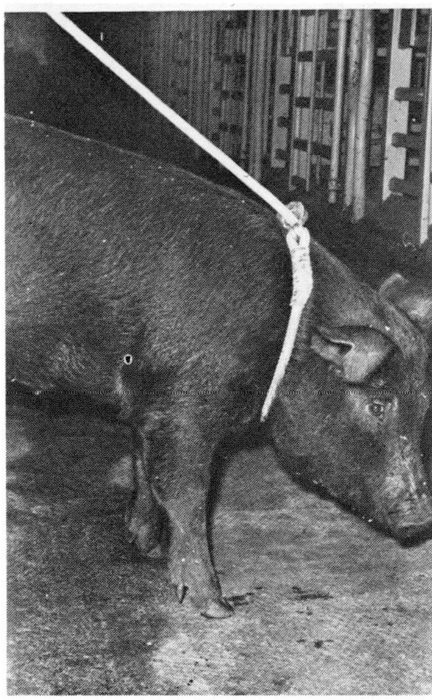

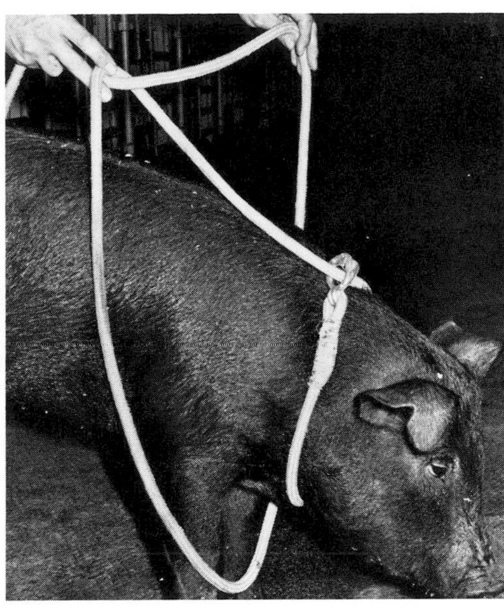

FIG. 11.3. Applying rope harness to a pig: Loop is placed over the head and gently tightened from behind *(left)*. Pig is allowed to walk through a half-hitch loop *(right)*.

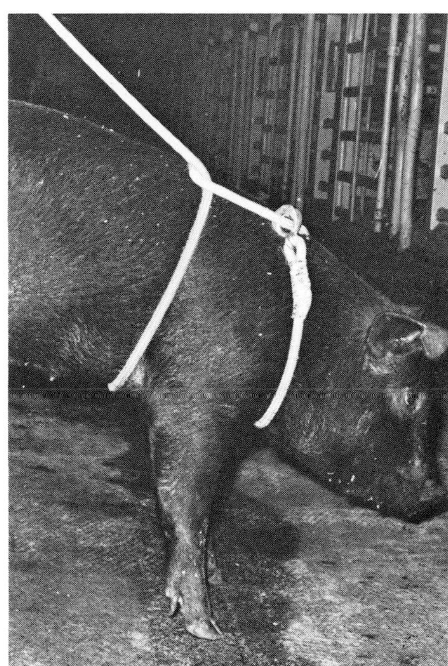

FIG. 11.4. Rope harness *(cont.)*: The second loop is tightened behind the forelegs *(left)*. Pig is tied to restrict movement *(right)*.

snugly over the upper jaw, ropes sized from 3.2 mm (⅛ in.) manila to 13 mm (½ in.) nylon can be used. A small honda should be formed and the running end pulled through the honda to form a loop. The honda should be small enough so that it covers no more than one-third of the arch over the top of the nasal bones. For an adult boar or sow, a 6.5-10 mm (¼-⅜ in.) nylon rope is suitable as a snout rope. Smaller pigs require cord from 3.2 mm (⅛ in.) to 6.5 mm (¼ in.).

The technique for applying the snout rope varies with the location of the animal. Ideally, if the pig is restricted in a chute or panel arrangement, one can approach from the side and rear. The enlarged loop is worked over the top of the snout until it rests in the mouth (Fig. 11.5). The rope is then pulled back into the mouth with a sawing action. It is usually not difficult to pull the rope between the lips.

If the animal is not confined in a chute but is quiet, it can be approached from the rear and to the side with the loop extended. The loop is held in both hands. The rope is draped over the upper jaw and brought into the mouth near the lateral commissures by a reciprocal sawing motion. Pressure is directed toward the tail at all times. The animal may try to go forward or back up, and the operator must be free to move with the animal.

Once the rope is placed in the mouth, pull the loop tight (Fig. 11.6 left). When the loop is tightened around the upper jaw the animal characteristically pulls back. Thus if the end of the rope is tied to a suitable post or ring in front of the pig, it pulls back, maintaining tension on the loop (Fig. 11.6 right).

It may be difficult to apply such a rope to a dangerous boar or a sow with piglets. Stay in close to the head to pre-

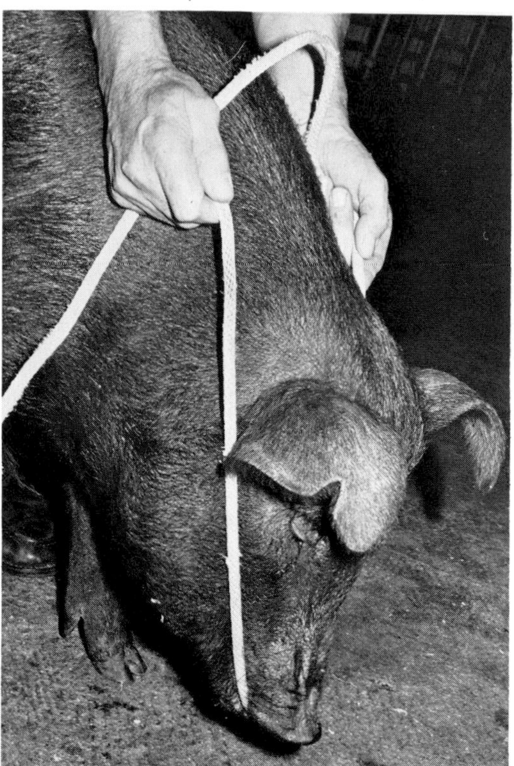

FIG. 11.5. Snout rope placement: Loop is pulled into the mouth.

vent the animal from turning to the side and raking the hand or the leg. It is sometimes difficult to thread the rope loop over the enlarged tusks of a big boar, but it is essential that the rope be placed behind the tusks since this serves as the anchor point for a large animal.

The cable snare is commonly used to capture swine and is safer and easier to manipulate than the snout rope. Usually the snare is formed on the end of a pipe or hollow tube. The handle that pulls the snare closed is on the opposite end, which is grasped by the handler. The cable maintains a previously formed loop, making it unnecessary for the hand to approach the head of the animal. Maneuver the snare over the top of the pig's head and into its mouth, sawing it back and forth until it is in place behind the incisors. The snare is closed by clamping the handle. The animal is held in the same manner as with the snout rope (Figs. 11.7, 11.8).

Once the snout rope or cable snare is in place, constant pressure must be applied forward lest the animal shake the loop from the upper jaw. Making use of the natural tendency of the animal to pull backward, the rope can either be held manually or secured to a post or ring. The snout rope or snare serves the same purpose for swine as the halter does for other species.

Other devices, based on the same principle, are used to hold swine (Fig. 11.9). Hog holders are essentially cable snares. Some have elaborate catches and quick releases. The cable obstetrical snare, designed to deliver a fetus, may be satisfactory for use on swine if made with a firm cable so that the loop stays open while being manipulated onto the upper jaw.

Do not leave a pig tied with a snout rope or cable snare for more than 15–20 minutes. A pig will pull back with sufficient force to produce a tourniquet effect around the upper jaw. In addition, the animal may chew through a rope and free itself. Baby pigs do not respond well to this technique and are likely to refuse to pull back if such a snout rope is applied. A pig will become wary after a snare has been used repeatedly; placement of the snare will become difficult.

Long-handled neck tongs can be employed to grasp a pig in order to place the snout rope (Fig. 11.10). The pig is approached from the rear and the tongs placed just behind the ears. The pig automatically pulls back and squeals, allowing the loop to be inserted in the mouth; if the animal is in a relatively confined position, the tongs fix the head, making it easier to place the snout rope or snare. The tongs are then removed. Tongs are used for medium-sized swine up to approximately 40 kg (90 lb).

Although even a tiny pig is capable of inflicting a severe

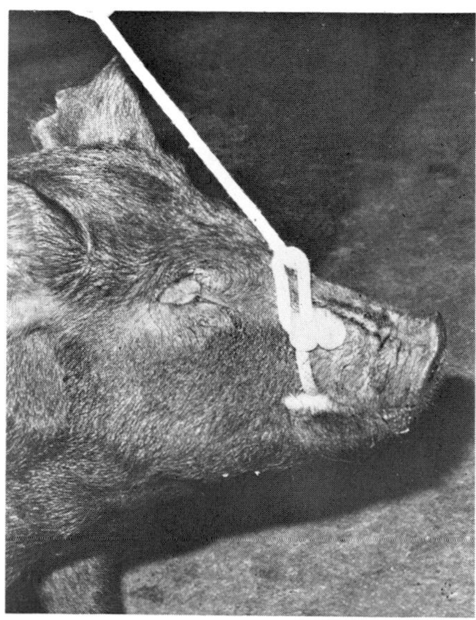

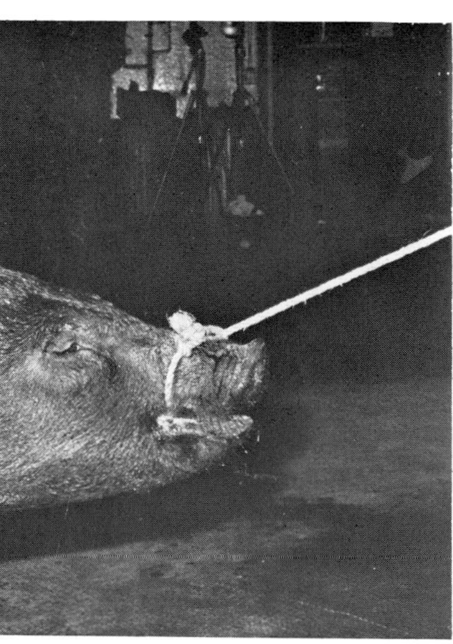

FIG. 11.6. Snout rope placement *(cont.)*: Loop is pulled tight from behind *(left)*. Free end of rope is brought forward to anchor the animal *(right)*.

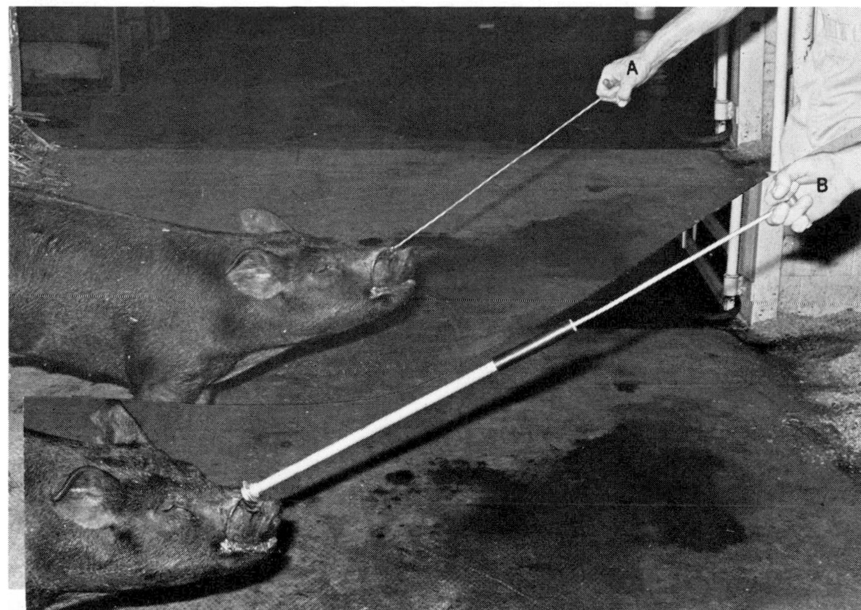

FIG. 11.7. Different types of snout snares: **A.** Obstetrical snare. **B.** Special guard to keep the loop open until it is in the mouth.

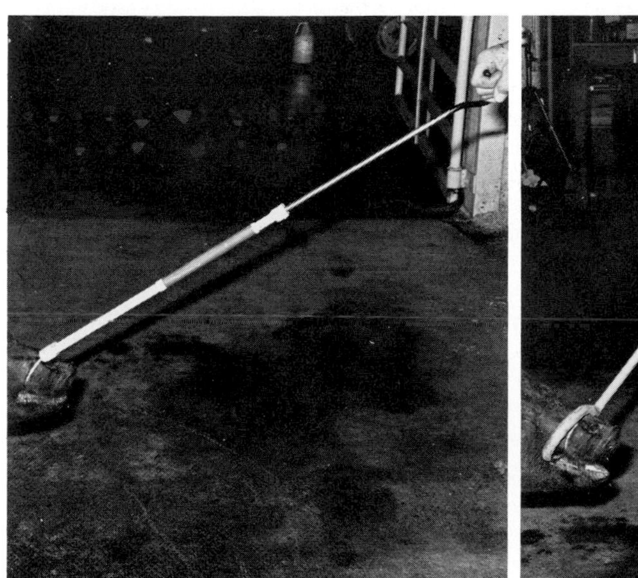
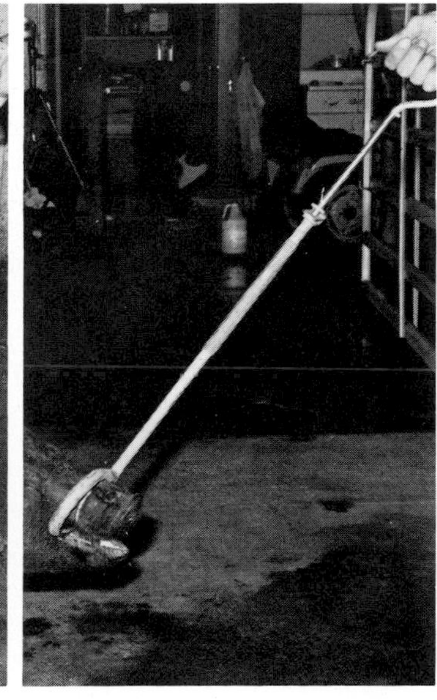

FIG. 11.8. Additional snout snares.

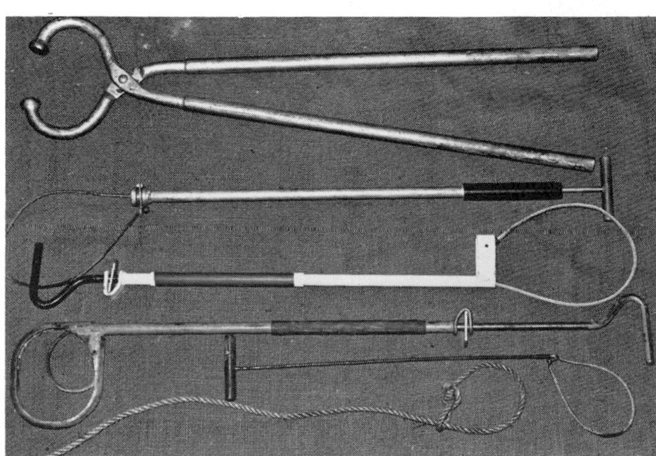

FIG. 11.9. Pig tongs and snares.

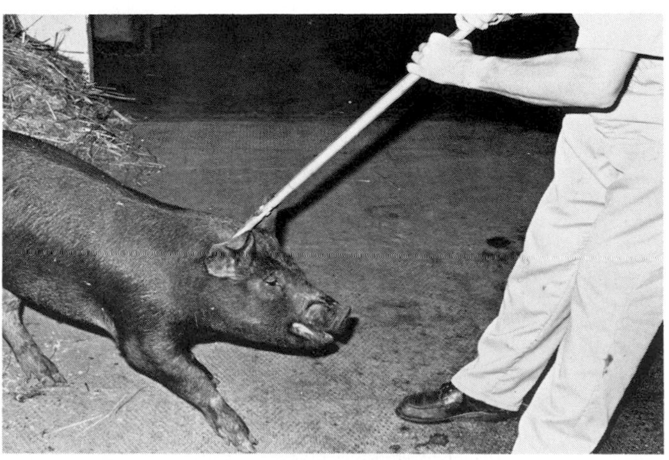

FIG. 11.10. Hog tongs placed behind ears of a pig will temporarily immobilize the animal.

bite, most newborn piglets do not resist handling if grasped gently. It is important to remember that a small pig squeals when taken from its mother. The squealing elicits a dramatic response from the sow, and if she can reach the handler she is likely to inflict serious injury. Therefore piglets should be separated from the sow prior to being picked up and handled. This is most easily accomplished by driving the sow into a farrowing chute where the piglets are free to move away from the sow because the fence does not reach to the ground. The sow is confined to the chute and cannot attack a handler picking up the small pigs. To minimize the sow's agitation and distress, move the piglets out of her hearing before manipulating them. A newborn piglet can be picked up by the tail (Fig. 11.11 left), which minimizes squealing. After the piglet gains 1-2 kg (2-4 lb), include one hind leg to prevent tail damage (Fig. 11.11 right). Larger piglets are handled by supporting the body under the chest (Fig. 11.12).

Pigs less than 28 kg (60 lb) should be confined to a small pen or alleyway before individual handling is attempted (Fig. 11.13). Pick up a pig by grasping above the hock and lifting the pig off its feet (Fig. 11.14 left). Be careful not to twist the leg as this may dislocate the hip joint. The pig can be held with the head up as well (Fig. 11.14 right). This is a suitable position for minor surgical procedures such as castration or vaccinations. The handler should keep in mind that a pig held in this way is able to bite. However, if the animal's head is confined by the handler's legs, this is not an undue hazard.

Larger pigs (45-56 kg) (100-125 lb) can be handled in a similar manner, except that it requires two people, each grabbing one of the hind legs, to lift it off the ground and hold it for a short time. Pigs over 68 kg (150 lb) are usually not captured or held in this manner. They should be placed in a small enclosure and handled with a snout rope or snare. If it is necessary to cast a pig or place it in lateral recumbency, one of the following techniques can be used. All require securing the pig with a snout rope before beginning.

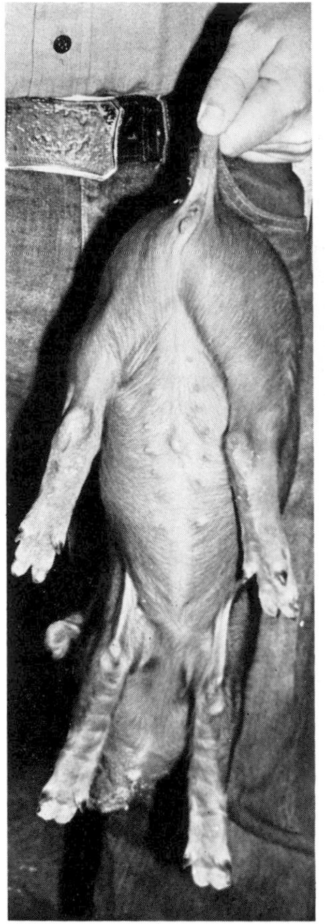
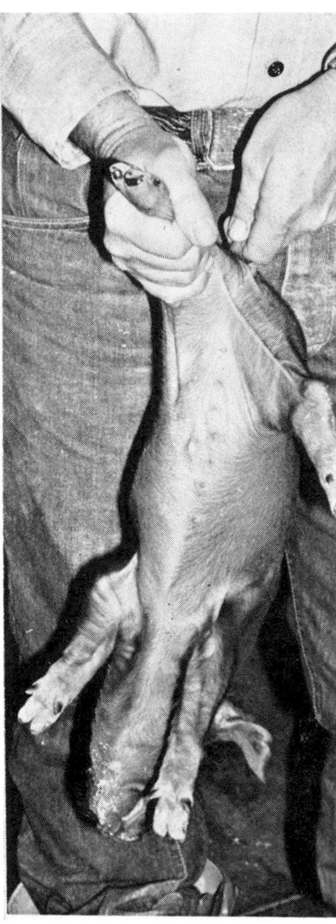

FIG. 11.11. Alternate methods of lifting a baby pig: Tail only (used only on newborn piglets) *(left)*. Hind leg and tail *(right)*.

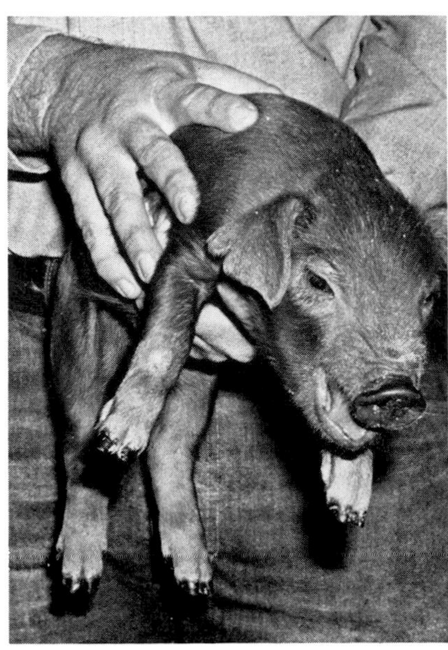
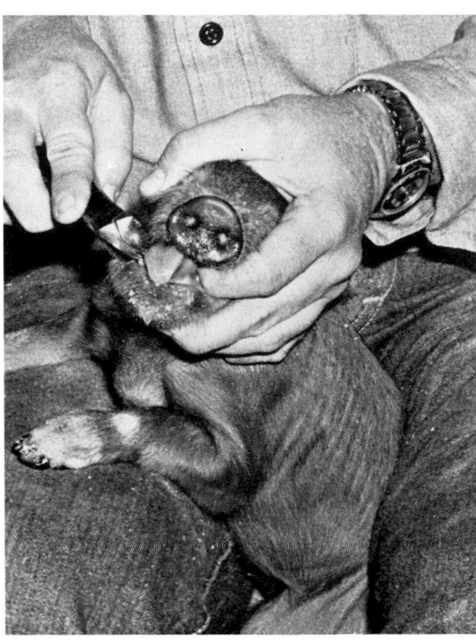

FIG. 11.12. Handling baby pigs: When lifting, the hands should be under and over the body *(left)*. Restraint for clipping "needle" teeth *(right)*.

FIG. 11.13. Squeezing pigs into small area with panel gates and shields.

1. The first technique utilizes a shackle. The shackle can be temporarily constructed of rope and a 50 × 100 mm (2 × 4 in.) board. A permanent shackle can be made by welding rings to an iron pipe through which loops of chain attached to the shackle are threaded (Fig. 11.15). Place the loops on the hind limbs, above the hocks if possible. Each hind leg is lifted in turn and placed through the loops

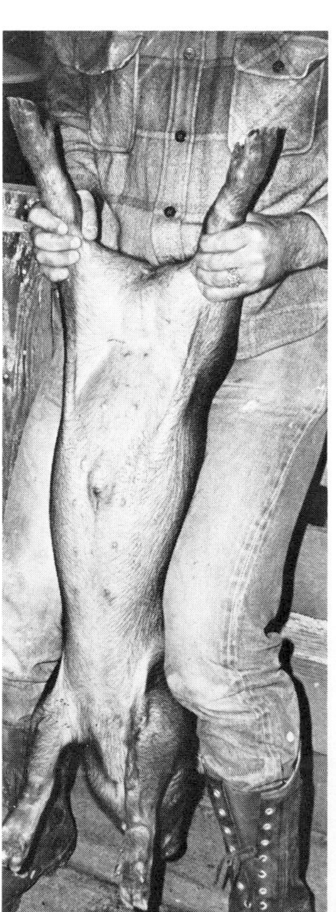

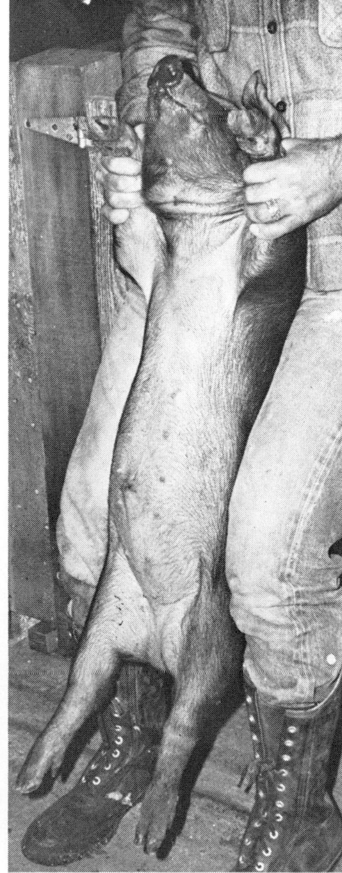

FIG. 11.14. Holding a medium-sized pig.

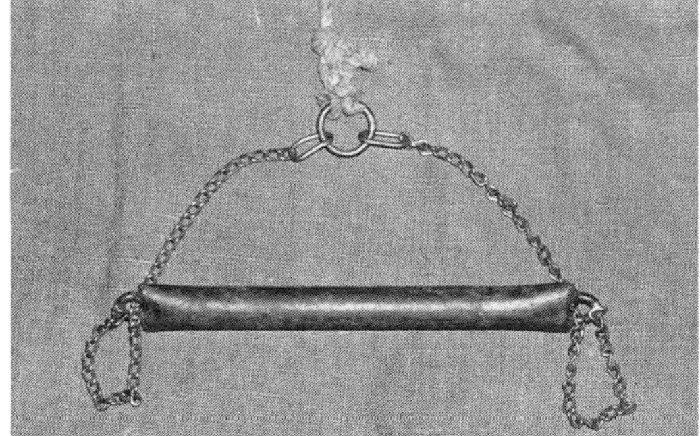

FIG. 11.15. Hog shackle.

(Fig. 11.16). It is not difficult to lift the leg of even a large boar for a sufficient length of time to do this. When both loops are in place, attach another rope to the center pull and stretch the animal. The pig will roll over on its side and can be maintained in either lateral or dorsal recumbency by applying tension (Fig. 11.17). A trucker's hitch is suitable to maintain tension. Once the animal is down, the legs can be maintained in the stretch position for minor surgery such as castration or repair of umbilical hernias.

2. A large pig can be cast by placing a hitch around the hock or using a hock hobble (see Fig. 8.53). Maintaining tension, bring the snout rope around through the hock loop and pull the hind leg up to the snout. The snout rope can be used to surround the hock as well (Fig. 11.18). The operator stands on the side opposite to the hobble and by pulling on the snout rope forces the animal to lie on its side. This is suitable restraint and positioning for castrating a boar.

3. For the third casting technique, place a short rope on both foreleg and hind leg on the side of the animal you wish it to lie when cast. Take the ropes beneath the body of the animal, up the opposite side, and over the back (Fig. 11.19). The handler stands on the side of the animal to which the ropes are secured and pulls on the rope over the top of the body. This procedure pulls the legs out from under the pig. The animal lies on its side but is otherwise relatively free and must be further secured (Fig. 11.20). An animal can be held by placing a knee on the neck. Use a local anesthetic for painful procedures. For prolonged operations, the legs must be secured and stretched to prevent them from flailing and injuring people.

Various types of troughs or cradles are used to restrain pigs in dorsal recumbency (Fig. 11.21). A bar or rope placed over the snout keeps the head down; the hind limbs are secured by a short length of sash cord. This method of restraint, coupled with either local or general anesthesia, is suitable for abdominal surgery in swine. The animal can also be bled from the anterior vena cava while in this position.

Commercial squeezes are available for handling swine. Pressure may be exerted on the neck, body, and buttocks to keep swine in the squeeze (Fig. 11.22). These devices are useful when working with adult or market swine that must be bled either from the ear or from the anterior vena cava.

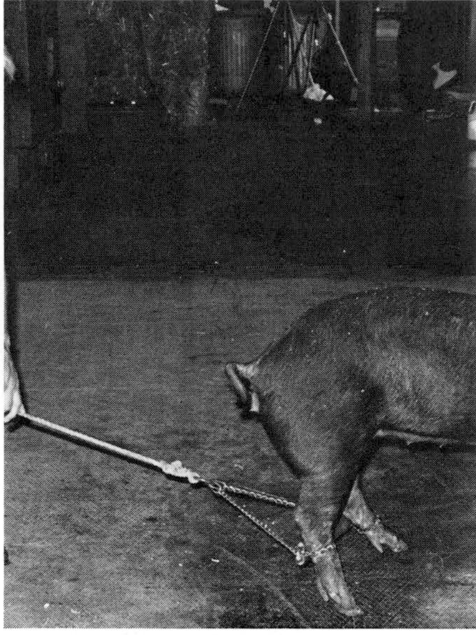

FIG. 11.16. Casting technique using a shackle: Form loops at ends of shackle *(left)*. Lift each foot and place a loop around a leg *(right)*.

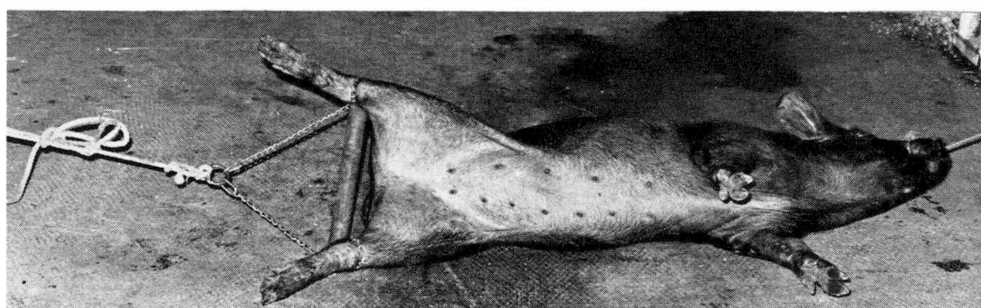

FIG. 11.17. Casting with a shackle *(cont.)*: Loops in place above the hock of a large pig *(upper)*. Animal cast and stretched, secure with a trucker's hitch *(lower)*.

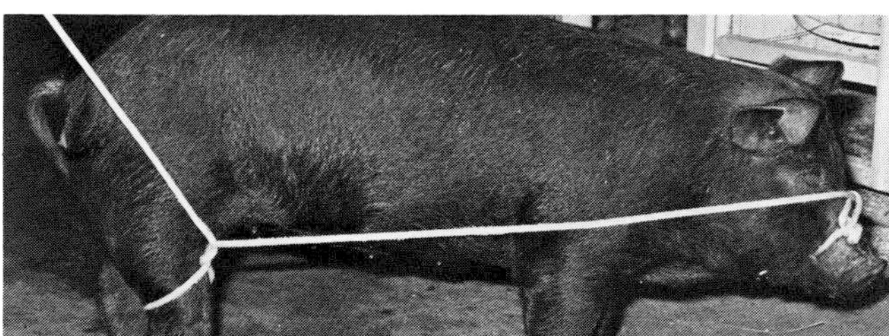

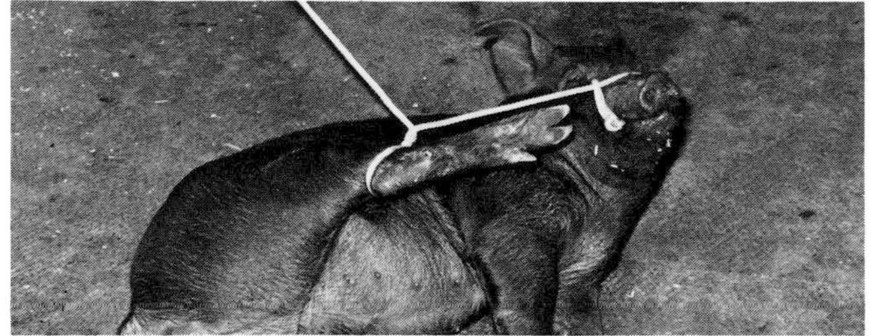

FIG. 11.18. Casting procedure utilizing snout rope and half hitch above the hock.

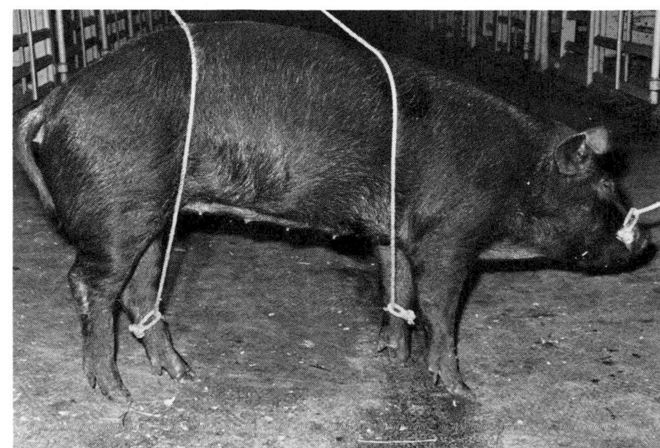

FIG. 11.19. Casting procedure for large swine: Secure with snout rope. Then place leg ropes.

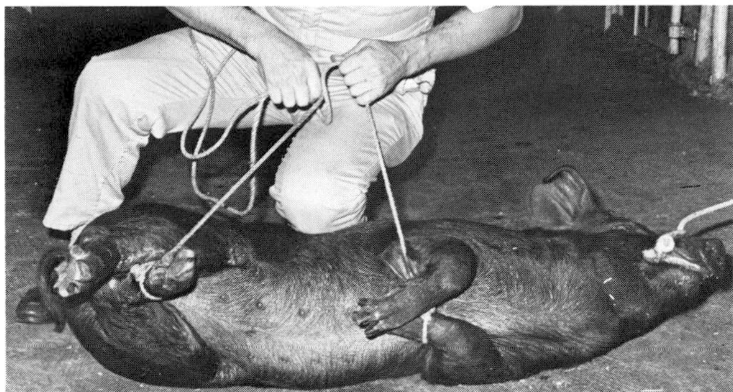

FIG. 11.20. Casting large swine *(cont.)*: Pull legs out from under the animal and roll it onto its side.

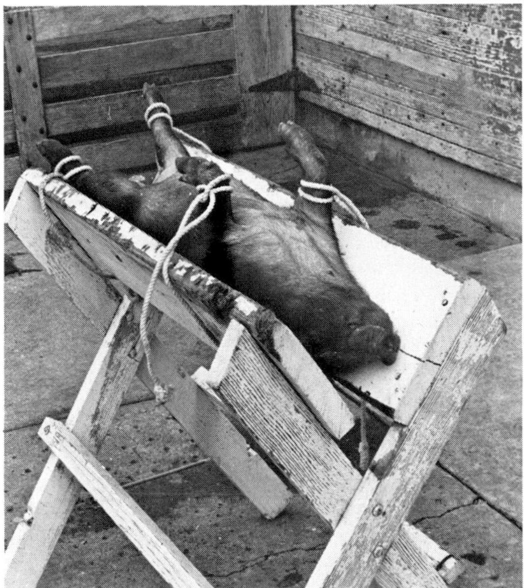

FIG. 11.21. Trough restraint for small pigs.

Examining the mouth of the adult pig is difficult. Place a snout rope on the animal and allow the pig to pull back against it. Open the mouth with a speculum and examine the mouth with a flashlight.

In a small pig, the oral examination can be conducted by inserting a speculum into the mouth from the position

FIG. 11.22. Portable squeeze chute for swine.

illustrated in Figure 11.14 (right). To control sideways motion of the head, grasp the forelegs and the ears together in one hand.

TRANSPORT

Swine are easily transported in crates, trailers, trucks, and trains. Hyperthermia is a constant threat.

CHEMICAL RESTRAINT

Ketamine hydrochloride (10-20 mg/kg) and phencyclidine hydrochloride (0.5-1 mg/kg) are excellent sedatives and restraint agents for swine. Precautions must be taken to prevent injection into fatty layers. Other agents used in swine include a combination of fentanyl (0.33 mg/kg) plus azaperone (10 mg/kg). Tilazol (2-8 mg/kg) is also used.

SUPPLEMENTAL READING

Dunne, H. W., and Leman, A. D., eds. 1975. Diseases of Swine, 4th ed. Ames: Iowa State Univ. Press.

12 DOGS

CLASSIFICATION

Order Carnivora
Family Canidae: dog

POSSIBLY the dog has been domesticated longer than any other animal. We have manipulated the gene pool of this species until greater physical and behavioral variations exist from breed to breed than in any other species of animal (Table 12.1). Names of gender are listed in Table 12.2.

TABLE 12.1. Weights of selected breeds of dogs[1]

Breed	Male		Female	
	(kg)	*(lb)*	*(kg)*	*(lb)*
Toy				
Pekingese	3.2-5	7-11	3.6-5.4	8-12
Toy poodle	< 2.3	< 5	< 2.3	< 5
Pug	6.3-8.2	14-18	6.3-8.2	14-18
Chihuahua	< 2.3	< 5	< 2.3	< 5
Pomeranian	1.8-3.2	4-7	1.4-2.7	3-6
Dachshund	5	11	5	11
Medium				
German shorthair pointer	25-32	55-70	23-30	50-65
Springer spaniel	22-25	49-55	20-23	45-50
Afghan	27	60	23	50
Labrador retriever	27-34	60-75	25-32	55-70
Greyhound	30-32	65-70	27-32	60-70
German shepherd (Alsatian)	36-57	80-125	< 45	< 100
Pointer	25-34	55-75	20-30	45-65
Giant				
Irish wolfhound	> 54	> 120	48	> 105
St. Bernard	68-90	150-200	< 80	< 175
Great Pyrenees	45-57	100-125	40-52	90-115
Great Dane	> 54	> 120	45	> 100

TABLE 12.2. Names of gender of dogs

Gender	Name
Male	Dog
Female	Bitch
Newborn and young	Pup, puppy

DANGER POTENTIAL

A dog's only weapons are its large canine teeth and to a lesser extent the toenails. Scratches from a dog are usually not serious, but a bite from a large Alsatian or St. Bernard can both disfigure and disable. Every precaution should be taken to prevent injury. Approach any strange dog with caution.

If a dog attacks, protect the face with a hand or an arm. Some dogs have trained offense responses (guard dogs). These animals should be handled only by competent personnel.

BEHAVIORAL CHARACTERISTICS

The behavior of a dog is determined by breed, training, previous disagreeable experiences, and degree of human association.

Canids have been studied in depth by animal-behaviorists; numerous texts elucidate various aspects of canine behavior [2,3,4]. Adverse behavioral patterns usually develop as the result of abuse or lack of understanding on the part of a dog owner.

In this book dogs are classified in the following categories for restraint purposes:

1. The stray or free-roaming dog that has little association with people except, perhaps, when fed. If owned, the owner may be afraid of the dog. These dogs must be handled as if they are wild canids because they are liable to bite with the slightest provocation.

2. The well-cared-for pet or working dog. Fortunately most dogs are in this category. They are docile and respond to a low-pitched soothing voice and slow, deliberate handling.

3. The extremely nervous, frightened dog. This dog can be recognized by an anxious expression, rapid movements of the head, and constant pricking of the ears in response to every sound or movement. The head will likely be ducked and the animal may cower in a corner. The lips may be pulled back in a grimace. These animals may also be boisterous and attempt to nip at the handler. Above all, they can be expected to bite in response to almost any type of approach. All these signs telegraph "beware!" to a perceptive handler.

There is little question that there are neurotic and even psychotic dogs which are unpredictable, to say the least.

4. The vicious, aggressive dog. Many of these dogs are large, are capable of inflicting serious injury, and will bite with little or no provocation. These dogs do not always exhibit aggressiveness in an obvious manner but signs of potential viciousness can be seen. The head is held low and these dogs will not look directly at you. They may attack without warning.

Rough handling may provoke adverse responses in the most amiable of dogs. Gentleness is wisest when dealing with any dog. Confidence and calmness are necessary for successful control.

It should be recognized that the temperament of a dog may change drastically when it is sick or injured. Such a dog is far more apt to bite than a healthy dog.

Kennel owners and others who must handle a variety of dogs day in and day out learn to understand dog behavior. They become adept at performing many procedures with no more restraint than talking to and soothing a dog into a relaxed state. The casual or infrequent dog handler will find it more difficult to carry out manipulative procedures and may have to resort to more stringent restraint practices.

A technique used by some veterinarians is to stand between the client or owner and the dog during an examination. The dog is not as likely to try to move away, but will tend to attempt to get closer to the owner and thus be moving toward the veterinarian.

Keep in mind the fact that dog owners often resent

anyone who handles an animal in what appears to them a rough manner, but many owners are incapable of holding a pet in such a manner that it cannot bite the person attempting to examine or treat it. Furthermore, some pet owners cannot be led to believe that a usually docile pet may respond to a strange or frightening situation by biting someone. Pet owners are seldom satisfactory assistants to the handler. A veterinarian may be liable if a dog bites its owner during an examination.

PHYSICAL RESTRAINT

The leash is an important device. The leather and chain leashes used by owners when they walk their dogs are rarely suitable for restraint. The light leather strap can be bitten through by a frightened or angry dog, and the swivel attachments are usually flimsy.

Collars vary greatly in design. Their function is to carry identification plates or tags, kill fleas, or look pretty (Fig. 12.1). It is virtually impossible to place a collar tightly enough around the neck to preclude a determined dog from slipping out of it. The only collar that is truly safe for restraint is a choke chain (Fig. 12.2). In lieu of a choke chain, a nylon cord 3 mm (1/8 in.) in diameter may serve both as a collar and a leash.

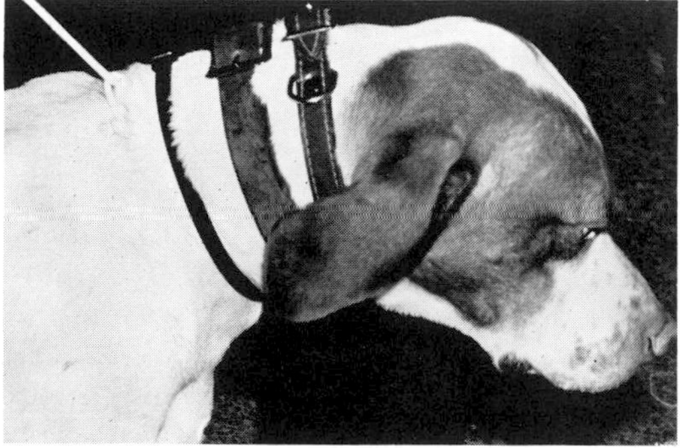

FIG. 12.1. Various collars for dogs. Second from the left is a flea collar.

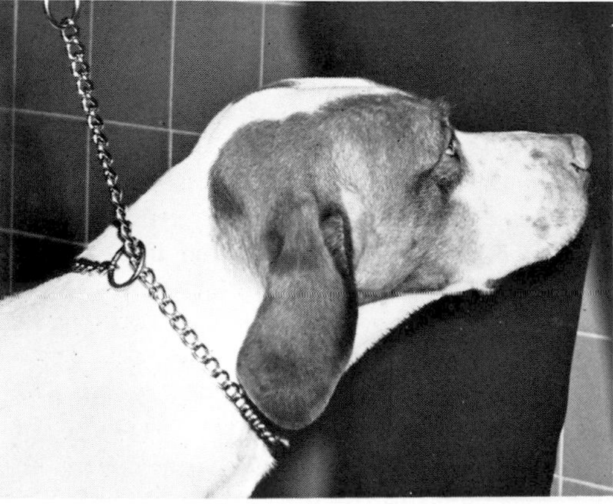

FIG. 12.2. Choke chain used to train and control dogs.

The collar and leash are used in many restraint situations. If a small dog is cowering in the back of a cage, a loop tossed over its head can be used to pull the dog off balance and to hold the head away from the handler while the dog is grasped. The aggressive, vicious dog can be subdued to administer a sedative by threading the leash through an eye bolt embedded in the wall and pulling the head up tight. Another person grabs the tail or a hind leg and makes a quick intramuscular injection. Do not wrap the leash around a table or chair leg; a medium-sized or large dog can easily upset a table. If an eye bolt is not available, the leash can be slipped through a partially opened door which is then lightly closed on the leash. The dog is pulled up tightly to the door which is held shut by the handler outside the door while the second person administers the sedative. Intravenous injections into the saphenous vein can be made with the dog held in this position.

Small dogs may be fitted with leather harnesses that surround both neck and thorax. These are excellent for pet owners whose dogs are attacked by other dogs when they are out walking. The owner can lift the pet into the air, away from another dog, without choking it.

Puppies of any breed can be handled with ease. Place one hand under the abdomen and chest to give support (Fig. 12.3). In almost all instances puppies respond quietly, offering no resistance. If the puppy squirms to free itself, place a hand over its back and bring it in close to your body for additional comfort and support.

Always talk to a dog as you approach it to pick it up. Approach from the side rather than directly from the front. Pick up small and middle-sized adult dogs as you would a puppy.

Removal of a dog from a cage is an art. The cage is its familiar territory, even if only temporarily. One should not play games with a vicious dog. Snare or net the dog to pull

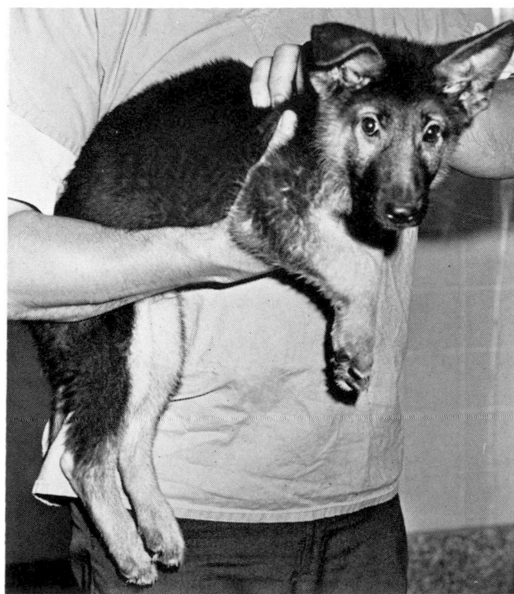

FIG. 12.3. Proper way to lift and carry a puppy or small dog.

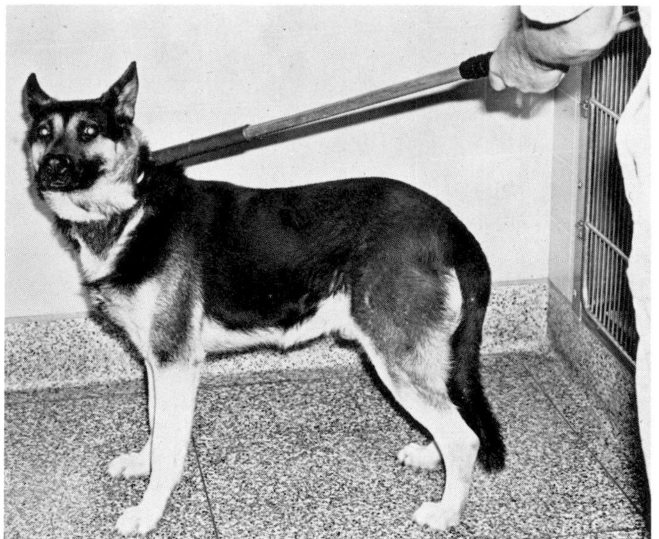

FIG. 12.4. Snare used to remove vicious dog from a cage.

it out of the cage, where a muzzle may be applied or a sedative given (Fig. 12.4).

The frightened or nervous dog is another matter. One should distract or confuse the dog to divert its intentions to bite. Do not reach out with a hand on first approach. Sit down close to the dog. The size of your body may intimidate it. You may then be able to soothe and calm the dog by speaking to it. Some handlers approach a small, nervous dog without looking directly at it. This confuses the dog. Another technique is to use the leash as described previously. One can also apply a muzzle or place a towel over the dog to partially blind it (Fig. 12.5).

To lift a medium-sized to large dog from the floor, squat next to the dog and place one arm around the dog's chest in front of the forelimbs. The other arm is placed under the abdomen. Pull the dog toward your body to shift the weight nearer to your legs. Then stand up (Fig. 12.6). If the dog struggles, pull it tightly against your chest or set the animal down on a table (Fig. 12.7). Reverse the process when releasing the dog; do not drop it or let it jump from the table.

Vicious dogs cannot be handled in this manner lest the handler be bitten. The owner or the customary handler of the animal may be able to muzzle it. Leash the animal before attempting to place the muzzle. A muzzle can be constructed from a piece of 5 cm (2 in.) gauze bandage or a small cord. Form a loop large enough to drape over the mouth of the animal, keeping the hand some distance away (Fig. 12.8). The leash prevents the dog from backing away when approached.

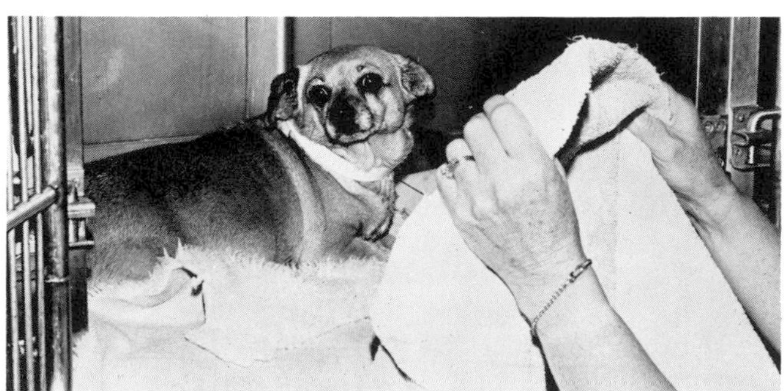

FIG. 12.5. Nervous or frightened dog may be calmed by placing a towel over it.

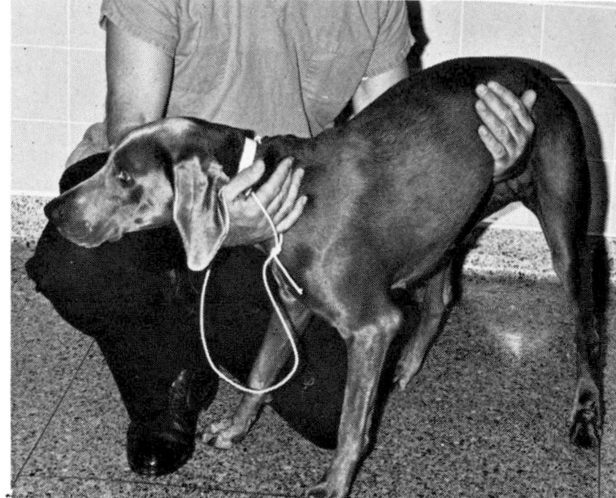

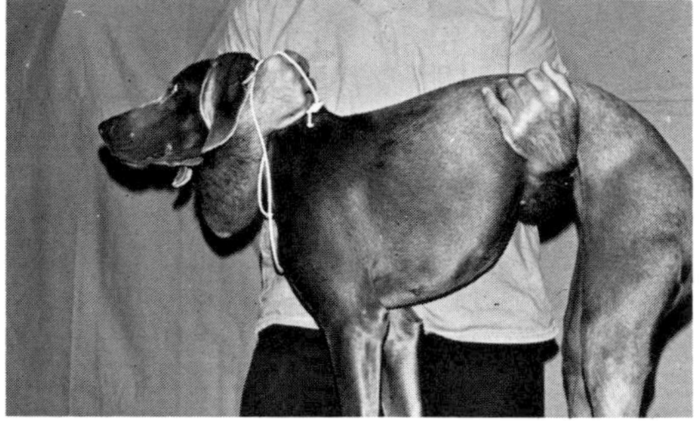

FIG. 12.6. Proper method of lifting a gentle dog.

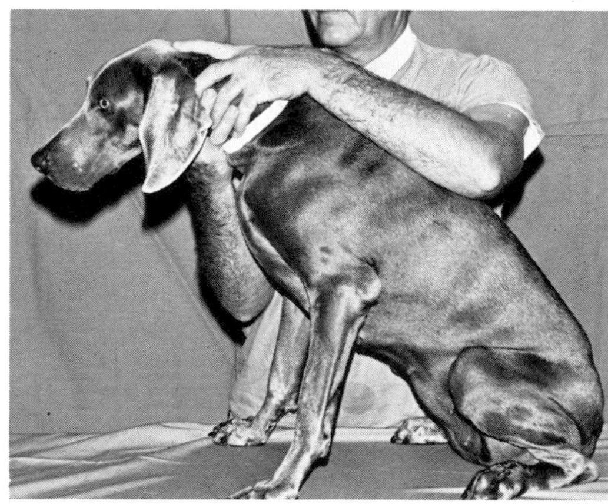

FIG. 12.7. Dog restrained on a table.

Dogs that are muzzled repeatedly may learn the trick of blowing on the light gauze loop as one attempts to place it over the jaws. Counter this by wetting the gauze.

With the muzzle properly in place, the dog cannot bite. Snug the knot sufficiently tight to preclude partial opening of the mouth. If the loop is anchored too close to the nostrils, the dog may paw the loop off or the nostrils may be clamped shut. Tie the muzzle at the back of the neck with a quick-release bowknot (Fig. 12.9). A dog may strug-

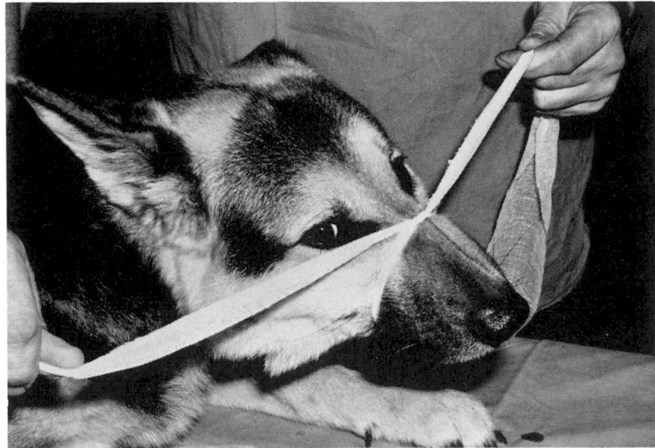

FIG. 12.8. Muzzling a dog with gauze bandage.

gle to the point of collapse from hypoxia or it may vomit. Both necessitate prompt release from a muzzle.

Remember that cooling mechanisms of the canid species involve panting and lolling of the tongue. Muzzling prevents both. Therefore do not work with a muzzled dog in a hot environment for prolonged periods.

Leather muzzles available commercially are not suitable for restraint unless fitted precisely to an individual dog. In most cases, there is enough play in the muzzle to allow partial opening of the mouth, and thus pinch biting. The wire muzzles used on racing greyhounds are suitable.

If a muzzle is unavailable and the manipulative procedure is short, the dog can be hand muzzled. Long-nosed breeds of dogs are easily handled in this manner. Brachycephalic or pug-nosed breeds are more difficult to grasp. Place the thumb over the bridge of the nose with the hand cupped around the lower jaw (Fig. 12.10 left). The middle finger is flexed and inserted between the rami of the mandible (Fig. 12.10 right). With the hand firmly placed, the finger between the rami forms an additional securing lock which prevents the head from pulling free. If this technique is prolonged or harshly administered, the dog may resist as though it were being strangled. Use it judiciously.

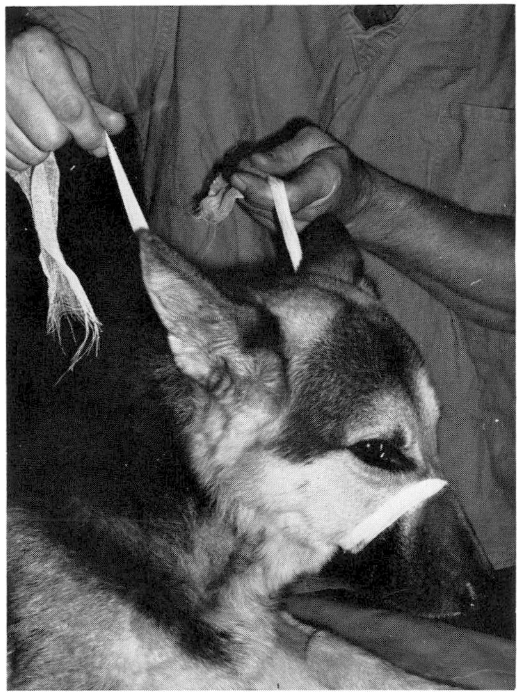

FIG. 12.9. Completing the muzzle: The strand ends are wrapped around the dog's muzzle, crossed beneath the jaw, and tied in a bow behind the ears.

To prevent the dog with a short muzzle from biting, surround the whole head with the hands. Be cautious not to cover the nares since this will inhibit breathing.

The dog's mouth can be easily opened for examination or medication. Place a hand over the bridge of the nose and pull the upper lips upward (Fig. 12.11). The head should be tilted up, which minimizes the dog's ability to bite. The other hand pulls the lower jaw open. If a tablet or capsule is to be given, it can be dropped at the base of the tongue. Gravity will aid in proper placement. A quick

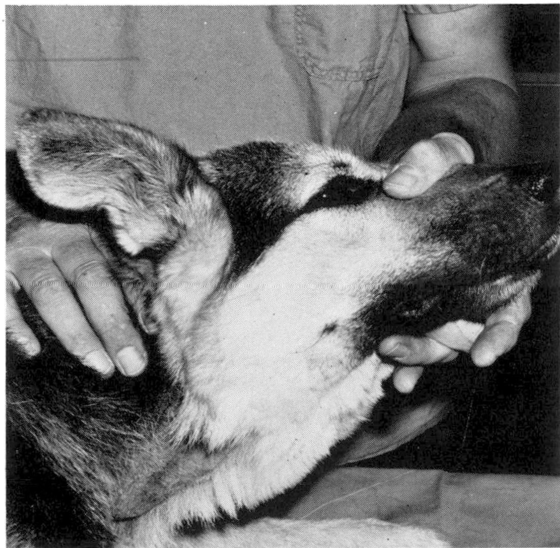

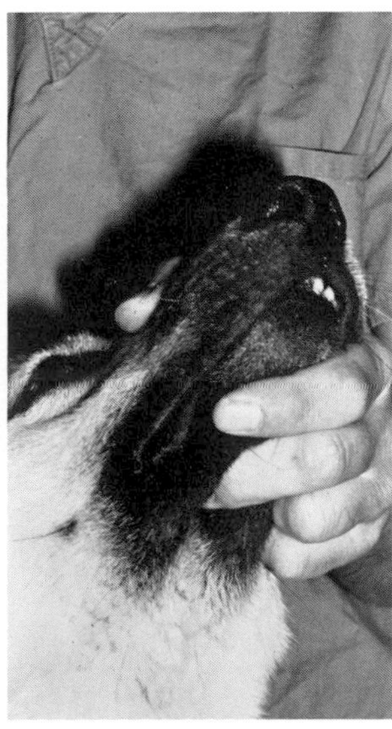

FIG. 12.10. Hand-muzzle control: Thumb over bridge of the nose and fingers beneath the jaw *(left)*. Finger inserted between rami of the mandible *(right)*.

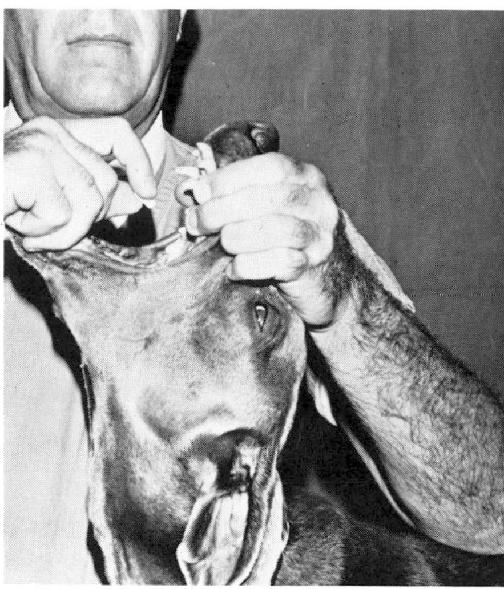
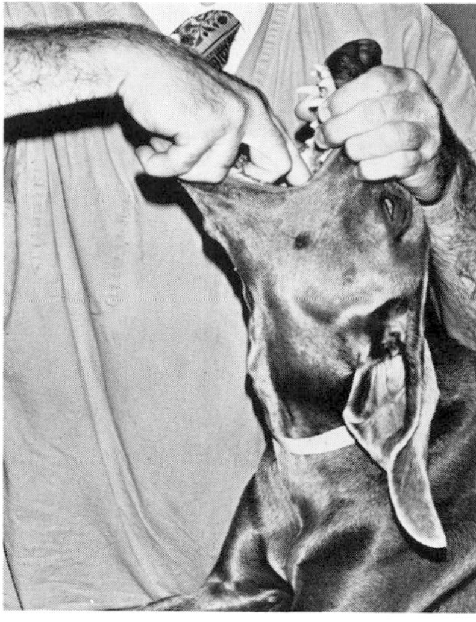

FIG. 12.11. Administering tablet medication to a dog.

thrust with the index finger will push the tablet over the tongue and initiate the swallowing reflex. Pull the fingers out quickly and release the hold on the jaws. If the dog swallows the tablet, the tongue will flick out between the teeth and lips as a flag that the mission is accomplished. If this does not occur, rest assured that the tablet will be spit out later.

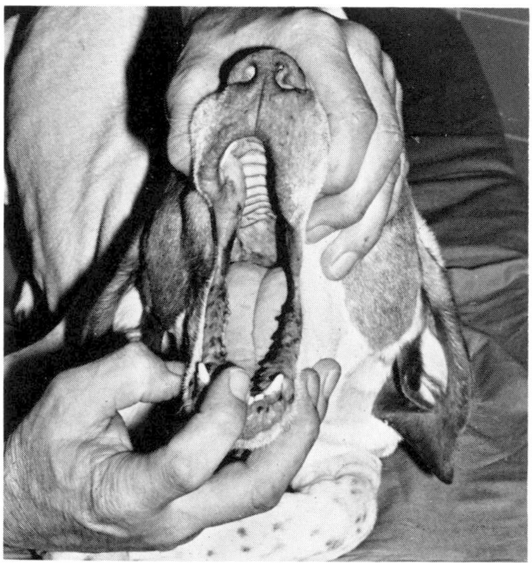
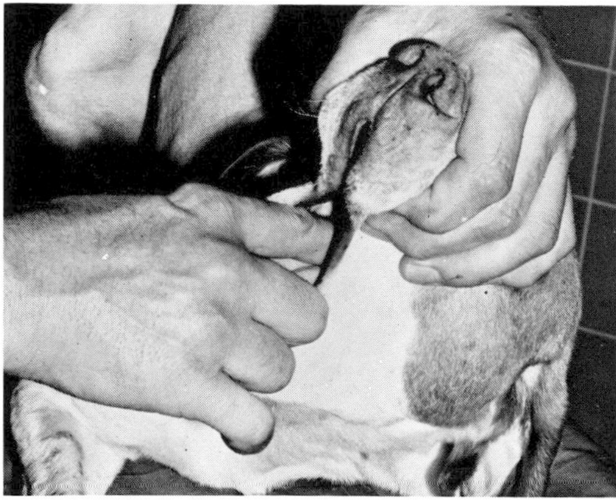

FIG. 12.12. Improper method of examining mouth and pilling.

A word of caution. Do not press the dog's lips over its teeth with your fingers (Fig. 12.12). The dog may bite its own lips and your fingers at the same time. Keep your fingers free of those crushing molars.

If it is necessary to keep the mouth open for an extended period, a special dental speculum (Fig. 12.13) or dowel speculum (Fig. 12.14) can be used.

Liquid medication can be given by depositing the fluid inside the lips at the commissure of the mouth (Fig. 12.15A). Tilt the head back slightly and form a pouch with the cheek. Deposit a small quantity of the medication to initiate licking and swallowing. This is signaled by the tongue flicking out of the mouth (Fig. 12.15B). If the dog fails to swallow, deposit some of the medication on the tip of the nose (Fig. 12.15C). Only after the dog is swallowing should a large quantity be deposited.

Elizabethan collars prevent self-mutilation (Fig. 12.16). These can be placed on the animal in either normal or reverse position. Similar collars can be improvised from sheets of plastic, large bottles, or buckets.

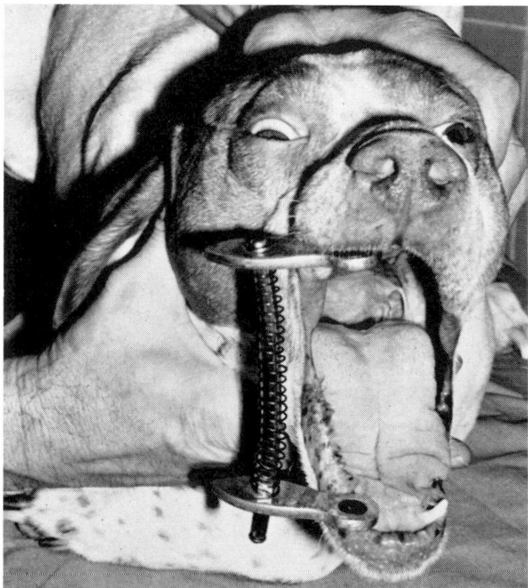

FIG. 12.13. Canine mouth speculum in place for oral examination.

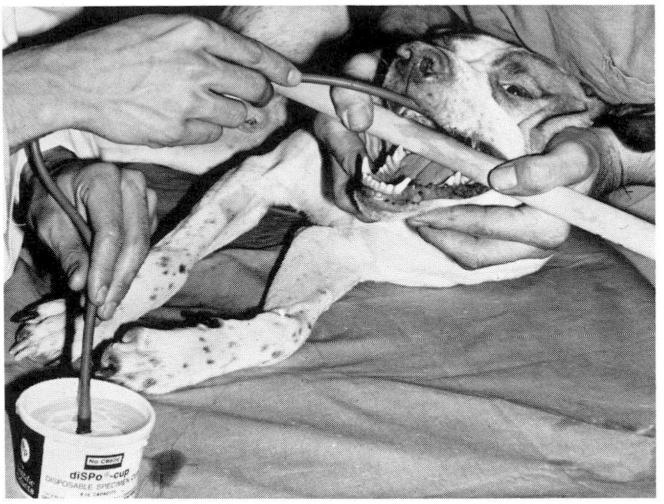

FIG. 12.14. Dowel speculum holding the mouth open for passage of a stomach tube.

Specimens of blood can be obtained from the cephalic vein (Fig. 12.17), the jugular vein (Fig. 12.18), or the saphenous vein, in which case the dog is restrained in lateral recumbency (Fig. 12.19).

Miscellaneous Procedures

Dogs vary tremendously in size. Accurate weight determination is necessary before medications are administered. Small dogs can be held and weighed on a bathroom scale (subtract your own weight). Larger dogs must be placed on a platform scale. Small to medium-sized dogs can be suspended and weighed from a hanging scale (Fig. 12.20).

Once anesthetized, the dog is secured to an operating table as illustrated in Figure 12.21. Short pieces of sash cord are placed over the limb above the knee and above the hock if possible. A half hitch placed over the limb below the loop secures the animal. This technique does not exert a tourniquet effect on the limb. Stretch the legs into the position desired. Keep in mind that an anesthetized

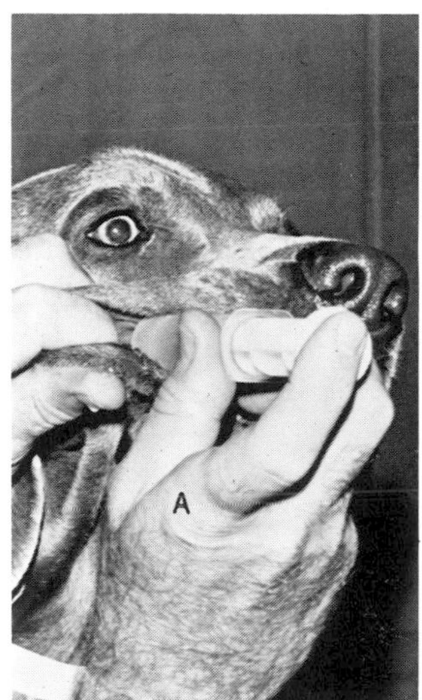

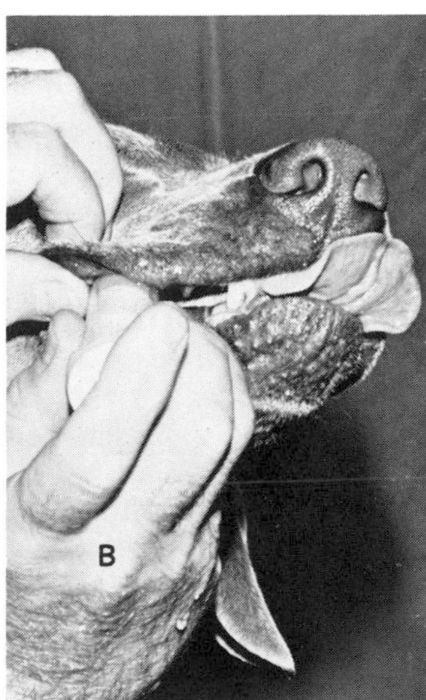

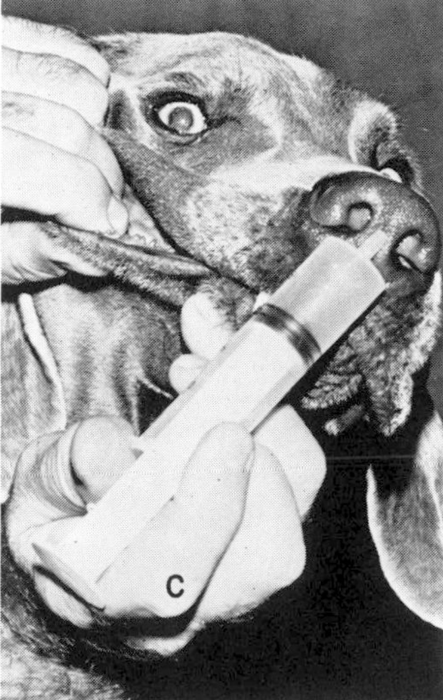

FIG. 12.15. Administering liquid medication to a dog: **A.** Form a pouch with the lips. **B.** Deposit small amount and wait for tongue to flick. **C.** If dog will not swallow, place a drop on the nose.

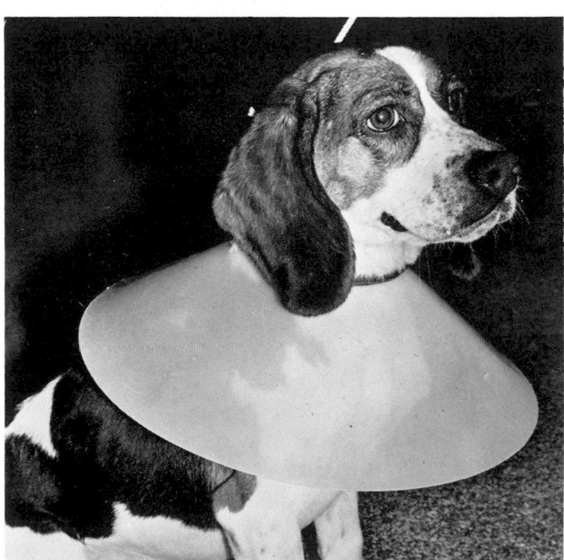

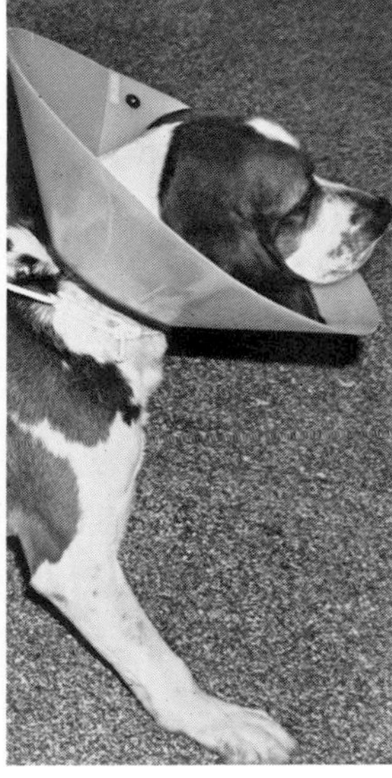

FIG. 12.16. Elizabethan collars.

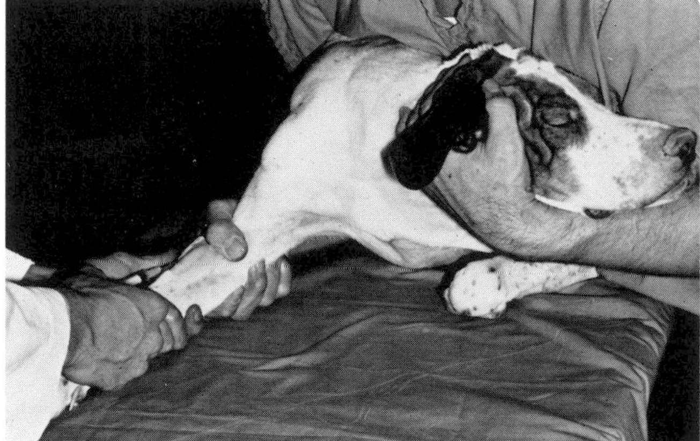

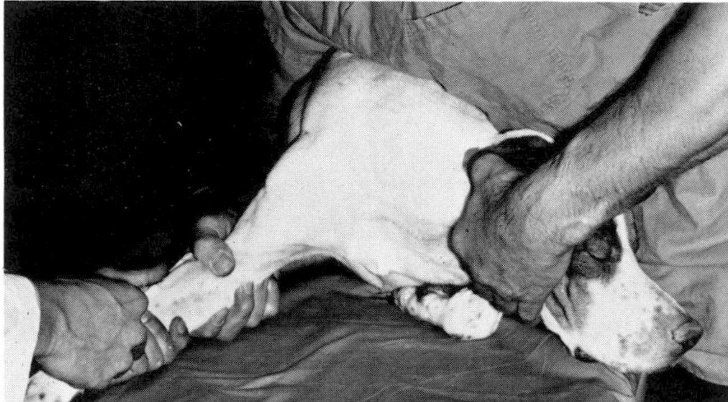

FIG. 12.17. Alternate methods for holding a dog for cephalic intravenous injections.

animal may be twisted into positions jeopardizing circulation of the limb or which overstrain joints and ligaments. Make the stretching as comfortable as possible. Deep-bodied animals can be secured more easily by placing sand bags on either side to prevent them from twisting over.

Cradles of various types are used to support the dog or to keep it in different positions to facilitate surgical procedures.

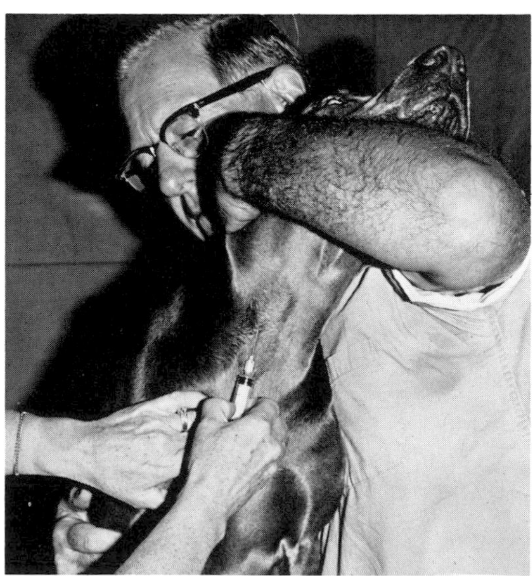

FIG. 12.18. Dog being restrained for collecting a blood sample from the jugular vein.

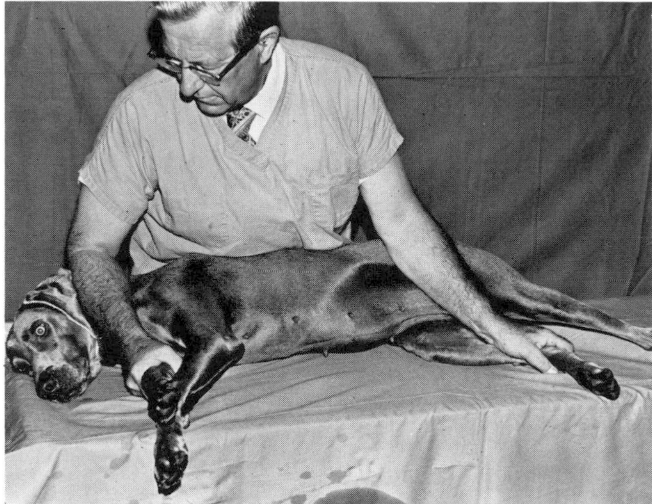

FIG. 12.19. Dog restrained in lateral recumbency.

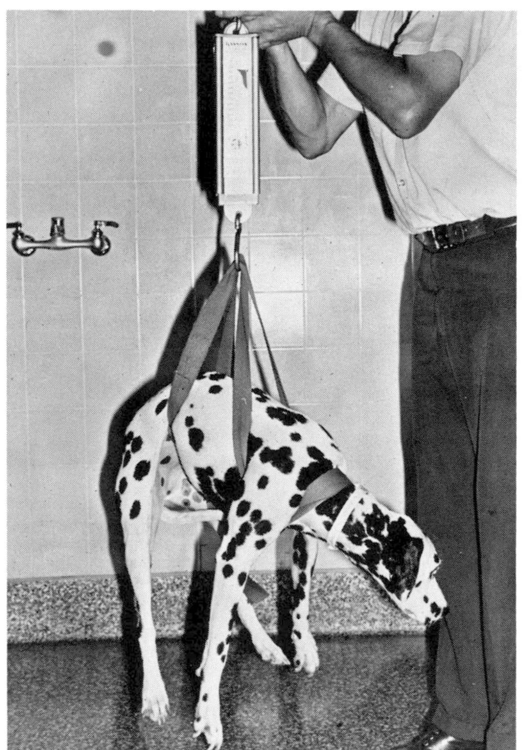

FIG. 12.20. Weighing a dog with a hanging scale.

Unfortunately dogs are frequently traumatized in automobile accidents. Be cautious when attempting to render first aid. The most docile dog is liable to bite when dazed or in shock. Try to leash the dog first, perhaps with a belt. A muzzle can be improvised with a necktie or shoelace. Try to ease the dog onto a coat or blanket. Then lift the dog by the coat to place it in a vehicle.

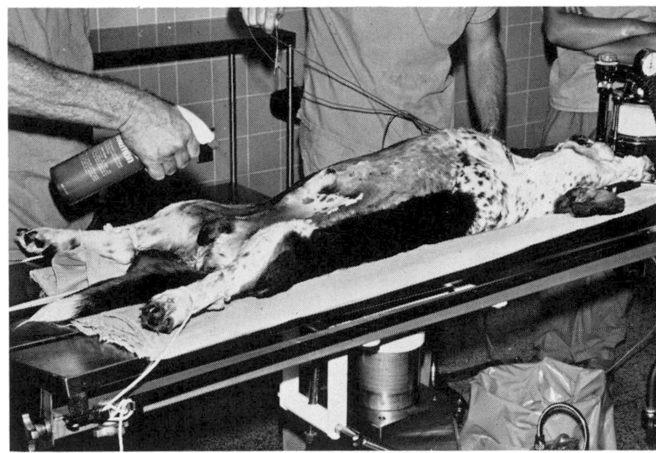

FIG. 12.21. Dog restrained preparatory to abdominal surgery.

TRANSPORT

Dogs are usually amenable to any type of transport. They customarily ride free in automobiles but must be confined in carrying crates for air shipment. Some dogs are susceptible to motion sickness. Appropriate preventive medication can be given prior to shipment. Chlorpromazine hydrochloride (0.55-1.1 mg/kg) is appropriate.

CHEMICAL RESTRAINT

Many tranquilizers are available to quiet the nervous or vicious dog. Each clinician establishes a usage pattern suitable to the type of dogs most often seen. Acepromazine maleate (0.55-1.1 mg/kg) is satisfactory. It has the added advantage in cases of epistaxis (nose bleeding) of lowering blood pressure slightly, which assists in achieving cessation of the hemorrhage.

The stray or feral dog presents a difficult handling problem. Animal control personnel are faced with a dilemma. They can rarely approach such dogs closely enough to net them, yet chemical restraint drugs are either unsafe or are restricted to use by licensed veterinarians only.

Popular literature and television have promulgated the idea that all one needs to do is point a tranquilizer gun at an animal and it is safely in hand. Some control programs have experienced up to 50% mortality when using the readily available nicotine alkaloids, especially when the drugs are administered by untrained personnel. Two suitable drugs are now available for remote restraint: xylazine hydrochloride (2.2 mg/kg) and a combination of tiletamine hydrochloride and zolazepam (5-10 mg/kg). Accurate weight estimation is essential for safe usage of these drugs. Both are restricted to use only by licensed veterinarians.

I would recommend that animal control program administrators limit the usage of chemical restraint to trained key personnel. These individuals could use one of the above drugs under the supervision of or in consultation with a veterinarian.

Training programs in the use of these drugs are not currently readily available but should be set up by veterinary schools or professional associations.

REFERENCES

1. Dangerfield, S., and Howell, E., eds. 1971. The International Encyclopedia of Dogs. New York: McGraw-Hill.
2. Fox, M. W. 1965. Canine Behavior. Springfield, Ill.: Charles C Thomas.
3. ______. 1974. Understanding Your Dog. New York: Coward, McCann, and Geoghegan.
4. Fox, M. W., ed. 1968. Abnormal Behavior in Animals. Philadelphia: W. B. Saunders.
5. Mather, G. W. 1969. Restraint of the laboratory dog. Fed. Proc. 28(4):1423-27.

13 CATS

CLASSIFICATION

Order Carnivora
Family Felidae: cat

ALTHOUGH domesticated for thousands of years, the cat [4] has never completely subjected itself to people. Despite many examples of affectionate relationships between persons and cats, in most instances the cat retains a certain aloofness. Weights and names of gender are listed in Table 13.1.

TABLE 13.1. Weights and names of gender of cats[3]

	Name	Weight
Adult male	Tom	Average: 3 kg (6.7 lb) Range: 1.4–5.4 kg
Adult female	Queen	Average: 2.6 kg (5.8 lb) Range: 1.4–3.6 kg
Newborn	Kitten	100 g
Pampered, castrated male	Male	May weigh over 9 kg (20 lb)

DANGER POTENTIAL

Of all domestic animals, the cat is one of the most difficult to handle. The cat defends itself by biting and clawing. Cats are well equipped with needle sharp canine teeth capable of inflicting serious wounds; retractable claws become formidable weapons in an excited and/or angry cat. Both forefeet and hind feet must be reckoned with when restraining a cat.

ANATOMY, PHYSIOLOGY, AND BEHAVIOR

Cats are agile and seldom tolerate manipulation without response. They are individualistic and vary widely in response to handling. Cats tend to be less amenable to manipulation than dogs. Behaviorally, they are less inclined than dogs to develop extremely close relationships with their owners.

Cats often exhibit territorial characteristics. Territoriality may cause a cat to resent being picked up from its own cage as an invasion of territory. The same animal, in strange surroundings, may not be aggressive when approached.

Most domestic cats will allow handling by the owner and usually will permit a stranger to approach if quietly reassured that no danger is present. To reassure a cat, talk soothingly but confidently to the animal. Some handlers blow gently into the cat's face or stroke it beneath the chin or beside the ears. Some cats enjoy a firm stroke over the back, always moving from head to tail [2]. Recognize that each cat is an individual and that it will take some time before a cat will accept handling by a new person. Some cats will never tolerate manipulation by a stranger. The handler must work quickly. A cat will tolerate manipulation just so long. When the cat's patience is exhausted, it is virtually impossible to proceed. Use as mild restraint as possible.

Cats are unpredictable, demonstrating quicksilver changes in behavior. They are sometimes rather timid and apprehensive and may show evidence of depression when placed in strange situations such as a new cage or unusual surroundings. The depressed state may quickly give way to a hostile state if the animal is roughly handled at this point. A ferocious cat, confined in a strange cage for several hours, may become more docile, especially if it can be convinced that ferociousness will not be rewarded in an acceptable manner. Even when apparently docile, if such an animal is picked up and subjected to painful manipulations, it may revert to aggressive hostility. If the manipulation is particularly unpleasant, the animal's behavior may rapidly deteriorate into a fit of rage, and one must cope with a snarling, clawing, biting buzz saw. In such a fracas, attempts to pin or control the animal usually end with an injured cat, the result of application of excessive pressure. It is wise to back off from an enraged cat, let the animal relax until it is calm, and begin anew with a less stressful method of restraint.

Cats with impaired breathing often resist handling, and struggling may compound the oxygen deficit.

PHYSICAL RESTRAINT

When a cat in a cage is approached, it quickly gives evidence as to whether or not it is friendly. An unfriendly cat hisses and spits at the person nearing the cage. If closer contact is attempted, the cat will strike out with a paw or attempt to bite an encroaching hand. The person must then either assume the cat is bluffing and risk picking it up or, preferably, take steps to divert the cat's attention while capturing it.

A cat permitted to come out of a cage into strange surroundings (strange territory) is usually intimidated and seldom exhibits hostile behavior. Often, if such a cat is approached in a calm manner, restraint will not provoke undesirable behavior.

Pick up the cat by placing a hand over the top of the animal and around to the opposite side, with the palm of the hand supporting the sternum or chest area (Fig. 13.1).

Alternatively, grasp the cat by the loose skin over the back of the neck close to the head (Fig. 13.2). This is a natural handhold for lifting a cat. If the cat becomes unruly, the head can be tilted backward, which unbalances the cat and lessens the inclination to bite and scratch [2]. Most cats do not resent this, but some may strike out to claw with the front feet and may, if extremely excited, attempt to scratch with the hind feet as well.

Grabbing a cat by the loose fold of skin on the neck makes use of a phenomenon retained from kittenhood. When the queen picks up her kitten by the scruff of the

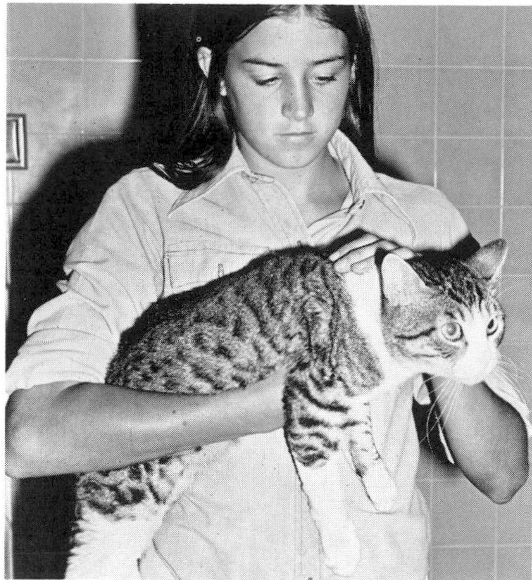

FIG. 13.1. Method for carrying a gentle cat.

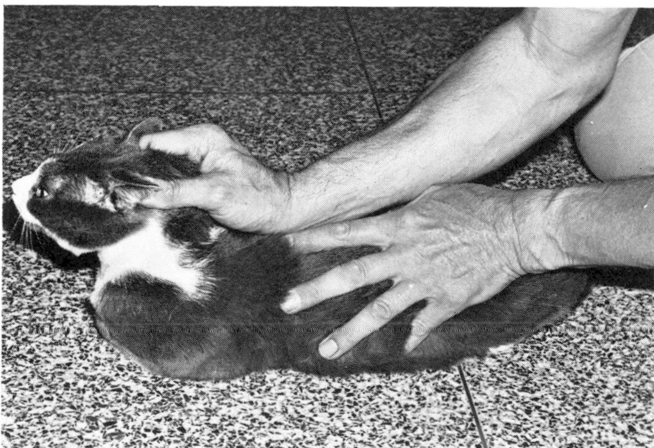

FIG. 13.2. Natural handhold for grasping a cat. Pick up skin close to the ears so head can be tilted up if cat begins to scratch.

neck to carry it, the kitten becomes limp, curling up the paws and tail, until it is released. This behavioral response prevents possible injury caused by the mother having to bite tightly or by the kitten twisting its neck in struggling. This response occurs in all species of domestic and wild felidae and can be utilized successfully to restrain kittens up to weaning age. Kittens grasped in this manner completely relax, and nonpainful manipulations can easily be carried out with this technique. If a cat becomes unmanageable during any restraint procedure, grasp it by the nape and hold it away from your body (Fig. 13.3). An obstreperous cat can be picked up and moved using both hands (Fig. 13.4).

It is seldom wise to wear gloves when handling cats, although gloves may minimize scratch injuries. One cannot rely on gloves to prevent penetration by the canine teeth; even a small cat can bite through heavy leather. Furthermore, wearing heavy gloves requires a harder squeeze

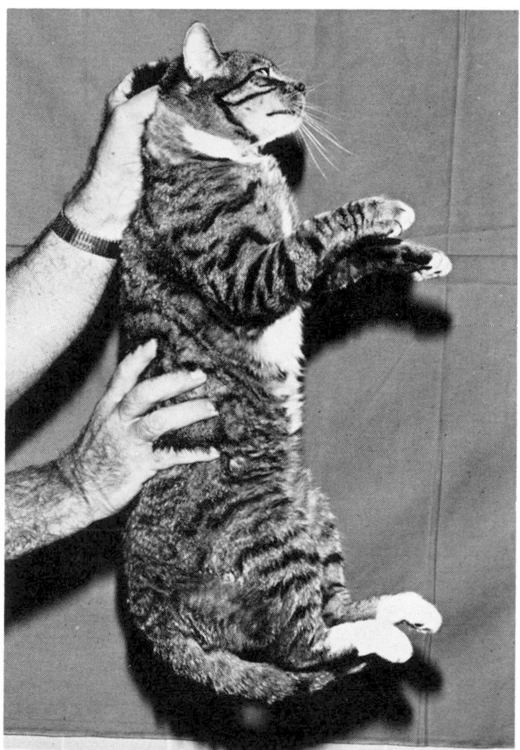

FIG. 13.3. Obstreperous cat can be grasped by the nape and pushed away from handler.

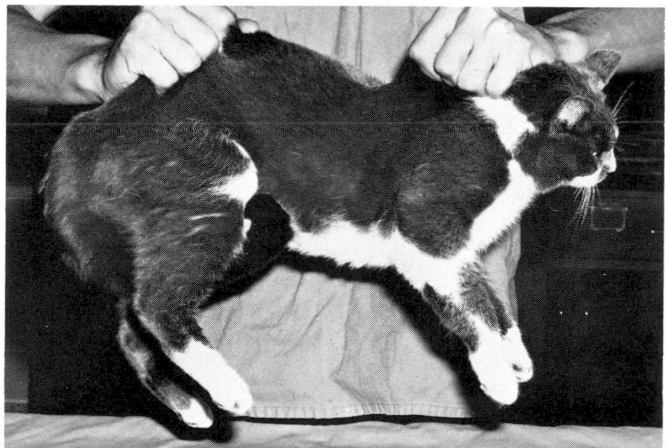

FIG. 13.4. Lifting an obstreperous cat.

to maintain the same degree of immobilization as can be achieved with a light grip of the bare hand. Squeezing hard may injure small cats. The danger is intensified because gloves reduce tactile discrimination, making it difficult to determine the degree of pressure applied. The pressure may be much greater than supposed and greater than necessary to maintain control.

When examining a cat in a veterinary hospital, have the animal placed on a smooth, slick table. The cat is distracted by the effort required to maintain stability and is less likely to use a paw for scratching. To carry out general examination and manipulation, gently hold the animal on the table with the hands cupped over the back.

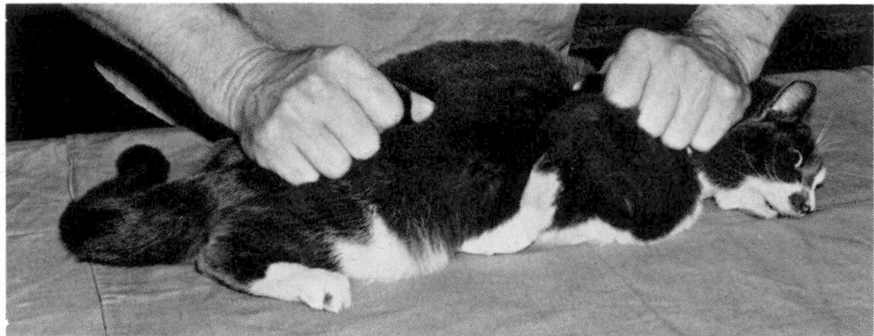

FIG. 13.5. Restraining obstreperous cat by pressing it onto a table.

If the animal must be held more firmly, grasp the loose skin over the back of the neck, pressing the cat onto the table with both hands, one over the neck and one over the loins (Fig. 13.5). This prevents the cat from lashing out with either front or back paws.

If more restricted activity is required, the cat can be held in lateral recumbency (Fig. 13.6). The paws are held immobile by the hands and the head is restrained by placing the forearm over the neck. It is necessary for an assistant to secure the head of an obstreperous cat during this procedure.

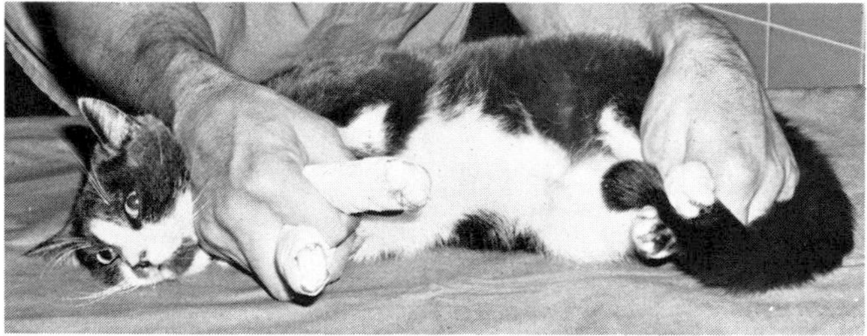

FIG. 13.6. Controlling feet and legs of a cat while in lateral recumbency.

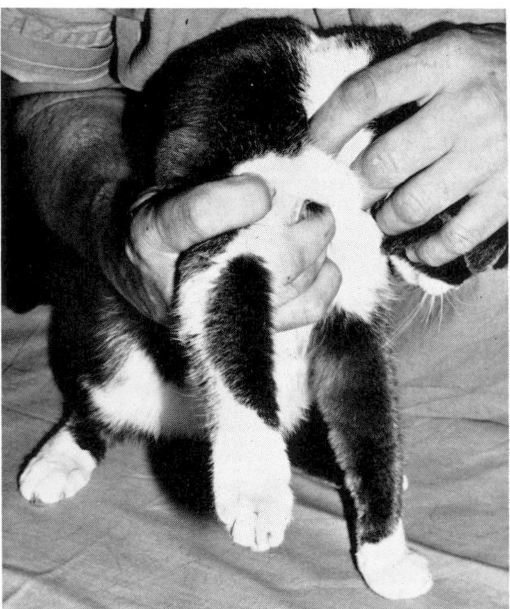

FIG. 13.7. Method of holding a cat for cephalic venipuncture.

If a procedure requires some time to complete, yet is not painful, the limbs can be immobilized by taping the front feet together and the hind feet together. The head can be easily controlled with one hand.

The mouth of the cat can be opened in the same manner as the dog's. If the front paws are not held by an assistant, the handler may be clawed. It is difficult to administer solid medication to a cat orally. Some prefer to use forceps to place a pill into the pharynx.

Intravenous injections can be given while holding the cat by a method similar to that used for holding the dog (Figs. 13.7, 13.8). The cat can be held on its back to withdraw blood samples from the jugular vein (Fig. 13.9).

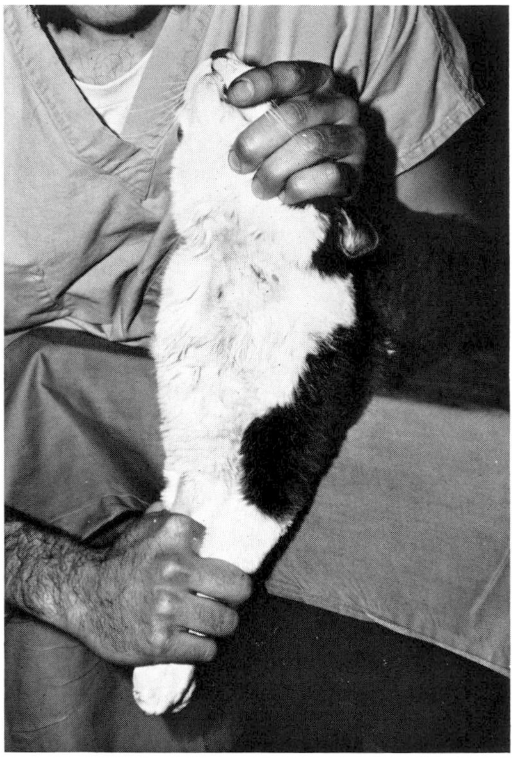

FIG. 13.8. Holding a cat for jugular venipuncture.

A cat can be controlled by wrapping it in a towel (Figs. 13.10–13.12). The extended claws entangle in the towel, keeping the paws within the wrap. A single limb can be withdrawn for examination or for exposure of a vein to administer intravenous medication or to withdraw blood for a laboratory sample. The head is controlled equally well and can be exposed for examination or treatment in the same manner as the legs.

A cat bag is a very satisfactory tool when restraining a cat for semiuncomfortable manipulations (Fig. 13.13). Pick the cat up by the nape to place it into the bag. Once the cat is inside the bag, either the head or any limb may be left outside for various manipulative procedures while the other limbs or the head are contained in the bag and cannot interfere. Cat bags can be obtained commercially, or they can be improvised.

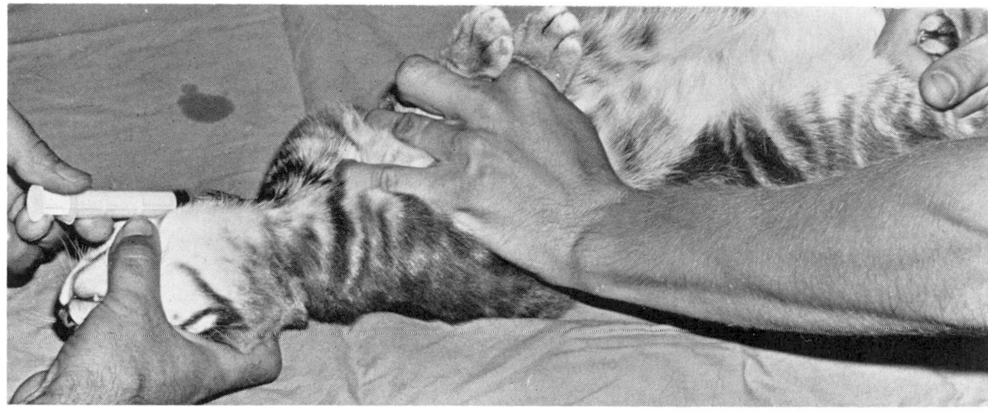

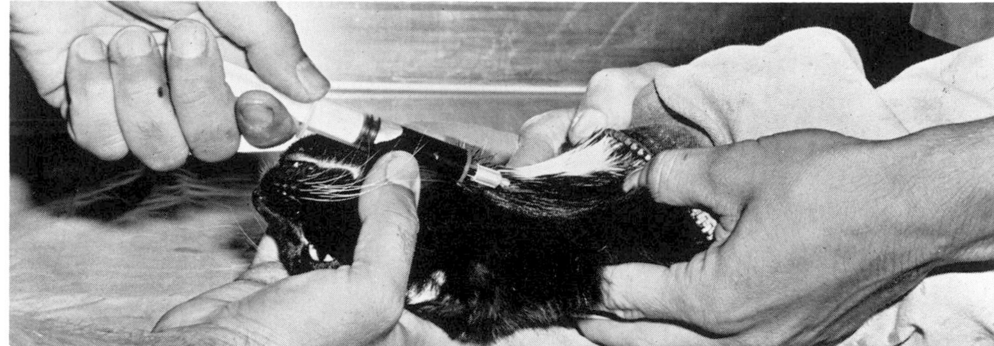

FIG. 13.9. Jugular venipuncture with cat on its back: Handheld *(upper)*. In restraining bag *(lower)*.

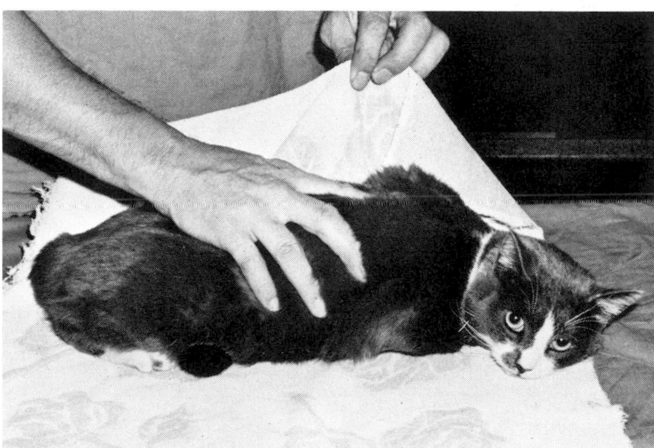

FIG. 13.10. Wrapping cat in a towel: Place cat in center of the towel.

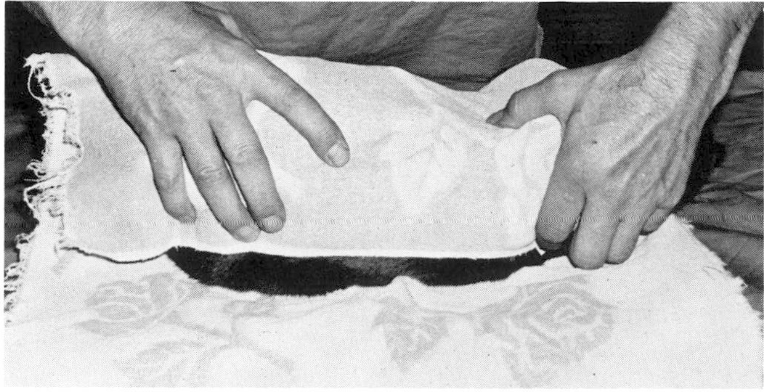

FIG. 13.11. Wrapping cat in towel *(cont.)*: Lift one end and wrap it quickly over top of the cat.

A net is another satisfactory tool for handling extremely nervous or unruly cats. The net can be introduced through a partially open door and placed over a cat in a small enclosure; a larger enclosure may be entered to place the net. Many manipulative procedures can be carried out on

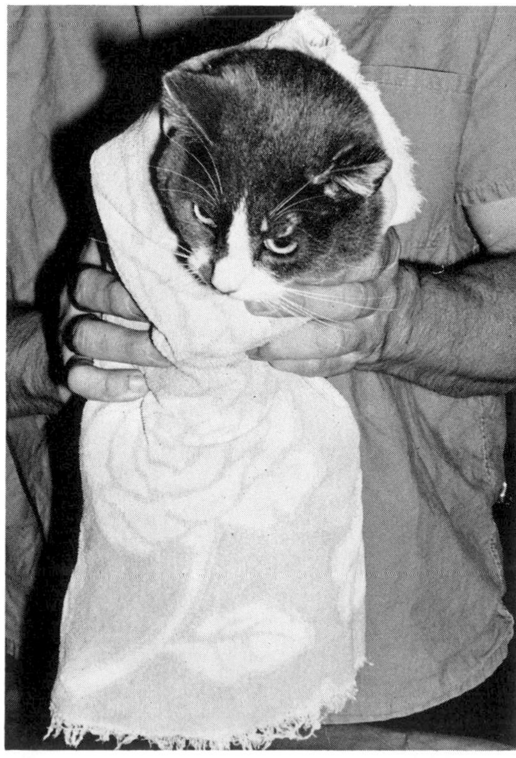

FIG. 13.12. Wrapping cat in towel *(cont.)*: Roll cat over to complete the wrap.

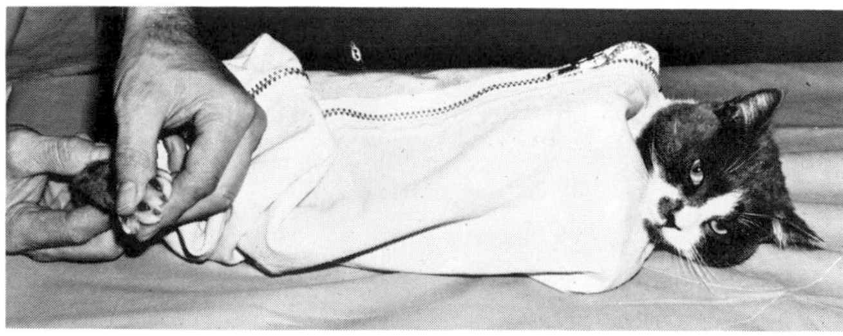

FIG. 13.13. Adjustable cat bag is a valuable tool for working with cats and many other small mammals.

a cat in a net. Almost any part of the body can be examined or a limb can be withdrawn through the mesh. The animal usually entangles its claws in the mesh, so scratching is not a hazard. The head can be restrained by grabbing the cat behind the back of the head with one hand.

It may be necessary to use a snare to handle extremely vicious cats. The snare is a less desirable tool because of the possibility of injuring the cat. Nevertheless, no one should be subjected to being bitten or scratched by an ill-mannered animal. It is not prudent to take excessive chances with any animal.

Experience in handling cats allows some individuals to approach seemingly vicious cats without difficulty. The inexperienced individual frequently must use special restraint techniques to minimize the chances of being scratched and bitten.

TRANSPORT

Cats are commonly carried or shipped to shows in small individual cages. Cats may vocalize loudly when confined.

CHEMICAL RESTRAINT

Ketamine hydrochloride is a frequently prescribed sedative and restraint agent for cats. Ketamine use obviates the necessity of prolonged, stringent physical restraint of the obstreperous cat. Dosages of 11-44 mg/kg provide excellent sedation to anesthesia for manipulative procedures in the cat.

REFERENCES

1. Fox, M. W. 1974. Understanding Your Cat. New York: Coward, McCann, and Geoghegan.
2. Joshua, J. O. 1971. Restraint of cats. Mod. Vet. Pract. 52:42-43.
3. Necker, C. 1970. The Natural History of Cats. New York: A. S. Barnes.
4. Pond, G., ed. 1972. The Complete Cat Encyclopedia. New York: Crown.

14 LABORATORY ANIMALS

CLASSIFICATION

Order Lagomorpha
  Family Leporidae: rabbit
Order Rodentia
  Family Muridae: rat, mouse
  Family Cricetidae: golden hamster, Chinese hamster
  Family Cavidae: guinea pig

NO SPECIAL names are given to male or female rodents. The young are called offspring, youngsters, or pups (mice and rats). The male rabbit is a buck and the female a doe. Newborn rabbits are kindling or youngsters. Weights are listed in Table 14.1

TABLE 14.1. Weights of laboratory animals

Animal	Male	Female	Newborn	Weanling
Guinea pig	1-1.2 kg	850-900	90	250
Rat	200-400	250-300	5-6	40-50
Mouse	20-40	25-90	1.5	7-15
Golden hamster	90-120	95-140	2	35
Gerbil	46-131	53-133	3	. . .
Rabbit				
New Zealand white	4-5 kg	4.5-5.5 kg	60	1.5 kg
Angora	2.7-3.6 kg			
Flemish giant	4.5-6.4 kg			
Dwarf	0.9 kg			

Note: Weight is in grams unless otherwise indicated.

DANGER POTENTIAL

All laboratory rodents and rabbits have large incisor teeth adapted for gnawing. All are capable of inflicting significant bite wounds. Rabbits may scratch a handler with the hind feet.

BEHAVIOR

Domestic laboratory animals are docile and easily handled unless previously subjected to unpleasant manipulation. Rats can easily climb up their own tails when suspended, but mice, less agile, cannot.

Free-living rodents and rabbits thermoregulate by escaping the heat through burrowing or nocturnal behavior. Domestic species will suffer from heat stress unless kept in a temperature-controlled environment.

Domestic laboratory animals are easily tamed and are frequently kept as children's pets.

PHYSICAL RESTRAINT

Hundreds of special devices have been developed to restrain various species of laboratory animals for certain specialized procedures. It is not my intent to describe or review all available techniques and equipment. Rather, general techniques suitable for the pet, zoo animal, or laboratory animal will be illustrated. Information about animals with specialized handling requirements can be obtained from the references listed.

Rabbits

Rabbits are usually kept in small cages or a hutch. A rabbit can be grasped by the loose skin over the back (Fig. 14.1). Lifting the hindquarters with the other hand provides added support (Fig. 14.2) and may prevent the rabbit from kicking backward as it is lifted, possibly fracturing the spine in the lumbosacral region. Small rabbits can be picked up in the same manner as puppies and kittens. Do not pick up a rabbit by the ears (Fig. 14.3) since this inflicts needless pain and causes the animal to struggle violently to free itself, even to the point of fracturing its neck. A docile rabbit can be carried in an arm (Fig. 14.4). When replacing a rabbit in a cage, set the rear quarters down first, with the head facing the door.

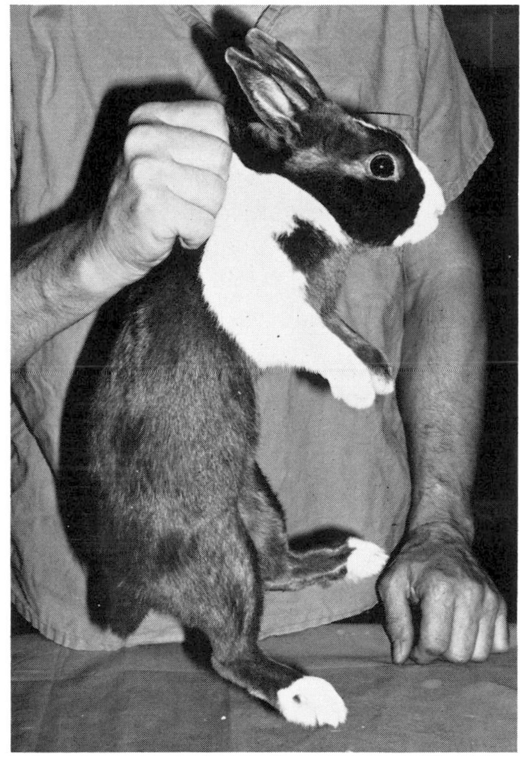

FIG. 14.1. Lifting rabbit by the loose skin over the back and neck.

To examine the abdomen, genital organs, or vent, hold the legs together (Fig. 14.5). The mouth can be held open with a dowel (Fig. 14.6).

Rabbits easily enter the torpid state (hypnotized, mesmerized) if placed and held on their backs for a few seconds (Fig. 14.7). Blindfolding may prolong the effect. A tap on the table is usually sufficient to reverse the trance. Occasionally a rabbit sitting upright will go limp while being examined. This eliminates the need for restraint, but the torpidity interferes with accurate clinical evaluation of a diseased rabbit.

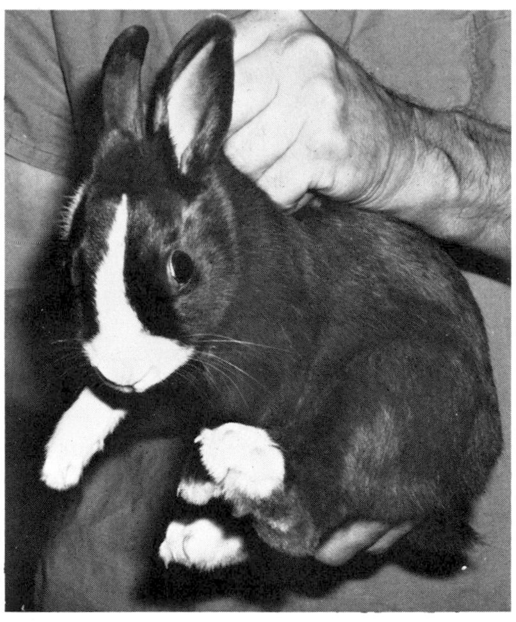

FIG. 14.2. Lifting rabbit by grasping neck skin while supporting the hindquarters.

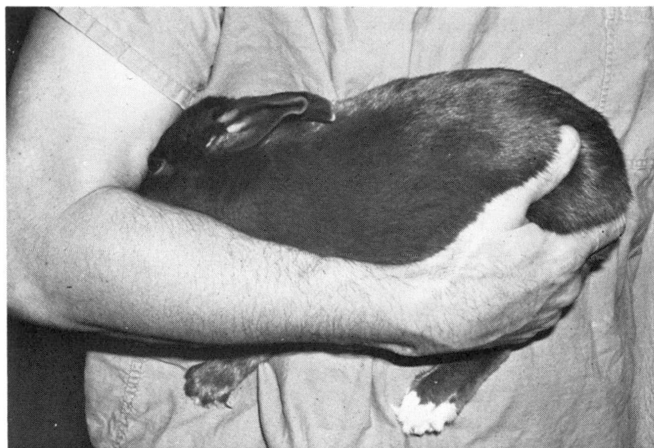

FIG. 14.4. Proper way to carry a docile rabbit.

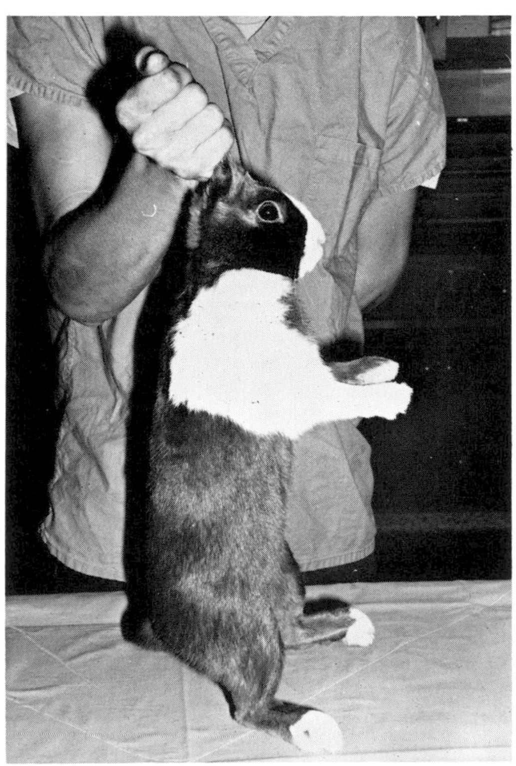

FIG. 14.3. Improper restraint. *Never* lift a rabbit by the ears.

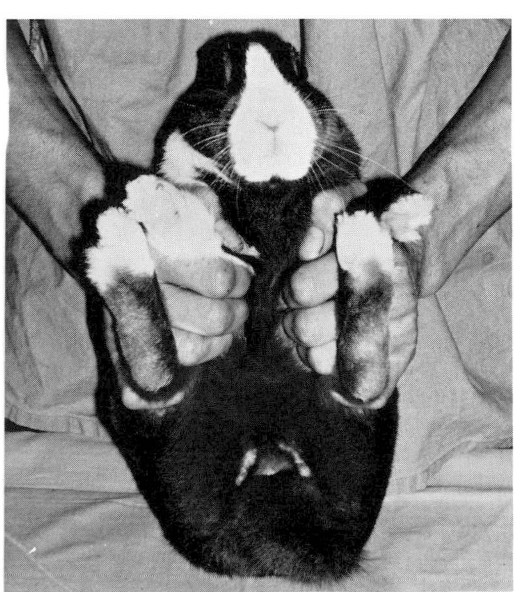

FIG. 14.5. Restraint for examination of abdomen and perineal area.

FIG. 14.6. Using a dowel to examine teeth or pass a stomach tube.

A rabbit wrapped in a towel cannot scratch the handler (Fig. 14.8). More effective restraint is achieved by placing the rabbit in a cat bag (Fig. 14.9) or a restraint box (Figs. 14.10, 14.11).

Blood samples are usually obtained from the prominent ear veins. The vein is raised by applying a tourniquet of string or umbilical tape to the ear (Fig. 14.12) or by clamping the vessel (Fig. 14.9).

Mouse

Mice vary greatly in the ease with which they can be handled. The pet mouse used to being handled by a child is easily picked up and held. It will seldom jump from your hand or bite. Some pets, although willing to accept manipulation by the owner, may not allow the same degree of familiarity by a stranger, particularly a veterinarian who is examining the animal.

Mice in laboratory colonies may be accustomed to being handled, but the relationship is not intimate. The animal is not tamed. In such institutions, devices or techniques that save time are often chosen over those that are easiest on the animal.

Mice are frequently picked up by the tail and removed from a cage. This can be done with the bare hand by simply grasping the tail close to the base with a finger and thumb. A thumb forceps should be used only as a last resort because it tends to make the rodent mean (Fig. 14.13).

If more restrictive restraint is necessary, grasp the tail, allow the animal to crawl away slowly, and reach quickly with thumb and finger alongside the neck to grasp a skin fold. Hold the animal as illustrated in Figure 14.14. The catch must be made quickly to prevent the mouse from twisting its head and biting the approaching fingers. The tail can be tucked under a little finger to contain the animal in one hand (Fig. 14.15).

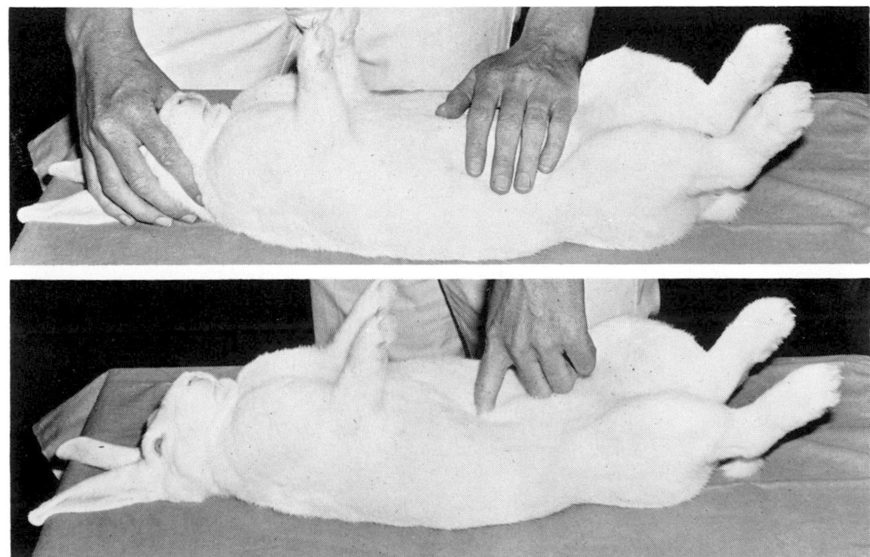

FIG. 14.7. Mesmerizing a rabbit.

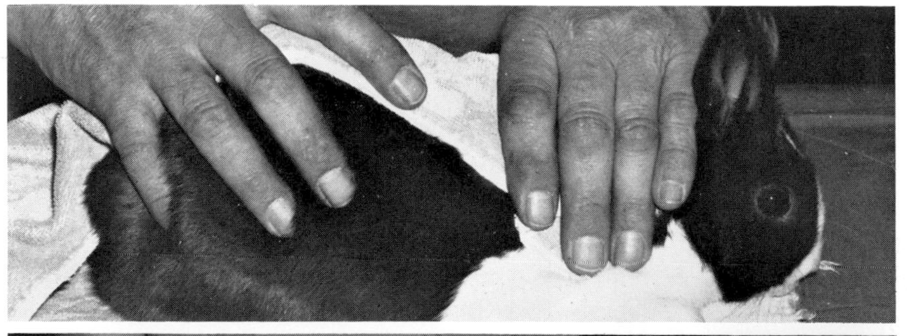

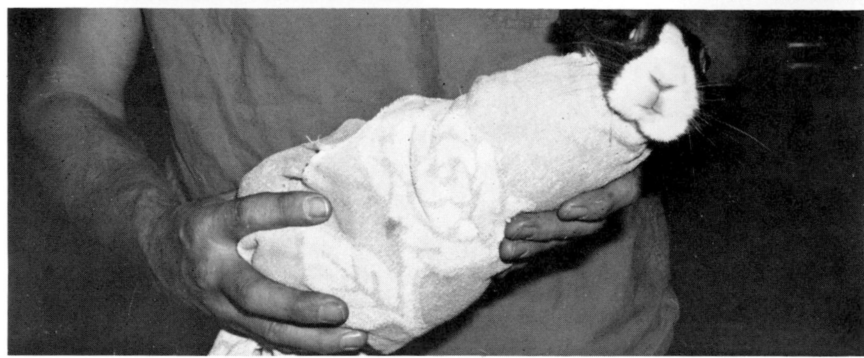

FIG. 14.8. Wrapping rabbit in a towel.

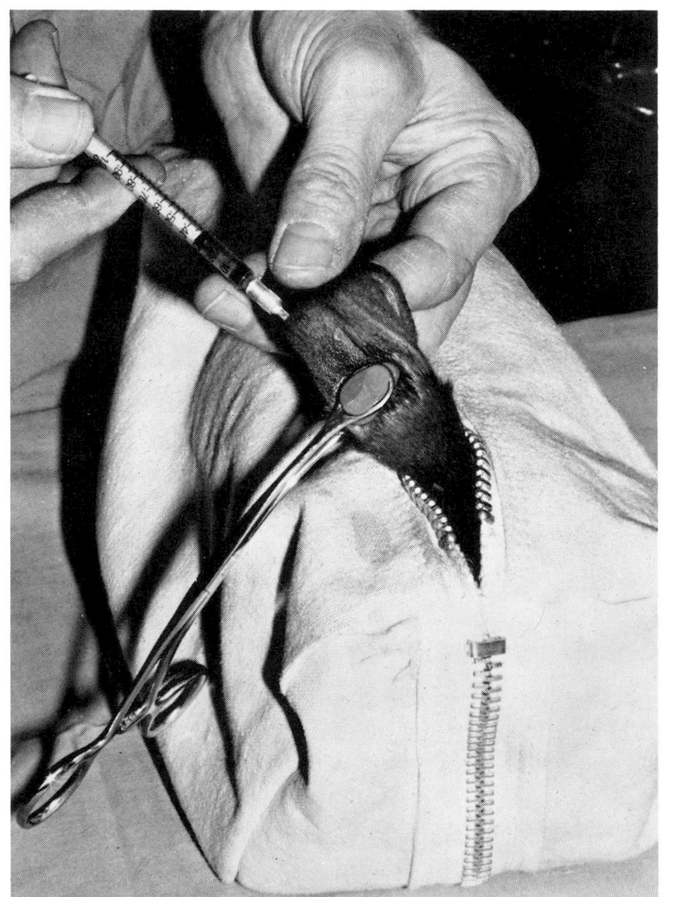

FIG. 14.9. Using tongue clamp to raise vein on rabbit in a bag.

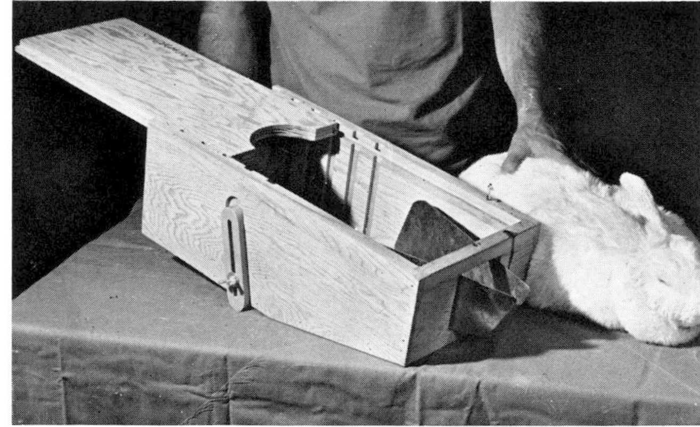

FIG. 14.10. Adjustable rabbit restraint box.

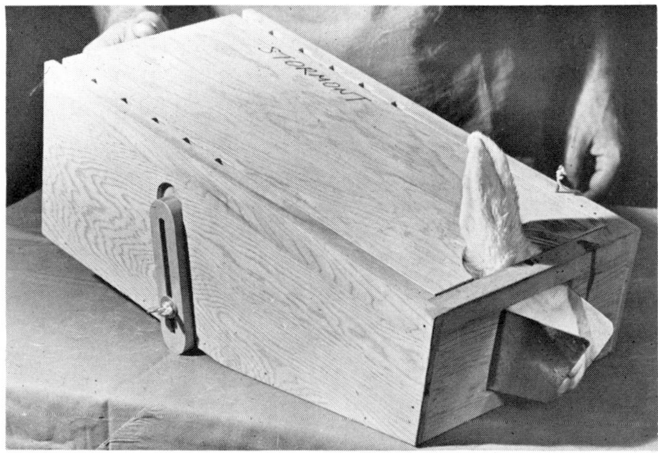

FIG. 14.11. Rabbit in restraint box.

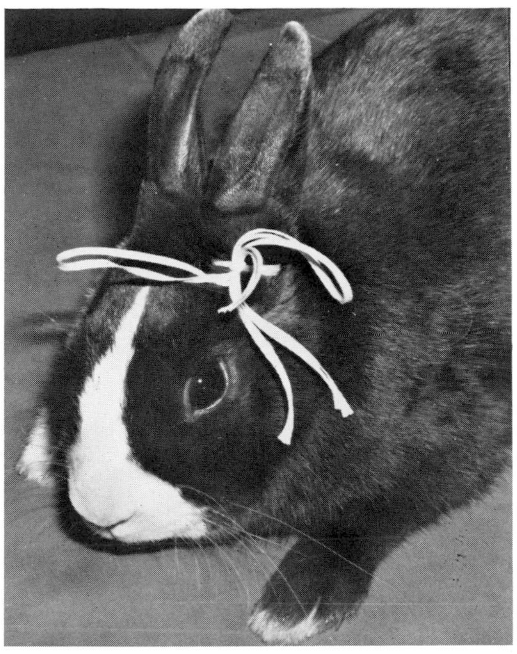

FIG. 14.12. Using umbilical tape or string to raise vein.

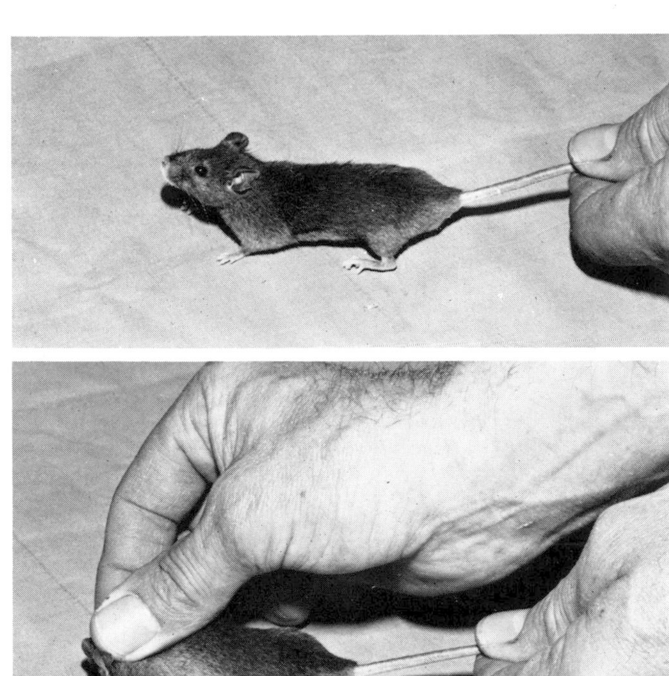

FIG. 14.14. Grasping a mouse.

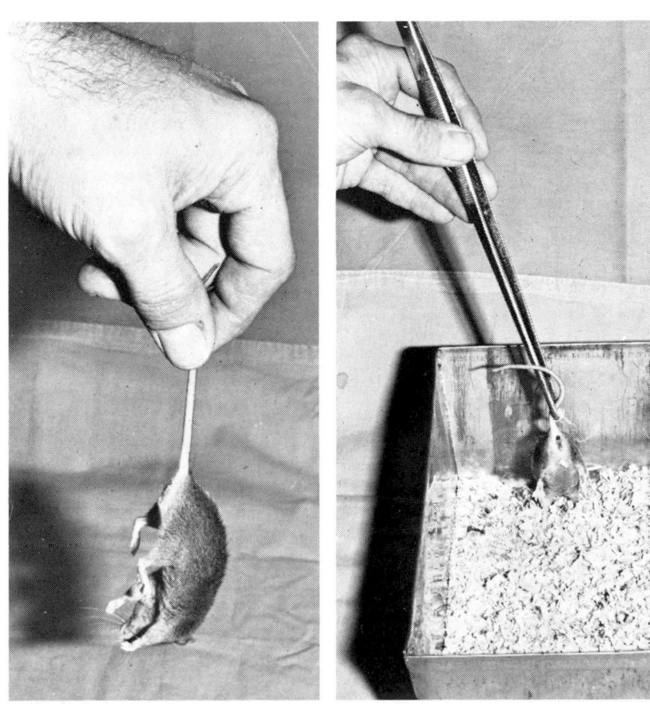

FIG. 14.13. Mouse can be lifted by grabbing the tail with bare hand or forceps.

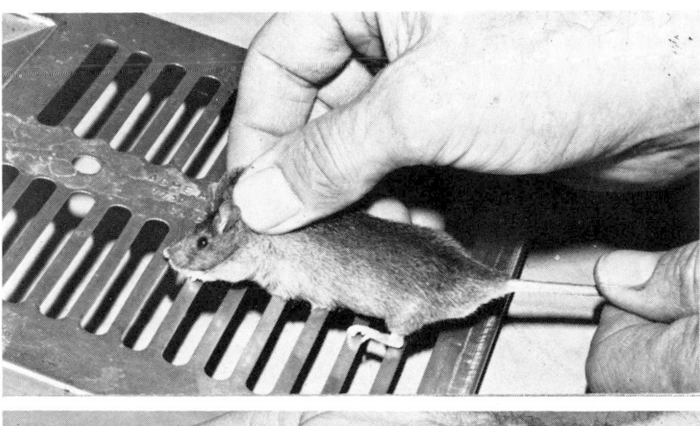

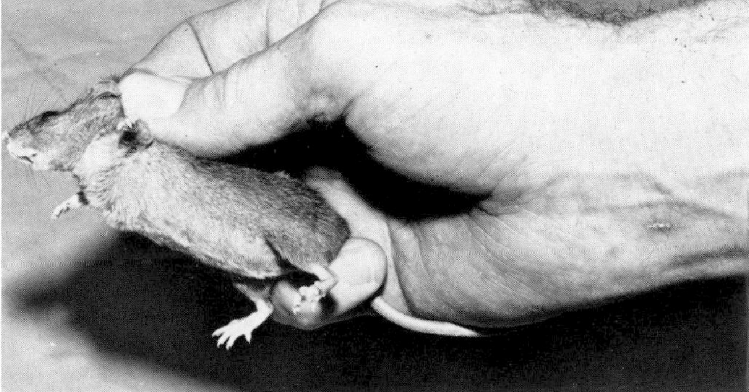

FIG. 14.15. Using a cage top to divert mouse's attention *(upper)*. Holding mouse by loose skin around the neck. Tail is anchored under little finger *(lower)*.

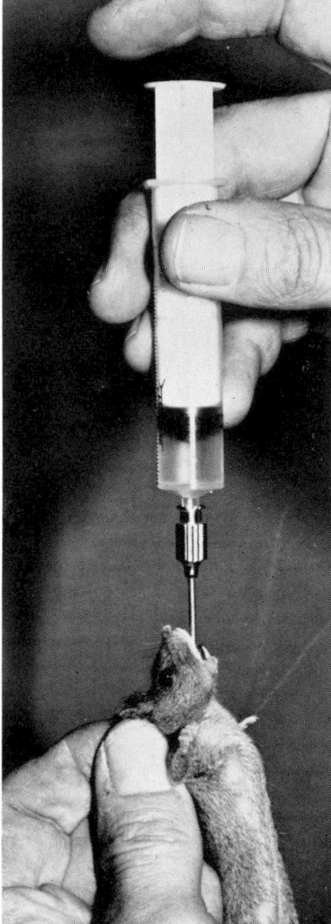

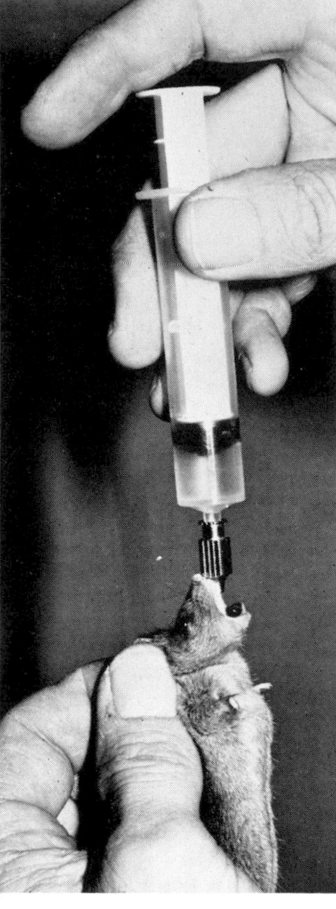

FIG. 14.16. Esophageal intubation with a 16 gauge, 5 cm needle with a ball soldered to the tip. Head and neck must be held straight.

Mice can also be held in the cupped hand by keeping the head between the thumb and forefinger. This is slightly more hazardous for the bare hand. Usually a light cotton or leather glove is worn to hold the animal. This technique is suitable when making quick injections or examinations.

Wild mice like *Peromyscus* spp. are jumpers. Place a holding cage in a tall garbage can before removing the top.

Oral medication can be given to a mouse with a needle stomach tube (Fig. 14.16). The stomach tube is a 16 gauge, 5 cm straight or curved needle, modified by soldering a ball to the tip to prevent damage to the esophagus. Insert the needle gently through the esophagus into the stomach.

If such a needle is unavailable, pass a small plastic or rubber stomach tube. The top of a plastic needle holder, special forceps, or a small dowel drilled with a hole may serve as a speculum to aid in passing the tube.

Obtaining blood samples from small laboratory rodents is challenging. Postorbital sinus bleeding is widely accepted as a satisfactory technique for use in all rodents, but it is most successful in mice and hamsters. It is more difficult to employ this technique with rats and very difficult with guinea pigs.

The animal is anesthetized by placing it in an ether jar. When anesthetized, grasp it, fixing the head firmly. Proptose the eyeball slightly by stretching the upper eyelid. Insert a suitable capillary tube slightly dorsal to the medial canthus of the eye with a slight twist (Fig. 14.17). Direct the tube toward the medial side of the bony orbit. Blood may well up in the tube during insertion. If not, once the bone is touched, withdraw the tube slightly to fill it. Rodents have been repeatedly bled by this method without producing ill effects; 0.1 ml of blood can be safely withdrawn from a 50 g mouse and 0.5 ml from a 200 g rat. These quantities are sufficient for complete hemograms and some blood chemistries if microtechniques are employed. Slight hemorrhage may occur following removal of the tube. This is not cause for concern; however, it would be inadvisable to use this technique to bleed a pet in front of an anxious owner.

Rat

Immature rats are handled much like mice. However, recall that a rat is capable of climbing up its own tail, so take care when lifting a rat from a cage by the tail barehanded. It is advisable to use a forceps (Fig. 14.18). Large adult rats should be grasped with a gloved hand (Figs. 14.19, 14.20). Once the head and neck are secured, the rat can be held with gloved or bare hands (Figs. 14.21, 14.22).

A rat can be wrapped in a towel. Plastic tubes partially restrict movement for induction of volatile anesthesia or the taking of whole body X rays (Fig. 14.23).

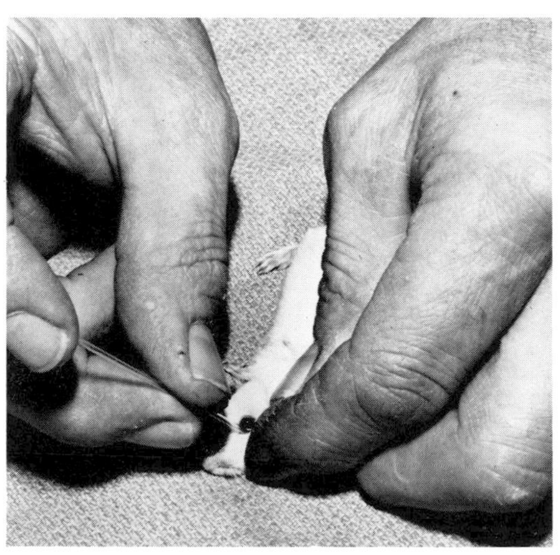

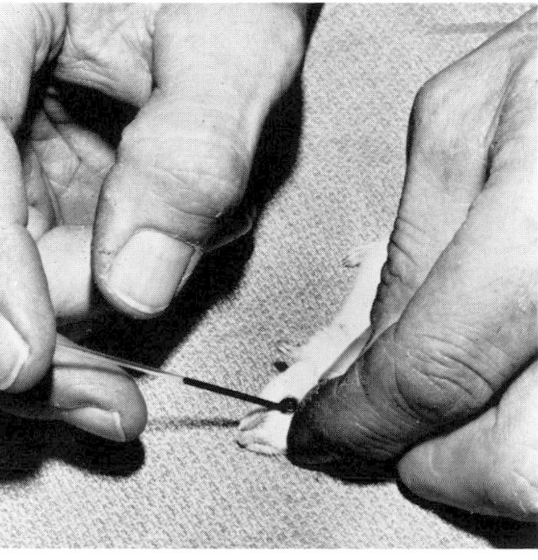

FIG. 14.17. Bleeding a mouse with capillary tube inserted retrobulbarly.

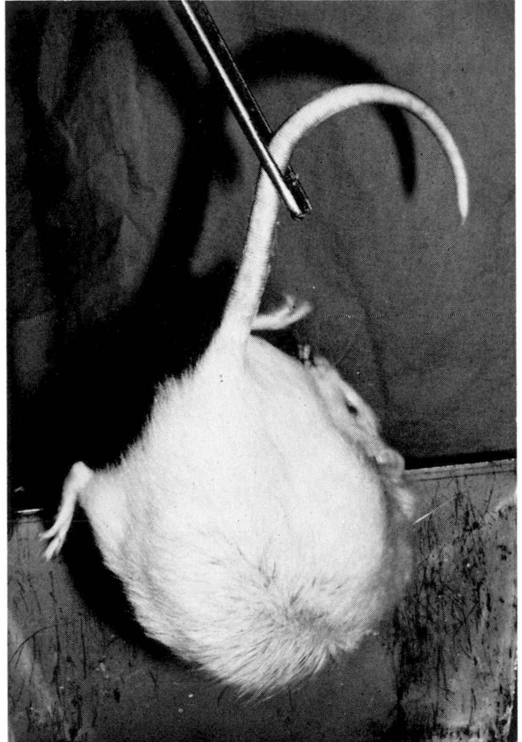

FIG. 14.18. Lifting rat by its tail with forceps.

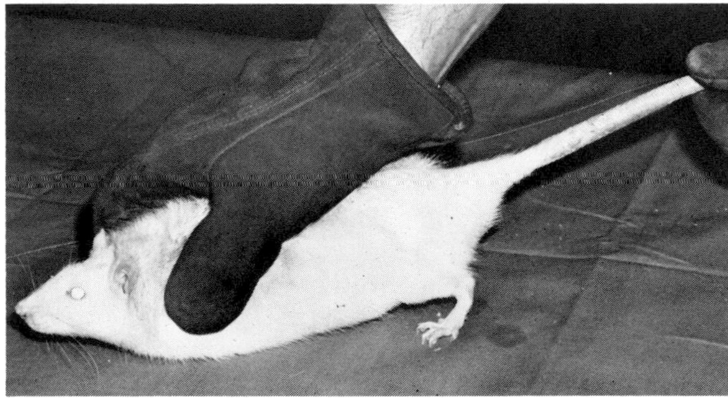

FIG. 14.19. Grasping an adult rat.

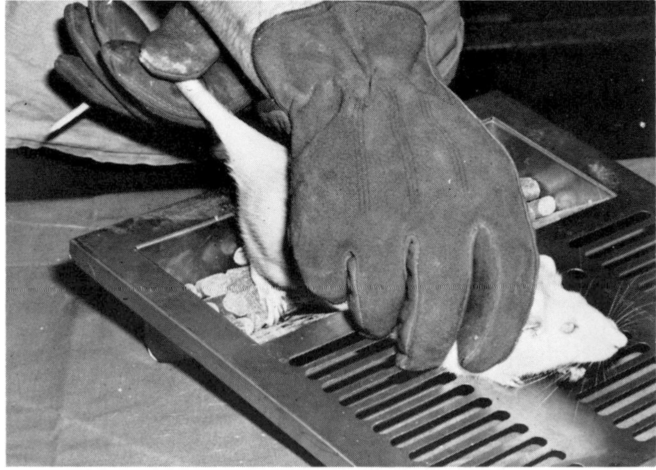

FIG. 14.20. Allowing rat to cling to a cage top to divert its attention while it is grasped.

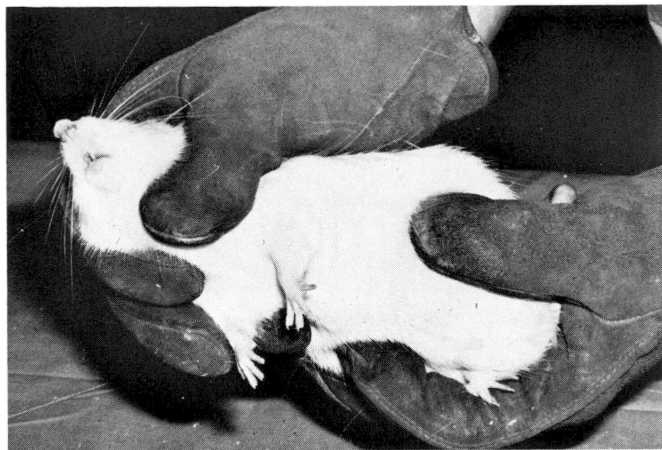

FIG. 14.21. Holding rat after grasping it, wearing gloves.

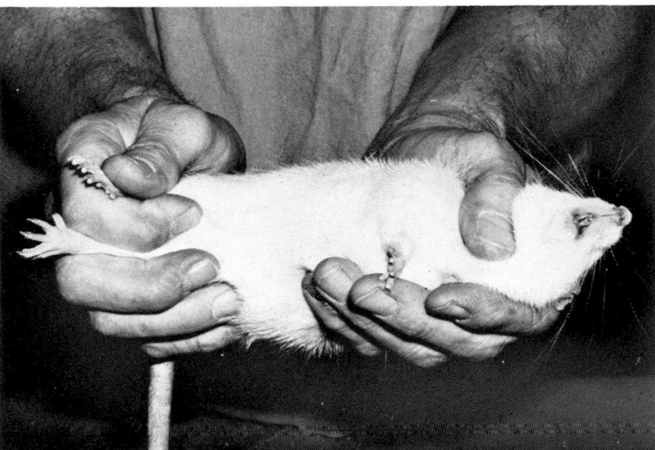

FIG. 14.22. Holding rat bare-handed for better control.

Esophageal intubation can be carried out by means of a dowel speculum or with forceps (Fig. 14.24). Needle stomach tubes (16 guage, 6.5 cm) are also used for medicating rats.

Guinea Pig

Pet guinea pigs are usually handled frequently and can ordinarily be picked up and examined with ease (Figs. 14.25, 14.26). Laboratory animals are more excitable and may resist being picked up, but usually relax when grasped gently. However, one must remember that all are rodents, capable of inflicting a serious bite if they are mistreated or suffer pain. If the restrainer is unfamiliar with the animal, a lightweight glove is indicated.

Dental problems are common to all laboratory rodents, especially guinea pigs. The mouth can be examined while the animal is under sedation or while it is physically restrained by the handler grasping its head or wrapping it in a towel (Fig. 14.27). A special speculum can be constructed or improvised to hold the mouth open and expand the cheeks to visualize the teeth and oral cavity (see Fig. 18.21).

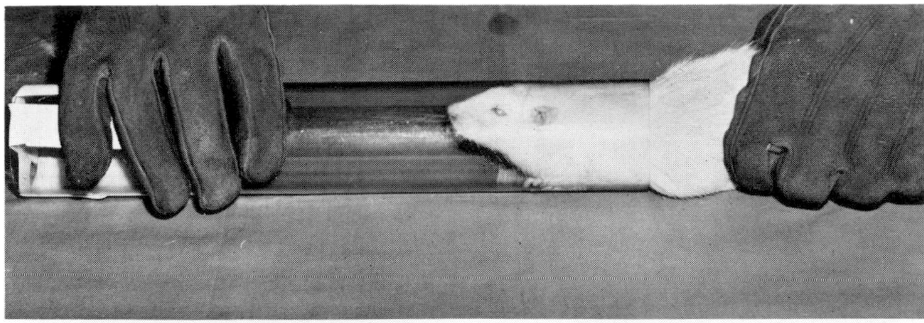

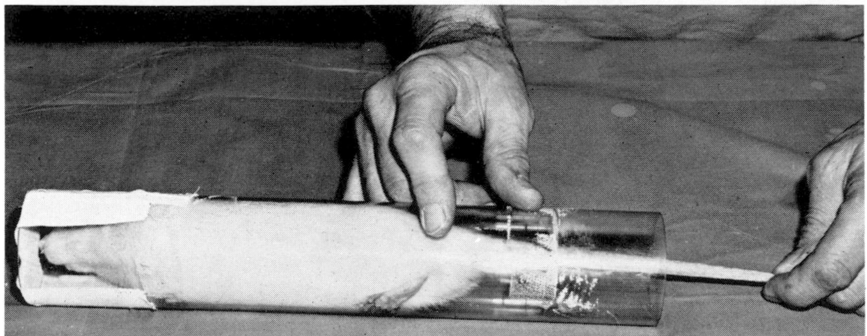

FIG. 14.23. Placing rat in a plastic tube.

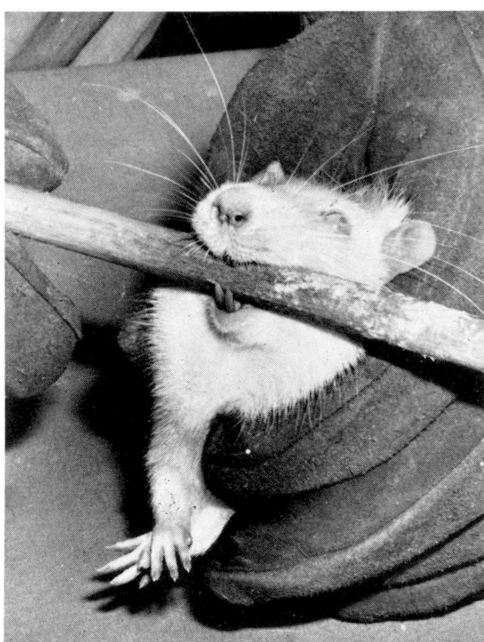

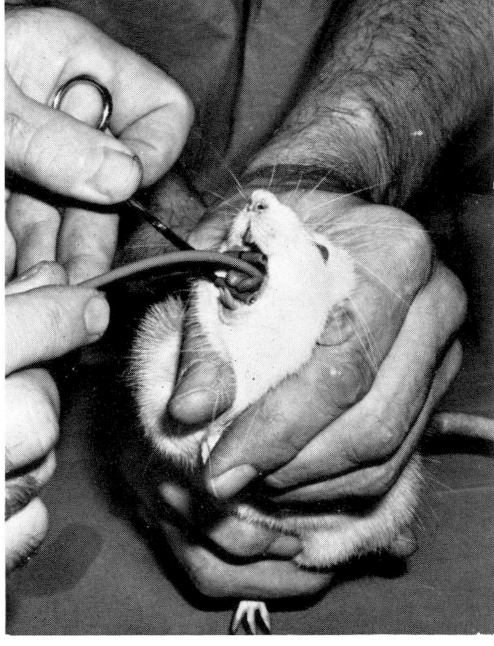

FIG. 14.24. Rat specula: Doweling *(left)*. Rubber tubing-coated forceps *(right)*.

Oral medication can be administered through a stomach tube as described for the mouse, or it may be given by inserting a plastic eyedropper through the commissure of the mouth and dropping the fluid into the pocket.

Hamster

Hamsters are slightly more nervous and more difficult to handle than mice or rats. In the hands of pet owners they are usually quite gentle. However, they are easily frightened if strangers try to grasp them and they may be difficult to capture. An animal may roll over on its back in a submissive posture, but will bite anyway. Hamsters have both speed and agility; they lack a tail to grab. If one is handling an unfamiliar animal, a light glove should be worn when the animal is picked up from the cage (Fig. 14.28). Some handlers pick up hamsters as they would rats.

Miscellaneous Species

Many other species of rodents are kept as pets or laboratory animals. The general principles governing restraint and handling of these rodents are similar to those mentioned above. Any unique problems will be discussed in Chapter 18.

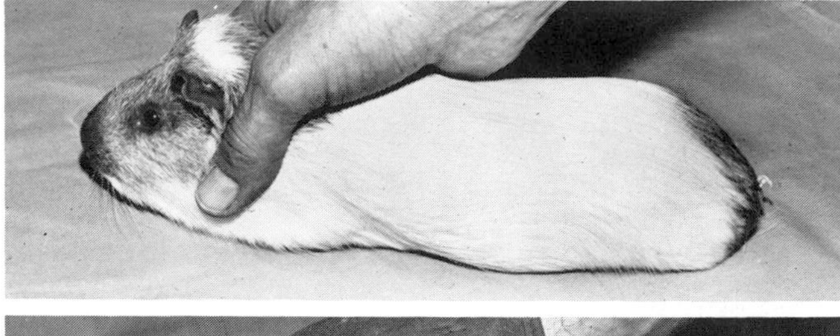

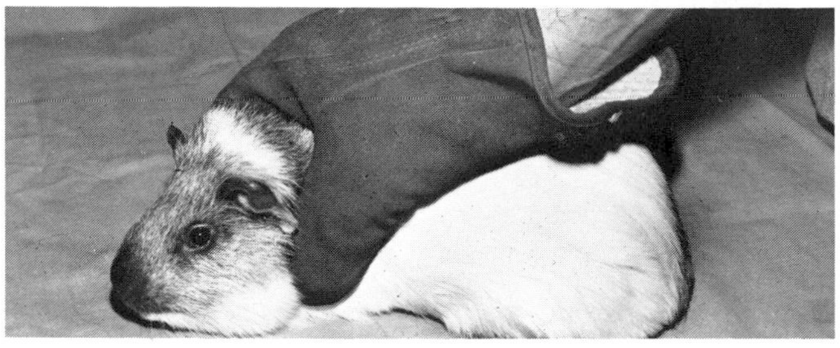

FIG. 14.25. Grasping guinea pig with bare and gloved hands. Only the rare aggressive animal requires use of gloves.

TRANSPORT

Laboratory animals are transported more frequently than any other group of animals. Special cages have been designed for air and truck shipment. Appropriate precautions must be taken to prevent overheating.

Pets can be moved in their own cages or in special shipping boxes. Rodents must not be left in cardboard containers because they can quickly gnaw their way out.

FIG. 14.26. Proper restraint and support of a guinea pig.

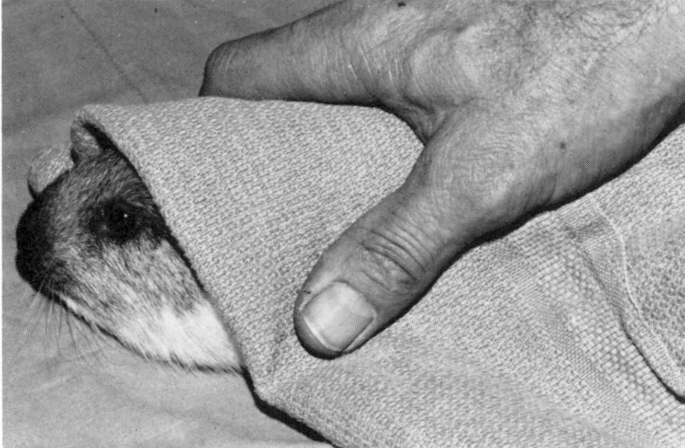

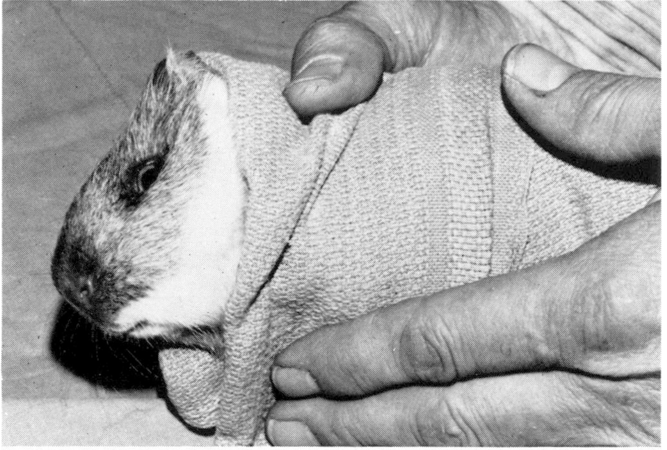

FIG. 14.27. Wrapping guinea pig in a towel.

CHEMICAL RESTRAINT

Numerous drugs are used to sedate and anesthetize laboratory animals. Table 14.2 gives dosages. The standard technique is to place the animal in a closed jar containing a volatile anesthetic agent such as ether or halothane. Anesthesia can also be induced with halothane administered via a mask, plastic tube, or anesthetic chamber. Tracheal intubation is difficult because of the restricted space in the oral cavity.

Intramuscular sedatives and restraint agents are useful in some species but not in others. Anesthetic techniques are described in detail by Gay [1]. Rabbits are refractory to many restraint agents. Many different drugs and combinations have been used. One suitable injectable anesthetic regimen is as follows: give acepromazine maleate (0.2–0.5 mg/kg), and xylazine hydrochloride (2–5 mg/kg). Wait 10 minutes, then give ketamine hydrochloride (30–50 mg/kg) intramuscularly. Anesthesia will last for 20 minutes at low dosages or up to an hour at higher dosages. Additional doses of 50–75 mg of ketamine hydrochloride can be given as required. Ketamine used alone will not provide suitable analgesia even with doses of 50 mg/kg.

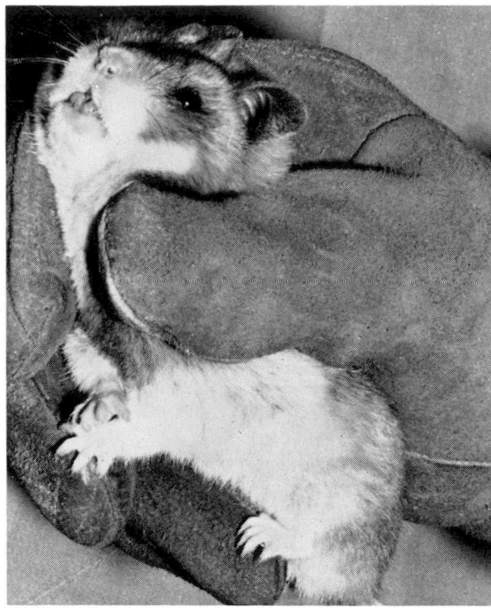

FIG. 14.28. Restraint of a hamster.

TABLE 14.2. Dosage of sedatives and restraint agents for laboratory animals

Drug	Rabbit	Mouse	Rat	Guinea pig	Hamster
			(mg/kg)		
Fentanyl-droperidol (Innovar-Vet)[2]	0.15-0.17	0.8-2	0.13	0.08	. . .
Tiletamine-zolazepam (Tilazol)*	10-25	. . .	. . .	10-30	10
Ketamine hydrochloride (Vetalar)[3]	20-44	44	44	20-44	. . .

*Unpublished data.

REFERENCES

1. Gay, W. I., ed. 1965. Methods of Animal Experimentation, pp. 89-100. New York: Academic Press.
2. Lewis, G. E., and Jennings, P. B., Jr. 1972. Effective sedation of laboratory animals using Innovar Vet®. Lab. Anim. Sci. 22:430-32.
3. Weisbroth, S. H., and Fuden, J. H. 1972. The use of ketamine hydrochloride as an anesthetic in laboratory rabbits, rats, mice, and guinea pigs. Lab. Anim. Sci. 22:904-6.

SUPPLEMENTAL READING

Beary, E. G. 1968. Laboratory Animals: Their Care and Use in Research: A Checklist. Bibliography. Natlick, Mass.: U.S. Army, Natlick Laboratories.

Cass, J. 1971. Laboratory Animals, an Annotated Bibliography of Informal Resources. New York: Hafner.

Clifford, D. 1971. Restraint and anesthesia of small laboratory animals. In L. R. Soma, ed. Textbook of Veterinary Anesthesia, pp. 369-84. Baltimore: Williams & Wilkins.

Hafez, E. S. E. 1970. Reproduction and Breeding Techniques for Laboratory Animals, p. 3. Philadelphia: Lea & Febiger.

Schuchman, S. M. 1974. Individual care and treatment of mice, rats, guinea pigs, hamsters and gerbils. In R. W. Kirk, ed. Current Veterinary Therapy V, pp. 588-614. Philadelphia: W. B. Saunders.

Universities Federation for Animal Welfare. 1975. The UFAW Handbook on the Care and Management of Laboratory Animals, 4th ed. London: E & S Livingstone.

15 POULTRY AND WATERFOWL

CLASSIFICATION

Order Galliformes
Family Phasianidae: chicken
Family Meleagrididae: turkey
Order Anseriformes
Family Anatidae: Pekin duck, Muscovy duck, goose, Canada goose, mute swan

CHICKENS, turkeys, ducks, and geese are frequently kept as pets or in small home flocks as well as in large commercial operations. Techniques for handling individual birds vary with the circumstances, although all are based on common principles. Names of gender of poultry and waterfowl are given in Table 15.1, weights in Table 15.2.

TABLE 15.1. Names of gender of domestic fowl

Bird	Male	Female	Young
Chicken	Cock, rooster	Hen	Pullet (female), cockerel (male), chick
Duck	Drake	Duck	Duckling
Goose	Gander	Goose	Gosling
Swan	Cob	Pen	Cygnet
Turkey	Tom	Hen	Poult

TABLE 15.2. Weights of poultry and waterfowl

Bird	Male		Female	
	(kg)	*(lb)*	*(kg)*	*(lb)*
Chickens				
Plymouth rock	4	9.5	3	7.5
Jersey black giant	6	13	5	10
White leghorn	3	6	2	4.5
Cornish	5	10	3	7.5
Turkeys				
Large strains	21	45	12	26
Medium strains	16	35	9	20
Small strains	8	17	5	11
Ducks				
Pekin	4	9	4	8
Indian runner	2	4.5	2	4
Muscovy	5	12	3	7
Geese				
Toulouse	12	26	9	20
Chinese	5	12	5	10

DANGER POTENTIAL

Chickens can peck and scratch. Roosters may develop large spurs on the legs which can seriously injure the unwary person. Roosters raised for fighting (illegal in most states) may be aggressive. Turkeys may peck, particularly at objects like rings and other jewelry, but rarely injure. Large tom turkeys can deliver painful wing blows and may scratch. Domestic ducks are easily handled, although some species have claws adapted for perching in trees and may scratch a handler. Ducks have blunt bills, so although they may peck, they rarely injure (Fig. 15.1).

Geese are large. They often peck viciously with heavy, pointed bills and may also strike an intruder by rapidly flapping strong wings. A small child can be severely injured by an angry goose. I vividly recall running to the house chased by an irate gander beating me with his wings and pecking at my backside.

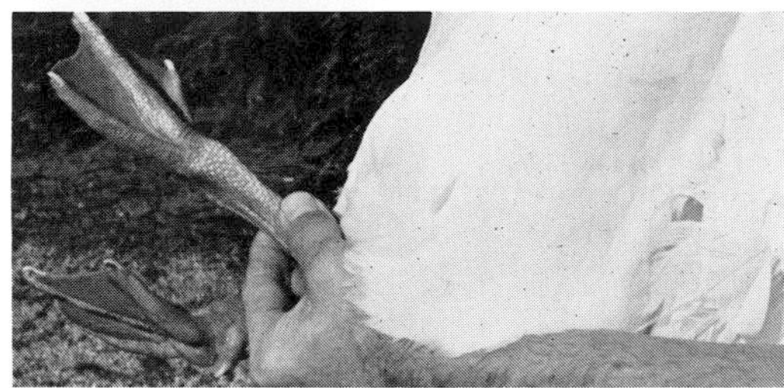

FIG. 15.1. Toenails of the Muscovy duck are adapted for perching in trees *(upper)*. Toenails of geese and most ducks are short and blunt *(lower)*.

BEHAVIOR AND PHYSIOLOGY

Most domestic birds are docile and amenable to handling. They are gregarious and stay bunched together when approached rather than scattering. This flocking trait enables the handler to single out and capture one individual without causing it to panic or frightening the other members of the flock.

Waterfowl will retreat to water if harrassed, so access to ponds must be prevented in order to capture them. Domestic birds rarely struggle once captured.

The trachea of birds is composed of a series of complete cartilaginous rings which prevent collapse of the trachea. This allows a handler to grasp the neck without danger of suffocating the bird.

All birds breathe by means of a bellows-type respiratory system (see Fig. 24.1). Be cautious when exerting pressure over the sternum.

FIG. 15.2. Grasping chicken simultaneously by wing and leg.

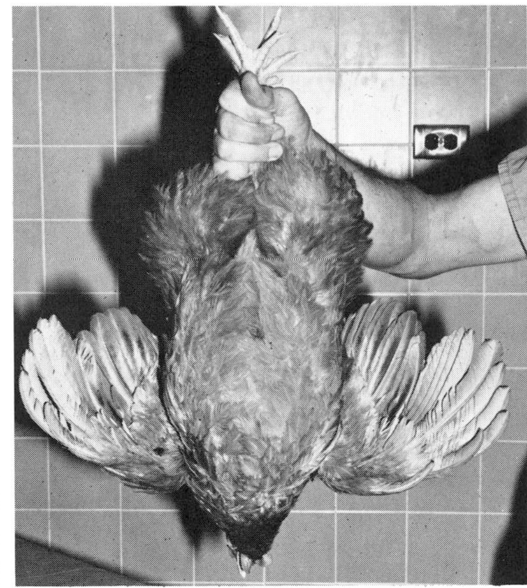

FIG. 15.4. Carrying chicken by the legs.

PHYSICAL RESTRAINT

Chicken

A chicken can be grasped by the leg and wing simultaneously (Fig. 15.2). It can also be captured with a net or a hook. Once the chicken is captured, it can be held by the legs (Figs. 15.3–15.6) or the wings can be subdued (Figs. 15.5, 15.6). The bird can be carried as illustrated in Figures 15.6 and 15.7 or placed on a table and restrained (Fig. 15.8).

A blindfolded bird will lie more quietly. When its vision is diminished the chicken behaves as if hypnotized. This is easily accomplished by placing a towel or cloth over the head (Fig. 15.9) or by tucking the head beneath a wing (Fig. 15.10).

The wings of a chicken can be interlocked to temporarily prevent flight. Carefully wrap the wings around each other as illustrated in Figures 15.11 and 15.12. Do not use this technique on peafowl or large turkeys, since it may fracture the wing or dislocate the joint.

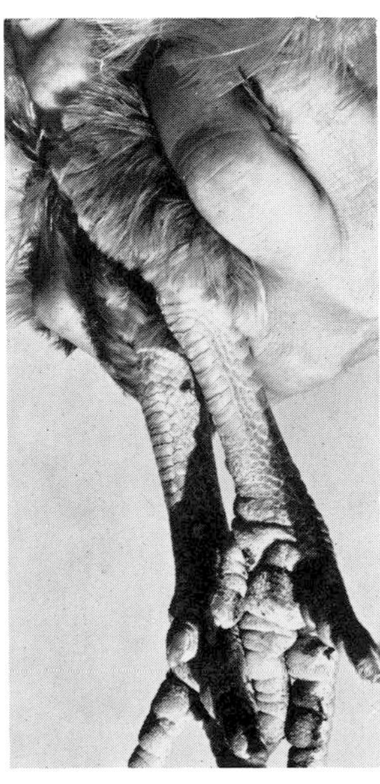

FIG. 15.3. Holding chicken by the legs while supporting body in the hands.

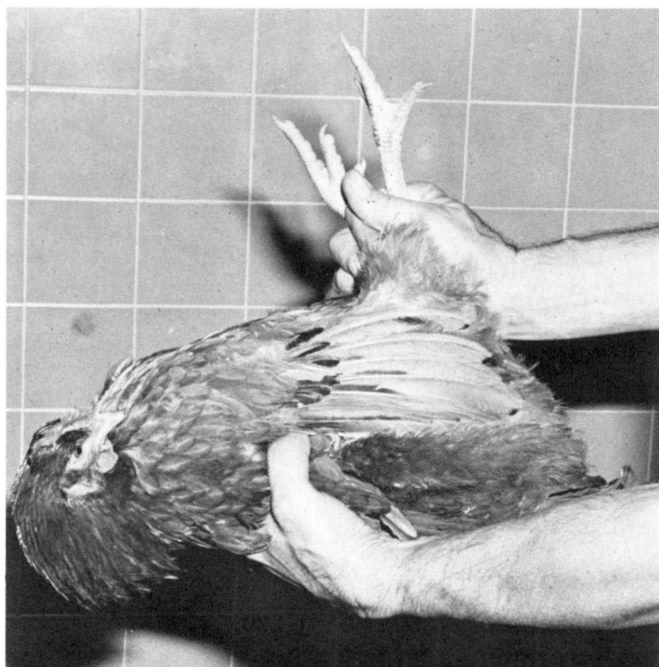

FIG. 15.5. Carrying chicken by holding both legs and supporting the back.

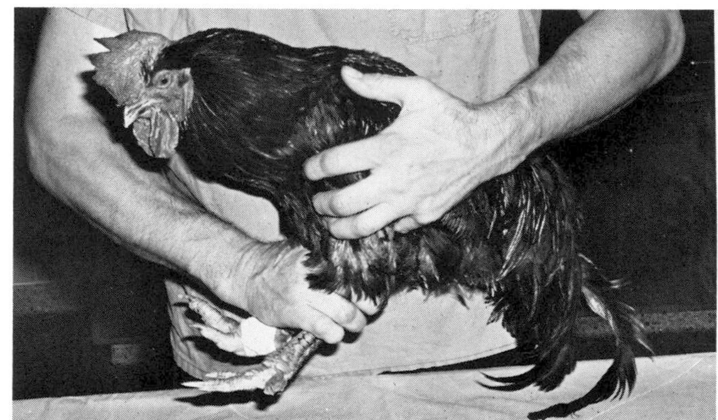

FIG. 15.6. Holding rooster, grasping both legs.

The beak of a chicken is easily opened and examined (Fig. 15.13). Blood samples are obtained from the brachial vein (Fig. 15.14). Masking tape is used for positioning an anesthetized bird for surgery (Fig. 15.15).

To restrain a chicken for caponization or abdominal exploration, stretch the legs backward and wrap a loop of rope or gauze around the wings, stretching the wings forward (Fig. 15.16).

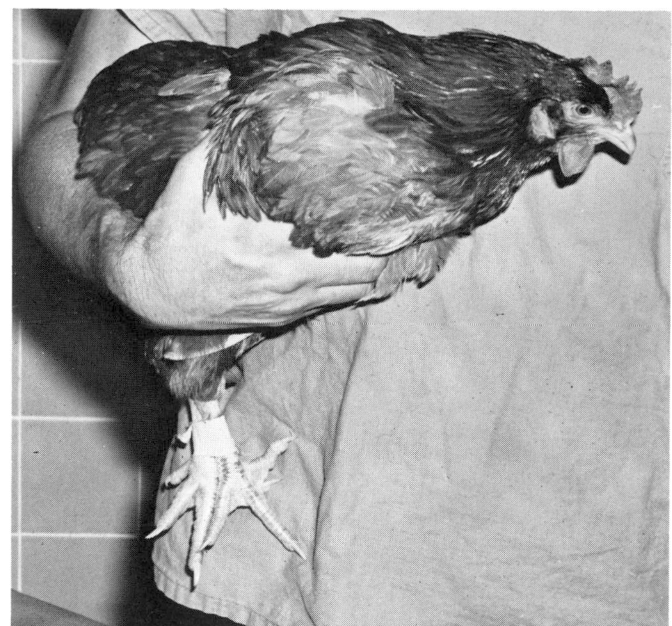

FIG. 15.7. A chicken can be carried with one hand if legs are taped together.

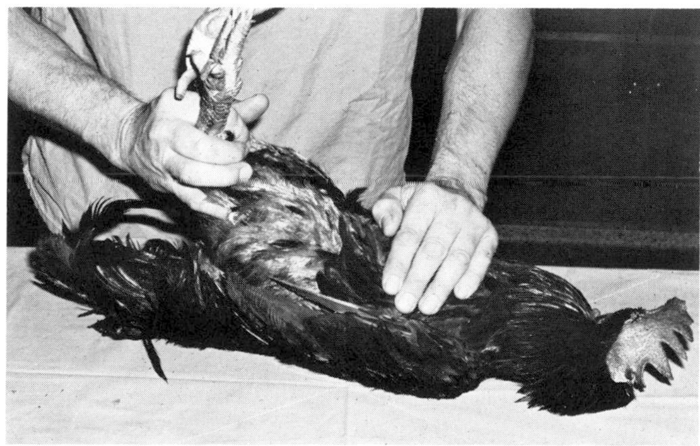

FIG. 15.8. A large rooster can be held on table with minimal struggling.

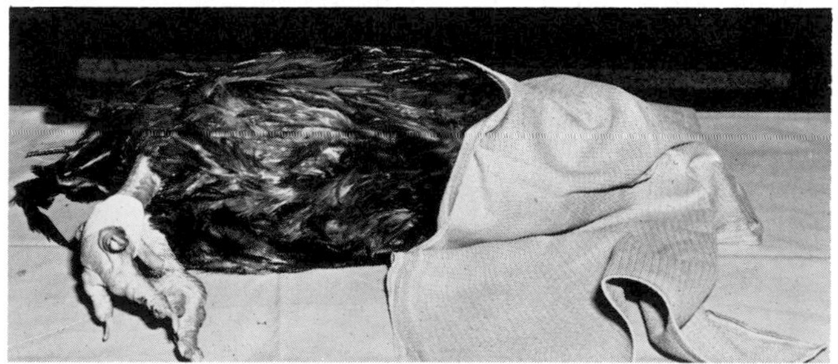

FIG. 15.9. A blindfolded chicken will lie quietly.

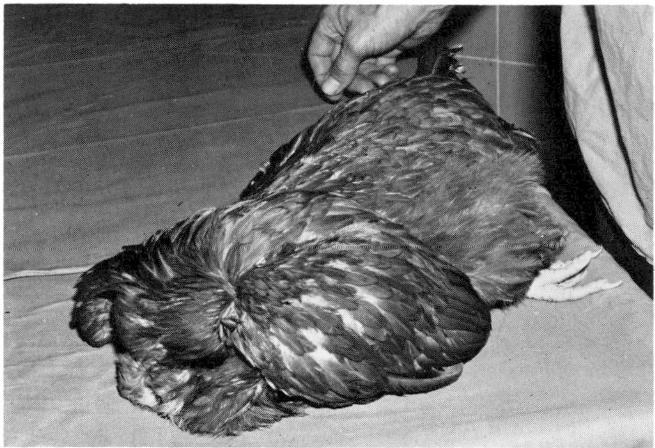

FIG. 15.10. State of hypnosis is induced by tucking head of chicken under its wing.

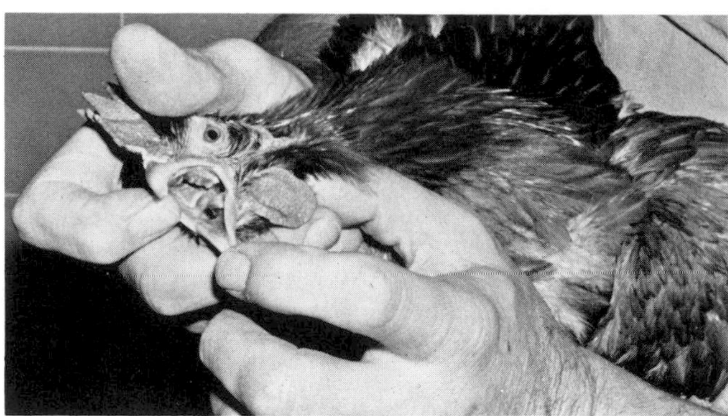

FIG. 15.13. Examining mouth of a chicken.

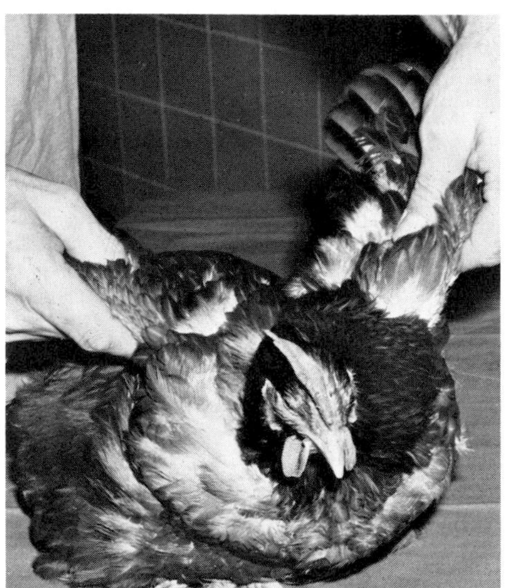

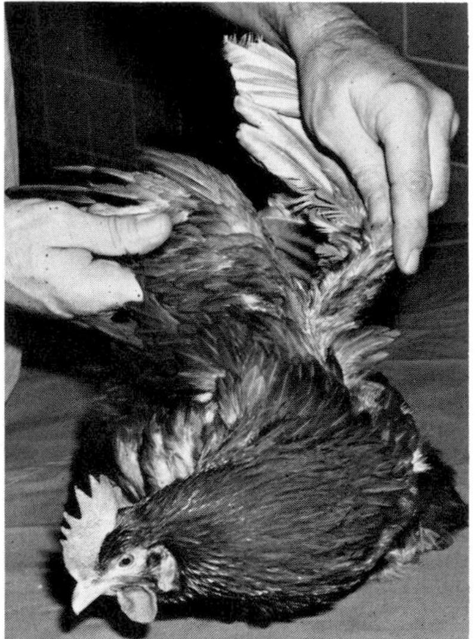

FIG. 15.11. Interfolding wings of a chicken: Grasp wings *(left)*. Cross wings and wrap them around one another *(right)*.

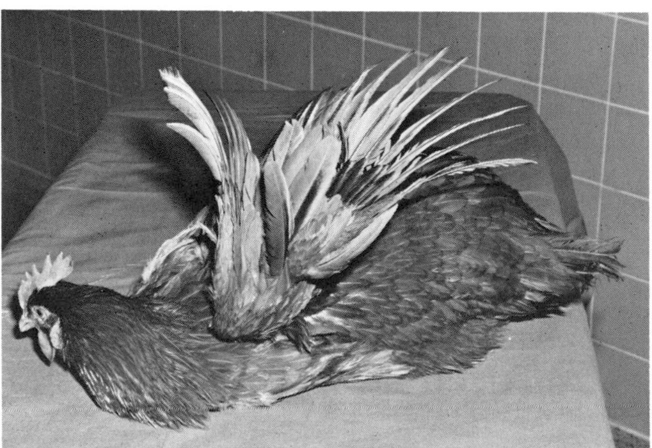

FIG. 15.12. Wings folded on a chicken.

Turkey

Adult turkeys are usually docile and can be approached while in a flock, especially by the usual caretaker. Slow but deliberate movements allow one to single out a bird and catch it without frightening the entire flock.

A large turkey can be captured as illustrated in Figures 15.17-15.19. A small bird can be grasped as one would a chicken (Fig. 15.20). If turkeys can be herded into a small pen, the handler can approach from behind in a kneeling position (Figs. 15.21, 15.22) to capture one. The heavy body of a large turkey cannot be safely suspended from the legs. Without body or wing support, the coxofemoral articulation of the hip may luxate.

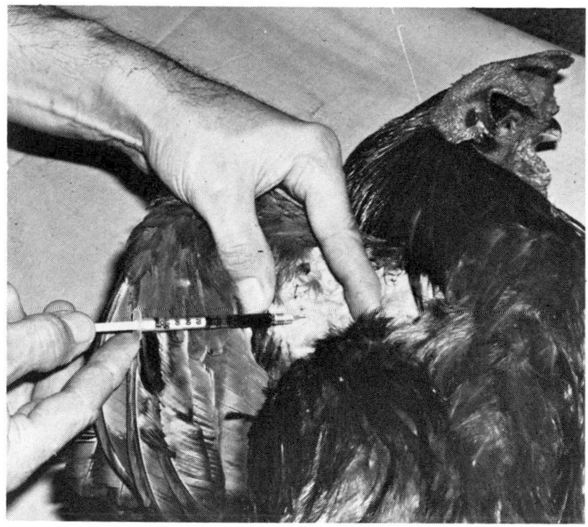

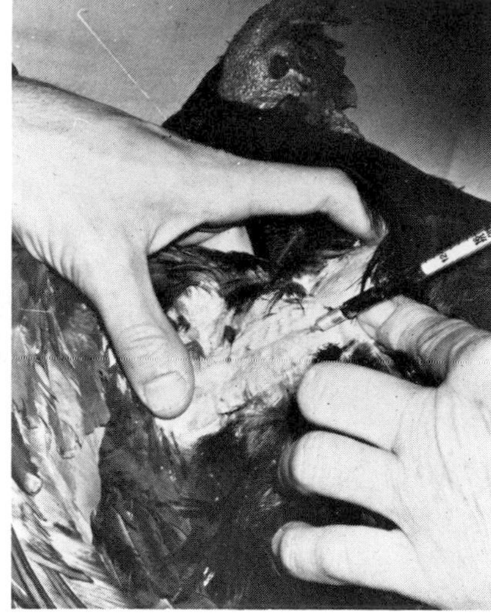

FIG. 15.14. Alternate approaches to venipuncture of brachial vein in a chicken.

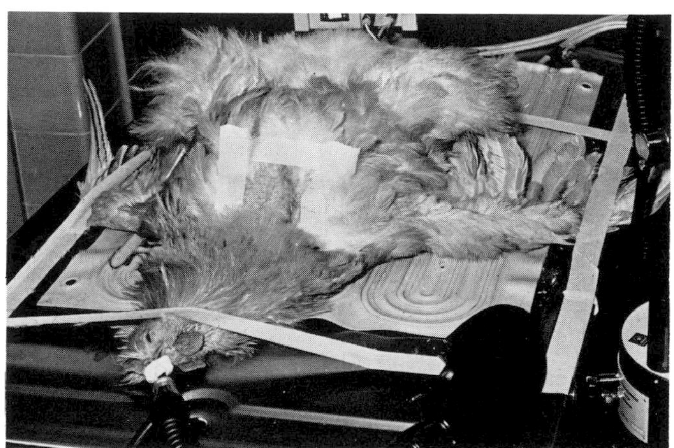

FIG. 15.15. Immobilization of anesthetized chicken using masking tape.

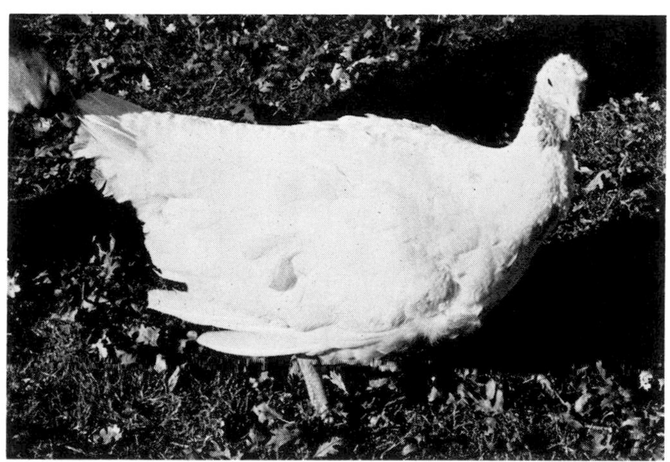

FIG. 15.17. Initial restraint of a turkey by grasping all the tail feathers.

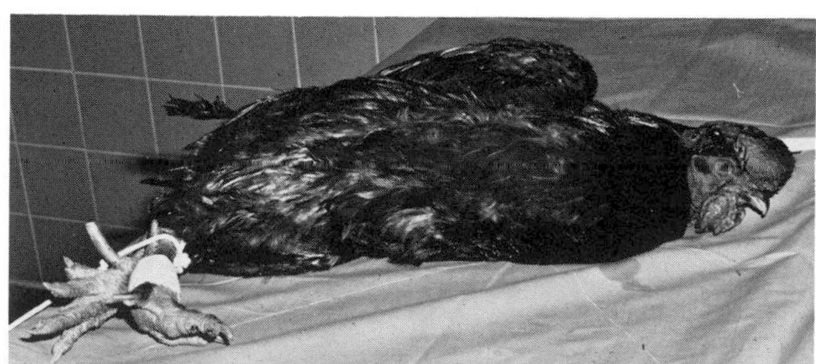

FIG. 15.16. Chicken can be stretched in lateral recumbency by securing both legs together. Another loop surrounds base of both wings.

FIG. 15.18. Reach under turkey and grasp both legs, allowing bird to fall on its breast.

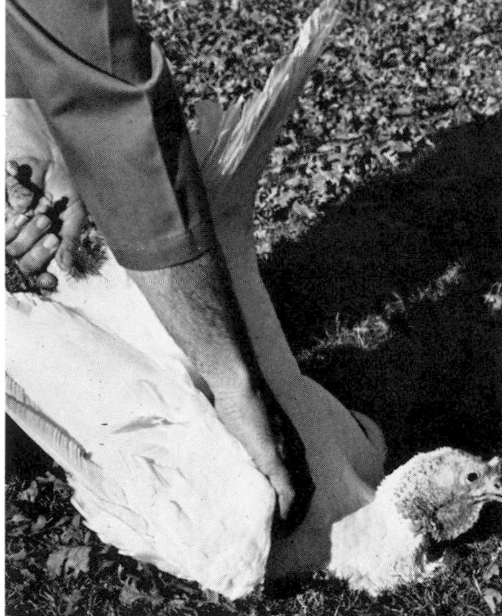

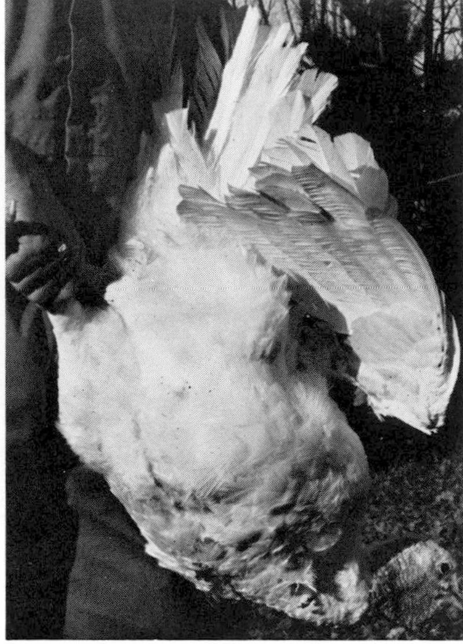

FIG. 15.19. Grasp a wing *(left)*. Lift bird by the legs and wing *(right)*.

FIG. 15.20. Alternate technique for approaching smaller birds. Grasp a wing, then reach under and grab the legs. Lift by legs and a wing as previously described.

FIG. 15.21. To capture turkey in close quarters, kneel behind bird and grasp both legs.

FIG. 15.22. With both legs in hand, lift small turkeys directly or, in addition, grasp a wing as before.

Blood samples are obtained via the brachial vein. Place the bird on a table and align the humerus to face directly at the person who will withdraw the blood (Fig. 15.23). A few feathers can be plucked to expose the vein.

Artificial insemination is a common practice in commercial turkey production. Adult breeder males (toms) must be ejaculated and females (hens) inseminated. When handholding a hen for artificial insemination, place her body between your legs with the breast toward you (Fig. 15.24). The vent is everted for the insemination by abdominal pressure. The tom can be held in a similar manner to force ejaculation, but since toms are heavier they are usually laid on a table (Fig. 15.25), in a trough, or on a handler's lap.

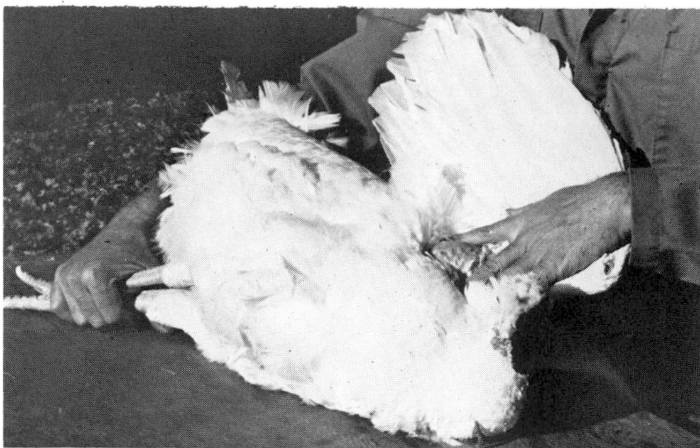

FIG. 15.23. Turkey held in position for venipuncture of the brachial vein.

FIG. 15.24. Hen turkey held for artificial insemination.

FIG. 15.25. Tom turkey in position for inducing ejaculation.

To administer oral medication, hold the turkey across the lap while sitting. Hold both of the turkey's legs near its body with one hand, stretching the head and neck straight out with the other hand [2]. An assistant then inserts a plastic tube containing the measured dose of medication through the mouth and into the crop. Hold a finger over the end of the tube to retain the fluid. When the tube is in place, remove the finger and gravity will empty the contents into the crop. The tube should be 25-27 cm (10-14 in.) long and 10 mm (⅜ in.) in diameter. Tablets can be given by pressing them into the esophagus, using the forefinger as a plunger.

Waterfowl

Ducks and geese can be captured with nets or hooks (Figs. 15.26, 15.27). Alternatively, they can be grasped by the neck (Fig. 15.28). If the duck is to be held for some time, or carried, grasp the wings (Fig. 15.29) or the legs and neck (Figs. 15.30, 15.31).

A large goose should be supported by the neck and wings when it is lifted (Fig. 15.32). If a heavy-bodied bird struggles, cervical injury may result. If the bird is to be carried, support the body (Fig. 15.33); the legs may flail, but they will not injure. Keep the head of a goose restrained at all times to prevent it from reaching around to peck at your face or eyes.

Birds frequently defecate when handled, so hold the bird in a manner that will prevent contamination of clothing.

Waterfowl can be sexed by everting the cloaca and prolapsing the penis (Fig. 15.34).

RELEASING BIRDS

Safe release from restraint is as important as safe capture. While holding a wing and both legs, bend over and, as the bird nears the ground, release first the legs, then the wing.

FIG. 15.26. Using hook to capture a goose.

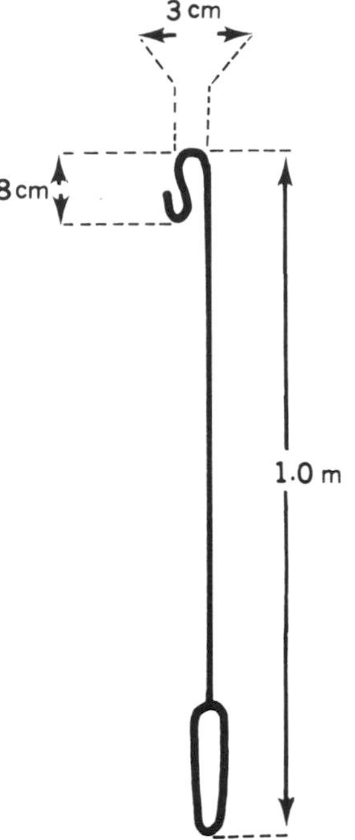

FIG. 15.27. Diagram of hook used to capture waterfowl.

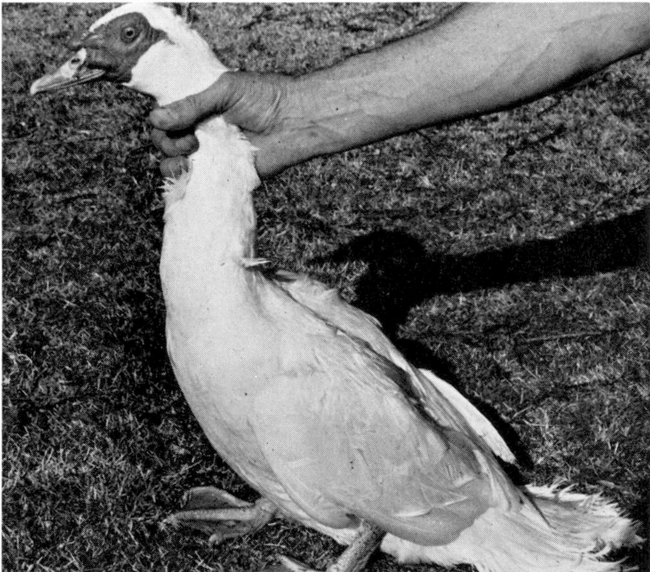

FIG. 15.28. Capturing duck *(upper)* and goose *(lower)* by grasping the neck.

FIG. 15.29. Lifting duck by grasping both wings. Keep a finger between wings.

FIG. 15.30. Controlling feet and neck of a duck.

TRANSPORT

Special crates are designed for transporting poultry and waterfowl to commercial processing plants. The individual pet bird can be placed in a cardboard box or put into a burlap or cloth sack. A hole is usually cut out of a corner of the sack to allow the head of a swan, goose, or turkey to protrude. This is not necessary for chickens or ducks.

FIG. 15.31. Carrying duck by controlling head, feet, and wings.

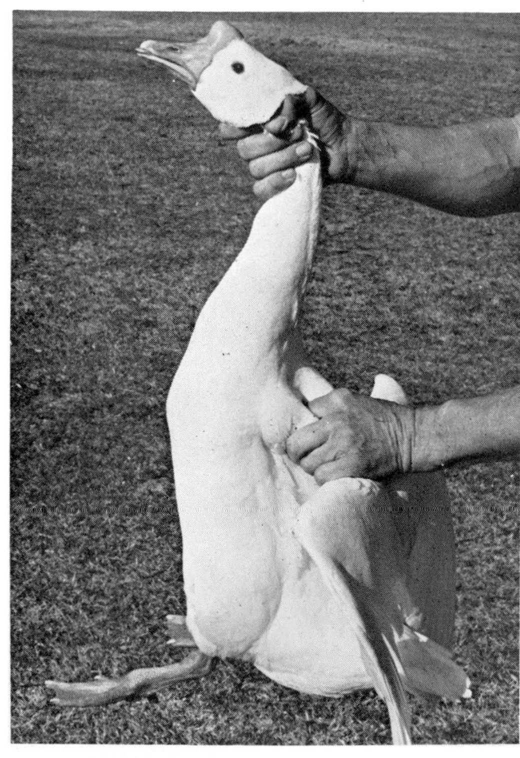

FIG. 15.32. Lifting a goose.

FIG. 15.33. Goose can be carried by supporting the body and controlling wings and head.

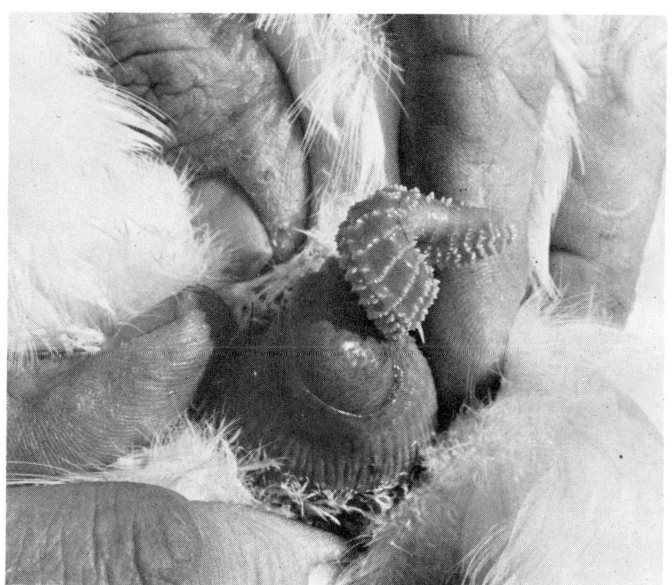

FIG. 15.34. Exposing penis of a gander.

A sacked bird transported in a closed automobile trunk on a hot day is threatened with hyperthermia and endangered from carbon monoxide fumes.

CHEMICAL RESTRAINT

Restraint drugs are not used for routine handling of domestic birds.

Ketamine hydrochloride (50 mg/kg) is used as a sedative and preanesthetic medication. Anesthesia may be induced and/or maintained with methoxyflurane and halothane.

REFERENCES

1. Grow, O. 1972. Modern Waterfowl Management. Chicago: American Bantam Association.
2. Leahy, J. R., and Barrow, P. 1951. Restraint of Animals. Ithaca, N.Y.: Cornell Campus Store.

3 Wild Animals

16 INTRODUCTION

THE BASIC principles of restraint apply to both wild and domestic animals. Nonetheless, some fundamental differences between wild and domestic animals must be understood and accounted for when wild species are handled.

DOMESTICATION [1-4]

Approximately 35 of the nearly 50,000 species of vertebrates have adapted to man's needs for food, fiber, work, sport, and beauty and are considered domesticated (Tables 16.1, 16.2). All but three or four species were living in harmony with man before the time of recorded history.

Domestication is an evolutionary process that involves a gradual (thousands of years) change in the gene pool of a species to allow adaptation to an artificial environment. Domestic animals must cope with buildings, fences, crowding, confinement, lack of privacy, changed photoperiodicity, altered climatic conditions, and different food.

TABLE 16.1. Domestic mammals

Common Name	Scientific Name	Family	Order
Mouse	*Mus musculus*	Muridae	Rodentia
Rat	*Rattus norvegicus*		
Guinea pig	*Cavia porcellus*	Cavidae	
Golden hamster	*Mesocricetus auratus*	Cricetidae	
Rabbit	*Oryctolagus cuniculus*	Leporidae	Lagomorpha
Dog	*Canis familiaris*	Canidae	Carnivora
Fox	*Vulpes fulva*		
Cat	*Felis catus*	Felidae	
Mink	*Mustela vison*	Mustelidae	
Ferret	*Mustela furo*		
Asian elephant	*Elephas maximus*	Elephantidae	Proboscidea
Horse	*Equus caballus*	Equidae	Perissodactyla
Ass (donkey)	*Equus asinus*		
Swine	*Sus scrofa*	Suidae	Artiodactyla
Bactrian camel	*Camelus bactrianus*	Camelidae	
Dromedary camel	*Camelus dromedarius*		
Llama	*Llama glama*		
Alpaca	*Llama pacos*		
Reindeer	*Rangifer tarandus*	Cervidae	
Cattle-European	*Bos taurus*	Bovidae	
Cattle-Zebu	*Bos indicus*		
Yak	*Bos grunniens*		
Banteng	*Bibos banteng*		
Gayal	*Bibos frontalis*		
Water buffalo	*Bubalus bubalis*		
Musk-ox	*Ovibos moschatus*		
Sheep	*Ovis aries*		
Goat	*Capra hircus*		

TABLE 16.2. Domestic birds

Common Name	Scientific Name	Family	Order
Pekin duck	*Anas platyrhyncos*	Anatidae	Anseriformes
Muscovy duck	*Cairina moschata*		
Goose	*Anser anser*		
Canada goose	*Branta canadensis*		
Mute swan	*Cygnus olor*		
Chicken	*Gallus gallus*	Phasianidae	Galliformes
Ring-necked pheasant	*Phasianus colchicus*		
Coturnix quail	*Coturnix coturnix*		
Peafowl	*Pavo cristatus*		
Guinea fowl	*Numida meleagris*	Numidae	
Turkey	*Meleagris gallopavo*	Meleagrididae	
Pigeon	*Columba liva*	Columbidae	Columbiformes
Budgerigar	*Melopsitticus undulatus*	Psittacidae	Psittaciformes
Canary	*Serinus canarius*	Fringillidae	Passeriformes

Genetic alteration during the evolutionary process took place by selection for specific characteristics that were economically or esthetically pleasing to man. Docile animals were selected over aggressive individuals. This may require only a single gene mutation. Other economically important characteristics included higher fertility, rapid growth, efficient food conversion, higher milk production, and disease resistance. Man has often selected polled cattle over horned breeds to minimize injury.

There was definite selection to reduce or eliminate undesirable wild characteristics such as territoriality, intraspecific dominance, elaborate food identification and gathering mechanisms, intricate courtship behavior, and fear of man. The constant selection yielded animals that are much easier to handle. They tolerate man's presence without a flight response. If physically restrained they rarely fight to the death, as do some wild species.

The wild animal restrainer must cope with a multitude of extremely sensitive behavioral responses. Wild animals react faster and are relatively much more powerful than domestic counterparts. Psychological, physical, and physiological stresses play more significant roles in the development of undesirable restraint sequelae.

It should be understood that we cannot domesticate the wild animals in our charge. Wild animals can be tamed to tolerate human beings. They can also be trained to peform tremendous feats. Trainers spend countless hours repeating maneuvers and gaining rapport with their animals. Trained animals frequently respond maximally only to the trainer. None of this results in domestication.

There are numerous instances where human-animal associations have been close and lasting. The lioness Elsa and Joy Adamson established a relationship that was superb, but such relationships are rare.

It must be indelibly imprinted on the minds of those who handle wild animals that even though tame and gentle, all wild species are capable of instantaneous reversion to the wild state. This is particularly important for restrainers when dealing with privately owned animals. No matter how faithful or well trained and affectionate a given animal is, the owner can never be absolutely sure that the animal will not turn on him or her under certain circumstances.

Some wild species kept as pets become extremely dangerous as they mature. A pet raccoon raised from infancy by a woman turned on her and disfigured her face. Most large solitary wild felids are gregarious while in the litter growing up, but as soon as they reach sexual maturity, they begin a solitary life that may be territorial. Owners who have raised such cats to adulthood may find that their loving, affectionate friend has become an adversary. The cat assumes that the home is its territory. The owner, a member of the family, or a visiting friend may suddenly be attacked as an intruder.

It is important to interject here that mild tranquilization, instead of rendering a tame animal more amenable to restraint, may abolish any inhibitions it has developed in regard to attacking a person.

In essence, restraint practices in wild animals must be absolutely necessary, well planned, and executed quickly and efficiently.

REFERENCES

1. Dembeck, H. 1965. Animals and Man. Garden City, N.Y.: Natural History Press.
2. Fox, M. W., ed. 1968. The influence of domestication upon behavior of animals. In Abnormal Behavior in Animals, pp. 64-75. Philadelphia: W. B. Saunders.
3. Hafez, E. S. F. 1968. Adaptation of Domestic Animals, pp. 38-45. Philadelphia: Lea & Febiger.
4. Zeuner, F. E. 1963. The History of Domestic Animals. London: Hutchins.

17 MONOTREMES AND MARSUPIALS

CLASSIFICATION (numbers in parentheses denote number of known species)

Order Monotremata
- Family Tachyglossidae: echidna (5)
- Family Ornithorhynchidae: platypus (1)

Order Marsupialia
- Family Didelphidae: New World opossum (65)
- Family Dasyuridae: marsupial mouse, Tasmanian devil, numbat (49)
- Family Notocryctidae: marsupial mole (2)
- Family Peramelidae: bandicoot (20)
- Family Caenolestidae: rat opossum (7)
- Family Phalangeridae: phalanger, koala (48)
- Family Phascolomidae: wombat (2)
- Family Macropodidae: wallaby, wallaroo, kangaroo (55)

ADULT males are referred to as males, bucks, or boomers; adult females as females, does, or flyers. The infant kangaroo is called a joey; the infant koala a cub.

Some weights of monotremes and marsupials are listed in Table 17.1.

TABLE 17.1. Some weights of monotremes and marsupials

Animal	Kilograms	Pounds
Echidna	2.5-6	6-13
Platypus	0.5-2	1.4
Tasmanian devil		
male	6.35-9.07	14-20
female	4.53-5.44	10-12
Bandicoot	0.55	1
Koala	5-15	9-33
Common wombat	15-35	33-35
Red-necked wallaby	4-24	14-24
Red kangaroo } Gray kangaroo }	23-70	51-154

MONOTREMES

Monotremes are primitive egg-laying mammals, represented by the platypus (restricted to the Australian continent) and the echidna (five species distributed throughout Australia, Tasmania, and New Guinea) [2].

Danger Potential

Venom of the male platypus is delivered through a hollow curved spur on the medial aspect of the tarsal (hock) joint of the hind leg. The spur is normally carried against the leg. When the animal becomes agitated, the spur erects. The platypus kicks with a jabbing motion, usually with both hind legs.

The venom gland is located on the medial aspect of the thigh and is emptied by a duct into a small reservoir situated below the spur. A second duct connects the reservoir to the spur.

The hind legs of the platypus are short. It is safe to pick up the animal by the tail while supporting the body with the other hand. The platypus is not particularly aggressive, but persons have been envenomated through careless handling. Human envenomation is not likely to be lethal. Signs reported are immediate intense pain, swelling at the wound, numbness around the wound, and a feeling of faintness.

The echidna possesses a diminutive spur and crural venom gland, but there are no reports of human envenomation by these animals. Echidnas cannot bite. Their diet consists primarily of small insects and other invertebrate species. The elongated snout is designed to probe in the soil or beneath objects, searching for prey.

Spines of the echidna are not barbed like those of the porcupine but are sharp enough to injure the ungloved hand. The echidna is an efficient burrower with strong claws, which can injure by raking if care is not taken.

Physical Restraint

The platypus is fragile and easily upset, but it can be netted from the aquatic environment and picked up gently. It responds adversely to excessive handling.

Echidnas can be picked up and handled without difficulty if a glove is worn (Fig. 17.1). If the animal is on a firm surface, place the gloved fingers underneath the abdomen to pick it up or grasp the tail or foot with one hand, placing the other beneath the animal to support the body.

The echidna rolls itself into a tight ball when disturbed, much as do hedgehogs and armadillos, exposing only the blunted spines. If the animal cannot be induced to relax, it must be chemically immobilized for examination.

FIG. 17.1. Handling an echidna.

An echidna dug into the soil can be brought to the surface by filling the hole with water.

MARSUPIALS

There are 248 species of marsupials. Size variation is extreme, ranging from 25 g to 70 kg.

Danger Potential

Scratching with strong claws is the primary mode of defense for most marsupials. The large macropods can injure severely by kicking with the hind legs as well as by scratching with both forefeet and hind feet. Some large species, like the eastern gray kangaroo and the red kangaroo, sit back on their hind limbs and tails to kick and claw. The power of the kick can knock a man down, and the claws can disembowel an adversary. If an aggressive kangaroo attacks when you are unprotected, strike the animal in the neck with your fist to drive it off. Hitting the body is ineffective.

All marsupials are capable of biting. A number of species are carnivorous, with strong jaw muscles and sharp teeth. Herbivores are somewhat less inclined to bite than carnivores; since their teeth are not adapted for tearing, they inflict less severe damage.

Anatomy and Behavior

Marsupials exhibit diverse morphologic and physiologic characteristics. The marsupials of the Australian continent and contiguous islands evolved in the absence of other mammalian species. Thus various species developed adaptations that allow them to exploit different habitats. Some marsupials are rodentlike in all their behavioral repertoire; others have become carnivorous, serving in the role of predator; still others are herbivorous, existing on sparse grass and herbaceous material much as do the large grazers in the order Artiodactyla [1,2,5].

Marsupials are characterized by a unique reproductive biology that provides for the final development of the fetus in a specialized pouch called a marsupium. The structure and location of the pouch vary from species to species. In all cases this may present special problems during restraint, because the presence of a fetus in the pouch is not easily recognized.

Upon arrival in the pouch, the fetus attaches firmly to one of the nipples, establishing a semipermanent bond. Rough handling may cause the fetus to pull away from the nipple and even be extruded from the pouch. If the fetus cannot be immediately reattached, it will die. The female will not replace the expelled fetus on her own, and it may not be possible for the very young infant to reestablish a connection if jostled loose. More mature infants, if not killed outright, may be traumatized.

Physical Restraint

The only North American marsupial is the Virginia opossum [4,6]. This medium-sized omnivorous marsupial has a formidable array of 50 small sharp teeth [6]. A cornered opossum is dangerous, but it can be handled and captured quite easily with a net. If suspended from the tail, the opossum can climb up its own tail and bite a handler; if the animal's feet are on a firm surface, it will pull away from pressure on the tail [6]. An opossum can be held by placing a gloved hand on either side of the head and neck.

An oppossum in a restricted enclosure can be snared (Fig. 17.2). Once the snare is around the neck, grasp the tail and hold it alongisde the handle of the snare to help control the animal.

Free-living opossum frequently prey on birds and their eggs, making them unwelcome inhabitants of zoo grounds. Baited live traps are frequently used to capture them for relocation (Fig. 17.3).

FIG. 17.2. Virginia opossum in a snare.

FIG. 17.3. Trapped Virginia opossum.

Small rodentlike species such as the sugar glider and bandicoot are handled in much the same manner as wild rats [9]. Gloves, towels, tubes, and other special devices can be purchased or improvised for handling these small marsupials (see Chapter 18).

The Tasmanian devil is reputed to be a ferocious carnivore. The reputation is only partially deserved, since this animal can be handled and moved with some ease. When threatened, the Tasmanian devil opens its mouth wide, exposing an array of carnivorous teeth to frighten off the attacker. A household broom or long-handled scrub brush can be used to ward off the mouth attack while maneuvering the animal into a position that enables a handler to grasp the tail and lift the animal off the ground. If suspended, the Tasmanian devil cannot turn on itself and climb the tail to bite the handler. If the animal must be further restricted, place a hand at the back of the head.

The Tasmanian devil is easily netted, but it is difficult to get a hand around the very short neck to extract it from the net. In a confined space it is more efficient and effective to place a snare over its head (Fig. 17.4) and stretch its hind legs (Fig. 17.5). One can then manipulate, examine, or collect blood samples from the Tasmanian devil as desired.

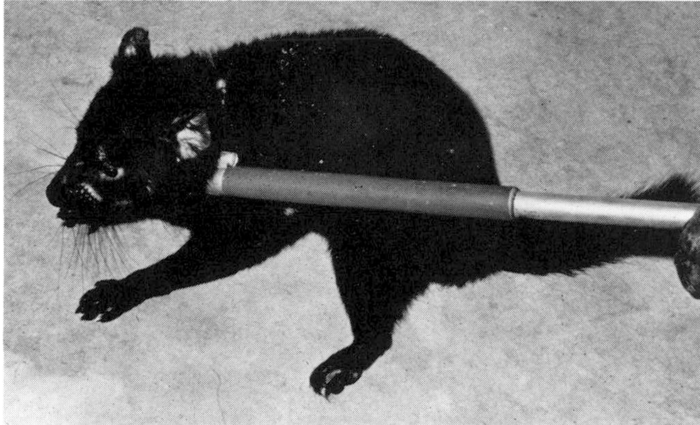

FIG. 17.4. A snare can be used on a Tasmanian devil.

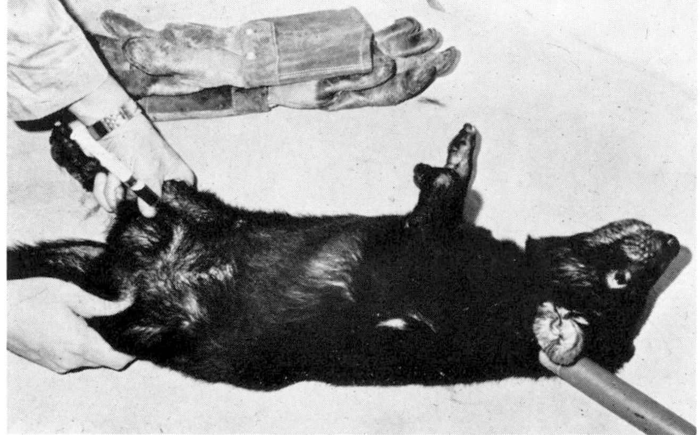

FIG. 17.5. Tasmanian devil being bled from the femoral vein.

Phalangers are small to medium-sized animals [7]. Many are arboreal and thus have sharp grasping claws. Nets and snares are suitable restraint tools. Some phalangers can be grasped with gloved hands. Bush-tailed phalangers can be grasped bare-handed around the back of the neck with one hand while grasping the rump or base of the tail with the other [7].

The koala, though rather phlegmatic, is capable of biting and scratching. Gloves may protect from bites and scratches. It likes to cling to soft furry objects and may tear clothing if the handler is wearing a sweater.

A koala in a tree may climb higher if someone taps on the tree trunk. If it is possible to climb above the animal, a light tap on the head will induce it to climb down. It is often captured in small timber by attaching a cloth to a pole and waving it over the animal's head. As the koala retreats, it is followed down the tree with this device. The collector at the foot of the tree must be alert, because koalas move rapidly on the ground.

Figure 17.6 shows a keeper approaching a koala sitting on a branch. The arms are grasped and pinned to the side of the body (Fig. 17.7). An experienced handler can carry out this restraint procedure bare-handed. If an extended period of restraint is required, the animal can be placed in a canvas sack or finely meshed net.

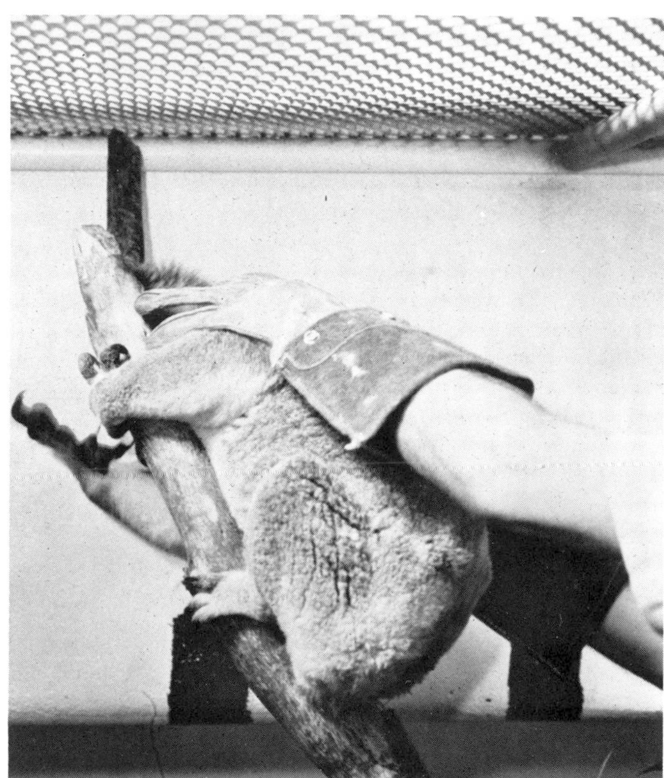

FIG. 17.6. Approaching a perched koala.

Do not grasp a koala around the middle. Young koalas can be held like rabbits by grasping the forelimbs and hind limbs on each side with one hand and fixing the head with the thumbs (see Fig. 14.5).

If properly conditioned as juveniles, koalas can be taught to permit the handler to pick them up and place them in the crook of his or her elbow to ride to another location. The koala clings to the handler's palm for support.

Wombats are heavy-bodied burrowing marsupials [13]. Although they are not overly aggressive, they possess strong limbs with claws designed for digging and burrowing. Captive wombats have been known to charge, bite, and grasp

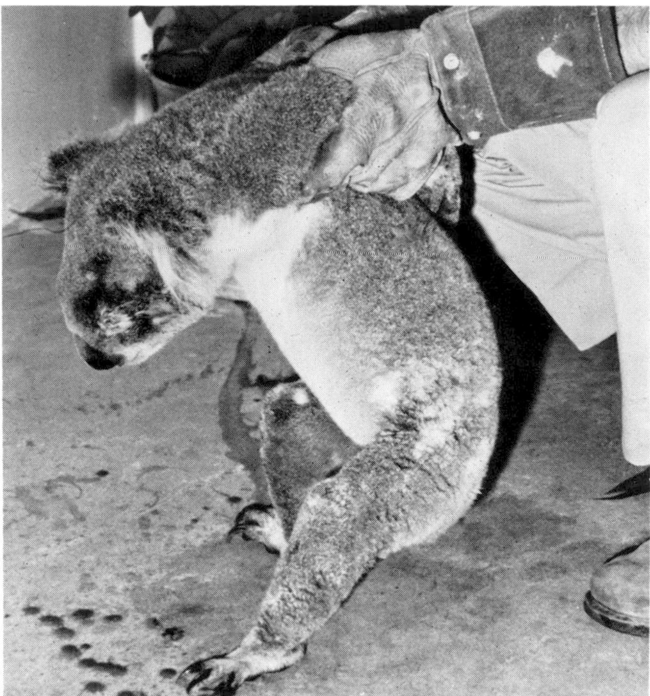

FIG. 17.7. Holding forelimbs of a koala behind the back. (This is also a suitable technique to control a small primate.)

handlers. The claws can inflict serious injury on another animal or a careless restrainer.

A tapering canvas bag with laced-up inspection ports is a useful device for handling wombats [13]. If the bag is placed over the head of the wombat, it will climb inside until it is firmly wedged.

Nets and squeeze cages are effective for moving wombats from one place to another or for restricting them for medication or examination. To pick up a wombat and put it into a crate, approach it from above and behind. Grasp the animal around the body just behind the front limbs and hold it against the chest. The wombat is not generally inclined to bite, but it may. If held in this position, the short neck and decreased mobility of the head prevent it from doing so.

All of the macropods are efficient jumpers. The hind limbs are highly developed, capable of carrying the animal through the air for great distances. The hind limbs are also used as weapons [1,2,5].

Use a sheet of burlap or opaque plastic to herd a mob (group) of macropods from one enclosure into another. Wallabies can be moved from one place to another by letting them crawl or hop along while directing them by the tail (Fig. 17.8).

Before capturing macropods, it is desirable to confine them in an enclosed solid-walled structure [8]. Capturing macropods from an enclosure surrounded by wire netting or cyclone fencing is hazardous; the animals may not recognize such a fence as a barrier and, in their excitement and fright, may jump against the fence, causing serious injury to themselves or to immature young in the pouches of females.

FIG. 17.8. Moving wallaroos and wallabies by directing them with an elevated tail.

FIG. 17.9. Netting a wallaroo.

Nets are useful for the initial capture of wallabies or other small species (Fig. 17.9). A net with a very fine mesh should be used to avoid entangling the claws. The macropod's agility and ability to jump may tax the skills of a person attempting to net it; once the animal is caught, the tail can be grasped and the animal removed from the net. After the tail is grasped, the limbs can be grabbed and the animal stretched (Fig. 17.10).

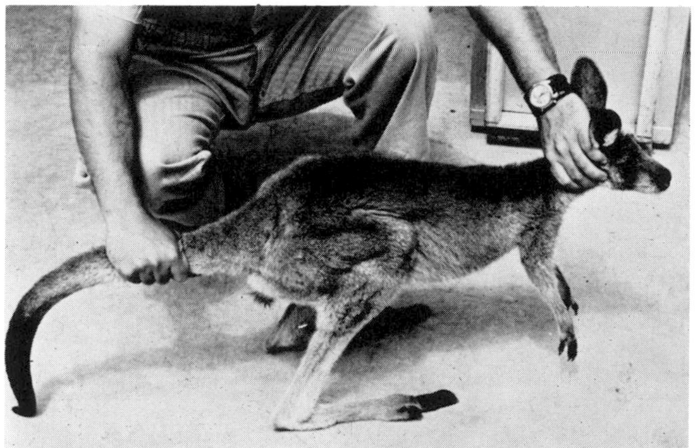

FIG. 17.10. Restraining a small kangaroo.

Small macropods may traumatize their spines if they are allowed to kick out with the hind legs while being captured or held by the tail. Agile wallabies in particular have suffered from posterior paralysis as a result of such injury. Large macropods can be held by the tail but should be turned to the side (Fig. 17.11).

Three experienced handlers can enter a group of large macropods to capture them by hand. Two, acting as a team, simultaneously grasp a selected animal by the tail, while the third person surrounds the animal's body with his arms or grasps and stretches the hind legs. A kangaroo may constantly turn to face and defy restrainers, preventing them from getting behind it to grasp the tail. A great degree of self-confidence is required for one person to close in and parry with such an animal, distracting its attention to allow the team to grasp the tail from behind.

A forked stick (see Fig. 24.10) similar to that used to push away an ostrich can be used to push a large male off balance until another person can grab the tail [1]. Once it is captured, a large kangaroo can be kept off balance by bringing the tail forward between the hind legs and taping it to the body. This technique is valuable if repeated handling is necessary within a few hours.

The length of time that anyone can hold a large macropod off the ground is limited. If prolonged examination or manipulation is required, the animal must be subdued and held recumbent on the ground. If at all possible, lay the animal on a padded surface such as an old mattress to prevent trauma during restraint (Fig. 17.12). The tail can be held by one person while another grasps the hind legs and a third person holds the front legs and restricts mobility of the head. A burlap bag can be placed over the head and forequarters to assist the person holding the head [1].

FIG. 17.11. Macropod is handled by lifting its hind feet off the ground, directing the hind legs away from the handler.

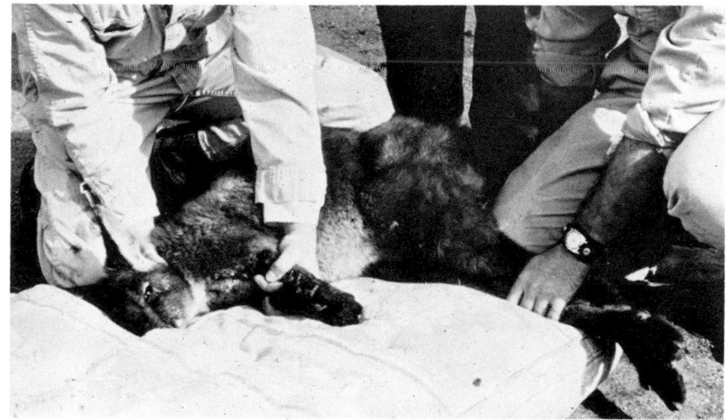

FIG. 17.12. Holding a kangaroo on a mattress.

A macropod that cannot be approached to grasp its tail can be squeezed into a corner with the use of solid shields, canvas, or a mattress. A mattress can be used to press the animal against the wall and hold it for grasping. A mattress is also desirable when a large animal must be chemically immobilized. During the induction state of immobilization, the animal may become hyperactive; unless activity is restricted, it may jump against walls and fences. When this type of activity develops, a mattress can be placed over the animal to hold it until it succumbs to the drug.

When a joey is mature enough to leave and return to the pouch, it can be handled and examined independently. To remove the joey from the pouch, grasp the youngster by the tail and give it a quick jerk.

Venipuncture in macropods can be carried out by a number of methods. Restraining the animal in lateral recumbency and stretching the head slightly exposes the jugular vein on the ventrolateral surface of the neck. In large animals the saphenous vein can be raised on the medial aspect of the tibia by applying pressure around the stifle. The femoral vein is accessible in all species, immediately posterior to and medial to the femur. Kangaroos have a medial tail vein which can be used to obtain blood samples, but it should not be used to administer intravenous medication.

Transport

Platypuses require extraordinary equipment and special care in transporting since they are extremely fragile and delicate. Marsupials can be crated and shipped like other mammals of comparable size. Macropods are subject to capture myopathy [10].

Crates with wire ends are hazardous for shipping or holding macropods. If they become frightened or excited, they will jump against the wire in an attempt to escape.

Chemical Restraint

Phencyclidine hydrochloride combined with acepromazine hydrochloride has been used successfully on red and western gray kangaroos [8]. The recommended dose is 1 mg/kg phencyclidine and 0.3 mg/kg acepromazine.

Etorphine hydrochloride and acepromazine maleate were used in another study to capture red kangaroos [15]. The dosages were 0.04 mg/kg etorphine and 0.16 mg/kg acepromazine. The animals were not immobilized but were sedated and handleable.

Xylazine and ketamine, 5-10 mg/kg and 10-20 mg/kg respectively, have been used in the Virginia opossum [14].

Marsupials have variable tolerances to general anesthetics. One study established the anesthetic dose of pentobarbital sodium to be 56-71 mg/kg [11,12]. The standard eutherian mammal dose is 30 mg/kg.

Ketamine has been shown to be a safe agent in red kangaroos (15 mg/kg) and wallaroos (19 mg/kg) [3]. My own experience verifies that ketamine (10-15 mg/kg) is satisfactory for handling wallaroos.

Effective anesthetic-immobilizing agents in the Virginia opossum include ketamine hydrochloride (20-25 mg/kg) and fentanyl-droperidol (0.4 mg/kg and 20 mg/kg respectively) [4]. Phencyclidine hydrochloride (5-6 mg/kg) was less effective.

REFERENCES

1. Calaby, J. H., and Poole, W. E. 1971. Keeping kangaroos in captivity; marsupials in captivity. Int. Zoo Yearb. 2:5-12.
2. Collins, L. R. 1973. Monotremes and Marsupials. Washington, D.C.: Smithsonian Institution Press.
3. Denny, M. J. S. 1973. The use of ketamine HCl as a safe, short duration anesthetic in kangaroos. Br. Vet. J. 129:362-65.
4. Feldman, D. B., and Self, J. L. 1971. Sedation and anesthesia of the Virginia opossum, *Didelphis virginiana*. Lab. Anim. Sci. 21:717-20.
5. Firth, J. H., and Calaby, J. H. 1969. Kangaroos. Melbourne, Australia: Cheshire.
6. Fritz, H. I. 1971. Maintenance of the common opossum in captivity; marsupials in captivity. Int. Zoo Yearb. 2:32-33.
7. Hope, R. M. 1971. The maintenance of the brush-tailed opossum in captivity; marsupials in captivity. Int. Zoo Yearb. 2:24-25.
8. Keep, J. M., and Fox, A. M. 1971. The capture, restraint and translocation of kangaroos in the wild. Aust. Vet. J. 47:141-44.
9. Lyne, A. G. 1971. Bandicoots in captivity; marsupials in captivity. Int. Zoo Yearb. 2:41-43.
10. Munday, B. L. 1972. Myonecrosis in free-living and recently captured macropods. J. Wildl. Dis. 8:191-92.
11. Watson, C. R. R., and Way, J. S. 1971. Anesthetics for kangaroos; marsupials in captivity. Int. Zoo Yearb. 2:12-13.
12. ______. 1972. The unusual tolerance of marsupials to barbiturate anesthetics. Int. Zoo Yearb. 12:208-11.
13. Wells, R. T. 1971. Maintenance of the hairy-nosed wombat in captivity; marsupials in captivity. Int. Zoo Yearb. 2:30-31.
14. Wiesner, H. 1976. Beuteltiere (marsupials). In Heinz-Georg Klos and Ernst M. Lang, eds. Zootierkrankheiten, p. 241. Berlin: Paul Parey.
15. Wilson, G. R. 1974. The restraint of red kangaroos using etorphine-acepromazine and etorphine-methotrimeprazine mixtures. Aust. Vet. J. 50:454-58.

18 SMALL MAMMALS

CLASSIFICATION (numbers in parentheses denote number of known species)

Order Insectivora
- Superfamily Tenrecoidea: solenodon, tenrec, other shrews (29)
- Superfamily Chrysochloroidea: golden mole (15)
- Superfamily Erinaceoidea: hedgehog (19)
- Superfamily Macroscelidoidea: elephant shrew (21)
- Superfamily Soricoidea: shrew, mole (284)

Order Dermoptera
- Family Cynocephalidae: flying lemur (2)

Order Chiroptera: bats
- Family Megachiroptera: fruit bat (150)
- Family Microchiroptera: insectivorous bat (831)

Order Edentata
- Family Myremecophagidae: anteater (4)
- Family Bradypodidae: sloth (7)
- Family Dasypodidae: armadillo (21)

Order Pholidota
- Family Manidae: pangolin (7)

Order Lagomorpha
- Family Ochontonidae: pika (14)
- Family Leporidae: rabbit, hare (52)

Order Rodentia
- Suborder Sciuromorpha: squirrel, marmot, chipmunk, gopher, beaver, kangaroo rat, springhaas (366)
- Suborder Myomorpha: rat, mouse, hamster, lemming, mole (1,183)
- Suborder Hystricomorpha: porcupine, cavy, capybara, chinchilla, agouti, guinea pig (180)

Order Tubulidentata
- Family Orycteropodidae: aardvark (1)

Order Hyracoidea
- Family Procaviidae: hyrax (6)

MOST SMALL mammal adults are referred to as male and female. The young and newborn are called infants, offspring, youngsters, or juveniles.

A few introductory remarks will indicate some techniques common to many species. Specialized problems or procedures will be described for taxonomic groups.

Biting, scratching, and clawing are common defensive and offensive actions by small mammals.

Many species can be handled with gloves. Nets and snares are useful, the type varying with the size and behavior of the animal.

All the insectivores have tiny sharp teeth. If agitated, they may bite at hard objects and injure their teeth.

INSECTIVORA

Tenrecs, shrews, and hedgehogs are primitive insectivorous mammals. Shrews are tiny rodentlike creatures with an extremely high metabolic rate. They cannot tolerate extensive manipulation because they may overheat and become hypoglycemic in a short time. Shrews are handled with techniques like those used for rodents, e.g., plastic tubes or other similar devices. These techniques minimize direct handling, which distresses this species. Light gloves protect the hands if it is necessary to grasp the animal.

The following species of shrews and solenodons are known to be venomous: American short-tailed shrew, European water shrew, masked shrew, British bicolored water shrew, and Haitian solenodon [3]. Probably other species are also venomous, at least to their prey species [3]. The venom apparatus in shrews and solenodons consists of modifed submaxillary salivary glands. The ducts of these glands open near the incisor teeth, which may be grooved or channeled to assist in the deposition of venom. Envenomation of a human being is usually of little consequence, but precautions should be taken.

Tenrecs and hedgehogs have teeth and may bite. Many individuals kept in captivity for a long period become docile and can be handled gently with the bare hand. However, if it becomes necessary to severely restrict movement, it is advisable to wear a glove (Fig. 18.1).

Modified hairs of hedgehogs and tenrecs are spiny. The spines or quills are not cast and are neither sharpened nor barbed like the quills of New World porcupines. Hedgehogs characteristically make a sudden jerking motion when touched, to make the intruder cognizant of its prickly nature.

Tenrecs and hedgehogs protect themselves by rolling up into a tight ball, exposing only quills. This is accomplished by a unique superficial muscle layer. These animals cannot be placed into a squeeze cage because as soon as they sense

FIG. 18.1. The quills of a hedgehog are not barbed nor extremely sharp. Nonetheless, gloves are usually recommended for handling them.

strange surroundings or pressure, they immediately roll into a ball. Nothing but spines is exposed for examination. Exerting more pressure crushes the animal. These animals can be examined only if they are completely relaxed. Otherwise, they must be chemically immobilized, using 20 mg/kg ketamine hydrochloride. This high dosage is required to abolish the sensitive reflex of rolling up in a ball when touched. Intramuscular injection into the superficial muscle can be made between the spines.

The sex of a tenrec or hedgehog can be determined by placing the animal in a box with a transparent bottom (plastic or glass). Once the animal has relaxed, the box can be lifted and the external genitalia viewed from below [11].

DERMOPTERA—FLYING LEMURS

Gloves and nets are used to handle these animals.

CHIROPTERA—BATS [6,9,15,16,20]

Danger Potential

All species of bats bite. It is necessary to control the head at all times. Vampire bats and some species of insectivorous bats are known to carry rabies virus. Bites from infected animals may prove lethal. Fruit bats have not been reported to carry rabies.

Anatomy and Behavior

Bats are the only mammals capable of sustained flight. The delicate bones of the forelimb support an elastic wing membrane called a patagium. Some species possess a highly refined echolocation sense which makes capture with nets extremely difficult. Most bats are nocturnal. Thus bright lights are useful to confuse bats while they are being grasped. Bats possess unique physiological adaptations that make them desirable for biomedical research. Some bats are hibernators, others become torpid or hypothermic daily. This heterothermic adaptation is unique among mammals and is utilized to conduct studies of infectious agents.

Physical Restraint of Fruit Bats

If the bat is in a cage that does not permit flight, it can be grasped with gloved hands. A towel or laboratory coat can be thrown over the bat to restrict its activity before grasping it. In flight enclosures a bright light can be shone on a perched animal and the animal grasped while it is dazed by the intense light. Nets can be used, but the hoop must be large enough to capture the bat without traumatizing the wings. The net should be made of a closely woven fabric; otherwise the claws will become hopelessly entangled.

Physical Restraint of Insectivorous Bats

Experienced handlers grasp most species at the nape of the neck with bare or rubber-gloved hands (Fig. 18.2 left). The novice should wear a leather glove (Fig. 18.3). Bats move rapidly, even when crawling, and may crawl past the handler's gloved hand to bite an exposed wrist or arm. A leather jacket and gauntleted gloves may obviate this action.

A towel or cloth can be dropped over insectivorous as well as frugivorous bats. Nets are also suitable capture equipment. Long-handled forceps can be used to grasp a leg or the skin over the neck in order to transfer a bat from one cage to another (Fig. 18.2 right). This technique can also be used to capture a bat, enabling the handler to grasp the animal as it struggles to free itself from the forceps.

Mist nets and elaborate devices have been designed for capturing free-living bats [7].

Physical Restraint of Vampire Bats

Vampire bats are less fragile and liable to injury than insectivorous bats. Extreme caution must be used when handling vampire bats because they may transmit rabies. Nets, forceps, and/or thick gloves offer some protection when handling these animals. Vampire bats can run and jump even when so engorged with food they are unable to fly.

Venipuncture to collect a blood sample from a bat is made possible by stretching the wing and locating the vein on the anterior edge between the carpus and shoulder.

Transport

Bats can be carried in light cloth sacks or cardboard boxes. They will not gnaw or chew their way out [16].

Chemical Restraint

Tranquilization and immobilization are seldom imposed

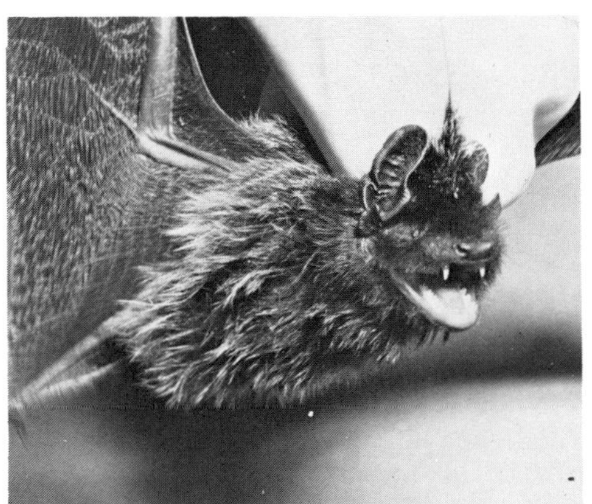

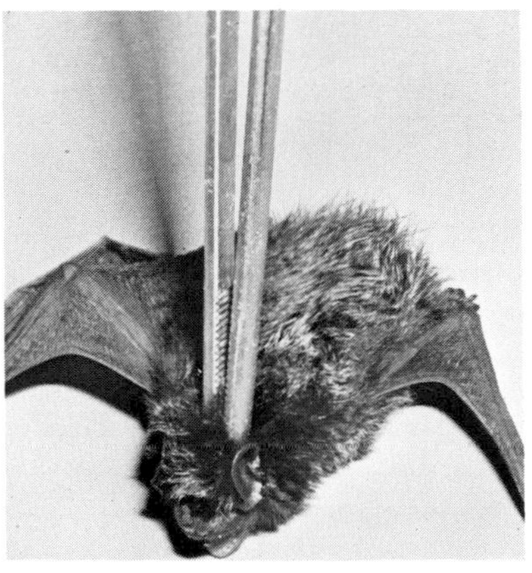

FIG. 18.2. Grasping a small bat with rubber-gloved thumb and forefinger *(left)*. Grasping with thumb forceps *(right)*.

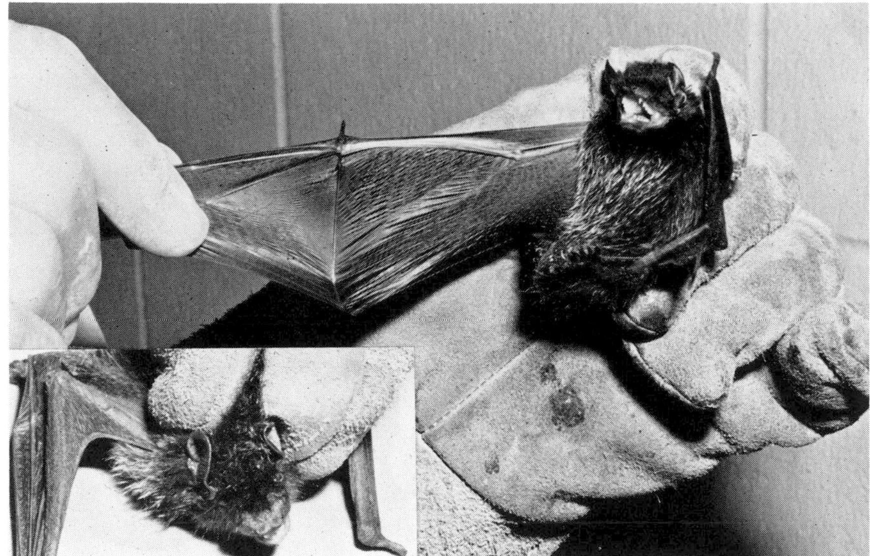

FIG. 18.3. Handling a small bat with leather gloves. Manipulating wings while restraining. Grasping loose skin behind the head *(inset)*.

on bats. Ketamine hydrochloride (5-15 mg/kg) has been used. Salivation and catatonia may be pronounced. Acepromazine alleviates catatonia.

Anesthesia has been carried out with pentobarbital sodium (30-50 mg/kg) given intraperitoneally [16].

EDENTATA—ANTEATERS, SLOTHS, ARMADILLOS [1,13,14]

Anteaters eat termites and ants. Some, such as the tamandua, are arboreal; others are strictly terrestrial,

The tamandua is a diminutive anteater with claws on the forefeet for opening termite nests and a prehensile tail (Figs. 18.4, 18.5). The claws are not so enlarged nor used in so aggressive a manner as the claws of the giant anteater. When a tamandua is frightened or threatened, it stands up with outspread arms. The enemy is grasped by the strong claws and held away from the body. The tamandua is handled by firmly holding the back of the head and tail simultaneously, preventing the animal from twisting around to grasp an arm with the front claws. Any manipulation of the ventral aspect of the animal requires grasping the front feet, preferably with a gloved hand. The tamandua is not aggressive, and minor procedures can be carried out with no physical restriction whatsoever. Intramuscular injections can be given by a quick jab. Examination can usually be conducted at close quarters without significant danger.

The giant anteater weighs up to 90 kg (200 lb) and is well known for its unpredictable and frequently aggressive actions. It has no teeth and is unable to bite, but it possesses extremely long, strong recurved claws on the forefeet that are especially designed for tearing open termite mounds (Fig. 18.6). If threatened, it grasps and pulls an enemy into its body. A person caught in such an embrace could not escape unaided. The dulled claws of an anteater were once driven into the bone of my wrist, so I can attest to the speed and strength of its forefeet. More than one zoo keeper has been chased from an enclosure by a giant anteater.

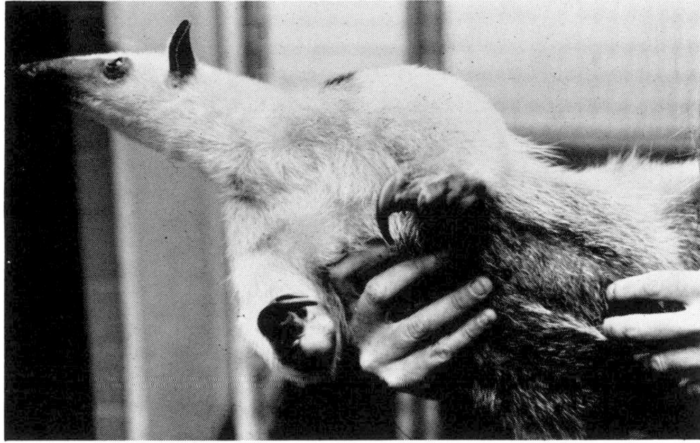

FIG. 18.4. The tamandua has strong claws but is not as aggressive as the giant anteater.

A giant anteater can be restricted by directing it into a narrow screened enclosure or box by means of a plywood shield and held in the box by bars placed in the opening. Such a box is suitable for sexing a giant anteater but does not allow sufficient exposure to conduct a general physical examination.

If blood samples or a comprehensive examination is required, the animal can be stretched by placing a snare on each foot [10]. This is not difficult to do, since these animals are somewhat slow moving. A handler can let the animal step into the loop of a snare with each foot. Nets are unsatisfactory for capturing a giant anteater; it is likely to tear the net to shreds with the forefeet.

The sloth appears to be slow and docile. On the ground it is clumsy or even helpless. The natural response of a threatened sloth is to curl up into a tight ball. When angered, however, it can move very quickly and slash accurately with the front claws. It also has a tendency to grasp an opponent and draw it close enough to bite.

The sloth's claws are highly adapted for grasping branches of trees. It is extremely difficult to detach a sloth that is clinging to a branch or post in an enclosure. If the

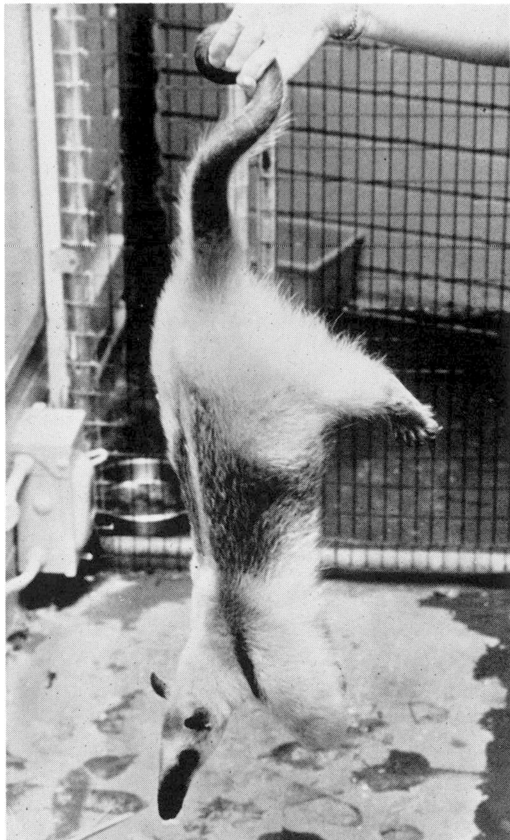

FIG. 18.5. Prehensile-tailed tamandua can be moved via the tail.

FIG. 18.6. Giant anteater.

sloth can be enticed to move onto a detachable pole, it can be carried on the pole to an area where the floor is smooth. Then a snare or catch strap can be attached to each of the limbs and the animal spread-eagled on the floor. If the animal is to be held with the hands, the handler should wear gloves to avoid being pinched by the curved claws [10].

Twenty-one species of armadillos, varying in size from 4 to 60 kg, are found in North and South America. Armadillos are insectivorous or omnivorous. They have short necks and heavy sharpened claws adapted for burrowing. The dorsal surface of the armadillo is covered by a segmented armored membrane.

The armadillo has molar teeth but no incisors and seldom bites. Its tail is useful for restraint practices since it can be grasped and held to prevent the animal from curling up into a tight ball.

Armadillos have poor eyesight and are primarily nocturnal. They are accustomed to a tropical habitat and susceptible to cold stress. Strict confinement in small squeeze cages can be detrimental to these animals since they will continually scratch and claw at the enclosure in attempts to escape.

The three-banded armadillo rolls into a tight, completely enclosed ball when frightened (Fig. 18.7). Complete examination of this animal is impossible without the aid of chemical immobilizers or the use of strong physical force to unwind the ball. The nine-banded armadillo and the giant armadillo are incapable of rolling into a complete ball and can be unrolled by grasping the plates near the midsection (Fig. 18.8). The armadillo has developed the ability to stop breathing for 5-10 minutes. Breath-holding is probably utilized during periods of intensive burrowing when the nostrils are buried in the earth [1]. This breath-holding technique enables the animal to enter complete torpidity, characterized by no apparent respiration, when it is disturbed. This phenomenon must be understood by the animal restrainer, since restraint may trigger the breath-holding reflex and subsequent torpidity, giving the impression that the animal has died. Cardiac function remains, and the animal will probably begin to breathe again within a few minutes. Obviously this phenomenon must be differentiated from unconsciousness resulting from other conditions. If injured during restraint, an armadillo should be separated from others; cannibalism of injured or disabled individuals is common within captive groups.

Chemical Restraint

The giant anteater has been successfully immobilized with phencyclidine hydrochloride (1 mg/kg) in combination with promazine hydrochloride (0.5 mg/kg) [2,12].

Sloths likewise have been handled with phencyclidine hydrochloride (0.5-1.65 mg/kg) [12,13].

Chemical immobilization and anesthesia have been carried out extensively on armadillos since they have become important in biological research. Ketamine (5-15 mg/kg) and tiletamine-zolazepam (2-10 mg/kg) are satisfactory for edentates.

PHOLIDOTA—PANGOLINS

Pangolins are insectivores without teeth. Their natural diet consists of termites and ants. Heavy claws of the forefeet are used for digging. All pangolins are nocturnal, and most species are arboreal. The pangolins occupy the ecological niche in Africa and Asia filled by the anteaters in the New World [8].

The scales of pangolins are razor sharp. When touched, the animal curls up into a tight ball. Wear a light glove when physical handling is required. Induce the curling

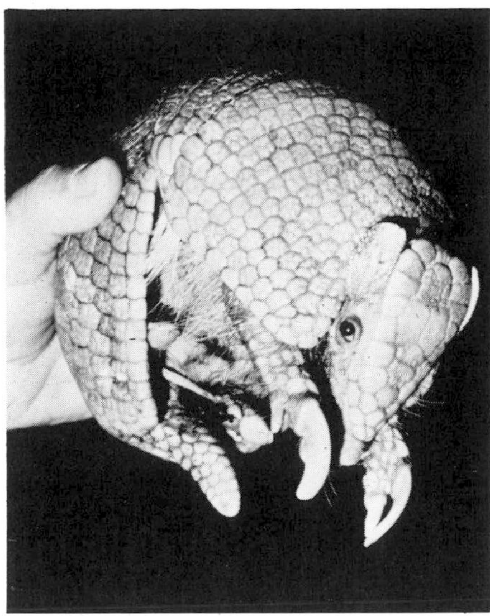

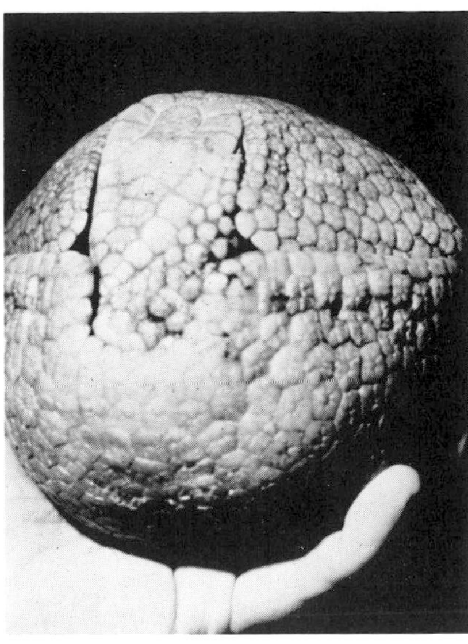

FIG. 18.7. Three-banded armadillo rolls up into a tight ball, difficult to open without injuring the animal.

FIG. 18.8. Grasping nine-banded armadillo near its midsection.

response; then gently uncurl the animal, taking care to avoid the claws. Pangolins are likely to spray urine in defense. Some individuals accustomed to people will permit touching and examination without rolling into a ball. Assumption of the defensive posture, however, is the usual response to being touched.

Suitable chemical restraint agents have not been identified for pangolins; however, based on the effects on other small mammals, it would appear that ketamine hydrochloride (5-15 mg/kg) or tiletamine-zolazepam (2-10 mg/kg) would be suitable.

LAGOMORPHA—RABBITS, HARES, PIKAS

Rabbits and hares are not aggressive animals and are unlikely to bite. They do have long, sharp incisor teeth and if annoyed sufficiently may bite. However, the prime danger to the handler is from scratching. A mature hare can lash out and scratch viciously with its powerful hind feet.

Rabbits are prone to injure themselves by bolting against walls and fences with sufficient force to contuse the brain or fracture the cervical vertebrae. A physically restrained hare may self-inflict back injuries, including fractures or dislocations, by simply flexing the loin muscles.

Manual restraint of the smaller rabbits and hares is simple. Gauntleted light gloves will prevent scratches. Techniques used for restraining domestic rabbits have some applicability, but caution must be used because the wild animal will probably resist vigorously and may injure itself. It is important to support the body weight when applying restraint techniques. Never suspend a rabbit or hare from its ears.

RODENTIA [4,5,12,21]

Rodentia, which includes over 1,500 species, is the largest order of mammals. Rodents vary in size from a mouse weighing 10-15 g to a capybara weighing 75 kg.

Danger Potential

All rodents have incisor teeth specialized for chewing and gnawing. Large species such as capybara, beavers, and porcupines can cause serious injury by biting.

Many species of rodents are burrowers, with sharpened claws which may scratch the handler. Specialized defense mechanisms will be described in discussions of individual species.

Physiology and Behavior

Rodents are usually small, with high metabolic activity. They frequently struggle violently during restraint, and one must resist the tendency to exert excessive pressure with gloved hands or nets which can interfere with respiration. Perhaps the most important fact to remember is that rodents possess no specialized thermoregulatory mechanisms. Thermal regulation is achieved by behavioral activity; they seek a source of heat when cold or retreat from a hot environment. Excitation and extreme activity during restraint procedures predispose them to hyperthermia inasmuch as they are unable to dissipate excess heat. The predisposition is intensified if the rodent is one of the many species with heavy fur pelts.

Physical Restraint

A person experienced in dealing with laboratory rodents should be cautioned that wild rodents cannot be handled with the same casualness permitted by laboratory animals. Gloves that provide protection from the bite of a laboratory species may be totally ineffective with a wild squirrel or a marmot. Wild rats, mice, and lemmings are generally handled with gloves (Fig. 18.9); however, the less they are subjected to handling, the better. When adapted to laboratory care, they may be handled bare-handed (Fig. 18.10). Squirrels should not be captured or held by their tails. The skin over the tail is friable and may pull from the tail [11].

Most rodent species can be netted, but the claws often become entangled in nets. A fine mesh that minimizes entanglement is best for capturing rodents. Small rodents are easily traumatized if struck by the hoop of a net. As with other species, the animal should be grasped securely around the head and neck through the net, and the net carefully worked off.

A large capybara can be netted and physically restrained, as depicted in Figure 18.11. Heavy-bodied marmots and woodchucks are difficult to remove from a net. A more effective method for handling these animals is a specialized squeeze cage (Fig. 18.12) that is adjustable for various sizes of animals. Plastic tubes are useful for transporting smaller rodents as well as for conducting examinations, taking radiographs, or administering anesthesia.

Snares are rarely used to capture rodents. Most have short necks, precluding safe use of a snare. A snare can easily tear the skin of an African crested porcupine. Specialized squeeze cages or boxes have been developed by those who work with captive rodent species in laboratories. The designs of squeeze cages are myriad and are usually

FIG. 18.9. Gloves should be worn when handling aggressive rodents or those of unknown behavior.

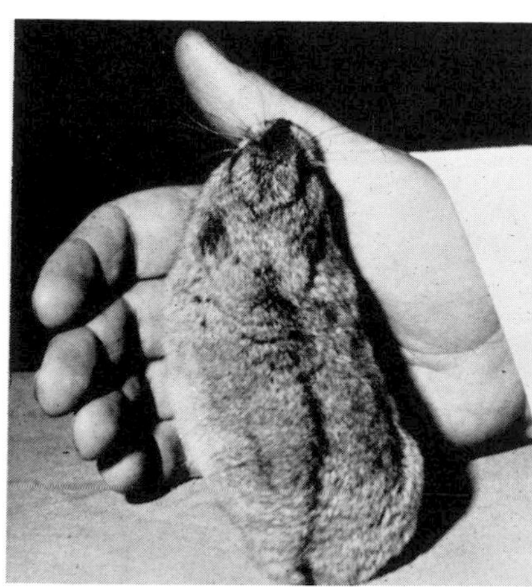

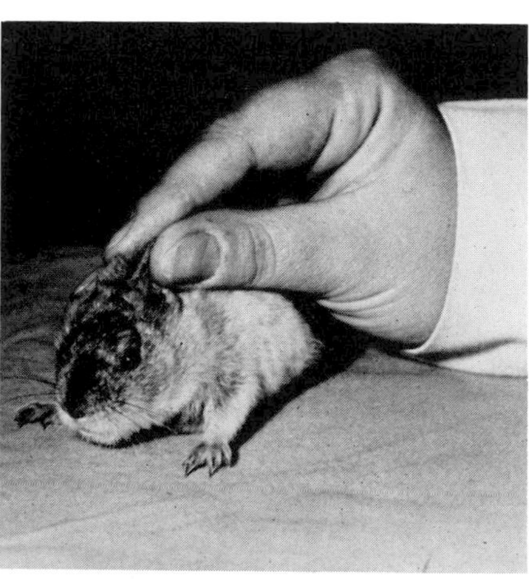

FIG. 18.10. Small docile rodents like this laboratory-reared lemming can be handled bare-handed.

FIG. 18.11. Manual restraint of a netted capybara.

FIG. 18.12. Special squeeze device for a woodchuck—designed by the Penrose Research Foundation, Philadelphia.

FIG. 18.14. Small rodents can be squeezed in special cages.

the result of personal observation of the behavior and anatomical structure of a particular species. The device illustrated in Figures 18.13–18.15 is adaptable for a number of small mammalian and reptilian species. It consists of a wooden frame with a plasticized wire bottom and a Plexiglas shield used to press the animal. Plexiglas permits visual communication with the animal, so excessive pressure is not applied. Rods keep the plastic shield in place once the animal is squeezed.

The box is inverted to expose the abdomen. If exposure of the dorsal surface is needed, the box is inverted and the squeeze is released slightly, allowing the animal to flip over. The squeeze is then reapplied, pressing the dorsal surface against the wire screen.

Do not lift a beaver by the tail alone [19]. They are heavy bodied, and the weight can cause spinal injuries. Restrain the beaver with a snare or preferably a catch strap around the neck. The animal can be lifted and held by a firm grip at the base of the tail as well for additional restraint.

Grasp the muskrat by the tail and pull the animal backward as it struggles to escape, placing the other lightly

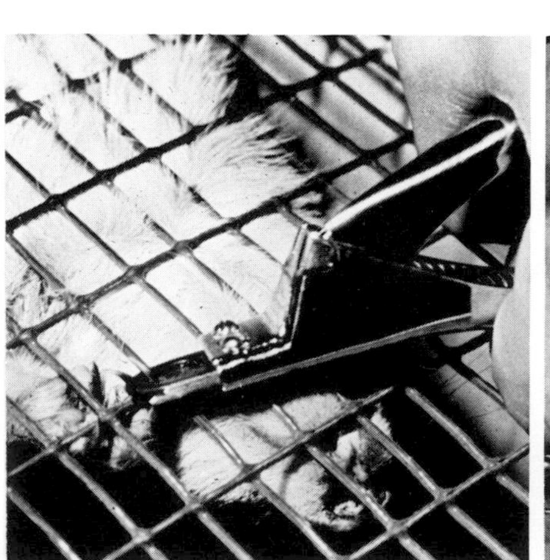

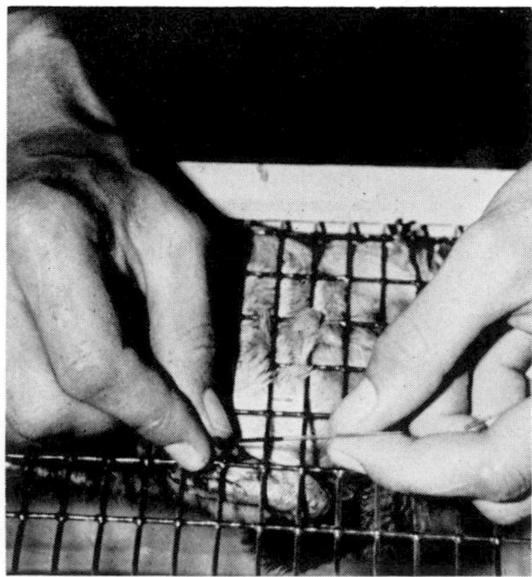

FIG. 18.13. Clipping a toenail *(left)*. Collecting blood from a prairie dog *(right)*.

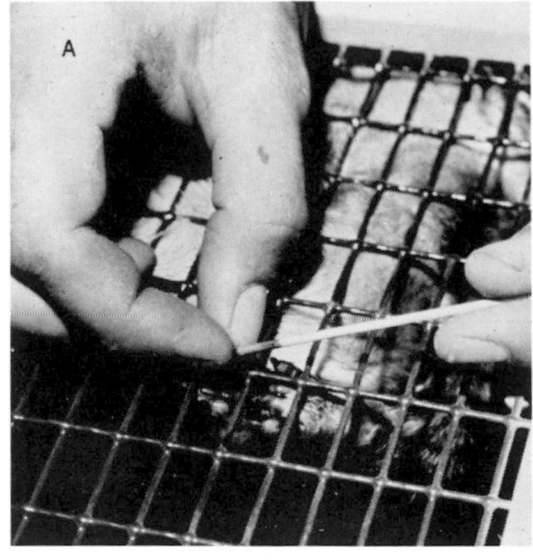

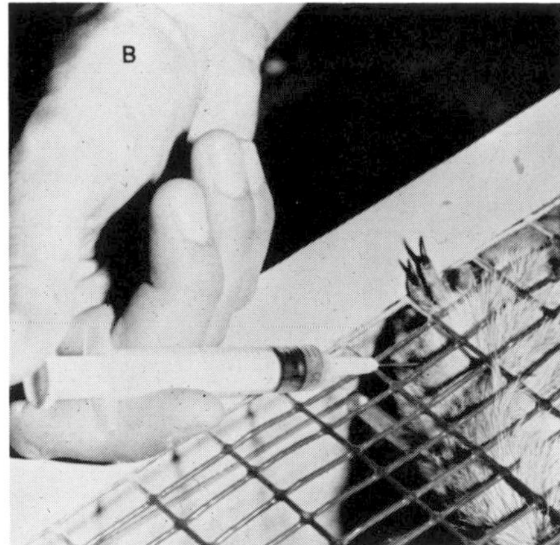

FIG. 18.15. **A.** Cauterizing clipped toenail with silver nitrate. **B.** Intramuscular injection.

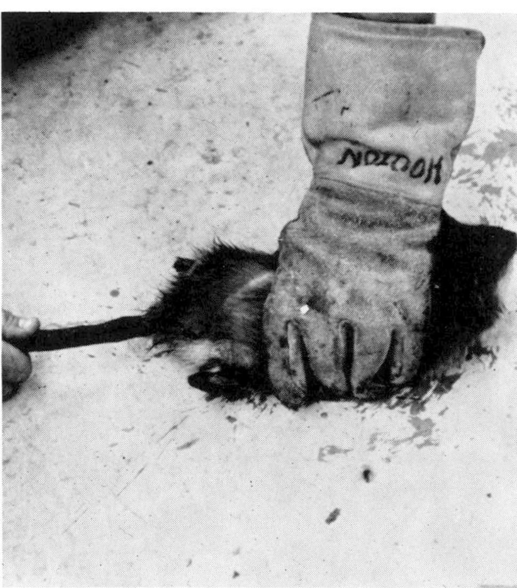

FIG. 18.16. Muskrat can be lifted by the tail or pinned, using the tail and a gloved hand.

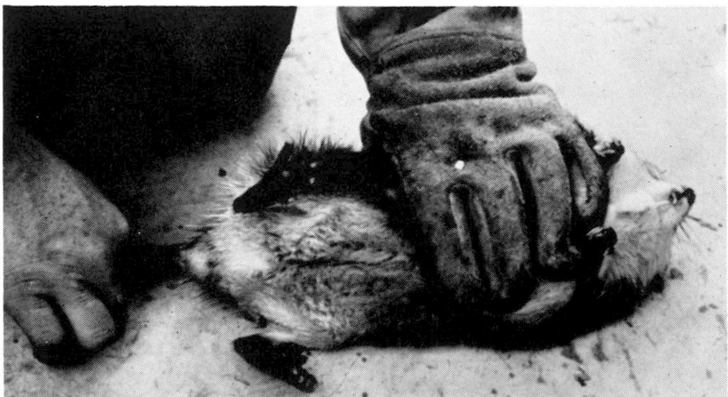

FIG. 18.17. Once muskrat is pinned, it can be manipulated into other positions.

gloved hand over the back to press the animal to the ground (Figs. 18.16, 18.17). To pick up the animal, surround the neck with thumb and forefinger, maintaining the pull on the tail and keeping the legs stretched to the front.

A chinchilla in a cage may exhibit defensive or even aggressive behavior when approached by a stranger. I had a personal pet which, when first acquired, would stand on its hind legs in the back of the cage and urinate on the hands of anyone who attempted to catch it. This individual would also use its forepaws to box with our pet cat through the bars of the cage.

The fur of the chinchilla is extremely delicate. Individual fibers can be pulled from the hide at a touch. Picking up the chinchilla by the body will result in the loss of some of the fur. The proper method of picking up a chinchilla is by grasping the animal at the base of the tail (Fig. 18.18A). If you are unfamiliar with the animal, wear a light leather glove. Once the chinchilla is grasped, it can be held on the hand, maintaining the grip on the tail (Fig. 18.19A).

If the coat requires examination, blow on the fur to separate individual fibers and look at the skin (Fig. 18.19B). By simply blowing in different locations, the whole body surface can be adequately examined.

If further manipulation or more restrictive restraint is necessary, grasp the animal by the ears as illustrated in Figure 18.18B. If a chinchilla must be securely held, wrap it in a towel to avoid damaging the fur (Fig. 18.20). Chinchillas are commonly plagued with dental problems. Dental specula are illustrated in Figure 18.21.

Porcupines have evolved quills to use as a protective

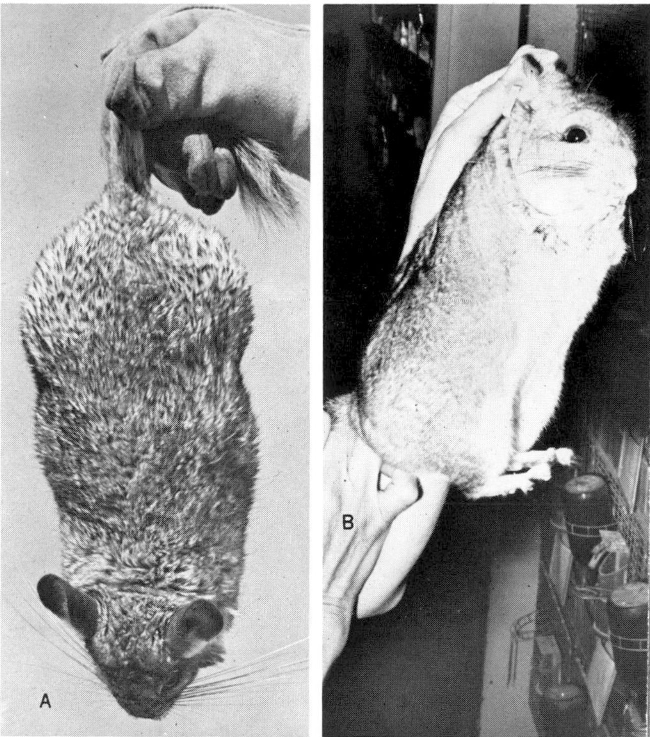

FIG. 18.18. Handling a chinchilla: **A.** Initial capture should always be via the tail. **B.** Additional restraint can be obtained by grasping the ears with thumb and finger.

mechanism against enemies. Old World crested porcupines have large smooth quills 30-45 cm (1-1.5 ft) long (Fig. 18.22). When frightened, they erect the quills and run backward, driving the long solid quills through the boots or legs of anyone approaching. The sudden rush can force the quills through a heavy leather boot or a broom (Fig. 18.23). The only adequate protection against such quills is a piece of heavy plywood or metal. An excited crested porcupine warns of impending attack by impatiently rattling the hollow tail quills [11].

Old World porcupines can be transferred quite easily; they are nocturnal and will readily enter a dark box in the cage. Close examination is possible if the box is constructed as a wire inner enclosure with removable covers.

Quills of the North American porcupine are much shorter than those of the crested porcupine. The quills are hazardous, both to the restrainer and to the animal itself. The porcupine is not capable of projecting the quills, but if it flips the tail and touches the handler, quills will impale the tissue. Furthermore, the quills have microscopic reverse barbs that catch in the flesh and cause the quills to migrate through the tissue unless they are quickly removed. Porcupine quills can be driven through heavy canvas and leather gloves, so neither of these provides adequate protection.

If two or more porcupines are in an enclosure, they should be separated before attempting to capture one; otherwise the excitement may cause one porcupine to bump into another, resulting in discharge of the quills by one or both and subsequent injury to the porcupines. Quills scattered on the floor can penetrate the feet of the porcupines.

Various techniques have been described for handling porcupines, most of them apparently by persons who have never handled one. Some have recommended throwing a canvas over the porcupine. Such a technique is not only

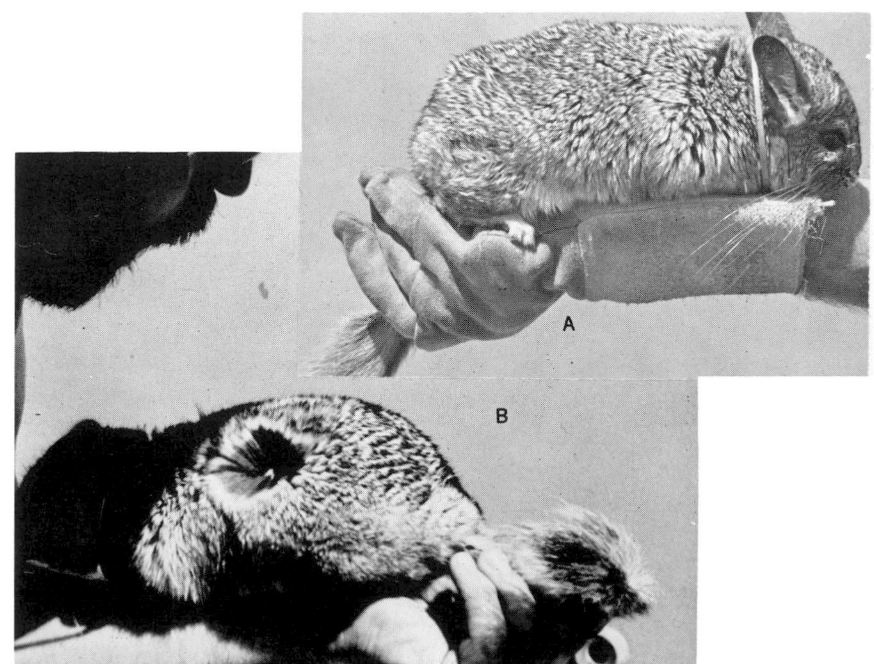

FIG. 18.19. **A.** Proper hold for a chinchilla. **B.** Fur can be examined by gently blowing to separate fibers.

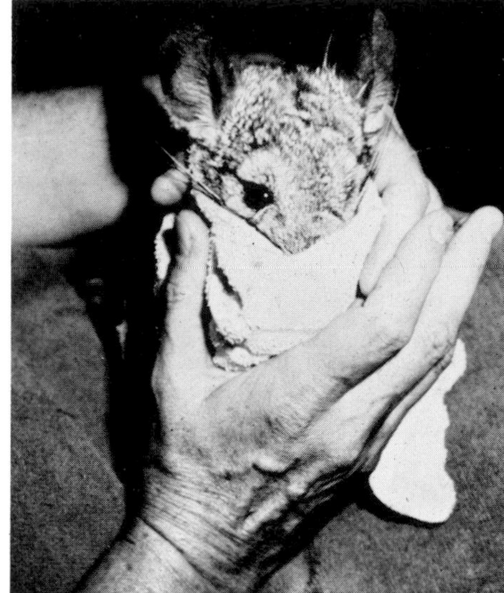

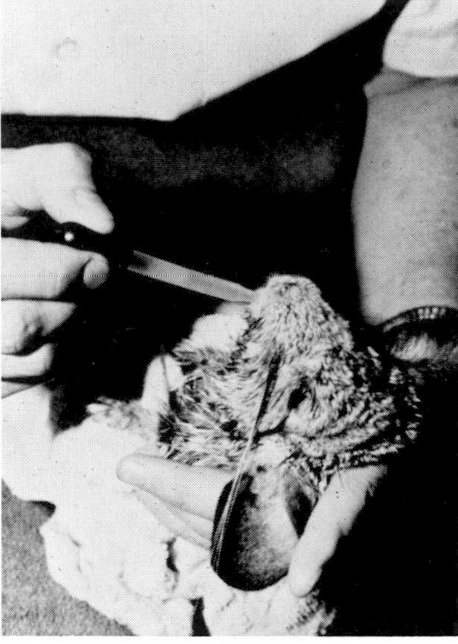

FIG. 18.20. Chinchilla wrapped in a towel for administration of medication.

FIG. 18.21. Chinchilla dental specula: Forceps on left is placed behind incisor teeth to open the mouth. Forceps on right is inserted into the cheek pouch to pull out cheek for better visualization of the teeth.

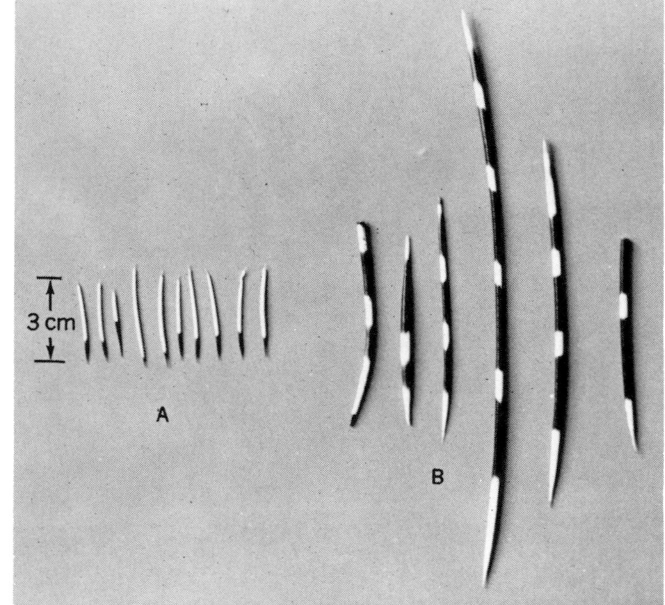

FIG. 18.22. **A.** Barbed quills of North American porcupine. **B.** Smooth quills of African crested porcupine.

dangerous to the handler but undesirable for the animal as well. All the quills that touch the canvas will be pulled from its body, causing dermal hemorrhage.

A technique successfully practiced by experienced persons is to approach the porcupine from the rear after it has headed toward a corner [17]. Usually the porcupine drags its tail on the ground, but when the animal is disturbed, the tail is flipped upward. Anything that comes in contact with that tail will receive an injection of quills. To guard against this, keep your hand on the ground at the same level as the tail and gently reach up to grasp some of the under hairs and quills and pull backward (Fig. 18.24). The animal will attempt to flip its tail upward as soon as it feels the touch, so maintain sufficient tension to keep the tail from pulling away. A wooden stick may be used to hold the tail down. It may be necessary to regrasp the tail with the other hand in order to achieve a firmer hold (Fig. 18.25). As long as the tail is grasped firmly, it cannot be flipped to discharge quills. Maintain pressure on the tail and gently lift the animal off its feet by sliding the other hand beneath the tail and up under its body (Fig. 18.26). The ventral surface is free of quills.

A squeeze cage with smooth inner walls can also be used to restrict a porcupine's movement. Be sure the quills are pressed in the right direction to lie smoothly against the body.

Manual restraint of a North American porcupine has been carried out with multiple snares—one snare around the base of the tail and others on each foot to stretch the animal. It can be placed in dorsal recumbency, exposing

FIG. 18.23. Quills of African crested porcupine driven through a broom.

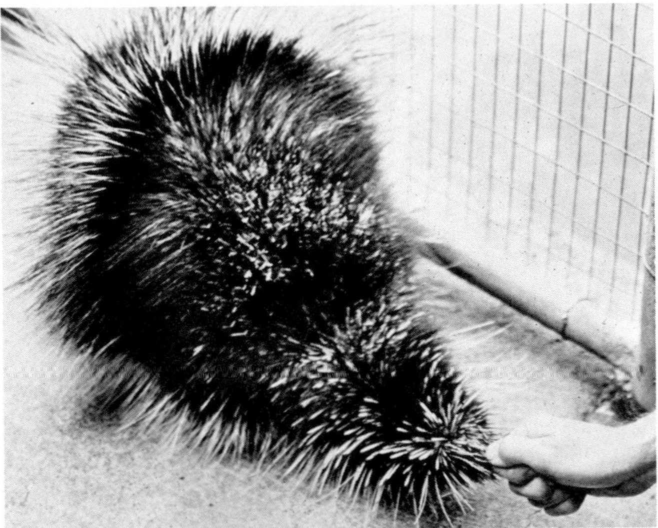

FIG. 18.24. Initial step in capturing a North American porcupine: Keep the hand low and grasp hairs at tip of tail.

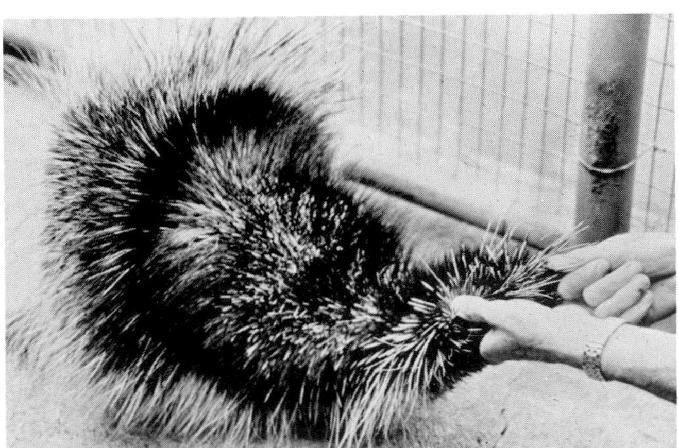

FIG. 18.25. Work the other hand carefully up the underside of tail until a firm grasp is possible.

FIG. 18.26. Final step in porcupine restraint: Insert a hand between hind legs underneath the abdomen to support the body.

FIG. 18.27. Capturing North American porcupine with snares.

the ventral surface for various procedures such as obtaining blood samples or tuberculin testing (Fig. 18.27).

The Brazilian tree porcupine has quills that are neither sharp nor barbed. The quills are more like those of hedgehogs or tenrecs. These animals can be handled quite easily with leather gloves.

TUBULIDENTATA

The aardvark is a medium-sized animal especially adapted for digging and burrowing. It has heavy claws, a thick body, and a short neck, making restraint rather difficult. When the animal's activity is restricted, it begins twisting and turning, expending great bursts of energy. The aardvark will begin to thrash wildly if snared; unless the restrainer is extremely careful, the animal will strangulate.

The aardvark has incisor teeth, but its mouth cannot open widely enough to be of any real danger.

The cone-shaped canvas bag described for restraining the wombat (see Chapter 17) is useful for controlling an aardvark. Phencyclidine hydrochloride (0.75 mg/kg) [10], ketamine hydrochloride (5-15 mg/kg), and fentanyl-droperidol (0.1 ml/kg) can be used for chemical restraint.

HYRACOIDEA

Hyraxes do not present any major restraint problems, but they are highly susceptible to stress and should be handled as little as possible. Strangers inside a cage may unnecessarily alarm the animals. It is desirable for a hyrax to be captured by the usual keeper.

Hyraxes can be gently directed into boxes or transfer cages without becoming alarmed. A net is satisfactory for capturing them from an enclosure (Figs. 18.28-18.30).

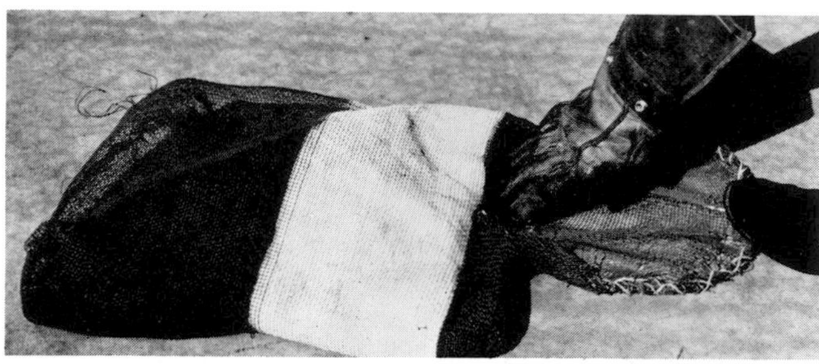

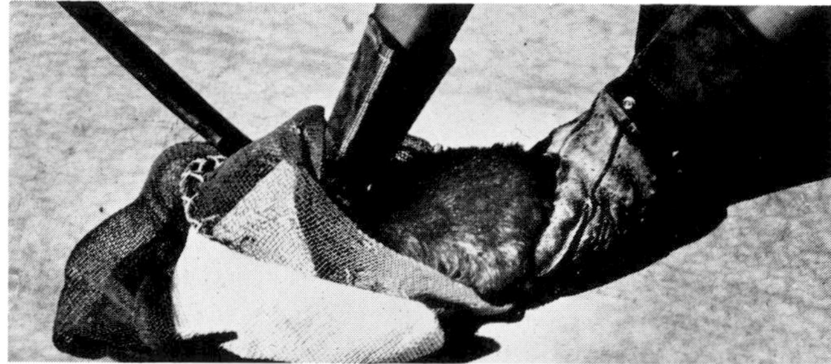

FIG. 18.28. Removing hyrax from a net: Grasp head through the net *(upper)*. Regrasp neck under the net with the other hand *(lower)*.

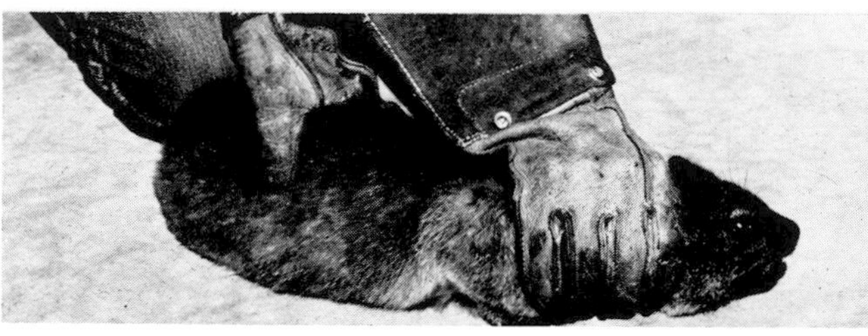

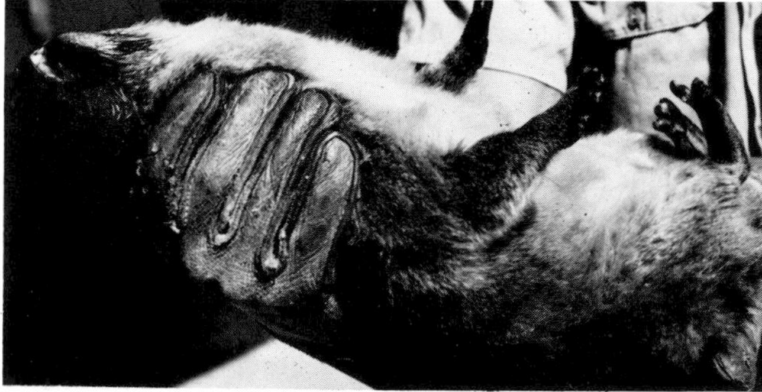

FIG. 18.29. Free of the net, hyrax can be manipulated into any position by maintaining neck hold.

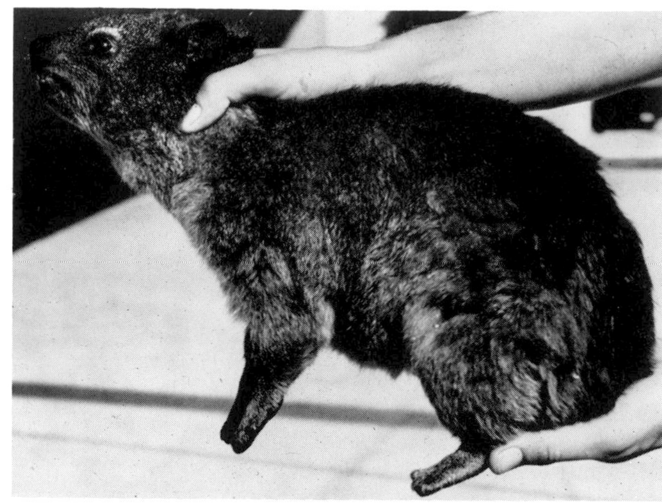

FIG. 18.30. The hyrax is not likely to scratch, thus it can be safely held bare-handed after capture.

REFERENCES

1. Anderson, J. M., and Benirschke, K. 1966. The armadillo, *Dasypus novemcinctus*, in experimental biology. Lab Anim. Care 16:202-16.
2. Beck, C. C. 1972. Chemical restraint of exotic species. J. Zoo Med. 3:3-66.
3. Caras, R. 1974. Venomous Animals of the World. Englewood Cliffs, N.J.: Prentice-Hall.
4. Carpenter, J. W., and Martin, R. P. 1969. Capturing prairie dogs for transplanting. J. Wildl. Manage. 33:1024.
5. Caudill, C. J., and Gaddis, S. E. 1973. A safe efficient handling device for wild rodents. Lab. Anim. Sci. 23:685-86.
6. Constantine, D. G. 1952. A program for maintaining the freetail bat in captivity. J. Mammal. 33:395-97.
7. ______. 1958. An automatic bat-collecting device. J. Wildl. Manage. 22:17-22.
8. Crandall, L. S. 1964. The Management of Wild Mammals in Captivity. Chicago: Univ. of Chicago Press.
9. Gates, W. H. 1936. Keeping bats in captivity. J. Mammal. 17:268-73.
10. Harthoorn, A. M. 1976. The Chemical Capture of Animals. London: Baillière, Tindall.
11. Karsten, P. 1974. Safety Manual for Zoo Keepers. Calgary, Canada: Calgary Zoo.
12. Melchior, H. R., and Iwen, F. A. 1965. Trapping, restraining, and marking arctic ground squirrels for behavioral observations. J. Wildl. Manage. 29:671-78.
13. Merritt, D. A., Jr. 1972. Edentate immobilization at Lincoln Park Zoo, Chicago. Int. Zoo Yearb. 12:218-20.
14. ______. 1974. A further note on the immobilization of sloths. Int. Zoo Yearb. 14:160-61.
15. Orr, R. T. 1958. Keeping bats in captivity. J. Mammal. 39:339-44.
16. Pye, J. D. 1967. Bats. In The UFAW Handbook on the Care and Management of Laboratory Animals, 3rd ed., pp. 491-501. London: E & S Livingstone.
17. Shadle, A. R. 1950. Feeding, care and handling of captive porcupines (Erethizon). J. Mammal. 31:411-16.
18. Skartvedt, S. M., and Lyon, N. C. 1972. A simple apparatus for inducing and maintaining halothane anesthesia of the rabbit. Lab. Anim. Sci. 22:922-24.
19. Whitelow, C. J., and Pengelley, E. T. 1954. A method for handling live beaver. J. Wildl. Manage. 18:533-34.
20. Wimsatt, W. A. 1970. Biology of Bats, vols. 1, 2. New York: Academic Press.
21. Zara, J. 1973. Breeding and husbandry of the capybara at Evansville Zoo. Int. Zoo Yearb. 13:137-39.

19 CARNIVORES

CLASSIFICATION (numbers in parentheses denote number of known species)

Order Carnivora
- Family Canidae: dog, fox, wolf (37)
- Family Ursidae: bear (7)
- Family Procyonidae: raccoon, kinkajou, panda (18)
- Family Mustelidae: skunk, otter, weasel (68)
- Family Viverridae: civet, mongoose (82)
- Family Hyaenidae: hyena (4)
- Family Felidae: cat (36)

ALTHOUGH most members of the order Carnivora are carnivorous, some species are omnivorous and/or insectivorous. The order has worldwide distribution. Carnivores are popular zoo exhibits. Names of gender are listed in Table 19.1. Weights are listed in Table 19.2.

Carnivores should not be handled immediately after they have ingested a meal. Regurgitation is a common reaction to fright or other emotional upsets and may result in aspiration pneumonia. Small carnivores can be shifted to or from swinging door cages using the technique described in Chapter 2.

TABLE 19.1. Names of gender for carnivores

Animal	Male	Female	Newborn or Young
Dog	Dog	Bitch	Pup, puppy
Fox	Reynard	Vixen	Kit, cub, pup
Wolf	Lobo, male	Female	Pup
Bear	Boar	Sow	Cub
Raccoon	Boar	Sow	Cub
Hyena	Male	Female	Pup
Skunk	Male	Female	Pup, kit
Otter	Male	Female	Pup
Ocelot, bobcat, mountain lion	Tom, male	Queen, female	Kitten
Tiger and other large cats	Male	Female	Cub, kitten
Cheetah	Male	Female	Kitten

TABLE 19.2. Weights of carnivores [17]

Animal	Kilograms	Pounds
Wolf	27-80	60-176
Gray fox	2.5-7	6-15
Coyote	9-12.7	20-28
American black bear	90-150	200-330
Grizzly bear	225-325	495-715
Polar bear	300-720	660-1,580
Sun bear	27-65	59-143
North American raccoon	15-22	33-48
Coatimundi	3-11.3	7-25
Cacomistle	0.87-1.1	2-3
Kinkajou	1.4-2.7	3-6
River otter	4.5-15	10-33
Sea otter	16-37	35-81
Striped skunk	0.75-2.5	2-6
Wolverine	14-27.5	30-60
Mongoose	0.34-4.5	0.75-10
Aardwolf	8	18
Spotted hyena	60-80	132-175
Cheetah	40-50	88-110
African lion	181-227	400-500
Tiger	225-340	500-750

Danger Potential

Teeth specialized for grasping and tearing prey are characteristic of all carnivores. The enlarged canine teeth are formidable weapons used for offense and defense as well as for food gathering.

The jaw muscles are well developed and tremendously strong. A hyena is capable of crushing the tibia of prey species with one snap. On one occasion I inserted a heavy stainless steel dental speculum, designed for a large cow, into the mouth of an immobilized bear. The bear suffered a convulsive seizure, bit down on the speculum, and completely collapsed it.

Wolves have extraordinarily powerful jaws. I have seen a medium-sized wolf collapse the bars of a cage especially designed to house rabid domestic dogs. Wolves can demolish a heavy stainless steel feeding bowl within a few moments.

The paws of most members of this order are fitted with claws that can rip and tear. All felids have dangerous claws. The larger carnivores are fully capable of killing a person who is careless when approaching or handling them.

In zoos large carnivores are customarily exhibited in an outdoor enclosure, with a bedroom as part of the exhibit area. A work area is usually constructed contiguous to the enclosure. When entering the work area, it is important for a handler to know exactly where the animal is in the enclosure and which doors are open or closed. Some tragic mistakes have been made by individuals who entered a work area under the assumption that the animal was barred from access to that area, only to find themselves confronted with a large cat or a bear.

Unless one has had extensive experience with members of this order, it is difficult to appreciate the speed and agility of these animals. Such speed and agility should be understandable, inasmuch as they must gather food by pouncing upon and grasping their prey, but many people fail to realize that most species can lash out with the claws much faster than a person can jump away. In addition, these animals have a mobile head that can reach forward or to the side quickly to bite.

It may be necessary to partially open the door to a carnivore cage to retrieve an object or to provide exposure for darting an animal. Be cautious! The animal may paw the door open wider and escape or attack. Keep a swinging door from opening too far by chaining it with a loose chain. Insert a rod through the netting or bars of a guillotine door so the door can be lifted only a specified distance.

Many carnivores have vesicular cutaneous invaginations on either side of the anus [7], opening to the surface through a short duct. Each sac contains an aggregation of glands or secretory epithelium. The secretions are exuded to define territorial limits by scent marking. Skunks and related mustelids eject the unpleasant secretion defensively.

CANIDAE—DOG, FOX, WOLF [8]

Various wild members of this family have been raised and trained by people, and many owners sincerely believe such an animal is as safe to handle as the domestic dog. This is far from the truth! Strangers who must handle such animals must assume they are wild and may suddenly revert to innate defensive behavior. Tamed wild canids are capable of inflicting serious injury.

A pet wolf brought to me for examination was docile as long as the owner was present. The animal required hospitalization. As soon as the client left, the wolf became vicious and could be given medication only when restrained with snares and a squeeze cage.

Semidomesticated foxes raised for fur are handled as shown in Figure. 19.1. Special tongs and snares are also used as needed. Zoo animals or captured free-living species should be handled carefully with nets (Fig. 19.2), snares, or squeeze cages.

FIG. 19.1. A ranch-raised fox can be grasped by the tail and gently swung to keep it off balance until the back of the head is grasped.

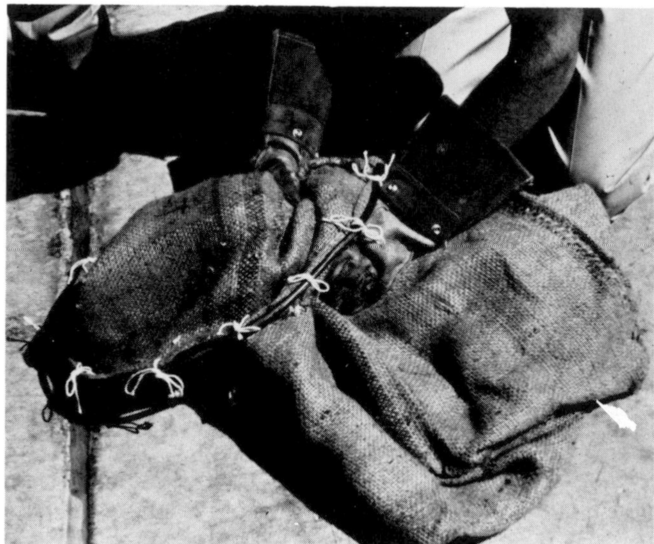

FIG. 19.2. Small canids are easily netted.

Gloves may be worn while grasping smaller species after initial capture with a net or snare (Fig. 19.3). All carnivores can bite through heavy leather gloves, which are worn primarily to guard against scratching.

A muzzle made of a small rope or gauze bandage will prevent a wild canid from biting. It may be wise to apply a muzzle as soon as the animal has partially recovered from chemical immobilization. However, caution must be exercised lest the animal overheat as a result of the inability to pant, the normal thermoregulatory mechanism for canids. Larger members of this family should be handled only with special squeeze cages or by chemical restraint.

URSIDAE—BEARS

Bears have tremendous strength. Mature bears are capable of bending bars and tearing off screens with their heavily clawed forepaws. Although bear claws are not sharpened and recurved like those of the large cats, the power of their paws and limbs makes bears dangerous to handle. Large bears such as grizzly or polar bears are capable of killing with a single swat of the paw (Fig. 19.4).

Immature bears can be hand held or controlled by nets or snares (Fig. 19.5). Mature bears can be handled only by the use of special squeeze cages [5] or by chemical restraint [16] (Figs. 19.6, 19.7). Squeeze cages used for bears must be especially strong; cages designed for large cats are usually not adequate. The use of chemical immobilization has minimized the need for such devices.

A large bear is usually transported in a heavy crate. The crate must be especially constructed of strong materials to prevent the bear from tearing it apart and escaping. Many bear crates are lined with galvanized sheeting, welded at the seams to prevent the animal from ripping it off by prying a claw beneath a seam.

To load a bear, the crate must be firmly secured to the cage opening with chains or ropes (Fig. 19.8). Otherwise the bear may run full tilt into the crate and, by the force of its body crashing against the opposite end, push the cage

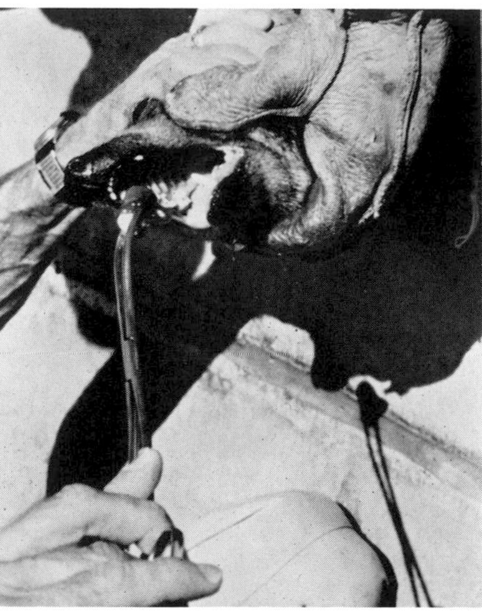

FIG. 19.3. Wild dog held tightly with gloved hands for examination and treatment.

FIG. 19.4. Although muzzled, this polar bear could still kill a person with a blow from a paw.

away from the opening and escape. A high-pressure water system is useful for directing a bear into a crate. It may be necessary to chemically immobilize a large bear before it can be placed into the crate.

Orphan cubs must be bottle fed from the proper posi-

FIG. 19.5. Small bear restrained with snares.

FIG. 19.6. Squeeze cages for large bears must be heavily constructed.

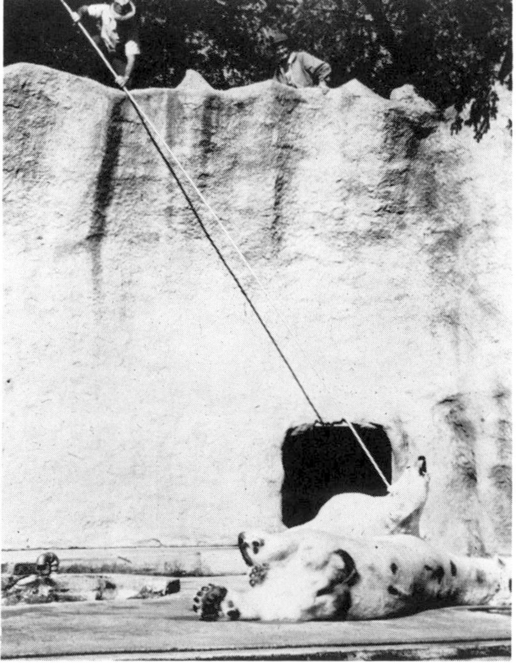

FIG. 19.7. Immobilized polar bear. (Check the stage of immobilization before entering an enclosure with this dangerous animal.)

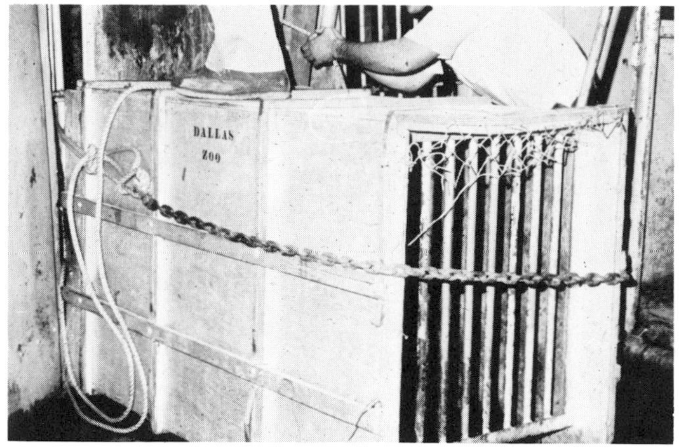

FIG. 19.8. Transferring large carnivore from cage to shipping crate. (Always chain or rope the crate to the door.)

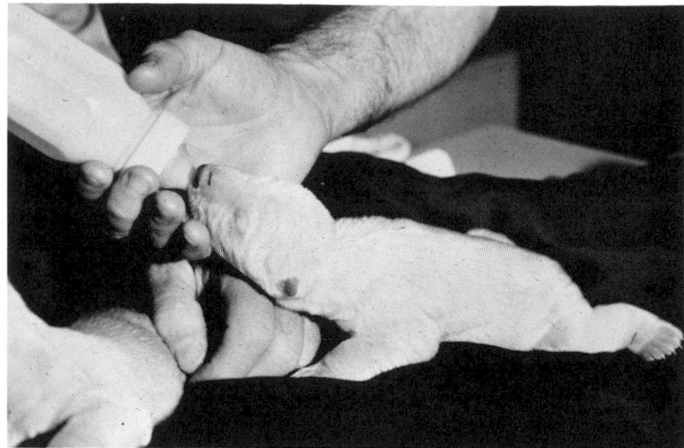

FIG. 19.9. Polar bear cub in proper nursing position.

TABLE 19.3. Chemical restraint agent dosages in carnivores

Animal	Etorphine HCl	Xylazine HCl	Tilazol* Combination: Tiletamine HCl / Zolazepam HCl	Phencyclidine HCl	Ketamine HCl	Ketamine and Xylazine*	Fentanyl-Droperidol†	References
Canid							0.04	
Wolf	0.03-1	2.5-5	2-6	1			(0.1 ml/kg)	2,3,4
Fox		2-3	3-8	1-1.5				12
Coyote			5-10	1				4,14
Ursid						4.4X		
Black bear	0.01	2-10	2.5-5.5	0.7-2.6	5-15	2.2K		2,4
Polar bear	0.01	2-10	6	0.8-3				2,4
Procyonid					20-29			
Raccoon			4-8	1.35				9
Kinkajou			2-4	0.85				
Mustelid								
Striped skunk			4-11	1	5-15			12
River otter			2-6	0.7				
Viverrid								
Mongoose			2-6					11
Civet cat			2-8	0.8-1.9				16
Binturong			1-2	1.25				3,11
Hyaenid								
Hyena		2.5-5	2	0.8	14			3,11
Aardwolf				0.75				
Felid								
Ocelot			2-6	0.5				
Mountain lion	0.006	8-10	2-4	0.85-1		0.5-1X 11K		2,3,11
Cheetah			2-6	0.77-1.1	8-10			4,11
African lion	0.006	8-10	2-6	0.5-2	5-7.5	0.5-1X 11K		2,3,4,6
Tiger		2-7	2-6	1		0.5-1X 11K		3,11
Bobcat		3-4	2-6		5.5-17			3,4

Note: Dosage is mg/kg unless otherwise noted.
*Unpublished data.
†Innovar-Vet.

tion (Fig. 19.9). If the animal is placed on its back, the milk may be inhaled into the lungs, causing aspiration pneumonia.

Chemical Restraint and Anesthesia

Many chemical agents have been used successfully to immobilize bears (Table 19.3). Anesthesia is handled in the same manner as for dogs (Figs. 19.10, 19.11).

MUSTELIDAE—SKUNK, WEASEL, OTTER

Mustelids are small to medium-sized carnivores that are dangerous to handle. They have needle-sharp teeth and are agile and aggressive.

All members of this family can be handled with nets, snares, or squeeze cages. Chemical immobilization can also be used. Pets can be manually handled with caution (Fig. 19.13).

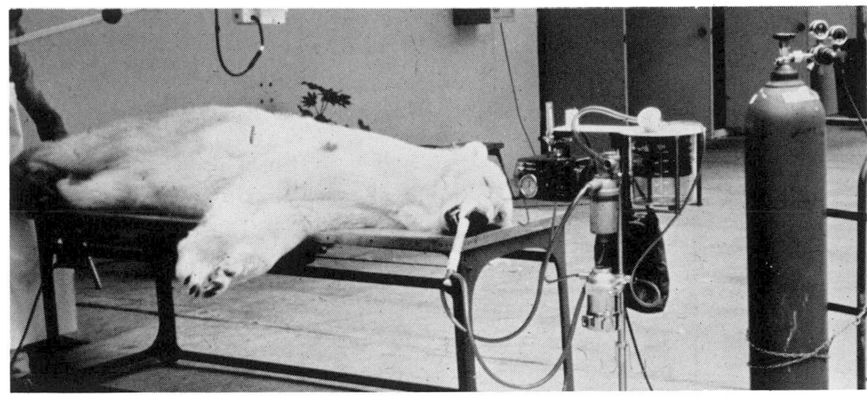

FIG. 19.10. Inhalation anesthesia for a bear is similar to that for a dog.

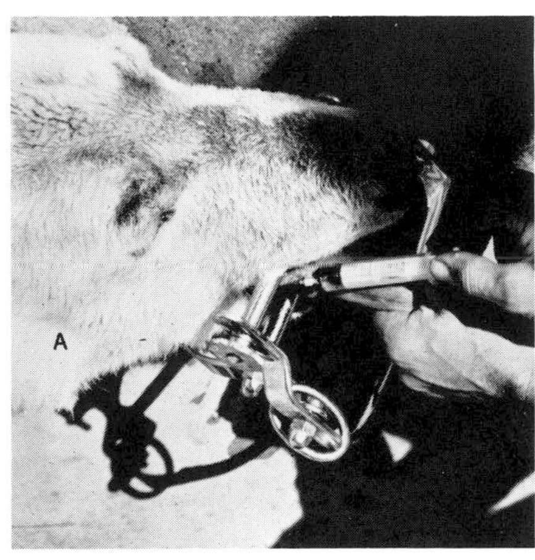

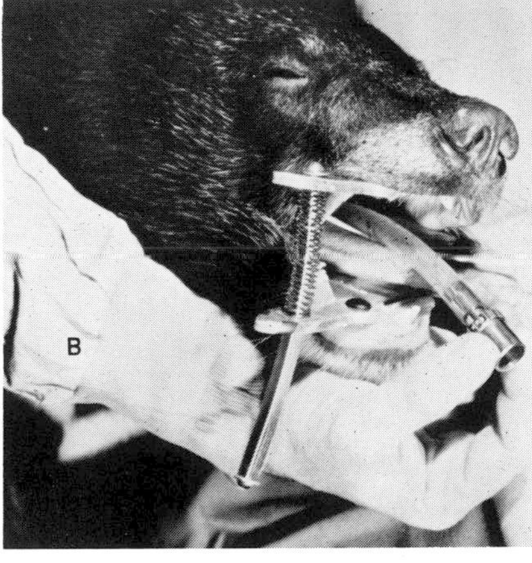

FIG. 19.11. **A.** Bovine speculum in use on anesthetized polar bear. Ventral glossal vein is being used to obtain blood sample. **B.** Canine mouth speculum used on small black bear.

PROCYONIDAE—RACCOON [9], KINKAJOU, COATIMUNDI

Members of this family are small to medium sized. Do not underestimate the danger of being bitten by a procyonid. Even hand-raised individuals may revert to wild behavior and inflict serious injury. I am aware of one North American raccoon that disfigured its owner's face for life in an unprovoked attack. Though raised from infancy, the raccoon laid open the lady's cheek as she held it lovingly in her arms.

Most procyonids are easily handled with nets or squeeze cages. Snares can also be used; however, most procyonids have highly prehensile forepaws and are capable of grabbing the loop of the snare and pushing it away or of pulling it off the head. An Elizabethan collar is used to prevent self-mutilation while wounds heal or during postsurgical care (Fig. 19.12).

Skunks bite readily [10]. Since they are one of the major reservoirs of rabies, they should be handled with great care [1,14]. Skunks defend themselves primarily by spraying the secretions of the anal sacs at enemies [7]. This sticky, irritating liquid can be projected accurately for a distance of up to 4 m (13 ft) and under favorable weather conditions can be smelled as far as 2 km (1 mi) away. The defensive position assumed by a threatened skunk is with hindquarters facing the enemy, feet planted firmly on the ground, and tail straight up in the air. It usually stamps with the front feet in warning before spraying. The spotted skunk lifts the hindquarters off the ground to spray. The musk sac is surrounded by heavy muscle layers. The spraying action is similar to that resulting from compression of a bulb syringe. Some believe that a skunk lifted up by the tail cannot spray. Not true! It can!

Immature skunks 8–10 weeks of age are not likely to

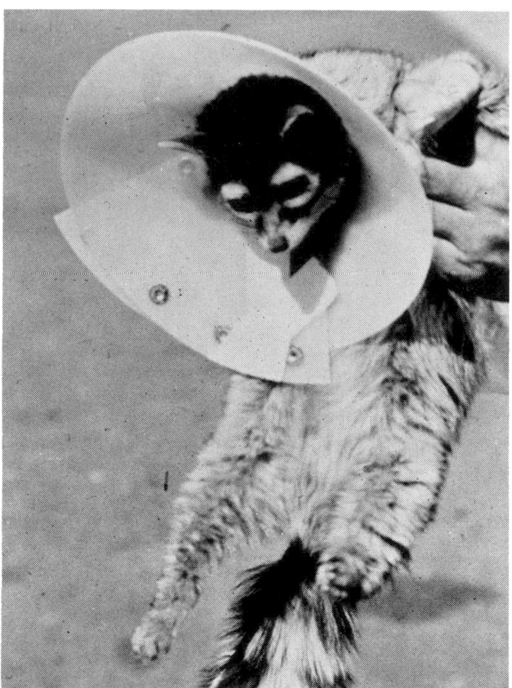

FIG. 19.12. Elizabethan collar on a cacomistle.

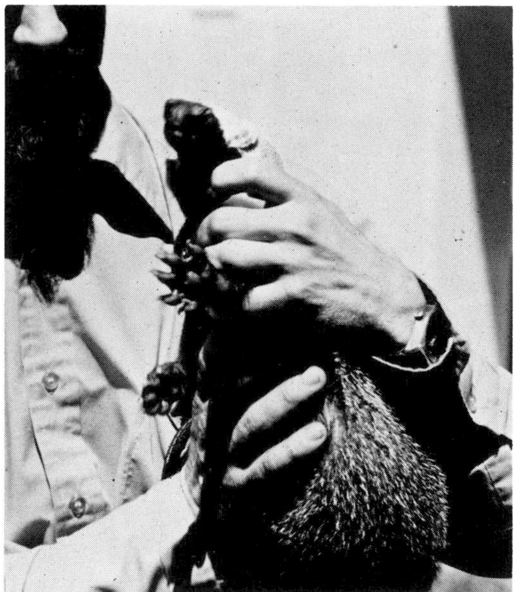

FIG. 19.13. Pet grisson—hand-held by owner. (Generally unsafe, since the owner rarely is qualified to maintain a proper grip.)

spray though capable of doing so. Capture intact skunks with a net [14] from behind a shield of plate glass or plastic, or wear goggles and protective rain gear.

The only safe way to handle an intact mature skunk is to sedate or anesthetize it. If the skunk is presented in a box, enclose the box in a plastic bag and insert a wad of cotton soaked with ether or halothane. Remove the skunk from the box only after it is sedated.

Ketamine hydrochloride (5-15 mg/kg) is an effective intramuscular anesthetic. Infant skunks can be wrapped in a towel for intramuscular injection. Mature skunks must be sedated first with a volatile anesthetic.

Clinical signs exhibited by those sprayed with skunk musk include skin burns, temporary blindness, nausea, convulsions, and loss of consciousness. If sprayed, wash spray from the eyes with copious amounts of water. Lacrimal secretions will also clear the eyes of the material within 15 minutes. Rinse the secretion from skin surfaces as well.

The pungent, acrid skunk musk is composed of many compounds, but the principal one is normal butyl mercaptan (butanethiol). The mercaptan can be made soluble, harmless, and odorless by applying strong oxidizing agents such as the alkaline sodium or potassium salts of halogens. Sodium hypochlorite (household bleach) is the most readily available, in a concentration of 5.25% chlorine. Dilute to approximately 500-1,000 ppm of available chlorine before applying. The chlorine oxidizes the mercaptan, breaking sulfur free from the carbon chain to form sulfate or sulfone compounds which are odorless. These compounds are water soluble and can be removed by repeated washings with copious quantities of water. Delicate fabrics and colored clothing may be damaged by this bleaching action. If clothing cannot be subjected to the bleach treatment, it is doubtful that odors can be removed. Remove odors from the skin surface by washing briskly with a carbolic soap (tincture of green soap) and warm water. Rinse the skin with a dilute bleach solution.

If the hair of an animal has been sprayed, a superficial clip may remove the bulk of the contamination. Close-clip the hair if necessary and rinse with a dilute bleach solution followed by repeated shampooing.

When a person's hair has been sprayed, the problem is compounded. A strong bleach solution will damage the hair, and one usually does not wish to close-clip the hair. Some have recommended repeated washing in tomato juice; thorough shampooing and rinsing, accompanied by application of a mild bleach solution, may prove to be the most practical method for removing skunk odor.

Skunks kept in captivity are usually descented by surgical removal of the anal glands [10]. Descented skunks can be handled like any other mustelid.

Some mink are semidomesticated for the production of fur. Although mink are small, they have relatively large canines, capable of inflicting severe lacerations.

Snares can be used to move mink, although special transfer cages are probably more satisfactory (Fig. 19.14). The experienced handler can remove mink from a cage as illustrated in Figures 19.15-19.17. Using a gloved hand, he grasps the mink by the tail and quickly pulls it from the cage. As it attempts to climb back into the cage, the handler's other hand grasps the animal behind the neck with the thumb and finger around the head. This sequence must be carried out quickly, because the animal is fully capable of biting the hand through the glove. Special restraint tubes have been utilized when repeated handling is required [13].

Chemical restraint of mink has become a standard practice and may produce less stress for the animal than

FIG. 19.14. Mink are transferred from cage to cage or transported to other facilities in wire transfer cages.

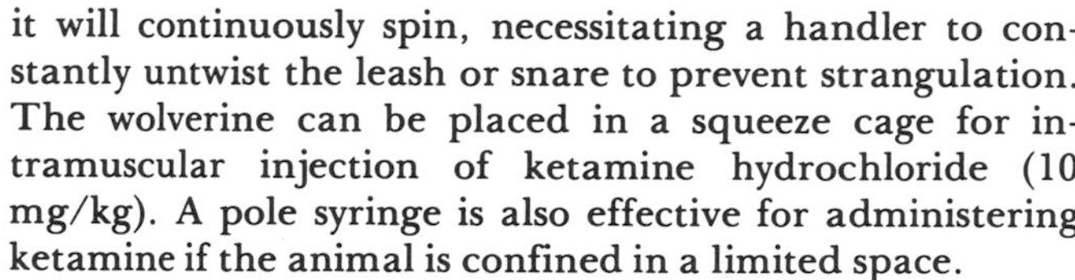

it will continuously spin, necessitating a handler to constantly untwist the leash or snare to prevent strangulation. The wolverine can be placed in a squeeze cage for intramuscular injection of ketamine hydrochloride (10 mg/kg). A pole syringe is also effective for administering ketamine if the animal is confined in a limited space.

River otters can be handled with nets or snares (Fig. 19.18). Sea otters require specialized handling techniques. Divers trap them from the wild in a basket net from beneath while the otters float on the surface of the sea. The net is closed with a drawstring to keep the otter inside and immediately lifted out of the water.

Sea otters are extremely susceptible to stress caused by handling and transporting. This unique marine mammal has no insulating blubber layer. Protection against

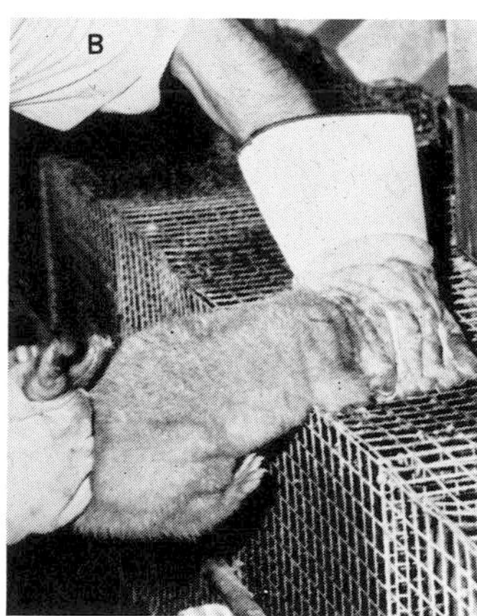

FIG. 19.15. Handling a ranch-raised mink: **A.** Remove mink from cage by grasping the tail. **B.** While it clings to the wire, pin the head.

manual restraint. Etorphine hydrochloride (0.1 mg total dose) and ketamine hydrochloride (5–10 mg/kg) are used.

The wolverine is the largest mustelid, weighing up to 18 kg (30 lb), and is about 38 cm (15 in.) high at the shoulder. The wolverine assumes a defensive posture by rolling over on its back, preparing to scratch and bite [15]. If tethered

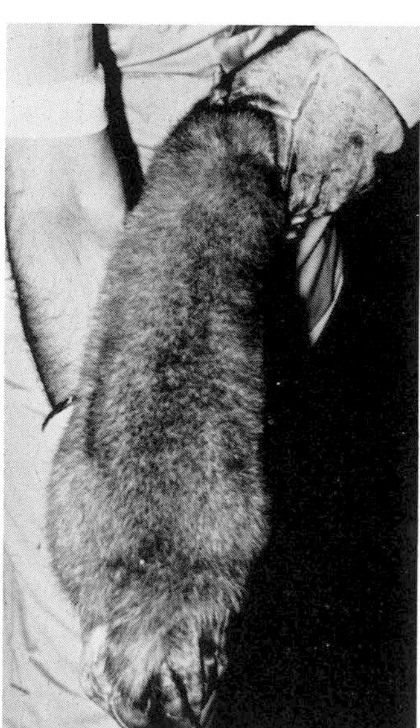

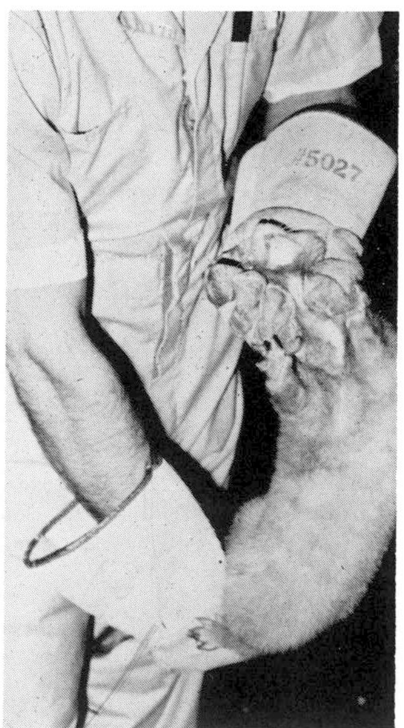

FIG. 19.16. Once head and tail are secure, the mink can be manipulated into various positions.

FIG. 19.17. Ranch-raised mink are usually docile enough to allow manual handling. Wild mink would be likely to bite through these gloves.

FIG. 19.18. River otter captured with a snare.

hypothermia is provided by the dense coat of fur. The underlying fine silky fibers next to the skin are kept dry by the perfectly groomed outer fibers. If the fur is soiled with feces, urine, or any other material during caging, the fur will mat and lose its water-resistance and hence its insulating qualities. Once the fur is matted, the entire coat becomes wet, heat is lost, and the animal becomes chilled. Seemingly irreversible pneumonia is the result. Great care must be taken to prevent soiling of the coat during any procedure involving sea otters.

VIVERRIDAE—MONGOOSE, CIVET CAT

Viverrids are similar in body structure, size, and habits to mustelids and can be handled in the same manner.

HYAENIDAE—HYENAS, AARDWOLF

Hyenas are efficient scavengers and active predators. Their primary weapons are their strong jaws and teeth. Handling mature hyenas requires the use of squeeze cages or chemical immobilization. They can tear through a net with little difficulty and are usually too strong to handle with a snare. Manual handling of larger than infant animals is unwise.

FELIDAE—CATS

Members of the cat family vary in size from those no larger than the domestic house cat to the Siberian tiger, which may weigh as much as 340 kg (750 lb). Obviously restraint techniques must vary with the size.

All felids have sharpened recurved claws, capable of lacerating flesh, and they are extremely fast in striking out with the front paws. Some species of cats claw only with their front feet; others, like the leopard, rake with the hind feet as well. Cats have great mobility of the head, allowing them to slash with the fangs from almost any position.

There is wide variation in behavior among members of the family Felidae. Some of the smaller cats are the most high-strung and vicious. The leopard cat is aggressive, even though it is often kept as a pet. The clouded leopard is one of the more docile species; if raised with human association, it is likely to respond well to mild restraint and can be handled by manual techniques (Fig. 19.19). It is seldom necessary to resort to snares, nets, or squeeze cages to restrain a hand-raised clouded leopard.

Other species respond to restraint in various manners. Once a cheetah is grasped with a snare and restrained, it usually lies quietly unless intense pain is inflicted. Even hand-raised cats will not tolerate any procedure inflicting pain, however slight. Contrarily, a leopard may fight vigorously the entire time it is being restrained.

Infant felids are easily handled manually (Figs. 19.20–19.22), or they can be wrapped in a towel or canvas (Fig. 19.23). More mature wild felids must be handled more carefully. Small individuals weighing up to 14 kg (30

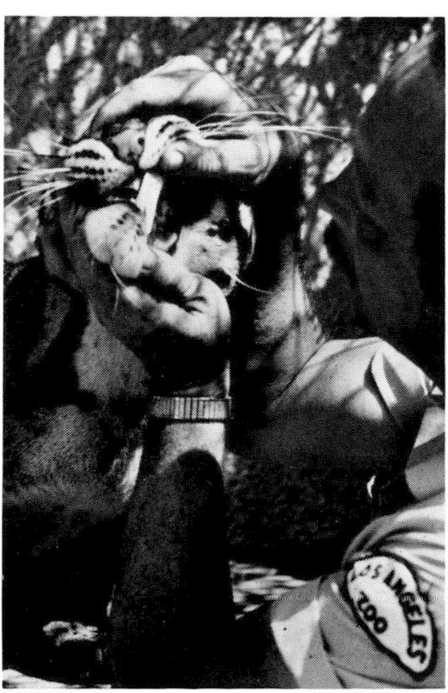

FIG. 19.19. Manual handling of wild felids is usually restricted to those who raised them. Clouded leopard *(left)*. Tiger *(right)*.

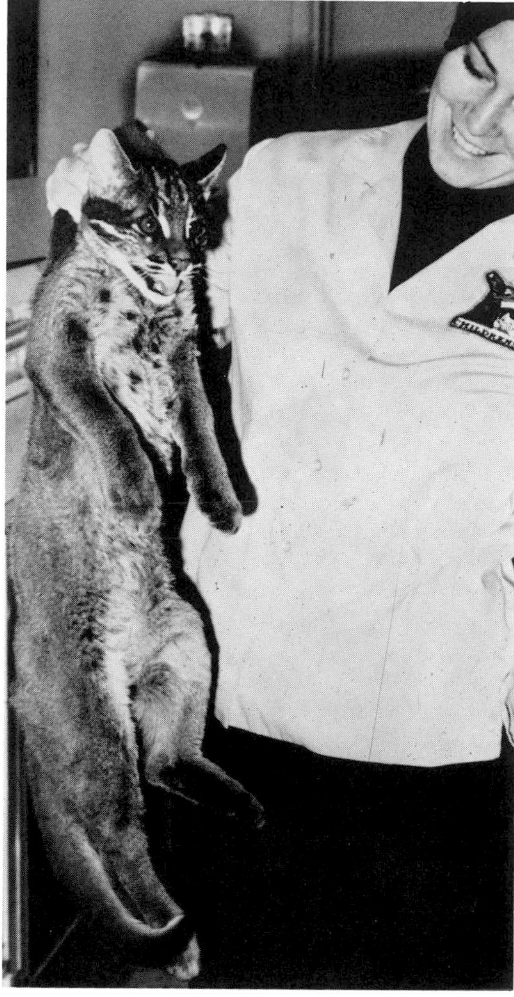

FIG. 19.20. Infant and juvenile felids usually become limp when grasped by the nape of the neck. This method is used by the mother to transport the kitten; thus relaxation is an expected behavioral pattern.

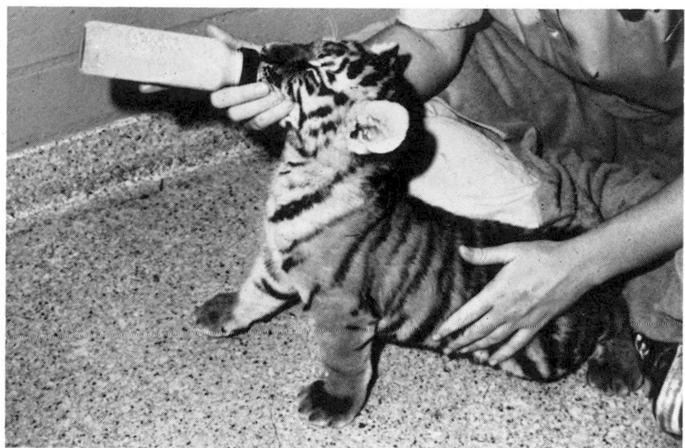

FIG. 19.22. Tiger cub in proper nursing position.

lb) can be handled with a net (Fig. 19.24); in some instances they can be grasped with a snare and then handled with gloved hands. Wide-gauntleted welder's gloves are useful when working with small felids, not so much to prevent biting (most of these animals can bite through the heaviest leather glove) but to protect the handler from severe scratching.

Trained cats are handled in a different fashion than truly wild or zoo cats. The trainer may be able to restrict the activity of even a large cat by the use of a snare (Fig. 19.25) or special chains (Fig. 19.26). These chains can be very useful in that the animal may be snapped or chained to a post and then grasped by the tail to administer a quick intramuscular injection in the hindquarter. The trainer handling an animal in this manner must be fully capable of restricting the animal.

Small cats weighing up to 15 kg can be grasped by a snare. The snare has the advantage of keeping the animal away from you as well as serving as an extension of your arm to catch the animal. Once the snare is in place, the

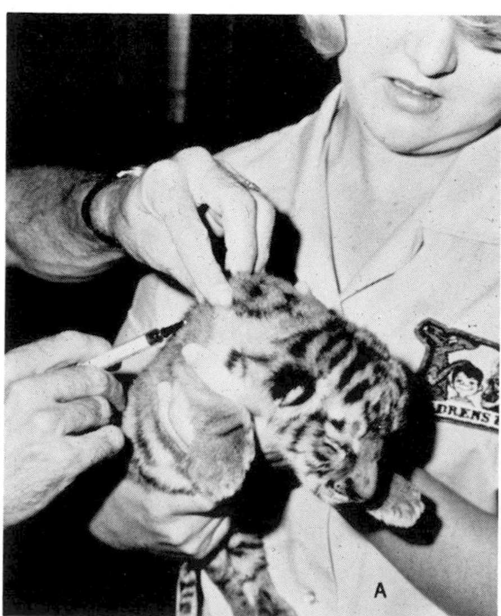

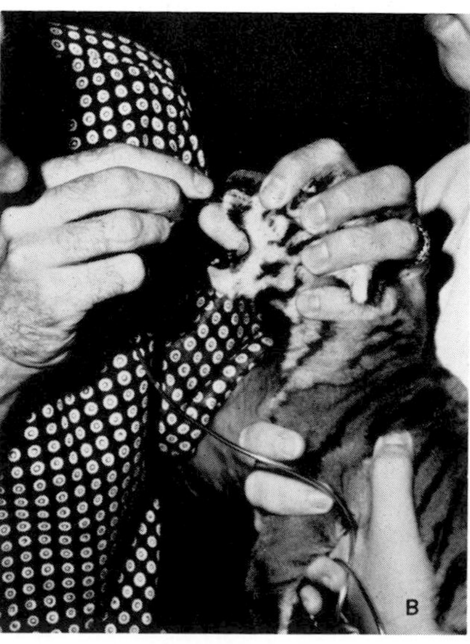

FIG. 19.21. Infant wild felids can be handled in much the same manner as domestic cats. **A.** Vaccination. **B.** Placement of stomach tube.

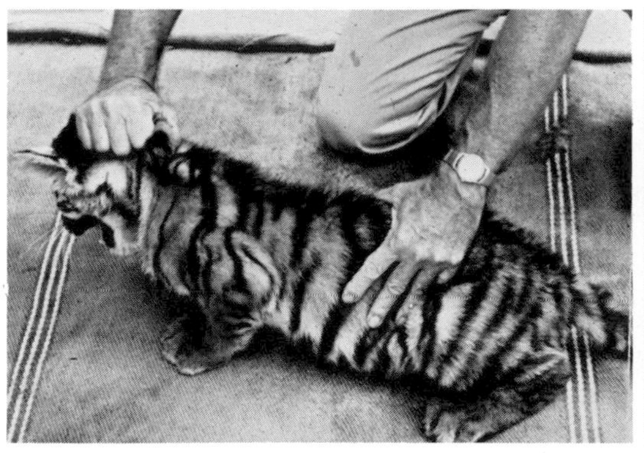

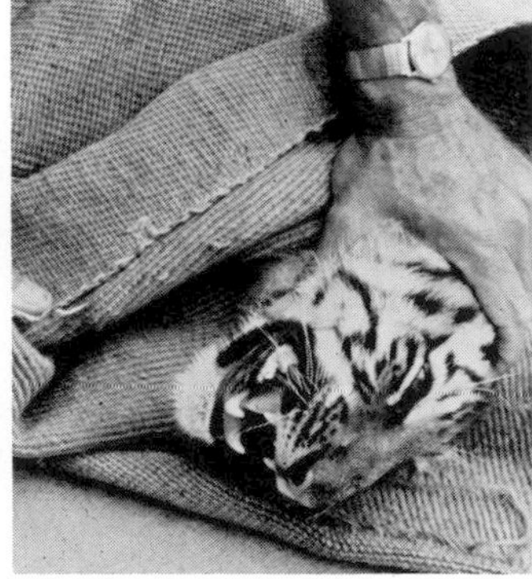

FIG. 19.23. Tiger cub can be wrapped in a towel or piece of canvas to eliminate scratching during examination.

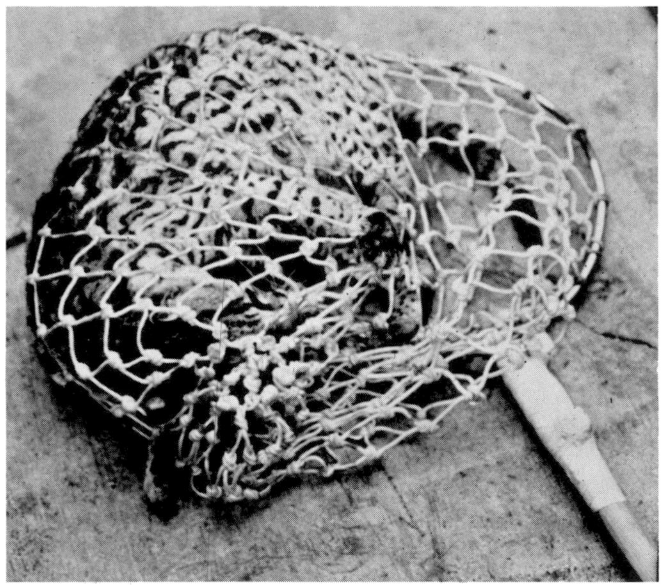

FIG. 19.24. Small to medium-sized wild felids are easily netted.

FIG. 19.26. Snap chain used to control a trained cat.

FIG. 19.25. Snare can be used on tame and semitame felids.

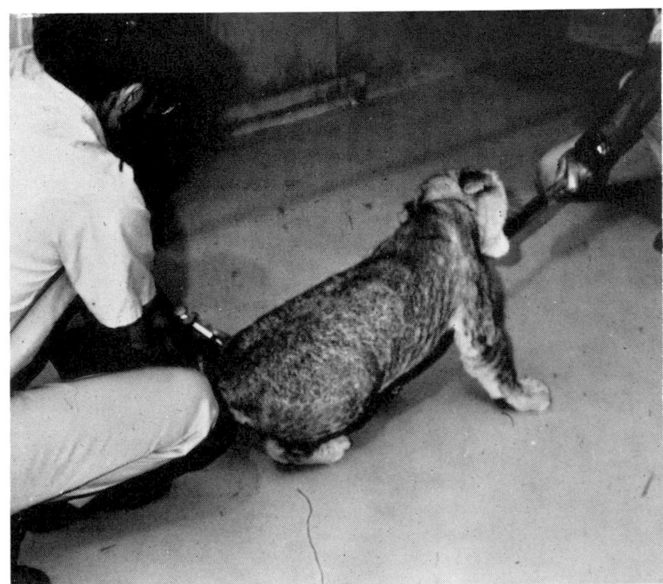

FIG. 19.27. Small wild felid stretched between snare and tail for administering medication.

tail can be grasped to hold the hindquarters facing toward the restrainer (Fig. 19.27).

If a net is used on cats it should be made of mesh fine enough to prevent the animal from poking its head or paw through the holes. The animal may strangle as it fights to extract its head from the mesh; if it puts a paw through the net, someone may be clawed.

Squeeze cages are frequently used for handling large cats. Numerous types and styles have been designed [12] (Figs. 19.28–19.33). Restraint cages are available commercially or can be custom-made. Because of the body conformation large cats such as the tiger, lion, jaguar, or leopard should be squeezed from side to side; smaller species from top to bottom. General manipulation and precautions for the use of squeeze cages are described in Chapter 20.

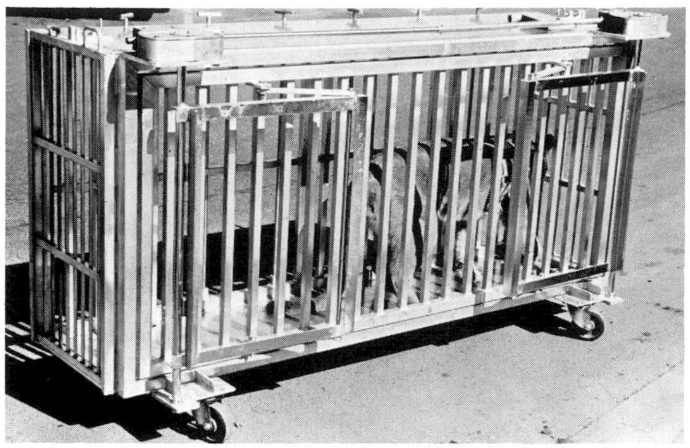

FIG. 19.28. Commercial carnivore squeeze cage.

When prolonged treatment is required, strap the immobilized animal to the table (Fig. 19.34) to preclude injuries to the animal (or handler) if it awakens from sedation unexpectedly.

Accessible veins for obtaining blood samples and administering intravenous infusions are both medial and

FIG. 19.29. Tiger in a squeeze cage.

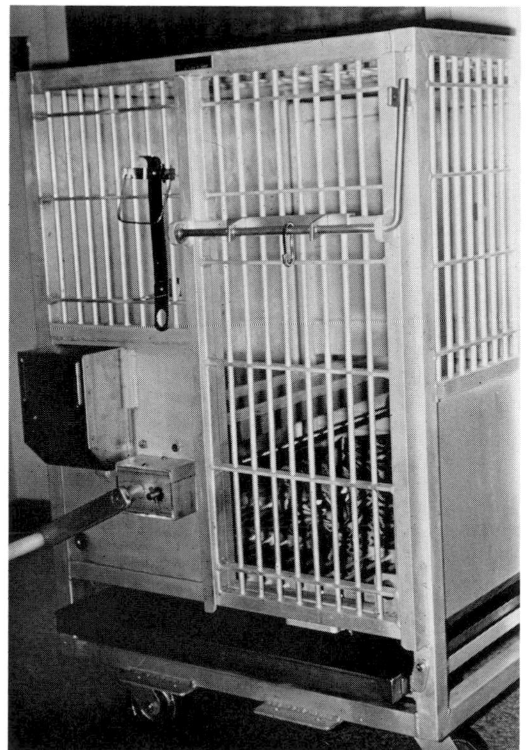

FIG. 19.30. Small felids can be handled in primate squeeze cages.

FIG. 19.31. Homemade squeeze cage for an ocelot.

FIG. 19.32. Homemade carnivore squeeze cage.

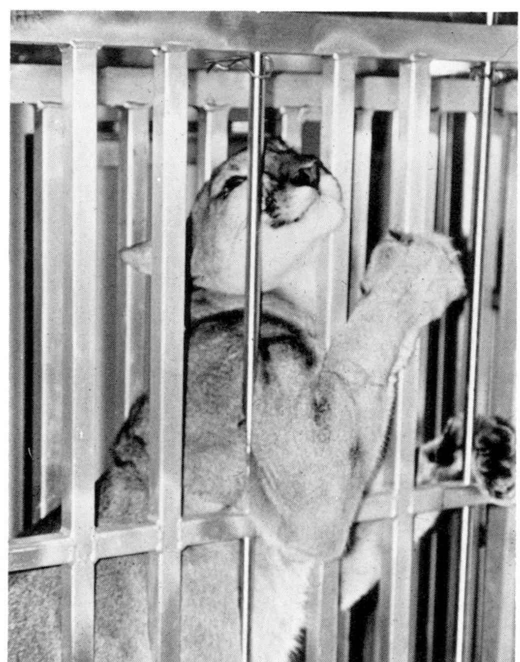

FIG. 19.33. Cats can reach out through bars. (Caution handlers and bystanders.)

lateral saphenous veins (Fig. 19.35) and the femoral, jugular, and cephalic veins. Felids also have a well-developed lateral tail vein which is easily penetrated.

The mouth of a felid can be opened for oral examination or surgery by using commercial or improvised specula such as a wooden block (Fig. 19.36).

Transport

Small cats can be directed into crates or transfer cages with a shield (Fig. 19.37). Larger cats can be baited into crates with food. Some must be immobilized before they can be crated. A plywood sheet, a human stretcher, or a sheet of canvas (Fig. 19.38) should be used to move an immobilized cat from one spot to another in preference to dragging it along the ground. Shipping crates and transfer cages must suit the size and behavior of the cat (Fig. 19.39).

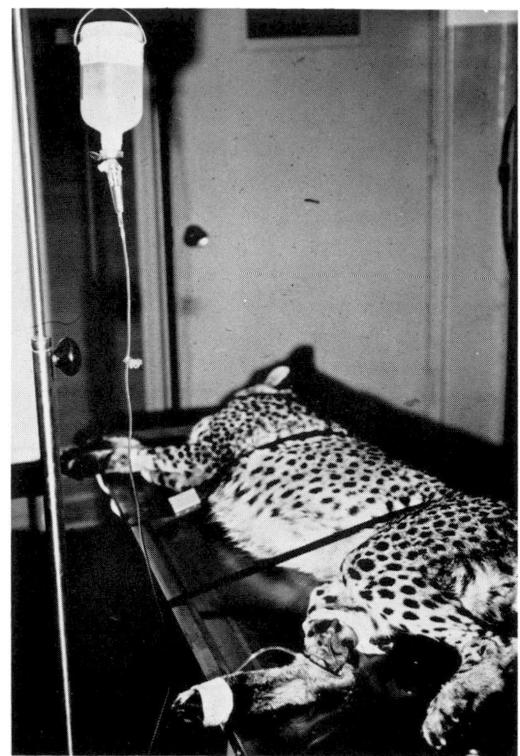

FIG. 19.34. Cheetah strapped to table while undergoing therapy.

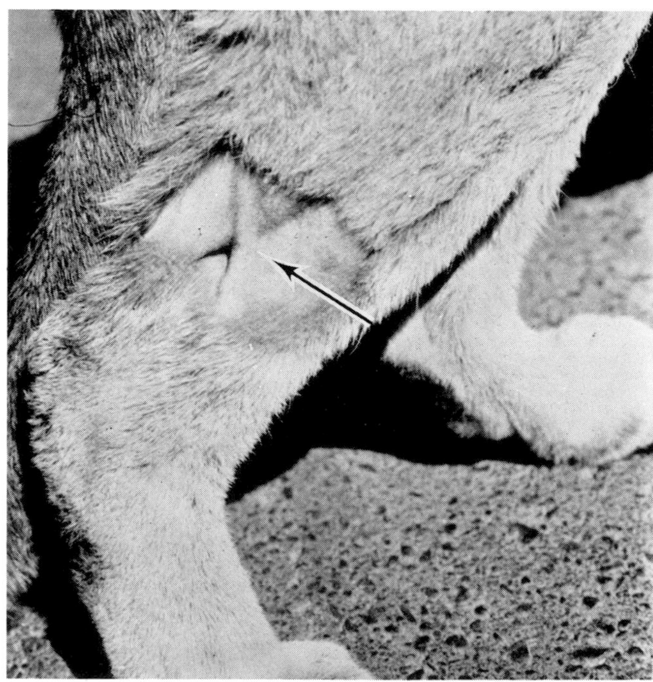

FIG. 19.35. Location of the lateral saphenous vein in a large felid.

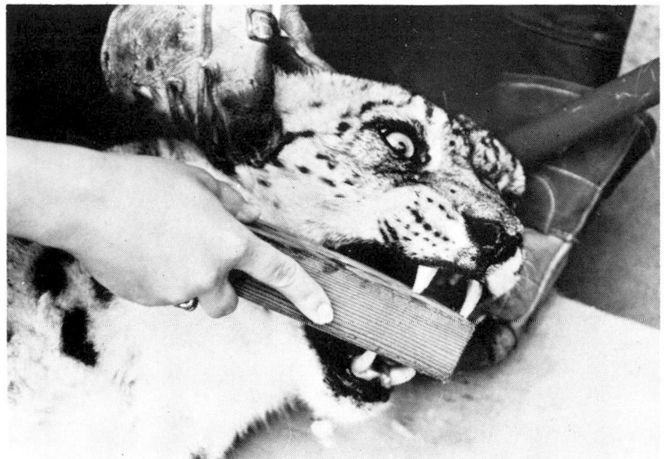

FIG. 19.36. Wooden block used to hold the mouth open.

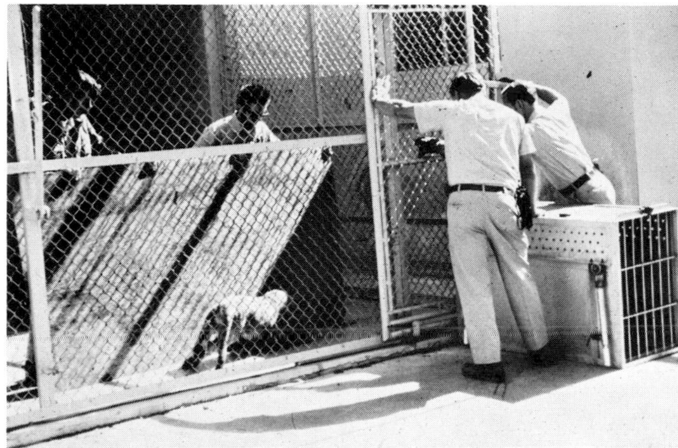

FIG. 19.37. Shield used to direct juvenile lion into transfer cage.

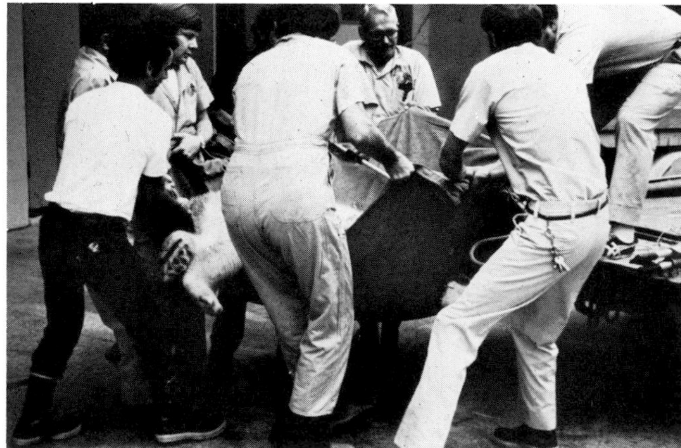

FIG. 19.38. Canvas sling for carrying an immobilized large carnivore.

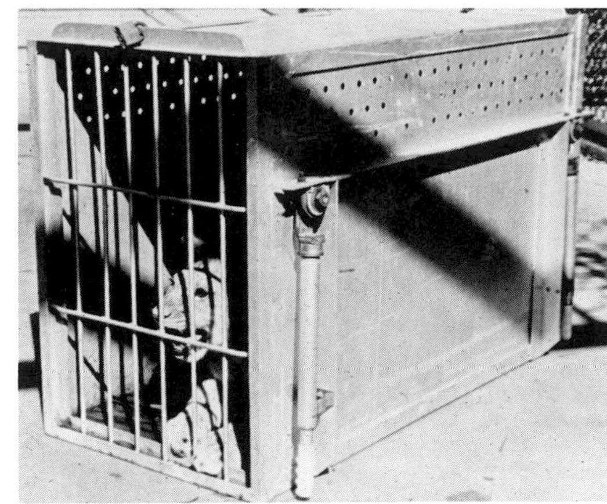

FIG. 19.39. Transfer cage for a large cat.

Chemical Restraint

Carnivores are frequently immobilized. Many chemical agents or combinations of agents have been used successfully. Table 19.3 lists reported agents with suggested dosages, to be used as a guide. If in doubt, contact an experienced zoo veterinarian. Always administer atropine sulfate (0.04 mg/kg) subcutaneously or intramuscularly when immobilizing or anesthetizing a carnivore.

REFERENCES

1. Adams, W. V.; Sanford, G. E.; Roth, E. E.; and Glasgow, L. L. 1964. Night-time capture of striped skunks in Louisiana. J. Wildl. Manage. 28:368-73.
2. Alford, B. T.; Burkhart, R. L.; and Johnson, W. P. 1974. Etorphine and diprenorphine as immobilizing and reversing agents in captive and free-ranging mammals. J. Am Vet. Med. Assoc. 164:702-4.
3. Bauditz, R. 1972. Sedation, immobilization and anesthesia with Rompun in captive and free-living wild animals. Vet. Med. Rev. 3:204-26.
4. Beck, C. C. 1972. Chemical restraint of exotic species. J. Zoo Anim. Med. 3:3-66.
5. Clifford, D. H., and Fletcher, J. 1963. Construction of a cage to confine, transport and treat bears. Int. Zoo Yearb. 3:121-24.
6. Ebedes, H. 1973. The drug immobilization of carnivorous animals. In E. Young, ed. The Capture and Care of Wild Animals, pp. 62-68. Capetown, South Africa: Human and Rousseau.
7. Ewer, R. F. 1973. The Carnivores. Ithaca, N.Y.: Cornell Univ. Press.
8. Fox, M. W. 1971. Behaviour of Wolves, Dogs and Related Canids. New York: Harper & Row.
9. Gregg, D. A., and Olson, L. D. 1975. The use of ketamine hydrochloride as an anesthetic for raccoons. J. Wildl. Dis. 11:335-37.
10. Hamann, J. R. 1969. Veterinary care for the pet skunk. Vet. Med. Small Anim. Clin. 64:409-11.
11. Harthoorn, A. M. 1976. Chemical Capture of Animals. London: Baillière, Tindall.
12. Hime, J. M. 1973. Mobile den wall installation for handling large mammals. Int. Zoo Yearb. 13:277.
13. Hummon, O. J. 1945. A device for the restraint of mink during certain experimental procedures. J. Am. Vet. Med. Assoc. 106:104-5.
14. Jacobson, J. O.; Meslow, E. C.; and Andrews, M. F. 1970. An improved technique for handling striped skunks in disease investigations. J. Wildl. Dis. 6:510-12.
15. Murrie, A. 1961. A Naturalist in Alaska, p. 125. New York: Devin-Adauer.
16. Seal, U. S.; Erickson, A. W.; and Mayo, J. G. 1970. Drug immobilization of the carnivora. Int. Zoo Yearb. 10:157-70.
17. Walker, E. P. 1968. Mammals of the World, 2nd ed., vol. 2. Baltimore: Johns Hopkins Press.

20 PRIMATES

CLASSIFICATION (numbers in parentheses denote number of known species)

Order Primates
- Suborder Prosimii (prosimians): tree shrew, lemur, loris, bush baby, tarsier (51)
- Suborder Anthropoidea: monkeys, apes
  - Family Cebidae: New World monkey (37)
  - Family Callithricidae: marmoset, tamarin (33)
  - Family Cercopithecidae: Old World monkey (58)
  - Family Pongidae: ape (11)
  - Family Hominidae: man (1)

ADULTS are called males and females. Newborn and young are called infants, youngsters, or juveniles.

Primates vary in size from a 60 g (0.15 lb) pygmy marmoset to a 275 kg (605 lb) gorilla (Table 20.1).

TABLE 20.1. Weights of primates

Animal	Kilograms
Tree shrew	0.1-0.2
Slow loris	0.5-1.5
Tarsier	0.08-0.15
Squirrel monkey	0.75-1.1
Capuchin	1.65-4
Spider monkey	6-8
Woolly monkey	5.5-6
Pygmy marmoset	0.06-0.07
Marmoset	0.1-1
Green monkey	7
Colobus monkey	12
Rhesus macaque	4.5-13
Baboon	14-41
White-handed gibbon	5-8
Chimpanzee	
male	56-80
female	45-68
Orangutan	
male	75-100
female	< 40-150
Gorilla	< 275

DANGER POTENTIAL

All species defend themselves by biting. All primates have large teeth and strong jaws with well-developed canine teeth. In gibbons and langurs—particularly the adult males—the canine teeth are vicious weapons, but the animals are by no means helpless without them. Owners of pet monkeys often have the canine teeth extracted, under the mistaken belief that this removes the danger of injury from biting. This is far from the fact. Even with the canines removed, a 20 kg macaque is capable of biting the finger from an adult person with one snap. Baboons and apes are particularly dangerous to handle because of their exceptionally large canine teeth and their extreme aggressiveness [5]. Research facilities housing such animals use special cages with a movable wall for animal restraint.

A secondary defense is scratching. Primate hands are able to grasp with strong fingers and hard fingernails, and scratches can be deep and painful. Medium-sized and large primates can also severely pinch and contuse any tissue within reach.

When working around primates, handlers should wear clothing that is neat and trim, with no pockets to provide a fingerhold and no full-length necktie; a primate may grab the tie and choke the handler. Primates also snatch glasses, pencils, or any other available object.

Unless one has had some experience working with primates, it is difficult to appreciate their tremendous strength. I recall a partially tamed 20 kg chimpanzee that could jerk its arm away from a man exerting the full strength of both hands on its arm to prevent it from escaping. To subdue this same animal required four adult men, one gripping each limb to stretch and hold it for examination.

The larger apes can maintain a grip on a hand or a limb that is impossible for a person to break. A man cannot pull his hand away from the one-finger grip of an orangutan or gorilla. With a full handhold, such animals can crush hands and dislocate joints with little effort. It is extremely important when working with primates that you not allow the animal to grab any part of your clothing or your body.

Primates are also able to throw objects at the handler, a trait possessed by no other group of animals and often resulting in significant injuries to people. Great apes have been known to pull a hypodermic dart from their bodies and throw it back at the marksman with tremendous force. When excited, a chimpanzee or a gorilla may pick up dirt, feces, water, stones, plant debris, or anything else handy and throw it at the handler.

Primates can remember and identify persons for whom they have conceived a dislike. Restrainers are most likely to be remembered. When I enter the enclosure to examine a chimpanzee at a nearby zoo, I must don a special raincoat and use a garbage can lid as a shield to deflect the bombardment of debris.

Some New World species of primates possess a prehensile tail. Although the tail is unlikely to harm a captor, it complicates restraint because it is used almost like another limb.

ANATOMY, PHYSIOLOGY, AND BEHAVIOR

The anatomical characteristic most important in restraint is the dexterity of the animal's hands and arms. A primate can prevent capture with a net by pushing away the hoop, or ensnarement by throwing off the loop. Primate hands can also grasp the handler to prevent him or her from carrying out the contemplated procedure and can seriously injure the handler as well.

All primates have tremendous relative strength compared to human beings, and this must be taken into account when selecting a restraint procedure. Most primates have short necks and relatively long limbs, allowing them to reach great distances. Nonhuman primates lack a human being's highly developed ability to dissipate heat by sweating, so provision must be made for temperature control when restraining these animals.

In nature most primates live either in family groups or in larger social units. A troop of baboons may include forty to fifty individuals. Groups have a definitive hierarchial social structure. The dominant individual is usually a male, but in some societies it is a female. The alpha animal is the most difficult to subdue and restrain and may interfere with the restraint of subordinates. In addition, the whole troop may attack, especially if one is attempting to separate out an infant or remove a dead infant.

A primate that has previously been restrained will be extremely difficult to approach a second time. They have excellent powers of recall, which make them wary of unpleasant experiences; a successful ruse can almost never be repeated.

When an individual is removed from a group of primates for examination, surgery, or other activities, the social structure is altered; reestablishment may require some time. During the interim, conflict is relative to the status of the individual who was removed. If it was the alpha animal, the subordinate animals will vie with one another to determine a new dominant individual. When the removed animal is replaced, it may find it impossible to reestablish itself at the previous level in the social structure. An individual can usually be safely reintroduced within three days of the removal; after that, success will be questionable. After a week the "newcomer" will almost certainly be attacked and possibly killed unless special techniques are used to disguise the introduction. One way to do this is to move the whole group into a new area and, during the confusion, introduce the newcomer. Another technique is to place the newcomer into the enclosure at night. Some feel that it is important to remove medicinal odors before reintroducing the animal.

It is essential for those who maintain captive populations of primates to understand these behavioral characteristics. The restrainer must be aware of and prepared to deal with the consequences of removing individuals from a social group.

PHYSICAL RESTRAINT

Prosimians

Small tarsiers and similar species can be grasped bare-handed (Fig. 20.1). Usually, however, the initial contact is made while wearing leather gloves (Fig. 20.2). Nets are universal tools. The animal can be extricated from the net and held bare-handed (Fig. 20.3).

Keepers of colonies of small primates (marmoset, tamarins) develop techniques for routine handling. One

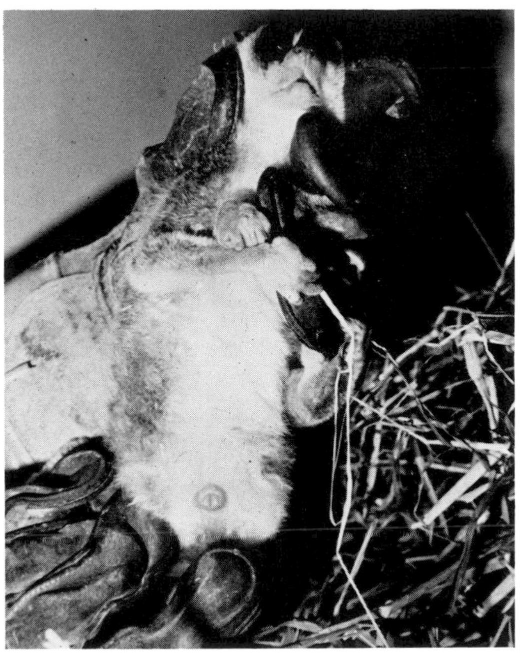

FIG. 20.2. Grasping a galago with gloved hands.

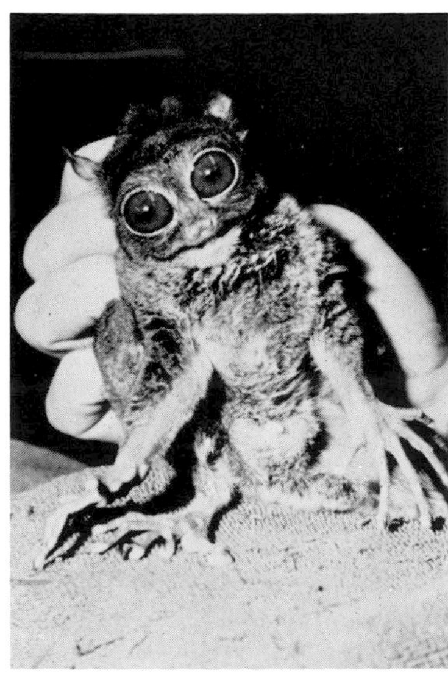

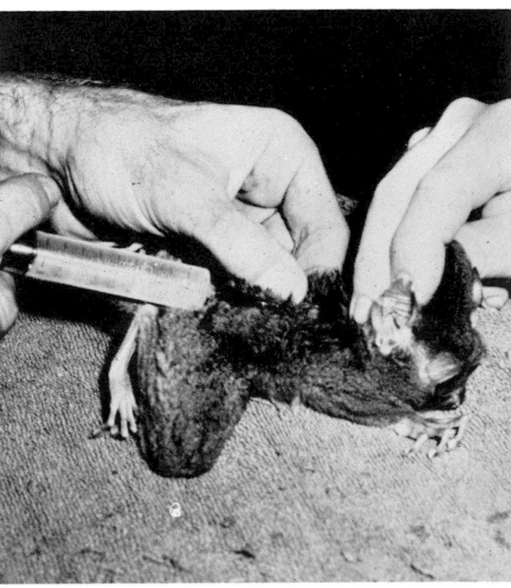

FIG. 20.1. Grasping a small tarsier bare-handed.

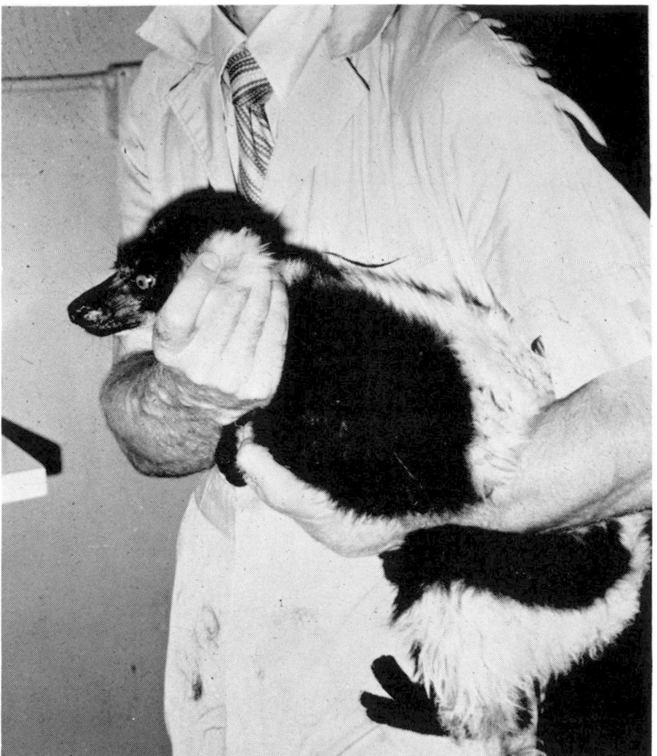

FIG. 20.3. Proper method of holding a lemur.

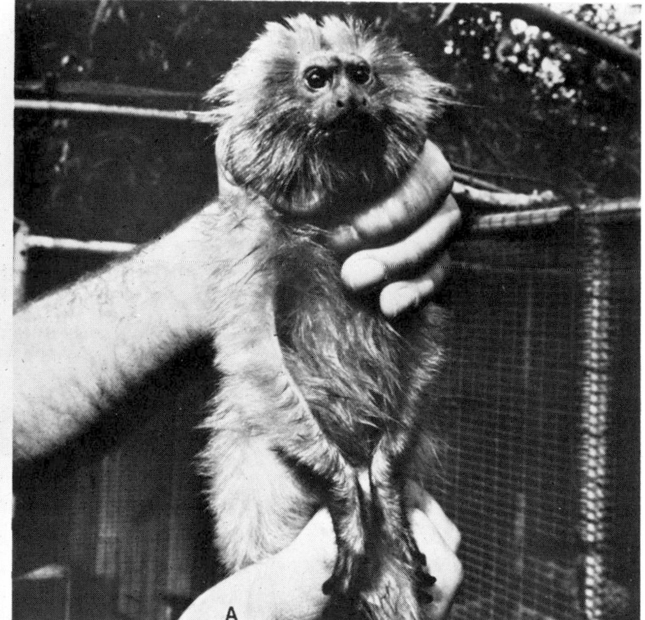

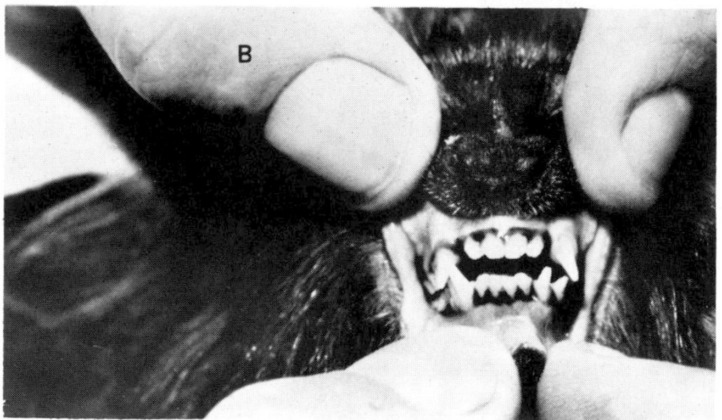

FIG. 20.5. **A.** Marmosets can be removed from the stockinette and held bare-handed. **B.** Marmoset teeth.

such technique, in use at the San Diego Zoo, incorporates a special night box with a sliding trap door. A hoop covered with a stockinette is placed over the opening (Fig. 20.4). When the night box is opened, the animal can be frightened into the stockinette. Once in the stockinette, it can be weighed or restrained and handled in any desired manner (Fig. 20.5). To remove the animal, grasp it through the stockinette behind the head at the nape of the neck, taking care to avoid the mouth. Slowly work the stockinette off, regrabbing the animal beneath the stockinette. The animal can then be manipulated bare-handed. Do not

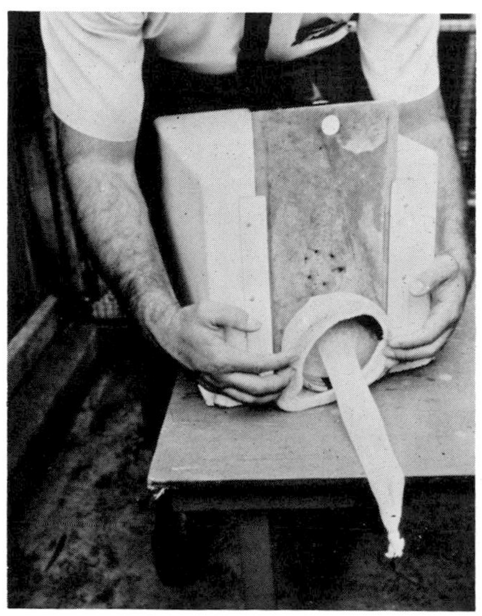

FIG. 20.4. Marmosets can be captured from their night box by placing a segment of stockinette over the door and encouraging the animal to crawl into it. (Technique of Dr. Charles Sedgwick)

disregard the sharpness of the teeth or the quickness of movement of these animals. Keep your hands away from the animal's face.

Chemical immobilization and medication of small species of primates should be preceded by accurate weight determination (Fig. 20.6).

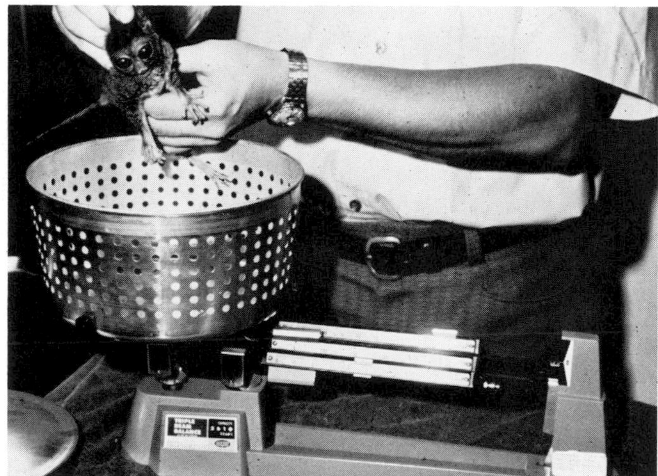

FIG. 20.6. Small species like the tarsier must be weighed to accurately determine dosage for chemical immobilization or for medication.

Simians—Small to Medium-Sized

Nets are commonly used to capture primates weighing up to 15 kg (33 lb). The diameter of the hoop and the size of the mesh are determined by the species to be captured. Mobile fingers and hands will entangle themselves if the mesh is too large. In no case should the animal be able to stick its head through the mesh, and preferably it should be unable to put a limb through the mesh.

Special precautions must be taken when entering a cage containing a group of monkeys, because the alpha male may attack. It is desirable to have a backup handler present, armed with a broom to ward off attack. In group cages it is not unusual for one or more monkeys to jump directly at the handler. This may be a defensive or an offensive maneuver, or the monkey may merely use the head or body of the handler as a springboard en route to some other vantage point. When entering a cage of animals known to jump at the handler, wear a face mask similar to those used by fencers (Fig. 20.7).

A small monkey (up to 5 kg [11 lb]) that has been netted may be grasped behind the head at the nape. This manipulation must be carried out swiftly to prevent the animal from turning its head to bite the handler. The arms of a larger monkey may be gripped above the elbows and pulled behind the back.

Some simple examinations and medications can be given with the monkey held in the net; if more complex procedures are required, it must be removed. This may be an arduous task, because the monkey will cling to the net with its mouth, all four feet, and its tail, if prehensile. Patience is required of all primate handlers. Once the monkey is out of the net, hold it as illustrated in Figures 20.8-20.11.

FIG. 20.7. A fencer's face mask should be worn when attempting to capture primates that may jump at the handler.

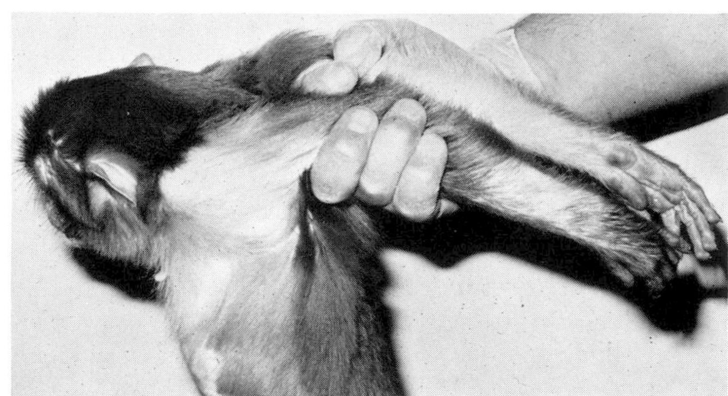

FIG. 20.8. Proper method for handling a primate weighing up to 10 kg. Note that a finger is kept between the arms of the monkey.

Gloves are an important tool for working with primates. Special double-thickness heavy leather gloves are made to help prevent mutilation of the handler's hands (Fig. 20.12). The extra thumb or mitten is offered to the animal to chew on while the other hand grasps the head. Some handlers prefer to wear gloves; others feel that bare hands are desirable because tactile discrimination and grip are enhanced. Macaques and baboons can crush the fingers even through a heavy leather glove. Some handlers use chain-mail butcher's gloves for added protection (Fig. 20.13).

Large primates should be handled in squeeze cages or by chemical restraint. Many types of squeeze cages have been designed for working with primates. It is not the purpose of this presentation to analyze and rate the myriad cages that can be fabricated; however, a few are illustrated (Figs. 20.14-20.16; see also Chapter 2). To remove an unsedated monkey from a squeeze cage, position the animal with its back to the door so it cannot bite the handler (Fig. 20.16). Then pull the arms behind the monkey's back as previously described.

Figure 20.17 illustrates a method of passing a stomach

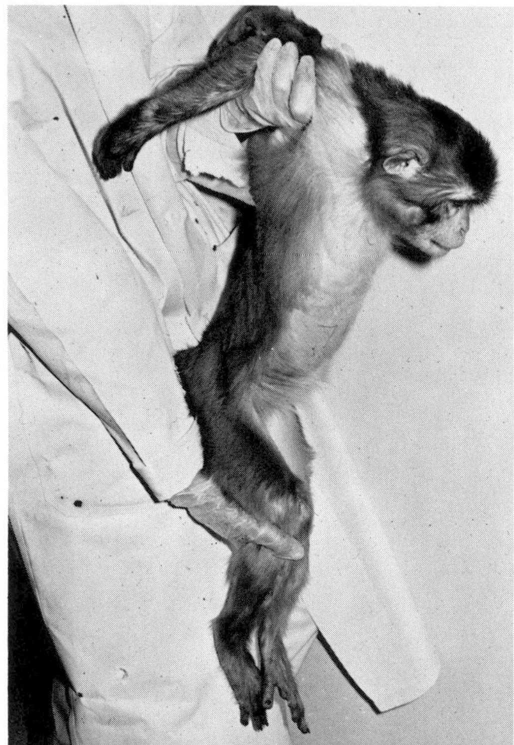

FIG. 20.9. Further restraint can be obtained by grasping the legs.

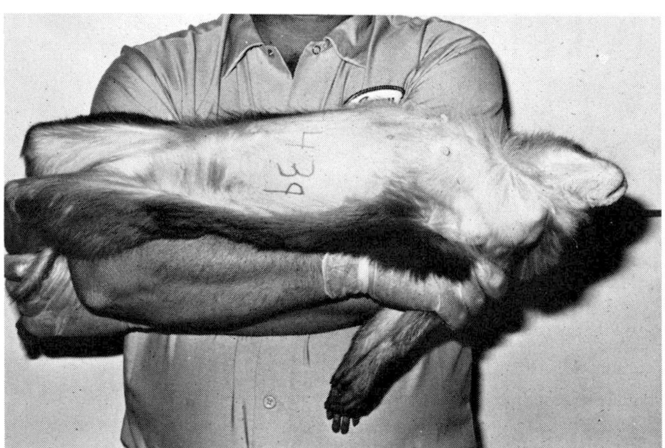

FIG. 20.10. Abdomen of a primate is exposed for examination or tattooing.

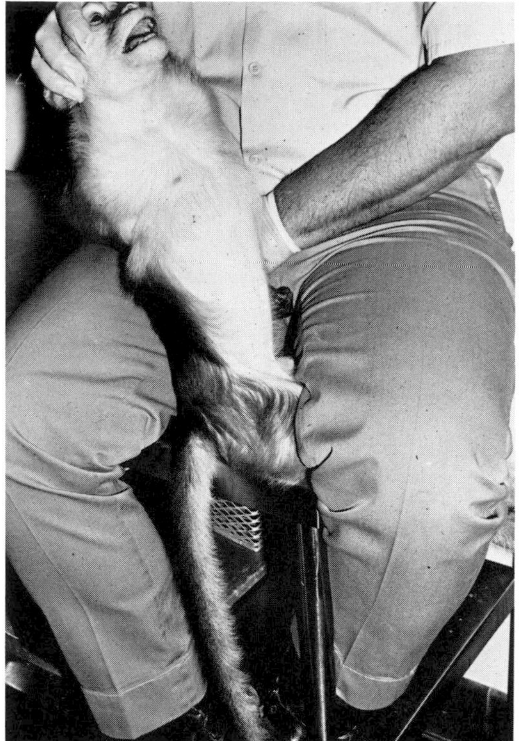

FIG. 20.11. Medium-sized primate, firmly secured.

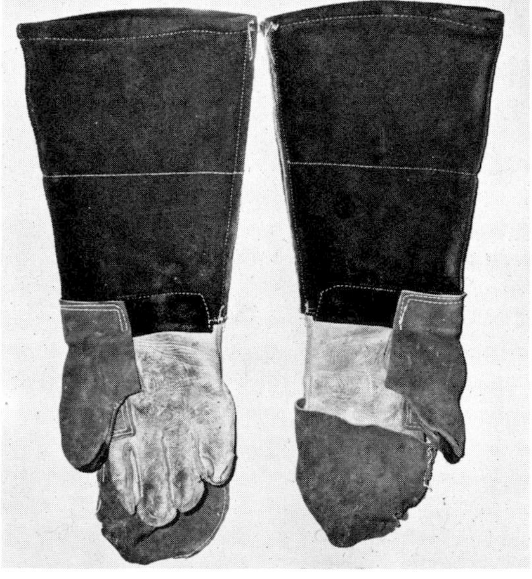

FIG. 20.12. Gauntleted, double-layered heavy leather gloves for working with primates.

tube in a physically restrained animal. The head is grasped as indicated and a piece of wooden doweling is inserted in the mouth. An appropriately sized stomach tube can be inserted over the top of the doweling or through a hole drilled in the doweling. The tube should be large enough to inhibit its passage into the trachea. The stomach tube is lubricated and inserted into the mouth. Gentle pressure on the tube will induce the animal to swallow it. After the tube has been inserted, the placement should be checked by immersing the other end of the tube in a pan of water. If the tube has been misplaced into the trachea, bubbles will be emitted with each expiration. A few bubbles may issue even though the tube is in the stomach, because of pressure on the stomach or perhaps some gas in the stomach. However, air bubbles from the stomach should not coincide with expiration. After proper placement has been assured, medication may be placed into the stomach.

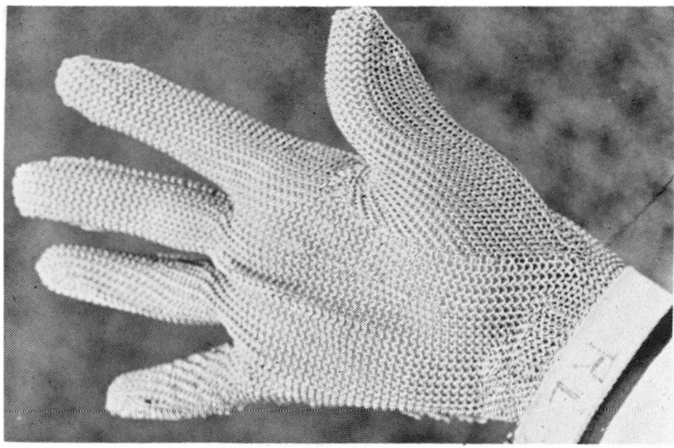

FIG. 20.13. Chain-mail (butcher's) glove to prevent serious injury from primate canine teeth.

FIG. 20.14. Commercial primate squeeze cage for animals up to 25 kg.

Blood samples from primates can be collected from numerous superficial vessels. The cephalic vessels of the arms are similar to those of human beings and dogs. In small to medium-sized New World monkeys the saphenous vein on the posterior aspect of the calf is an excellent vessel from which to obtain samples. In small animals or in moribund animals with low blood pressure, it is necessary to cut alongside the vessel to see it and direct the needle into it. When large quantities of blood are required, the femoral vein is penetrated just prior to the point of entrance into the pelvis (Fig. 20.18). It is desirable to slightly sedate the animal in order to avoid movement while the needle is in the vessel.

FIG. 20.15. Primate transfer cage.

FIG. 20.16. Removing monkey from a transfer squeeze cage.

Apes

In nearly all instances the manipulation of gorillas, chimpanzees, and orangutans requires chemical immobilization. Some zoos have constructed massive squeeze cages for apes, but most of these become inoperable or are not sufficiently adaptable for universal use. I do not recommend that squeeze cages be built into new facilities. The chemical immobilization agents presently available are generally safe, efficient preparations and impose much less stress and harm on the animals than any method of physical restraint adequate to hold them.

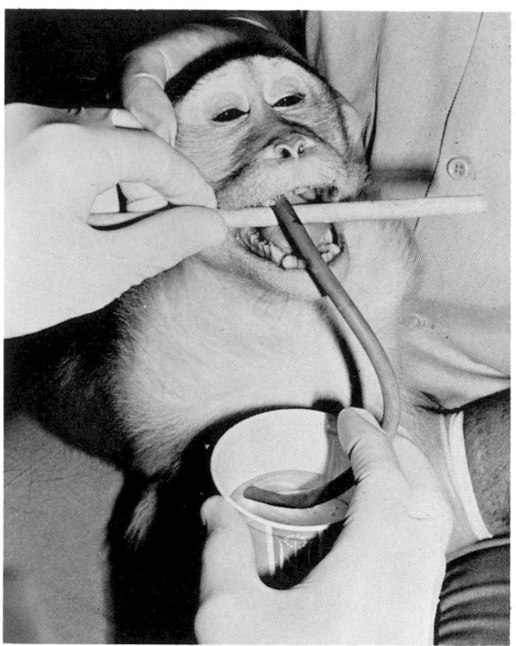

FIG. 20.17. Passage of a stomach tube using a dowel as a speculum.

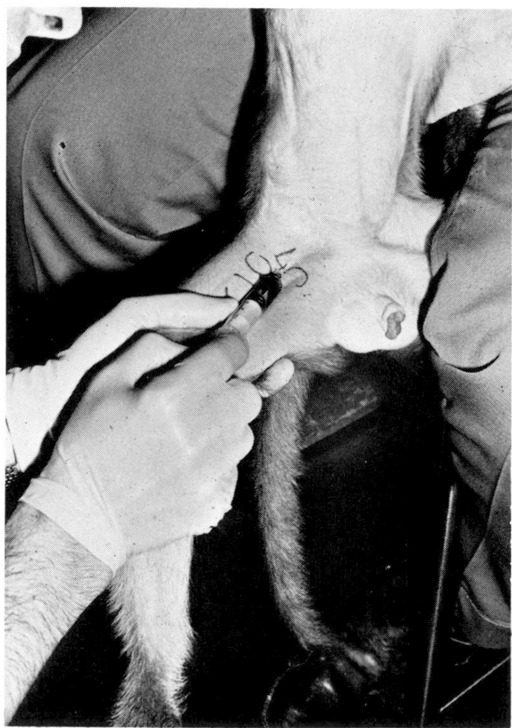

FIG. 20.18. Bleeding primate from femoral vein.

Special Procedures and Techniques

Tuberculin testing is an absolutely essential program for the management of captive primates. The preferred sites for intradermal injections of tuberculin are the eyelid, the abdomen, or the thorax. For the eyelid site the animal is restrained in a suitable manner, the upper lid is stretched,

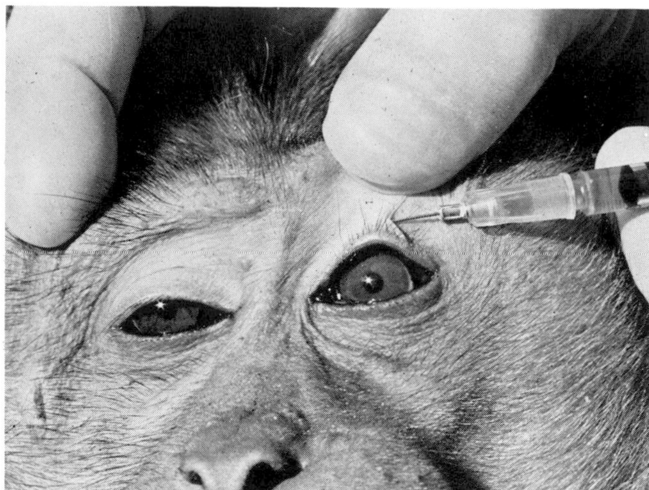

FIG. 20.19. Intrapalpebral intradermal injection of tuberculin. Notice stretch of the eyelid.

and the material is injected with a fine 26 gauge short beveled needle (Fig. 20.19).

Any relatively hairless area can be used, such as the forearm or the area around the nipple; or the hair can be carefully clipped from a chosen site on the abdomen. Palpation is required to read the test at all sites other than the eyelid. Results of injecting the eyelid can be seen. The test site should be inspected at 24, 48, and 72 hours.

To transfer a small primate from one swinging-door cage to another, use the technique described in Chapter 2.

Primates have become important as laboratory animals. Large colonies of various species are maintained in primate centers of university and other research laboratories throughout the world. A great deal of experience relating to the care and handling of animals has been gained in such institutions, but many of the techniques practiced are not really desirable for general primate restraint. For instance, in one primate center, large macaques are allowed out of individual cages to run free in a small narrow room. The handler captures one animal by grasping it by the loins as it runs by. The handler must be strong enough to maintain a grip on the animal and keep swinging it until the arms can be grasped, one at a time, and pulled behind the animal's back. It takes special skill to accomplish such a capture, and this technique is not recommended for the majority of handlers. Some acceptable special restraint devices used in primate centers include the wooden cross (Fig. 20.20) and plastic tubes for small species (Figs. 20.21, 20.22) [4].

One particularly perplexing problem faces the veterinarian who must handle a pet primate. The pet may arrive in the arms of its owner, but when it becomes necessary for the veterinarian to grasp the animal, the owner is completely incapable of assisting in the operation [3]. Attempts to take the animal or initiate restraint while the animal is held in the arms of the owner may result in the owner being bitten. If at all possible, the animal should be transferred to a squeeze cage and quickly immobilized with a

FIG. 20.20. Primate secured on a plywood cross (used for continuous medication or other treatment).

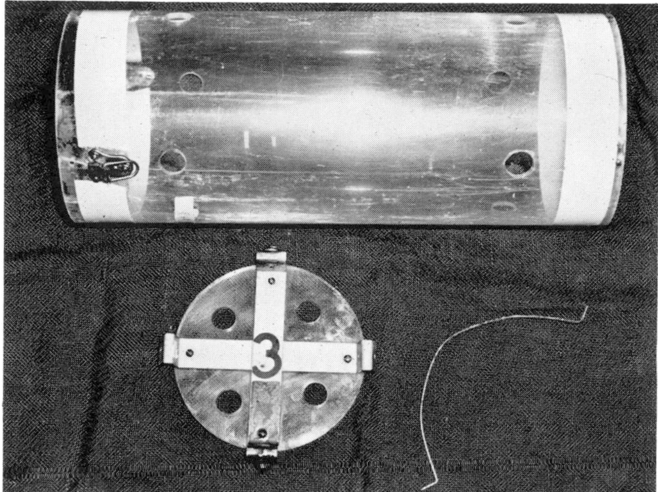

FIG. 20.21. Plastic tube used to transport and/or manipulate small primates.

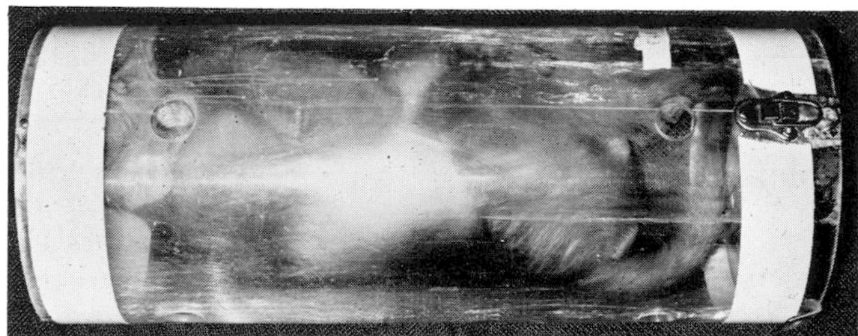

FIG. 20.22. Small primate in a plastic tube.

FIG. 20.23. Sheet of plywood used as a stretcher for a large gorilla.

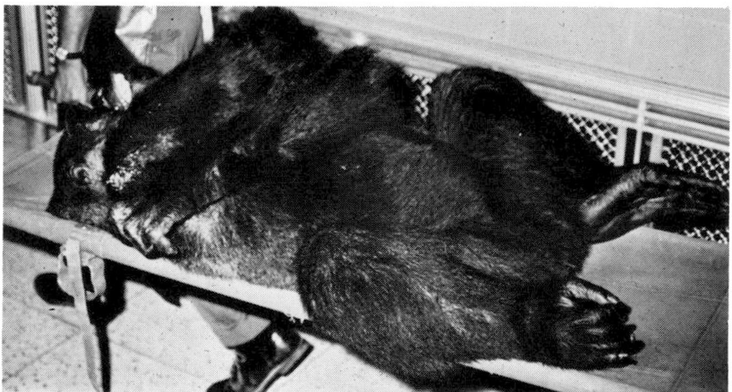

FIG. 20.24. Transporting gorilla on a human stretcher.

chemical restraint agent. Alternatively the animal should be handed to a third party who is capable of handling it. In nearly all instances pet primates have been fed a deficient diet and are likely to suffer from demineralization of the bone, predisposing them to fractures; extraordinary care must be taken to avoid injuring them by the restraint practice.

TRANSPORT

Small to medium-sized primates are placed in small cages and easily moved from one place to another. Large animals such as the gorilla may be transported on heavy plywood sheets or human stretchers following chemical immobilization (Figs. 20.23, 20.24).

CHEMICAL RESTRAINT

Primates are easily handled with chemical agents now available. Fortunately the dosage levels of each drug are essentially the same throughout the order.

Ketamine hydrochloride in doses of 5-20 mg/kg is an excellent immobilizing agent for all classes of primates [1]. Mild sedation to anesthesia is accomplished within this dosage range. A tiletamine-zolazepam combination in a dose range of 2-10 mg/kg is the newest member of the cyclohexanone group of chemicals and provides excellent anesthesia for primates.

Phencyclidine hydrochloride in a dose range of 0.7-1.5 mg/kg was one of the first drugs used and is still used for immobilizing large apes or when prolonged anesthesia is desirable [2]. Phencyclidine can also be given orally (2-3 mg/kg), disguised in grape juice or orange juice if the animal is trained to drink from a cup.

Animals sedated with phencyclidine remain under the effects of the drug for hours. A potentially fatal condition, known by medical primatologists as post sernylan sleep, may develop when the drug is given late in the day and the animal is left alone to sleep off the sedation. Sometimes handlers have returned the next morning to find a dead ape. The cause is unknown, but it is theorized that the animal rouses in a darkened enclosure providing little stimulation to continue recovery. With no incentive to move about, the animal goes back to sleep, and possibly a thermoregulatory upset or metabolic change develops which leads to the animal's death.

The chemical restraint of a primate in a large open cage is difficult. Small species have limited muscular areas available as injection sites. Even the chimpanzee or orangutan may provide a small target for the dart. Apes are extremely wary, expecially if they have been previously immobilized. They will create a violent disturbance when anyone attempts to point a gun or blowgun at them. The animals may charge the bars, reaching out and grabbing at the marksman (Fig. 20.25). They may spit or throw feces, debris, or water. They may keep running about the cage, preventing a stationary target; or they may station themselves in a corner, folding their arms and legs into positions that leave essentially no muscle mass visible as a target site. Patience, experience, and readiness to take advantage of fleeting opportunities are necessary to achieve success in such cases.

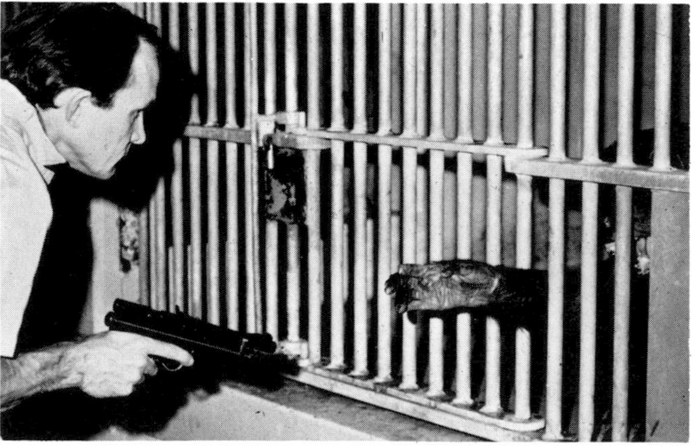

FIG. 20.25. Chemical immobilization of the large apes is difficult. Here the animal is close to the bars, grabbing at the handler.

Inhalation anesthesia is accomplished with primates ranging in size from the tiniest marmoset to the gorilla (Fig. 20.26). Small primates require supplemental heat during recovery from anesthesia. A human infant incubator is useful for this purpose (Fig. 20.27).

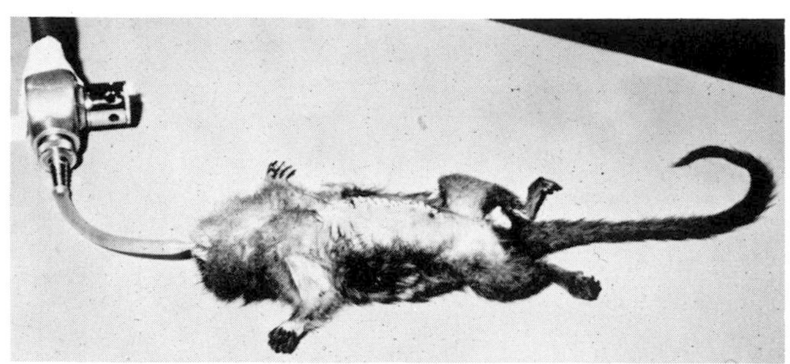

FIG. 20.26. It is difficult to intubate tiny primates, but Cole endotracheal tubes can be used effectively.

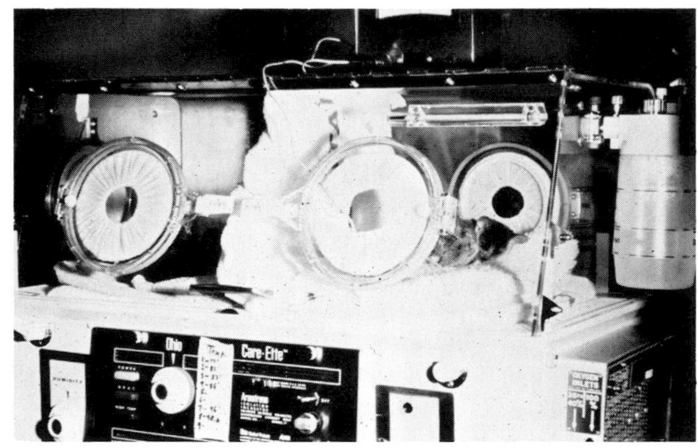

FIG. 20.27. Human incubator provides extra warmth and increased oxygen—vital to the recovery of sick small mammals.

REFERENCES

1. Bonner, W. B.; Keeling, M. E.; Van Ormer, E. T.; and Haynie, J. E. 1972. Ketamine anesthesia in chimpanzees and other great ape species. In G. H. Bourne, ed. The Chimpanzee, vol. 5, pp. 255-68. Baltimore: University Park Press; Basel: Karger.
2. Foster, P. A. 1973. Immobilization and anesthesia of primates. In E. Young, ed. The Capture and Care of Wild Animals, pp. 69-76. Capetown, South Africa: Human and Rousseau.
3. Harris, J. M. 1974. Restraint and physical examination of monkeys and primates. In R. W. Kirk, ed. Current Veterinary Therapy V, pp. 586-88. Philadelphia: W. B. Saunders.
4. Swan, S. M. 1970. An apparatus for the transportation and restraint of non-human primates. Lab. Anim. Care 20:1131-32.
5. Whittingham, R. A. 1971. Trapping and shipping baboons. J. Inst. Anim. Tech. 22:66-82.

21 MARINE MAMMALS

CLASSIFICATION (numbers in parentheses denote number of known species)

Order Cetacea: whales
- Suborder Odontoceti: toothed whale, dolphin (80)
- Suborder Mysticeti: baleen whale (12)

Order Sirenia: dugong, manatee (4)

Order Pinnipedia
- Family Otariidae: eared seal, sea lion (13)
- Family Odobenidae: walrus (1)
- Family Phocidae: seal (18)

THE POLAR bear and the sea otter are also marine mammals, but they are discussed in the chapter on carnivores.

Marine mammals are adapted to an aquatic habitat and present special problems to the animal restrainer. In the United States they are all under the regulation of the Marine Mammal Protection Act.

Marine mammals vary in length from 1 to 30 m (3–100 ft) and may weigh up to 81,000 kg (89 t) (blue whales) [3]. Names of gender are listed in Table 21.1.

TABLE 21.1. Names of gender of marine mammals

Animal	Adult Male	Adult Female	Newborn and Young	Group Name
Whales	Bull	Cow	Calf	Pod
Dolphin	Bull	Cow	Calf	
Sea lion	Bull	Cow	Pup	Herd
Walrus	Bull	Cow	Pup	
Seals	Bull	Cow	Pup	

DANGER POTENTIAL

Toothed whales are carnivorous, but fortunately most of them are not aggressive. Divers can usually enter their pools or tanks without fear of injury.

A dolphin may throw its head about and smash it into a handler. Dolphins often bump objects with their large and powerful mandible and can easily break the arm of a person.

All cetaceans flip the flukes up and down. A large killer or pilot whale could crush a person with a slap of the tail. Baleen whales do not bite, but the tail fluke may injure a handler.

The dugong and the manatee are mild-mannered herbivorous animals that are unlikely to be dangerous.

Seals and sea lions are efficient carnivorous predators. They have strong teeth capable of tearing flesh. The large bulls of various species are able to badly bruise or crush an individual who approaches too closely, but such behavior is unusual.

ANATOMY AND PHYSIOLOGY

Marine mammals lack readily accessible limbs that can be grasped for restraint. Just beneath the skin of most species is a blubber layer of insulation that must be reckoned with during restraint procedures. Marine mammals have a highly specialized thermoregulatory system, eliminating heat by conduction from the appendages as they swim through the water. The single most important precaution to take for the safety of a captive animal is to make sure it does not overheat. A continual spray of water should be played on restrained animals.

The massive bodies of large marine mammals are buoyantly supported by water displacement in their oceanic habitat. When they are "dry-docked" or beached for examination or treatment, tremendous pressures are exerted on anatomical systems, particularly on the flexible thorax. Such burdens may embarrass respiration if the animals are kept out of water too long.

Cetaceans breathe through a blowhole on the top of the head. Never obstruct this orifice.

PHYSICAL RESTRAINT

Cetaceans—Whales and Dolphins

Cetaceans are popular attractions at zoos and oceanaria. They are intelligent and can be trained to perform spectacular feats (Fig. 21.1). Their trainers can usually enter the pools and administer intramuscular injections while in the water. Killer whales have been trained to lay their tail fluke on the side of the pool while blood is withdrawn from a fluke vein.

A calm dolphin can be removed from a tank by means of a stretcher sling (Figs. 21.2, 21.3, 21.4). If the animal must

FIG. 21.1. Killer whale clearing a rope.

FIG. 21.2. Small docile cetaceans can be removed from a small pool by handler gently directing the animal to the bank.

FIG. 21.3. Sling is slowly moved around dolphin to lift it onto the bank.

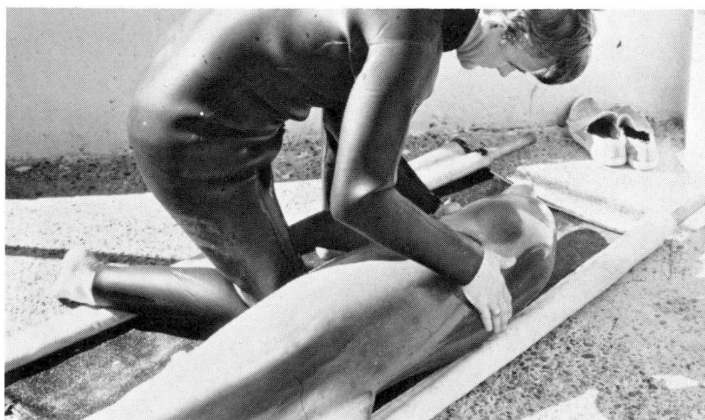

FIG. 21.4. Examination and treatment is carried out quickly to minimize heat stress.

be dry-docked for more than a few minutes, it should be placed on a foam rubber pad or a mattress. A cetacean can be examined while lying on its abdomen, or it may be gently rolled over onto its side (Fig. 21.5). Be certain to pull the flipper back against the body before rolling it to preclude

FIG. 21.5. Cetaceans can be partially immobilized by draining their pools.

injury to the forelimb as the posture change is made [5]. No marine animal should be left unattended while out of the water. Watch carefully for signs of heat stress and possible injury from flipping over or flopping into dangerous positions.

Sunlight and wind are added hazards to beached cetaceans. The skin is highly susceptible to sunburn, and the eyes can be damaged as well. Water sprayed over the body continually is helpful in preventing hyperthermia (Fig. 21.6), or the animal may be covered with a wet sheet that is rewet as necessary. A blindfold may serve the double purpose of protecting the eyes from exposure to sunlight and diminishing visual stimulation.

These air-breathing mammals can stay submerged for

FIG. 21.6. Hosing down beached cetacean.

only a limited time. When several animals are in a large pool, one can be separated and partitioned off for capture with a net. The separation net must extend completely across the pool from top to bottom. Handlers or divers should be in the pool to aid the animal should it become entangled and to assist in maneuvering the net into position. When one animal is captured and lifted from the pool, the net should be removed so others do not become entangled.

The release of a restrained dolphin should be as painstaking as the capture. Replace the animal on the sling and lower it gently into the water. When the animal is buoyant, free it from the sling. An attendant should be in the water to head the animal toward the center of the pool—never toward the periphery. As soon as the animal is released from the sling and has taken one complete breath, it can be allowed its freedom.

Generally large cetaceans are captured by emptying the tank. Attendants should stay beyond the arc of the tail fluke to avoid serious injury. Blindfolding the animal reduces activity, although this does not guarantee that the animal will not flip around and bite. Close to the side of the animal at midbody is likely to be the safest place to stand (Fig. 21.7).

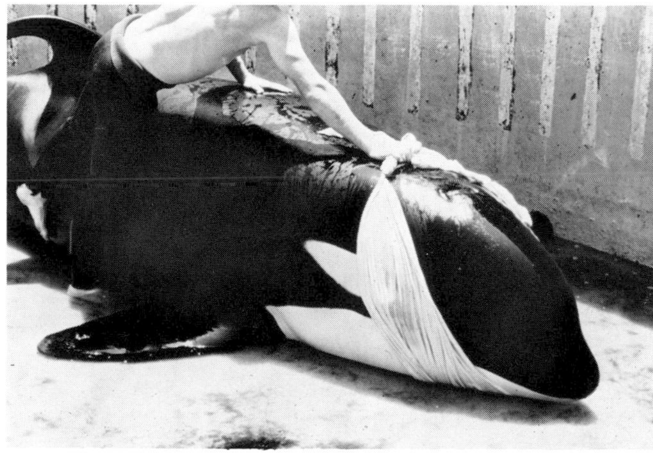

FIG. 21.7. Killer whale that has been drydocked by draining the pool. A blindfold minimizes struggling.

Blood samples for laboratory evaluation can be withdrawn from the tail fluke (Fig. 21.8). The vessels are located approximately one-third of the way posterior to the leading edge of the fluke. A groove indicates the site location on both dorsal and ventral surfaces [6]. The flipper of smaller species is also a suitable site (Fig. 21.9).

Intramuscular injections are given in the large muscle mass just ahead or on either side of the dorsal fin (Fig. 21.10). In a dolphin, use an 18 or 20 gauge needle, 3.5 cm (1.5 in.) long. In cetaceans over 180 kg (396 lb), use a 5 cm (2 in.) needle [6].

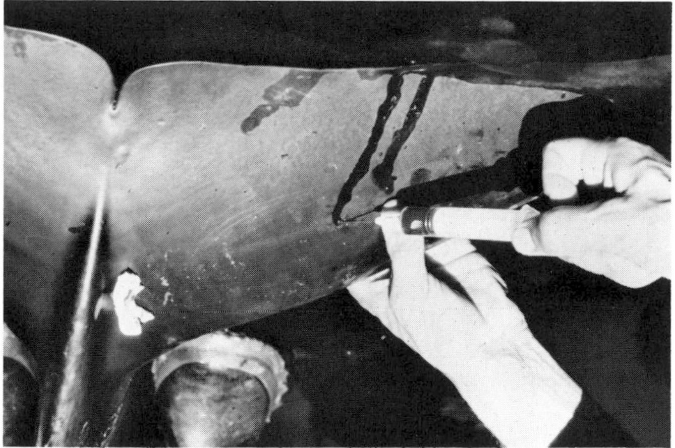

FIG. 21.8. Withdrawing blood from vein in the fluke of a cetacean.

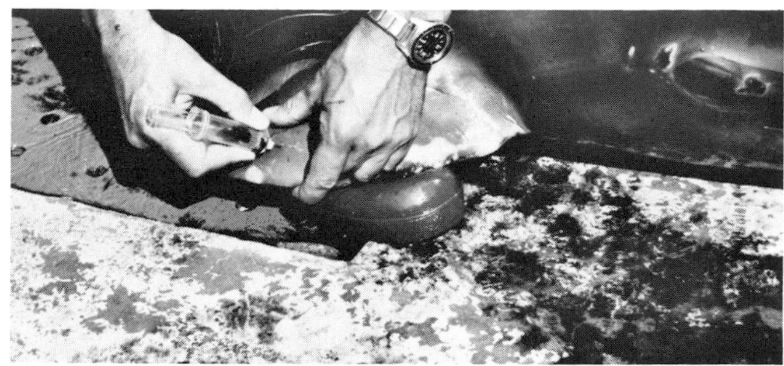

FIG. 21.9. Withdrawing blood from interdigital vein on the foreflipper.

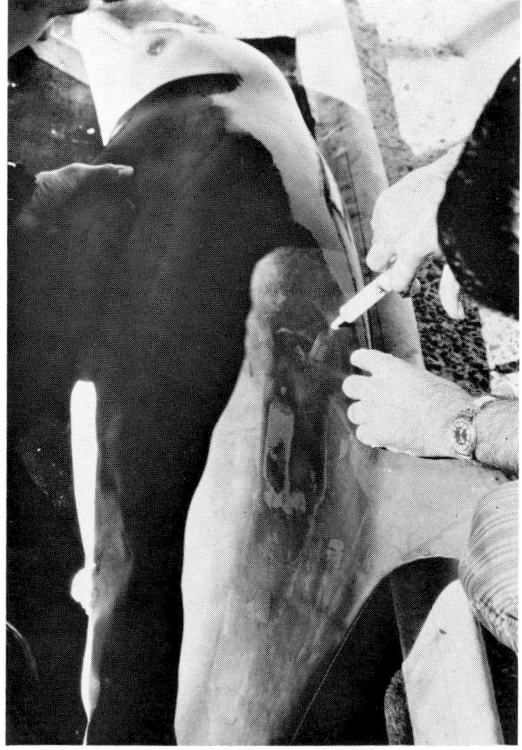

FIG. 21.10. Giving intramuscular injection on side of dorsal fin.

Sirenians—Manatees, Dugongs

These inoffensive animals are handled by the same methods as cetaceans.

Pinnipeds—Seals, Sea Lions, Walrus

Pinnipeds are amphibious. It is virtually impossible to capture a sea lion in a pool; it can elude or jump over nets. The animal should be enticed onto the bank with food or the pool should be drained before attempting capture.

The sea lion's head is extremely mobile; it can reach out effortlessly to the side and to the back. Be extremely cautious when working with these animals, because their bite wounds may be severe. Handle small pups with gloved hands by grasping them over the neck from the back. Encircle the neck with a come-along or a leather strap (Figs. 21.11, 21.12) to restrict mobility while grasping the neck and head. Further restriction can be imposed by placing a burlap bag over the animal and fixing it tightly against the body with the handler's feet (Fig. 21.13).

An experienced handler can seize a small to medium-sized pinniped by the tail and quickly grasp the back of the head to manually restrain it for enough time to induce anesthesia (Figs. 21.14–21.18).

FIG. 21.11. Close-up of pinniped neck strap.

FIG. 21.12. Restraining a seal with a neck strap.

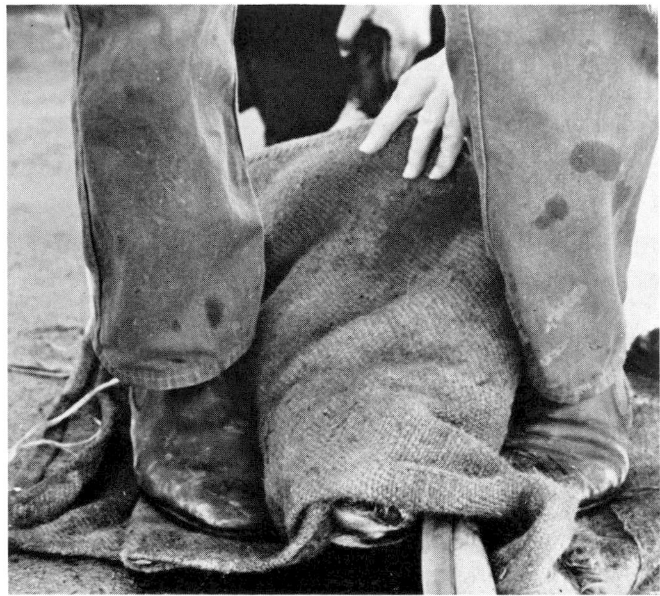

FIG. 21.13. Pinniped with burlap sack secured tightly over the head; body is restrained by feet of the handler.

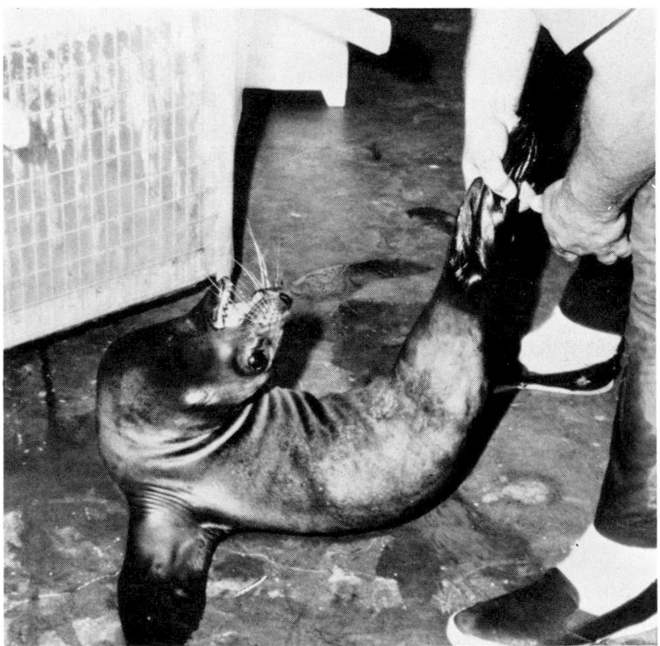

FIG. 21.14. Manipulating juvenile sea lion by grasping the pelvic flippers. (The head is extremely mobile; avoid the sharp teeth.)

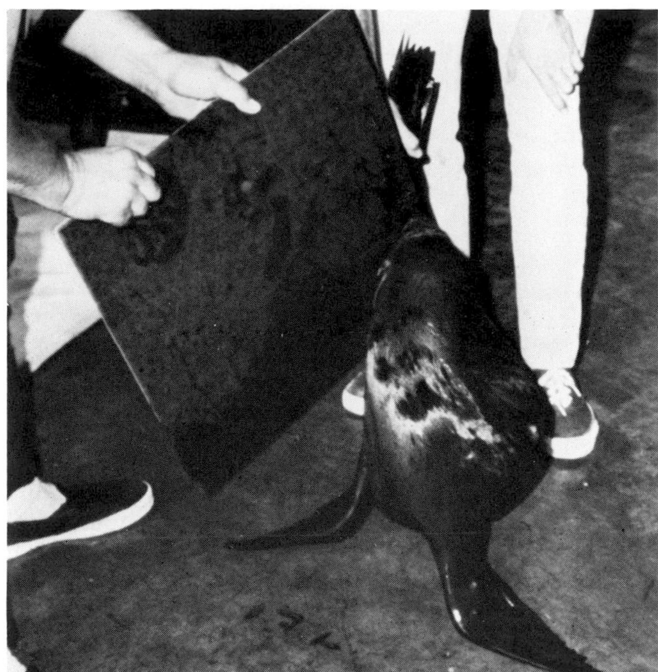

FIG. 21.15. Shield can be used to closely approach a juvenile sea lion.

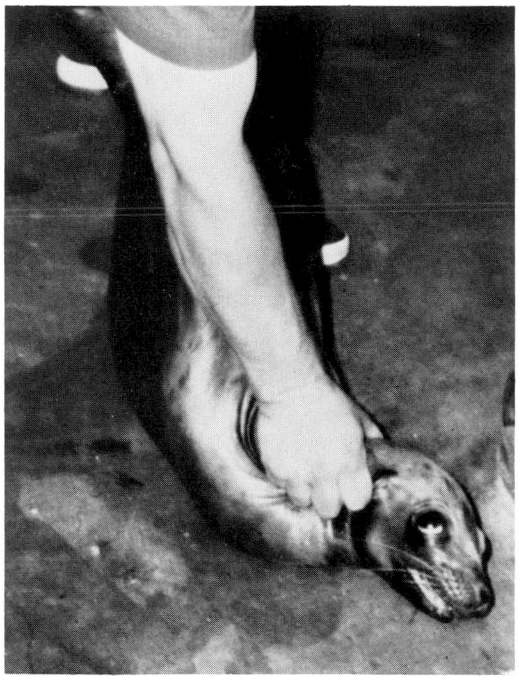

FIG. 21.16. Sea lion can be grasped firmly by loose skin at the base of the head. (This must be done quickly and completely in one motion or animal can twirl its head to bite.)

Small seals and sea lions can also be captured in nets for examination or intramuscular injections (Fig. 21.19). A special heavy mesh net on a 2.5 cm (1 in.) pipe hoop is effective for handling small pinnipeds (Fig. 21.20). I have netted adult female California sea lions as they rushed from their night quarters when the door was opened.

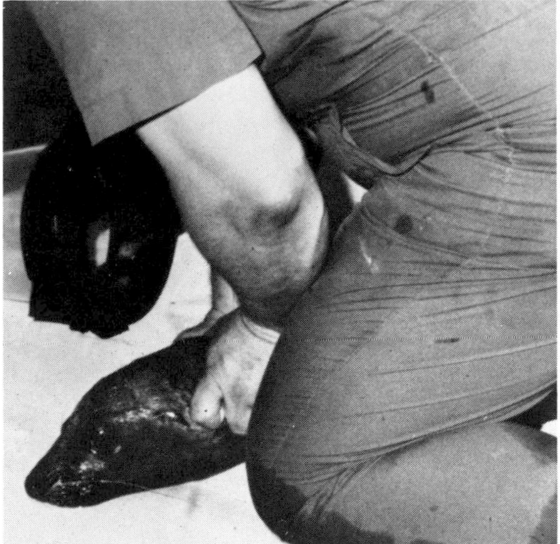

FIG. 21.17. Once the head is controlled, further restraint can be accomplished by straddling the body.

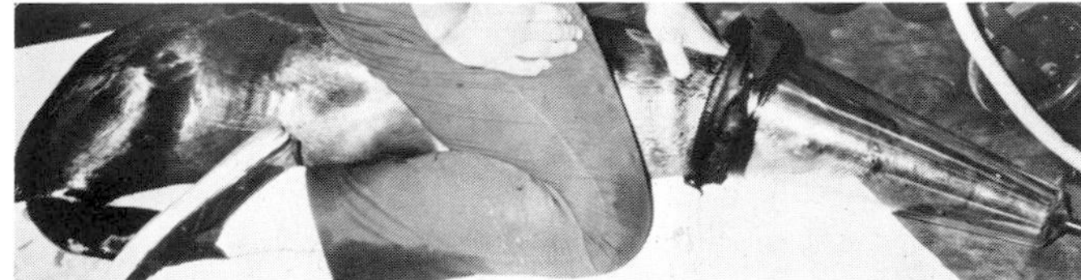

FIG. 21.18. Halothane anesthesia via face mask.

Special squeeze cages have been designed for handling pinnipeds (Figs. 21.21, 21.22). The cage can be lowered into a pool and baited with fish to entice the animal inside. The cage is then lifted to the pool edge, and the squeeze is applied. The animal can be strapped to the board on the floor of the cage and removed.

FIG. 21.19. Netting small sea lion. (Photo courtesy of J. Sweeny)

FIG. 21.20. Special hoop net used to restrict movement of a small sea lion.

FIG. 21.21. Commercial pinniped squeeze cage. (Photo courtesy of Research Equipment Co., Bryan, Texas)

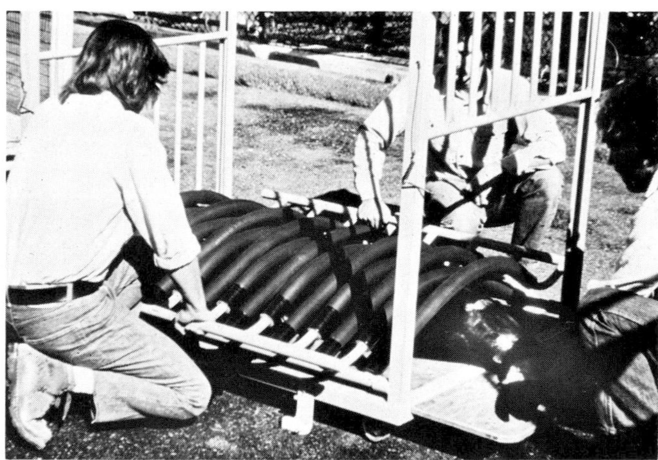

FIG. 21.22. Another commercial sea lion squeeze cage. (Photo courtesy of J. Sweeny)

Usually pinnipeds must be sedated or anesthetized to obtain blood samples. Blood vessels are deep to preclude excessive cooling in the oceanic environment. Three locations permit access to veins. In sea lions the caudal gluteal vein is penetrated on either side of the coccygeal vertebrae through the notch formed where the ileum and sacrum unite. Blood specimens can be obtained from a venous sinus formed by the confluence of the internal and external jugular veins and the internal thoracic vein [4]. Sedate the sea lion and place it in dorsal recumbency. Extend the head. Establish a line from the manubrium of the sternum (forward end) to the point of the shoulder. Insert an 18-20 gauge, 3.5-5 cm (1.5-2 in.) needle at a site 1.9 cm (0.75 in.) lateral to the sternum on the previously described line. Insert the needle downward in front of the first rib. Maintain a negative pressure on the syringe while inserting the needle.

Phocids (true seals) have an extradural venous sinus. A spinal tap in the lumbar area after proper surgical preparation can yield large quantities of blood.

Young sea lions, susceptible to verminous pneumonia, can be nebulized in a combination therapy-transport cart (Fig. 21.23).

FIG. 21.23. Crate completely enclosed to allow nebulization medication for respiratory tract diseases.

TRANSPORT

The transportation of marine mammals has become an art during recent years because of the popularity of commercial oceanaria [1,7]. Marine mammals are difficult to obtain and expensive to purchase and maintain. Although general principles are outlined here, no one should attempt to ship one of these valuable animals without first consulting a commercial oceanarium or aquarium.

Dolphins and whales can be lifted from the water and transported in special slings (Fig. 21.24). Slings for cetaceans must be constructed with attention to detail. The sling should be long enough to support the flukes or such that the flukes just clear the end of the sling [5]. If the trip is short and the animal small, fold the flippers back alongside the body. Protect the eyes from exposure to straps, buckles, netting, or anything else that might injure them.

Cut holes for the foreflippers in a sling for larger animals to be shipped long distances. The sling can be anchored to the walls of a box for long-distance truck or air shipment (Fig. 21.25) [7]. A special Santini box with a double-pad-

FIG. 21.24. Lifting killer whale in a sling. (Photo courtesy of L. Cornell of Sea World, San Diego)

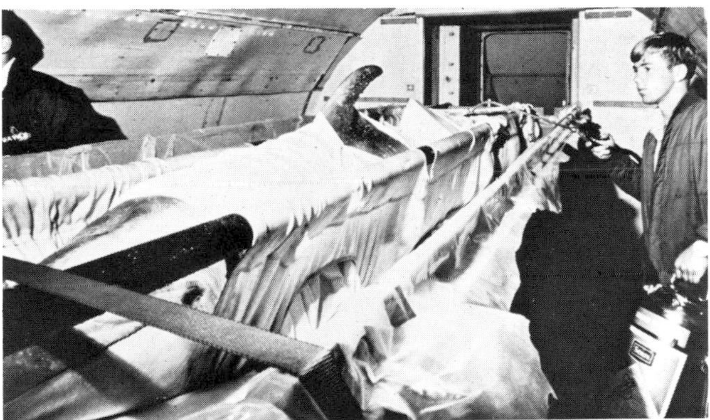

FIG. 21.25. Sling for transporting a cetacean secured inside an airplane. (Photo courtesy of L. Cornell of Sea World, San Diego)

ded floor has been successfully used for shipping dolphins [1].

Overheating is an ever present hazard for marine mammals during transport. The skin must be constantly moistened and the body temperature monitored throughout the voyage. Ice can be added to shipping boxes to retard overheating.

Individual variations in animal behavior will greatly affect the success of transport. If the animal continually thrashes about during the journey, abrasions and/or pressure sores are almost sure to occur. The flippers are also liable to damage. An improperly designed sling or a thrashing animal may traumatize the base of the flipper or impair circulation by continued pressure.

Pinnipeds can be shipped in rather simple crates (Figs. 21.26–21.28). However, they also must be carefully monitored to prevent hyperthermia.

Sirenians are transported by much the same methods as cetaceans.

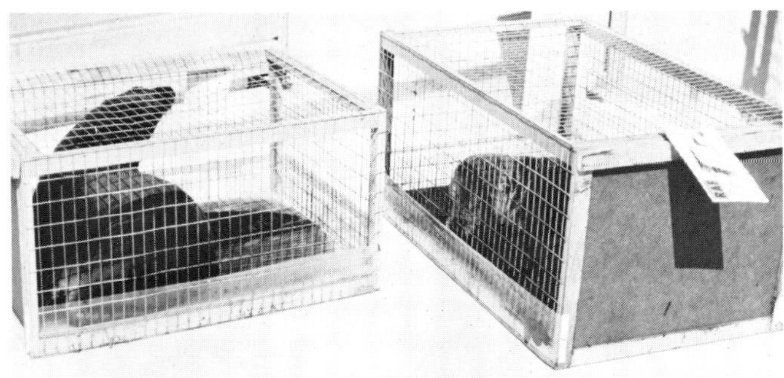

FIG. 21.26. Transfer crates for shipping pinnipeds. (Keep these animals in a cool environment. Periodic wetting may help.)

FIG. 21.27. Larger crate for moving a small elephant seal.

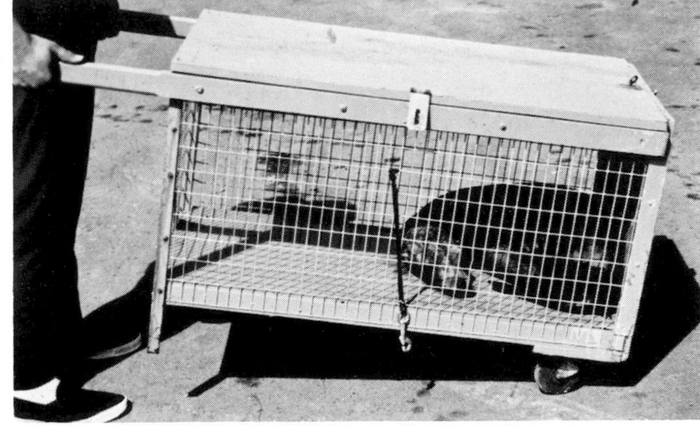

FIG. 21.28. Wheeled crate for moving small pinnipeds within a research facility.

CHEMICAL RESTRAINT

It is difficult to generalize about the application of chemical restraint to marine mammals. Be cautious and seek counsel before attempting it.

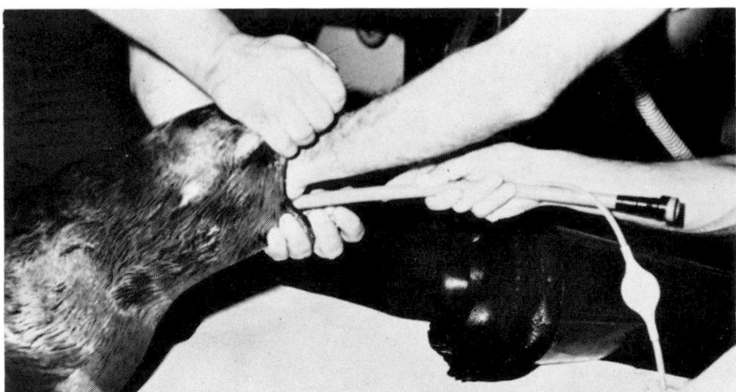

FIG. 21.29. Manual insertion of endotracheal tube after anesthetizing animal with halothane.

Reactions to restraint drugs are extremely variable in marine mammals. Their unique cardiovascular and pulmonary physiology makes it difficult to accurately predict drug effects.

Attempts to immobilize free-living animals in their aquatic environment have been less than satisfactory. California sea lions, elephant seals, and Weddell seals have been immobilized with tiletamine-zolazepam (0.5-1.5 mg/kg).[1] Ketamine hydrochloride has been used on a variety of pinnipeds in dosages of 4.5-11 mg/kg [2].

Marine mammals can be anesthetized with intravenous injections of barbiturates or by inhalant anesthesia. Small pinnipeds can be induced with halothane and a cone while physically restrained (see Fig. 21.18). As soon as possible, insert the endotracheal tube to provide respiratory assistance during periods of apnea. The mobility of the larynx may necessitate manual placement (Fig. 21.29).

1. Unpublished data.

Before instituting general anesthesia in cetaceans, administer thiamyl sodium (Surital) (10 mg/kg) in the fluke vein. After the animal relaxes, open the mouth and place the endotracheal tube into the pharynx. Pull the larynx from its intranarial position and insert the endotracheal tube. Continue anesthesia with halothane and oxygen with a positive pressure respirator-anesthetic unit. Details of the technique are described by Ridgeway [5].

REFERENCES

1. Dudok Van Heel, W. H. 1972. Transport of dolphins. Proc. Conf. Transportation of Exotic Animals, Zool. Soc. Lond.
2. Geraci, J. R. 1973. An appraisal of ketamine as an immobilizing agent in wild and captive pinnipeds. J. Am. Vet. Med. Assoc. 163:574-77.
3. Nishiwaki, M. 1972. General biology. In S. H. Ridgeway, ed. Mammals of the Sea, p. 19. Springfield, Ill.: Charles C Thomas.
4. Odend'hal, S. 1969. Possible venipuncture site in the California sealion. Proc. Semin. Medical Care and Husbandry of Marine Mammals, Stanford Res. Inst., Menlo Park, Calif.
5. Ridgeway, S. H., ed. 1972. Homeostasis in the aquatic environment. In Mammals of the Sea, pp. 689-94. Springfield, Ill.: Charles C Thomas.
6. Sweeney, J. C., and Ridgeway, S. H. 1975. Procedures for the clinical management of small cetaceans. J. Am. Vet. Med. Assoc. 167:540-45.
7. Wilkie, D. W.; Bell, G. B.; and Coles, J. S. 1968. A method of dolphin transport and its physiological evaluation. Int. Zoo Yearb. 8:198-202.

22 ELEPHANTS

CLASSIFICATION

Order Proboscidea
Family Elephantidae: Asian elephant, African elephant

THE ELEPHANT has captured attention since prehistoric eras. The caveman drew pictures of extinct elephant relatives on the walls of his home, and undoubtedly used the flesh of the animals for food.

The Asian elephant (Fig. 22.1 left) is one of the oldest domesticated animals. Both Asian and African species have carried human beings and moved their heavy loads for hundreds of years. The African elephant (Fig. 22.1 right) is much less tractable than the Asian and thus has not been as extensively used as a beast of burden.

A mature male elephant is called a bull, the female a cow. Newborn and young are called calves.

ANATOMY AND PHYSIOLOGY

Table 22.1 lists some approximate sizes of elephants. Precise data are not readily available, and estimates vary [2,11,12].

The respiratory apparatus of the elephant is unique. In most mammals the lungs are situated inside the chest cavity in a vacuum, as illustrated on the right side of Figure 22.2. Air is pressed out of the lungs by respiratory muscles, and negative pressure in the chest refills the lungs with air. In the elephant the parietal movement depends on free movement of the chest wall in a bellowslike action. This

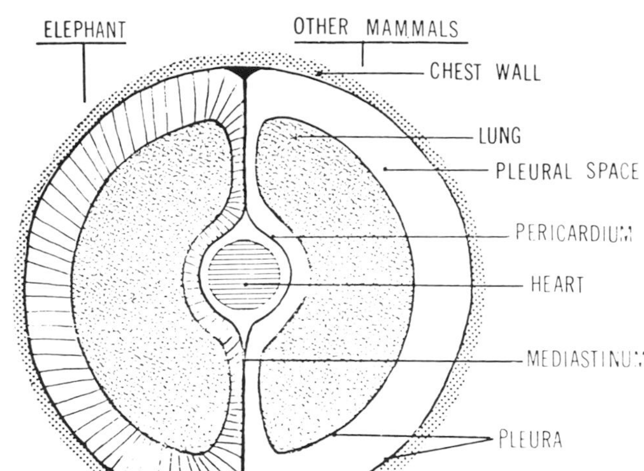

FIG. 22.2. Diagram illustrating adhesions of elephant lung to chest wall contrasted with usual mammalian morphology.

fact must be taken into consideration when the elephant is cast or put into positions that may inhibit respiration.

Literature citations indicate that the elephant cannot lie for a prolonged period either on its side or on the sternum (Figs. 22.3, 22.4). However, I have kept elephants in lateral recumbency for as long as 3½ hours during surgery. At other times elephants have been down on the sternum for some time without showing ill effects.

The trunk of the elephant is a vital organ. An injured or paralyzed trunk is a serious problem; if it cannot be cor-

FIG. 22.1. Elephants: Asian female *(left)*. African female *(right)*.

TABLE 22.1. Elephant size

Animal	Approximate Weight						Length		Height	
	Newborn		Adult female		Adult male					
	(lb)	*(kg)*	*(lb)*	*(kg)*	*(lb)*	*(kg)*	*(ft)*	*(m)*	*(ft)*	*(m)*
Asian elephant	198	90	7,920	3,600	9,900	4,500	18-21	5.5-6.4	8.2-10	2.5-3
African elephant	220	100	11,000	5,000	13,200	6,000	20-24.6	6-7.5	9.8-13	3-4

FIG. 22.3. Elephant should not be allowed to remain in sternal recumbency for extended periods; respiration may be compromised and elbows and knees suffer excessive straining and contusion.

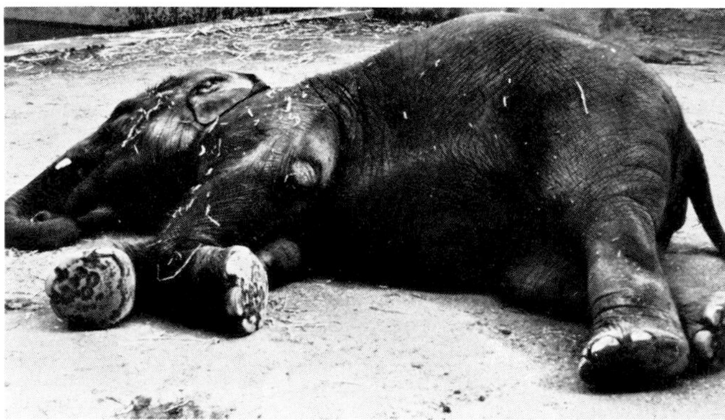

FIG. 22.4. Elephant can remain in lateral recumbency under chemical immobilization and anesthesia for up to 3½ hours and maintain normal respiratory function.

rected, the elephant will starve. The tip of the trunk is an extremely delicate prehensile organ. A tiny peanut or a large tree trunk may be grasped with equal ease. The trunk also places food into the mouth. Water is drawn up into the tip of the trunk and blown into the mouth. The trunk is involved in almost every activity of the elephant.

The trunk hangs limp at birth, and the calf nurses with its mouth as other mammals do. Within a few weeks the calf begins to develop dexterity with the trunk.

The tusks of the elephant are modified incisor teeth, not canines. Ivory is a unique form of dentine. When casting or restraining elephants, consider what the tusks will hit against or drive through.

Tusks are present in both male and female African elephants, but they grow larger and longer in the male. In the Asian elephant the female may either lack tusks or have small, underdeveloped tusks. The tusks continue to grow throughout life. The tusk is hollow at the base. This is the pulp cavity, containing nerves and blood vessels, and is the growing point. The pulp cavity ends approximately at the point where the tusk passes the trunk. A broken tusk is likely to abscess if the pulp cavity is exposed. Fortunately ivory usually breaks diagonally across the tusk, minimizing the chance of opening the pulp cavity. An elephant suffering from the pain associated with an abscessed tusk is extremely dangerous.

Elephants have highly vascularized ears which are the major organ of heat regulation. An elephant customarily stands flapping its ears on a warm day. During immobilization the elephant is incapable of moving its ears, and hyperthermia may develop with prolonged restraint when the ambient temperature is high.

Elephants are called "pachyderms" because the skin is thick. They share this characteristic with some other large herbivores such as the rhinoceros and hippopotamus. The thickness of the skin is primarily due to thickened dermis. The epidermal layer is approximately the same thickness as that of horses or cattle and is extremely sensitive. Insects such as flies or mosquitoes annoy elephants and cause them to blow dirt or mud over their backs to protect the delicate epithelial surfaces.

The elephant is also subject to scratches or cuts through the epidermis into the dermis. These bleed as readily as those of other mammalian species. The epidermal surface of the elephant is also susceptible to abrasion, so ropes pulled beneath or around the animal in restraint must be moved slowly to prevent burns.

BEHAVIOR

The elephant is intelligent and sensitive. It is highly responsive to a trainer who exercises proper judgment in obtaining mastery over it. One who works around these animals must have confidence and be capable of providing consistent discipline without inflicting prolonged pain. Otherwise the elephant will become either belligerent or unreliable, seeking occasions to catch the handler off guard. A belligerent elephant can inflict serious or fatal injury. The elephant is quick to pick up voice tones indicating that a handler is confident or, conversely, is frightened or uncertain.

It is important to be aware of an elephant's mood and take special precautions when it is annoyed or excited. The position of the ears is an important indicator of mood. When an elephant is excited, its ears are brought forward and extend out from the head; the trunk is curled upward.

As with most animals, elephants develop likes and dislikes for those who work with them. An elephant's dislike for an individual is difficult to overcome. Usually it is necessary to assign another person to care for that particular animal.

Physical examination of an elephant may prove

frustrating because elephants are incapable of standing still. Even though gentle and unconcerned about the manipulations in progress, these animals constantly shift weight from one leg to another. At the same time the trunk moves about, investigating the examiner's body, extremities, and anything else within reach. They are particularly fond of removing objects from pockets.

Sexually mature male elephants are extremely unpredictable and can be aggressive, particularly when in musth. Musth is a rut period similar to that seen in cervid species, but it occurs in irregular cycles.

Few facilities in the United States are adequately equipped to handle a mature male elephant. The Portland (Oregon) Zoo has been singularly successful in maintaining a large Asian male in a breeding herd. Zoo management has gone to great expense to provide this animal with individual stalls and shifting stalls controlled by huge hydraulically operated doors.

A few other zoos in the United States maintain mature male elephants, but in most instances the facilities are less than ideal. It must be possible to care for a mature male without the keeper ever coming near it, or the animal must be kept constantly chained to prevent its injuring keepers or destroying housing.

DANGER POTENTIAL

When watching elephants lumber around in a zoo, one may receive the impression that the elephant is slow. Their bulk belies the speed with which they can whirl and charge, slap with the trunk, or stomp with the feet. The elephant copes with enemies in many ways, and people are potential victims [3,8]. Serious students of elephant handling should read *Elephant Tramp* [9].

The trunk probably causes more injuries than any other weapon. It can be used as either an offensive or a defensive weapon. This strong grasping organ can grab a person by the arm, leg, or neck and pull him or her in. The elephant does not usually bite, but the victim can be pulled close by the trunk and kneeled on, stepped on, or banged against solid objects.

In addition to direct contact, the trunk can be used to throw objects such as feces, straw, dirt, pieces of wood, rocks, or other missiles at a handler. Also the elephant can draw water up into the trunk and spray it at any individual to whom it takes a dislike.

Some elephants acquire the bad habit of slapping people with the trunk. They become very clever and bait an individual into moving closer by extending the trunk for petting. A certain curve is left in the trunk; when the person moves closer, the trunk is flipped out. The force can fracture facial bones or ribs or knock a person over.

The tusks are an obvious hazard. Elephants have gored unwary victims and have also been known to crush people against walls with the tusks. The vast bulk of the elephant may also injure by pressing people against solid objects.

When working around elephants, be extremely careful; the elephant continually moves from one foot to another, and if due caution is not exercised, you may be stepped on—either by accident or on purpose.

Some elephants are adept at kicking with the hind legs. The kick is usually directed backward. The speed with which a kick is administered will astound persons who believe that the elephant is slow moving.

The elephant's tail is not usually used as a weapon, but it may become one if it is whirled in such a manner as to administer hammer blows, much as a baseball bat strikes a ball. The keeper chaining or shackling a hind leg or the person administering intramuscular medication in the hindquarters must be watchful of the tail.

It is unwise for anyone to work on an elephant alone. Some elephants become adept at maneuvering a person into a position where he cannot get free. The regular attendant should be present to control the head and command the elephant to move into positions suitable for examination and/or treatment. Even the presence of the keeper is sometimes insufficient to control an animal that is really annoyed or intent on injuring a person. The handler must be constantly aware of the position and activity of the elephant and must always have a planned escape route.

PHYSICAL RESTRAINT

The hook (bull hook or ankus) is an indispensable tool for working with elephants (Fig. 22.5). The hook should not be so sharp that it will tear the skin. Its primary purpose is to exert pressure to sensitive spots on the body, inducing the elephant to move away from the source of the pressure. Thus by pushing or pulling at various sites on the body, such as behind the ear or behind the front leg, one encourages the animal to move in a specific direction (Fig. 22.6). Immature elephants can be handled entirely by use of a hook if properly trained.

Elaborate maps of sensitive sites have been worked out by those who use the elephant daily as a beast of burden. A modified location map is illustrated in Figure 22.7. The handler of an individual elephant soon discovers where the animal can be touched to elicit the most satisfactory responses.

The hook handle should not be used indiscriminately as a club; an elephant resents harsh discipline. Not only is it inhumane, but unless the person using this type of discipline is capable of establishing and maintaining complete fear-mastery over the individual, it is both unwise and ineffective as a means of restraint.

A whip is used in place of the hook by some handlers,

FIG. 22.5. Variations of elephant or bull hook.

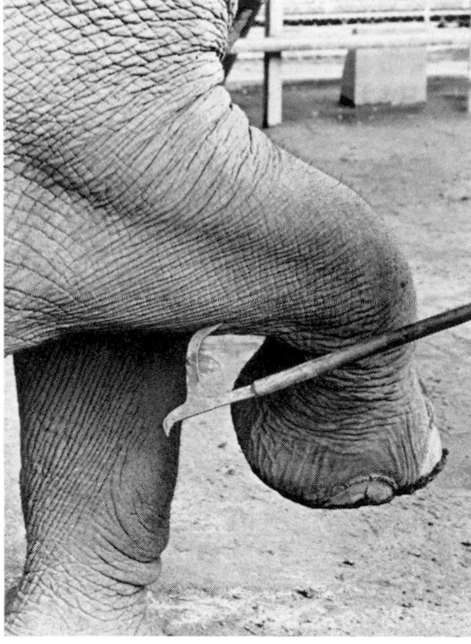

FIG. 22.6. Elephant hook used to make a trained elephant lie down and lift a leg.

FIG. 22.7. Adult female Asian elephant chained to a tree. Black dots represent sensitive sites for hook pressure. Arrows emphasize danger sources.

FIG. 22.8. Chaining using clevises.

particularly in circuses. Battery-operated stock prods and even electrified elongated hooks have been used to gain mastery over a particularly obstreperous elephant, but these techniques have no place in routine handling and restraint.

The hook is used on mature adult elephants, but if the elephant chooses not to obey, there is little the handler can do. Such elephants must be chained or shackled to examine or treat them.

All captive elephants should be trained to accept chains on all four legs, whether or not the animal is routinely chained. The chaining procedure should begin in infancy so that the elephant becomes accustomed to standing quietly when chained. It is imperative that an adult elephant be fully accustomed to the use of chains so that, if manipulative procedures must be carried out to administer medication or treatment, the animal can be adequately controlled by leg chains.

Numerous methods are used for attaching chains to the leg. Chains on the hind legs must be placed above the hock, lest they slip off. The bell shape of the front feet permits chaining just above the foot. Some handlers attach a short length of chain around the leg as a permanent bracelet or anklet. A clevis is used to attach the bracelet to the tethering chain (Fig. 22.8). Another method is to wrap the end of the chain around the leg and clevis it to an appropriate link (Fig. 22.9).

Both of these methods are rather slow, exposing the handler to prolonged opportunity for injury from a nervous or fidgety elephant. A rope loop is sometimes used for temporary restraint while the chain is being attached. Dangerous animals should be chained from behind a fence separating the handler and the animal.

A much faster but equally secure method of attaching the chain to the elephant's foot is with the use of Brummel hooks (Fig. 22.10). Brummel hooks may be obtained in various sizes (Table 22.2) and function in pairs. A permanent anklet is attached to the elephant's lower leg with one Brummel hook as part of the chain. The chain, which is permanently attached to a post or ring, includes a Brummel hook as the end link. When the elephant is brought into position, these two hooks can be joined rapidly since the

FIG. 22.9. Wrapping end of chain around leg.

hooks have no moving parts. The Brummel hooks are secure since it is unlikely that the elephant can shake the chain or manipulate it in such a way that the hooks line up in the proper position for disengagement.

Elephant chains should not be too long. In one instance an elephant was chained inside a truck with a chain sufficiently long that it could move partially out the door. An attendant mistakenly left the door ajar. As the elephant tried to escape, it fell, fracturing a leg.

Chains can serve another purpose. In one instance an elephant was transported to a veterinary clinic but refused to leave the truck. No amount of coaxing could lure him out, although he could easily step from the low truck bed. A chain was attached to a front leg, strung out the door, and attached to a metal post. The truck was then slowly driven forward, forcing the elephant to step from the truck.

Animals unaccustomed to chains may strain or jerk so hard on a leg chain that tendons, ligaments, or joints are injured. Training at an early age will obviate this.

If the skin of an elephant tends to abrade under the chain encircling the leg, the chain can be covered with a length of canvas or rubber hose. Chain size is important; 7.9 mm (5/16 in.) or larger should be used on adult elephants (Table 22.3).

TABLE 22.2. Brummel hook specifications

Size and Material	Bail Diameter		Length of Hook		Safe Load	
	(mm)	*(in.)*	*(mm)*	*(in.)*	*(kg)*	*(lb)*
00 Aluminum alloy	10	0.42	37.44	1.56	227	500
0 Manganese, bronze	10	0.42	37.44	1.56	454	1,000
1 Manganese, bronze	13	0.56	48.72	2.03	909	2,000
2 Manganese, bronze	15	0.69	63.12	2.63	1,818	4,000

TABLE 22.3. Chain size and strength

Diameter of Rod		Length of Link		Working Load* Proof coil		Working Load* High test	
(mm)	*(in.)*	*(mm)*	*(in.)*	*(kg)*	*(lb)*	*(kg)*	*(lb)*
4.8	3/16	31.7	1.25	341	750	. . .	. . .
6.4	1/4	38.1	1.5	568	1,250	1,136	2,500
7.9	5/16	44.4	1.75	852	1,875	1,818	4,000
9.4	3/8	50.8	2	1,193	2,625	2,318	5,100

*Working load is approximately one-half the breaking strength.

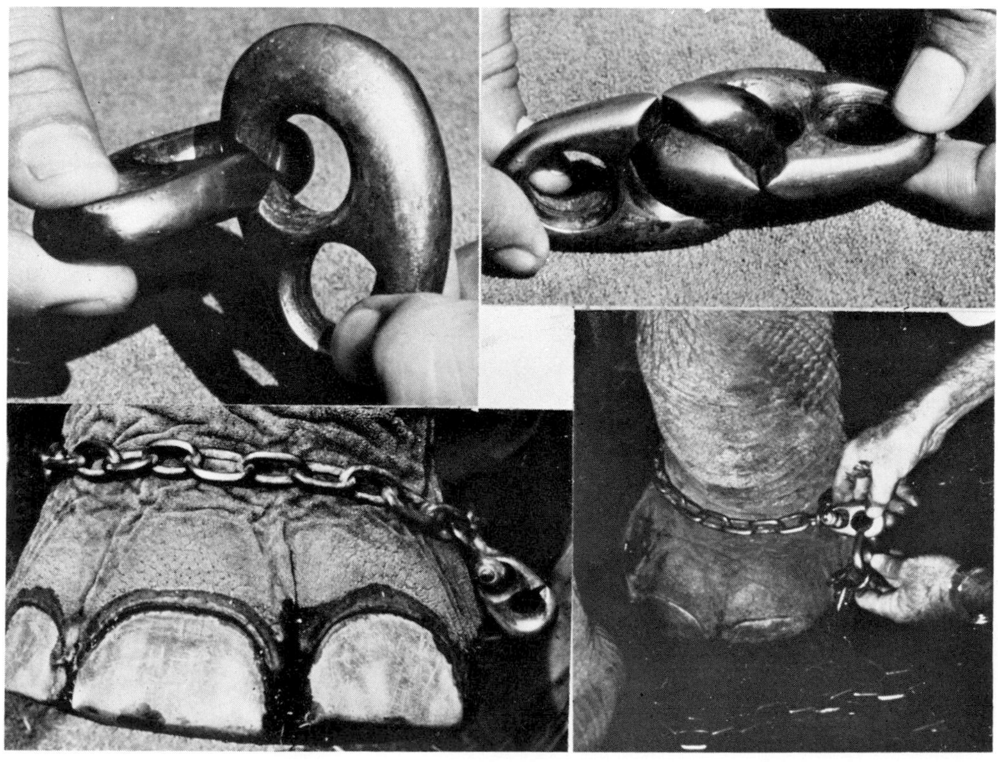

FIG. 22.10. Chaining using Brummel hooks: Manipulation of the hooks *(top)*. Bracelet on elephant's leg *(lower left)*. Attaching bracelet to anchoring chain *(lower right)*.

Well-trained docile elephants can be quieted slightly or even made to stand still momentarily if the handler grasps the ear (Fig. 22.11). An elephant can be taught to lie down on command and remain down until the handler signals it to rise. Examination and limited treatment can more easily be carried out with the elephant lying down.

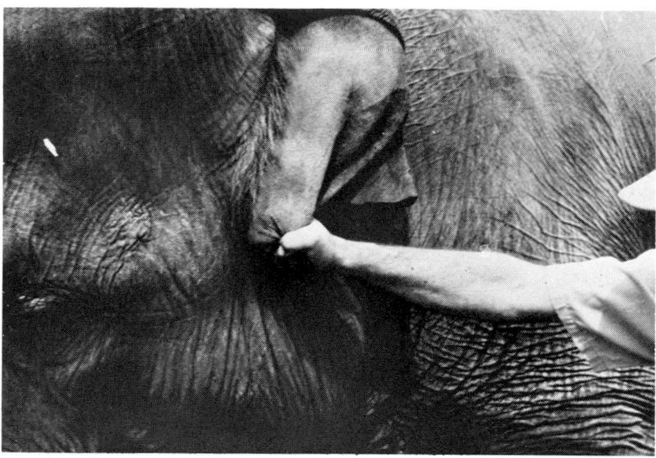

FIG. 22.11. Earing an elephant (a mild form of restraint).

An elephant's feet frequently require manicuring and/or examination to remove foreign objects, to check for excessive wear of the footpads, or to trim nails (Fig. 22.12). A wise keeper or handler will work with the elephant all the time so it becomes accustomed to lifting the feet. If toenails are rasped frequently there is no need to cut them too short; thus the best interests of both animal and handler are served.

Elephants can be taught to raise a foot on command and hold it in the air or rest it gently on the leg of a handler (Fig. 22.13) or on a pedestal (Fig. 22.14). It is obvious that a handler cannot keep an elephant's leg in the air without its consent. It is important that no painful procedures be carried out when the elephant is in this position, since it would then step down quickly, possibly injuring the attendant. In addition, it is highly unlikely that anyone would be able to induce that elephant to lift the foot again, for fear that additional pain might be inflicted.

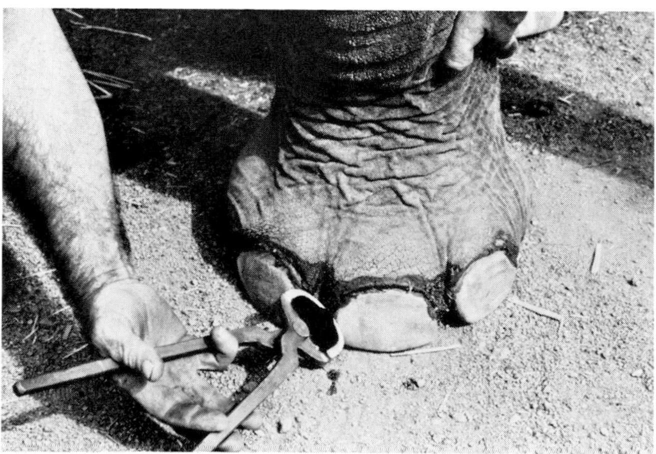

FIG. 22.12. Trimming "hangnails" with foot on the ground.

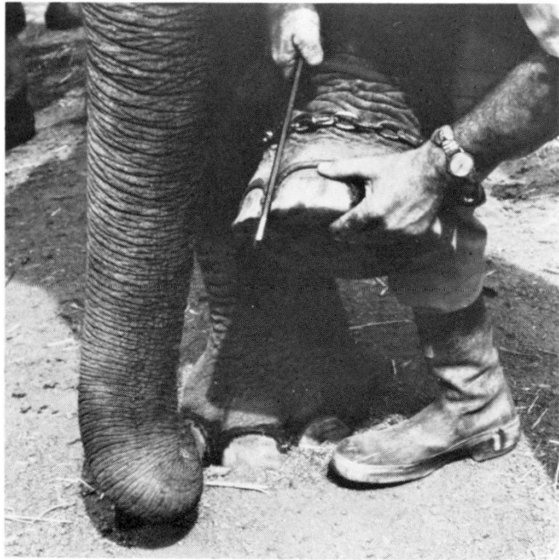

FIG. 22.13. Rasping a toenail while holding foot on a knee (possible only with well-trained animal).

Designers of facilities for elephants should remember that the trunk has extensive reach. All electrical fixtures, water pipes, and other loose objects must be kept well beyond trunk distance. When an elephant must be moved to new quarters or into a hospital area, attendants frequently overlook the fact that, though items are out of

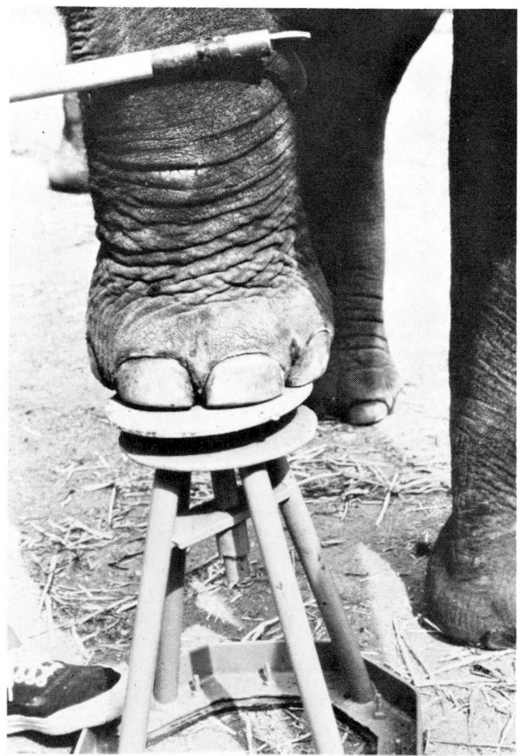

FIG. 22.14. Foot on pedestal for examining and trimming feet and nails.

reach of other animals, they are within range of the extended trunk. There is not only danger for the attendant when water pipes burst or electrical wires are torn loose, but the elephant may likewise be electrocuted in the process.

The animal's reaching ability creates other hazards, because elephants continually investigate; anything hanging loose or in the pockets of the examiner may be grasped and swallowed. This includes stethoscopes, thermometers, and other diagnostic equipment. I recall an elephant that was being treated for an abscess. The elephant grasped the protective plastic sleeve the examiner was wearing while cleaning the abscess. The plastic sleeve was torn off the examiner's arm and swallowed immediately. This represented a serious hazard for the elephant since it could have produced an obstruction within the intestinal tract. Fortunately the plastic was passed in the feces without incident.

Elephants are notorious for grasping purses, glasses, or other objects from patrons of zoos who are unwise enough to allow such items within reach of the highly prehensile trunks.

A small elephant (up to 1.5 m tall) can be cast using block and tackle (Fig. 22.15). This restraint allows performance of minor operative procedures such as toenail surgery or trimming hooves on somewhat obstreperous elephants. There is a slight hazard in that the legs, especially the joints of the hind legs, are stressed during this manipulation. The hobbles are constructed of 19-25 mm (¾-1 in.) rope loops, long enough to encircle the leg. The block and tackle must be pulled slowly. Frequently the elephant will go down on its sternum and remain there until pulled over.

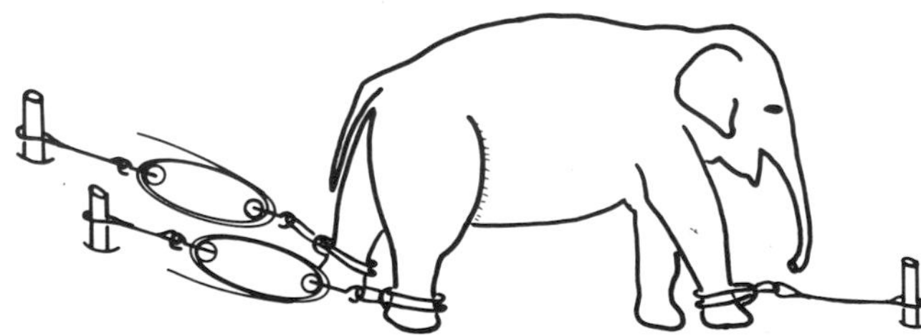

FIG. 22.15. Casting a small elephant with block and tackle (suitable for animals up to 1.5 m tall at the shoulder).

An elephant in sternal recumbency can be pulled over by placing a rope on the tusk or around the front leg. The rope is then passed up over the head or body and the animal pulled over. A truck or small tractor may be required to pull over a large elephant. Be certain that the head rope does not pass over the eye or ear. Position the legs before turning to minimize twisting.

CHEMICAL RESTRAINT

Etorphine hydrochloride (M99) is valuable when it is necessary to obtain complete control of an elephant [1,5,6,7,10,13,14]. It is a marvel of biology that 5 mg of a drug given to an elephant weighing 5,000 kg can immobilize it within 15-30 minutes.

The dose is 0.0022 mg/kg for the Asian elephant and 0.0017 mg/kg for the African. The dose expressed in different terms would be 1 mg/450 kg in the Asian and 1 mg/600 kg in the African. Prolonged immobilization and anesthesia can be induced by repeated administration of etorphine.

In two separate instances with the same elephant, anesthesia was prolonged up to 4 hours by intravenous administration of M99 [4]. After initial immobilization, 1 mg of M99 was given every 15 minutes. It was found to be most satisfactory to administer this as a continuous drip in physiological saline solution.

In a zoo the procedure for immobilizing an elephant should include draining of any pools in the enclosure and chaining the elephant so it does not fall into a moat or empty pool. If the elephant is tractable, the drug can be administered by a hand-held syringe. If not, a projectile syringe can be used.

Effects of etorphine will be observed within 10-15 minutes. The trunk hangs limp or loses some of its investigativeness. The animal will start to sway back and forth. Keepers or handlers must stay away from the animal from this point on because the elephant may fall suddenly. A nearby attendant can be fatally crushed if the elephant falls on him. Recumbency occurs within 20-30 minutes.

Special needles are required for injecting elephants. Needles should be 6-8 cm long with an outside diameter of 2 mm (15 gauge). Opinion varies as to whether or not the tip should be plugged and holes bored on the sides of the needle, to prevent the needle from cutting skin plugs. A conventional needle can be modified by placing a tiny bead of solder on the proximal lumenal side of the bevel [7]. A special needle can be constructed by plugging the end with solder, filing it to produce a symmetrical tip, and boring tiny holes behind the plug to disperse medication laterally into the tissue (see Fig. 4.15). Barbed needles should not be used on captive elephants because of the trauma associated with removal. Collared needles will remain in place sufficiently long to disperse the drug.

If the elephant must fall on a specific side, take appropriate steps before induction to guide the fall (Figs. 22.16, 22.17). An elephant that has fallen on the wrong side may be turned with ropes, using the parbuckle principles (Fig. 22.18). This device was used by the lumber industry to move huge logs in an era before modern equipment was available.

FIG. 22.16. Casting method to ensure that elephant falls on the correct side when chemically immobilized (side view).

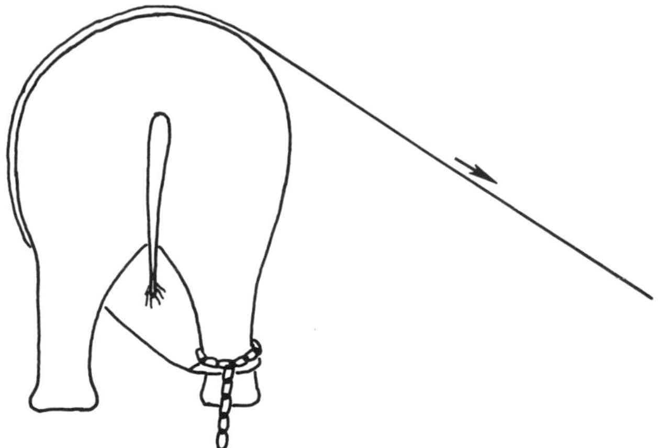

FIG. 22.17. Casting method (end view).

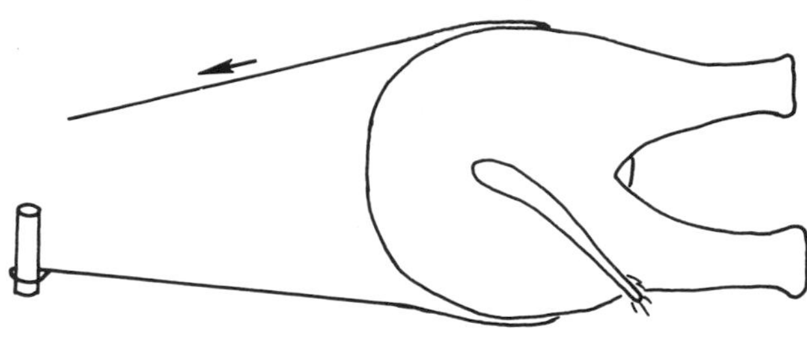

FIG. 22.18. Parbuckle principle used to turn large animals such as elephants and rhinos from one side to the other.

An immobilized elephant may fall into either sternal or lateral recumbency. There is a controversy regarding the proper recumbent position for a restrained elephant. Harthoorn indicates that an elephant should not be allowed to stay in sternal recumbency for longer than 20 minutes [7]. Lateral recumbency is also dangerous; however, I have maintained a 9-year-old Asian elephant in lateral recumbency for 3½ hours on two separate occasions. Blood P_{CO_2} and P_{O_2} values were normal throughout the procedure.

At the conclusion of the necessary period of immobilization the antidote, diprenorphine (M50-50), is administered intravenously in an ear vein. The dosage is double that of the etorphine. The elephant will begin to investigate with its trunk within 1 2 minutes. Shortly thereafter it will begin rocking the body and roll into the sternal position. The front legs are placed in front of the animal and the forequarters are raised. Slowly the elephant will rise all the way. This will occur within 2-15 minutes.

Although etorphine is the drug of choice when immobilizing elephants, other drugs may be used to calm them so that work can be done in the standing position, or to tranquilize them to the point where they will lie down and allow minor manipulative procedures to be carried out. Acepromazine in dosages of 0.04-0.06 mg/kg is highly effective. Xylazine in dosages of 0.08-0.14 mg/kg has been used. Various combinations of etorphine, xylazine, and acepromazine are used in practice by experienced clinicians. Inexperienced persons should not use any immobilizing agent without first consulting someone experienced in its use.

Etorphine is used both as an immobilizing and an anesthetizing agent. Inhalant anesthetics should be used with extreme caution. I am aware of three fatalities from three attempts to use halothane (fluothane) anesthesia on elephants. In one instance an elephant was immobilized with etorphine, and supplemental halothane was administered. Upon completion of surgery, the effects of etorphine were reversed with diprenorphine. While still under the partial effects of halothane, the animal injured itself so severely that euthanasia was required.

If an elephant is to be placed into a special padded recovery stall following surgery, be sure the tusks cannot reach the wall. Tusks can be devastating to the padding.

TRANSPORT

Young elephants can be crated and moved via truck, airplane, or ship if the crates are adequately constructed. Adult elephants require special facilities. Semitrailer or railroad cars must be reinforced inside with 6.4 mm (¼ in.) sheet iron to withstand the tusks and butting of elephants. Rings must be provided to secure both front and hind legs.

Chains must be kept short, and access must be provided for handlers to chain and unchain the elephant. Normally elephants are removed from trucks on short ramps or at docks where the elephant can walk out at truck-bed level. They can climb or step down a few feet if trained to do so.

To move elephants within a zoo, chain difficult elephants between two docile animals. This technique can be used with two pieces of heavy machinery such as large tractors by attaching chains in front and behind the elephant as depicted in Figure 22.19.

REFERENCES

1. Alford, B. T.; Burkhart, R. L.; and Johnson, W. P. 1974. Etorphine and diprenorphine as immobilizing and reversing agents in captive and free-ranging mammals. J. Am. Vet. Med. Assoc. 164(7):702-5.
2. Anderson, S., and Jones, J. K., Jr. 1967. Recent Mammals of the World, p. 360. New York: Ronald Press.
3. Caras, R. A. 1964. Dangerous to Man, pp. 69-79. Philadelphia: Chilton Books.
4. Fowler, M. E., and Hart, R. 1973. Castration of an Asian elephant using etorphine (M99) anesthesia, J. Am. Vet. Med. Assoc. 163:539-43.
5. Harthoorn, A. M. 1965. Application of pharmacological and physiological principles in restraint of wild animals. Wildl. Monogr. 14, p. 74.

FIG. 22.19. Moving a tame elephant.

6. ______. 1971. The capture and restraint of wild animals. In L. R. Soma, ed. Textbook of Veterinary Anesthesia, pp. 404-37. Baltimore: Williams & Wilkins.
7. ______. 1973. The drug immobilization of large herbivores other than antelopes. In E. Young, ed. The Capture and Care of Wild Animals, p. 52. Capetown, South Africa: Human and Rousseau.
8. Karsten, P. 1974. Safety Manual for Zoo Keepers, pp. 46-49. Alberta, Canada: Calgary Zoo.
9. Lewis, G. W. 1955. Elephant Tramp, An Autobiography as Told to Byron Fish. Chicago: Peter Davis. (Nontechnical. Should be required reading for all who work with elephants.)
10. Pienaar, U. de V. 1966. Capture and Immobilizing Techniques Currently Employed in Kruger National Park and Others. Skukuza, Ronco: South African National Parks and Provincial Reserves. Kruger National Park.
11. Sikes, S. K. 1971. The Natural History of the African Elephant. New York: American Elsevier. (A very extensive bibliography)
12. Walker, E. P. 1964. Mammals of the World, vol. 2, pp. 1321-23. Baltimore: Johns Hopkins Press.
13. Wallach, J. D., and Anderson, J. L. 1968. Oripavine (M99) combinations and solvents for immobilization of the African elephant. J. Am. Vet. Med. Assoc. 153:793-97.
14. Wallach, J. D.; Frueh, R.; and Lentz, M. 1967. The use of M99 as an immobilizing agent in capturing wild animals. J. Am. Vet. Med. Assoc. 151:870-76.

SUPPLEMENTAL READING

General

Kempe, J. E. 1950. The training of elephants. Ctry. Life 107 (January 27):228-29.

Williams, J. H. 1950. Elephant Bill. Garden City, N.Y.: Doubleday.

Anatomy and Physiology

Anderson, R. J. 1883. Anatomy of the Indian elephant. J. Anat. (London)27:491-94.

Benedict, F. G. 1936. The physiology of the elephant. Wash. Carnegie Inst. Publ. 474.

Benedict, F. G., and Lee, R. D. 1957. Further observations on the physiology of the elephant. J. Mammal. 19:140-43.

Dhindsa, D. S.; Sedgwick, C. J.; and Metcalfe, J. 1972. Comparative studies of the respiratory functions of mammalian blood. VIII. Asian elephant *(Elephas maximus)* and African elephant *(Loxodonta a. africana)*. Respir. Physiol. 14:332-42.

Engel, S. 1963. The respiratory tissue of the elephant *(Elephas indicus)*. Acta Anat. 55:105-11.

Forbes, W. A. 1879. On the anatomy of the African elephant *(Elephas africanus,* Blum.). Proc. Zool. Soc. Lond., pp. 420-35.

Rensch, B. 1957. The intelligence of elephants. Sci. Am. 196:44-49.

Restraint and Handling

Counsilman, J. W. 1954. Demerol hydrochloride as an anesthetic for an elephant. North Am. Vet. 35:835-36.

Fowler, M. E. 1974. Some thoughts on handling large animals. J. Zoo Med. 5:27-30.

Harthoorn, A. M., and Bligh, J. 1965. A new oripavine derivative with potent morphine-like properties for the restraint of the large wild African mammals. Res. Vet. Sci. 6:290-99.

Jainudeen, M. R. 1970. The use of etorphine hydrochloride for restraint of a domesticated elephant *(Elephas maximus)*. J. Am. Vet. Med. Assoc. 157:624-26.

Kodituwakku, G. E.; Dissanayake, K.; and Seneviratne, E. 1961. General anesthesia in an elephant. Ceylon Vet. J. 9:75.

Pienaar, U. de V.; Van Niekerk, J. W.; and Young, E. 1966. The use of oripavine (M99) in the drug immobilization and marking of wild African elephant *(Loxodonta africana blumenbach)* in the Kruger National Park. Koedoe (S. Afr.) 9:108-24.

23 HOOFED STOCK

CLASSIFICATION

Order Perissodactyla (odd-toed ungulates)
- Family Equidae (7 species): horse, ass, zebra
- Family Tapiridae (4 species): tapir
- Family Rhinocerotidae (5 species): rhinoceros

Order Artiodactyla (even-toed ungulates) (total: 79 genera, 194 species)
- Suborder Suiformes
  - Superfamily Suoidea
    - Family Suidae (5 genera, 8 species): wild pig
    - Family Tayassuidae (1 genus, 2 species): peccary
  - Superfamily Anthracotherioidea
    - Family Hippopotamidae (2 genera, 2 species): hippopotamus
- Suborder Tylopoda
  - Family Camelidae (3 genera, 6 species): camel, llama
- Suborder Ruminantia
  - Infraorder Tragulina
    - Family Tragulidae (2 genera, 4 species): chevrotain
  - Infraorder Pecora
    - Family Cervidae (17 genera, 41 species): deer
    - Family Giraffidae (2 genera, 2 species): giraffe, okapi
    - Family Bovidae (46 genera, 128 species): antelope, sheep, goat
    - Family Antilocapridae (1 genus, 1 species): pronghorn

TABLE 23.1. Names of gender of hoofed stock

Animal	Mature Male	Mature Female	Newborn and Young
Horse	Stud, stallion	Mare	Foal — colt (male) filly (female)
Zebra	Stallion	Mare	Foal — colt
Ass	Stallion, jack	Mare, jenny	Foal
Tapir	Male	Female	
Rhinoceros	Bull	Cow	Calf
Peccary	Boar	Sow	Piglet
Hippopotamus	Bull	Cow	Calf
Camel	Bull	Cow	Calf
Llama	Male	Female	Calf
Deer	Buck, stag, hart	Doe, hind	Fawn
Elk	Bull	Cow	Calf
Moose	Bull	Cow	Calf
Reindeer	Buck	Doe	Fawn
Giraffe	Bull	Cow	Calf
Okapi	Bull	Cow	Calf
Pronghorn	Buck	Doe	Kid
Cattle	Bull	Cow	Calf
Antelope	Buck	Doe	Kid, lamb
Water buffalo	Bull	Cow	Calf
Cape buffalo	Bull	Cow	Calf
Yak	Bull	Cow	Calf
Bison	Bull	Cow	Calf
Rocky mountain goat	Buck	Doe	Kid
Musk-ox	Bull	Cow	Calf
Goat	Buck	Doe, nanny	Kid
Ibex	Buck	Doe	Kid
Aoudad	Buck, ram	Ewe	Lamb
Tahr	Buck	Doe	Kid
Sheep	Ram, buck	Ewe	Lamb
Mouflon	Buck, ram	Ewe	Lamb
Bighorn sheep	Buck, ram	Ewe	Lamb

NAMES of gender are listed in Table 23.1; weights of representative species are listed in Table 23.2.

PHYSICAL RESTRAINT [10]

Hoofed animals do not recognize a chain link or wire net fence as a barrier, especially when newly placed in such an enclosure. Fences should be draped with plastic sheeting or burlap sacking until the animals are accustomed to the fence (Fig. 23.1).

Although difficult to herd otherwise (Fig. 23.2), hoofed animals can be moved effectively and efficiently by manipulating opaque plastic sheeting (Figs. 23.3, 23.4). The technique was first described by Oelofse [13] and subsequently has been used in numerous instances to cap-

TABLE 23.2. Weights of representative hoofed stock

Animal	Kilograms	Pounds
Grant zebra	350	770
Wild horse	350	770
Ass	260	572
Rhinoceros		
White	2,300-3,600	5,060-7,920
Black	1,000-1,800	2,200-3,960
Indian	2,000-4,000	4,400-8,800
Tapir		
Malayan	225-300	495-660
Mountain	225-300	495-660
Brazilian	225-300	495-660
Wart hog	75-100	165-220
Peccary, collared	16-30	35-66
Hippopotamus		
Nile	3,000-3,200	6,600
Pygmy	160-240	352-528
Camel		
Bactrian	450-690	990-1,518
Dromedary	450-650	990-1,430
Llama	70-140	154-308
Alpaca	70-140	154-308
Vicuna	35-65	77-143
Guanaco	48-96	106-211
Caribou	318	700
Wapiti	200-350	440-770
Moose	825	1,815
Deer		
Fallow	40-80	88-176
Sika	25-110	55-242
White-tailed	22-205	48-451
Mule	50-215	110-471
Giraffe	500-980	1,100-1,936
Okapi	250	550
Pronghorn	36-60	79-132
Eland	900	1,980
American bison	1,000	2,200
Antelope		
Roan	240	528
Sable	214	471
Saiga	23-40	51-88
Buffalo		
African	600-900	1,320-1,980
Water	1,000	2,200
Impala	99	218
Gazelle (Thompson)	40	88
Gnu	180	396
Sheep		
Mouflon	70-150	154-330
Aoudad (Barbary)	50-115	110-253

FIG. 23.1. Visual barrier over the wire for introducing ungulates into new fenced-in enclosures.

FIG. 23.2. Attempting to herd deer can be a frustrating experience.

FIG. 23.3. Opaque plastic sheeting used for directing hoofed animals.

ture and retain wild animals in an enclosed area. Hoofed animals recognize the sheeting as a barrier and can be persuaded to move through gates, into shipping crates, or into alleyways by judicious manipulation of this material.

This technique is effective with cervids, antelope, zebra, and rhinoceros, but dorcas gazelle and giraffe do not respond well. They reach their heads under and flip up the

FIG. 23.4. Plastic sheeting used to temporarily confine a herd of deer in a small area for close inspection.

plastic sheeting, and obviously it is difficult to obtain a sheet of plastic high enough to prevent a giraffe from seeing over it.

If an enclosure is too small to admit a truck or trailer, remove downed sick or injured animals by dragging them on a large 1.22 × 2.44 m (4 × 8 ft) piece of plywood sheeting 15 mm (⅝ in.) thick, with holes drilled approximately 4 in. from the forward corners. Thread wire or chain loops through the holes and attach a chain. Lay the plywood sheeting next to the animal and carefully lift or pull the injured animal onto the sheet. The plywood sheet can be dragged from the pen by the use of snatch blocks. Canvas tarp can be used similarly, but if the animal must be dragged some distance, the musculature, ribs, and bones may be bruised through the canvas.

Avoid the use of whips or clubs when working with any animal. A swat from a household broom or a flat scoop shovel is more effective, producing noise to scare the animal and encourage it to move rather than simply inflicting pain. When animals are excited by the stress of moving or during the alarm stage, pain perception is lessened. Mild pain will not be felt; if intense pain is inflicted, hostile reactions may ensue.

TRANSPORT [5,6,8,16]

The critical time for injuries during transport operations is at the time the animal is loaded into or released from the crate. Lack of preparation is the prime cause of failure when loading, unloading, or conducting capture operations of hoofed stock. Too often insufficient thought is given to the needs of the animal and the specific behavioral characteristics which must be accommodated to achieve the desired end with as little fuss as possible.

Most operations are carried out much too rapidly. The animals should be fed in open crates or capture pens over a period of several days to allow them to become accustomed to confining areas. Special chutes can be arranged to guide the animals into a new enclosure, or ramps can be constructed in alleyways or alongside a familiar barn or shed so animals do not feel harassed. It is possible to arrange the chutes to funnel an animal to a crate, chase it in, and close the door behind it. Other alternatives are to immobilize the animal and allow it to recover in the crate; or immobilize it, give an antidote, and as the animal regains mobility direct it into the crate.

Each group of animals has its own set of requirements for crating. Unfortunately there is no single source of information on crating for all species. The International Air Transport Association (IATA) has published a manual describing a variety of crates and cages, but it does not give enough detail [8].

It is important to select a crate or cage suited to the biological requirements of the animal [5]. The floor area should be based on that required by the animal when resting and recumbent and its special needs for space to lie down or arise. For instance, an antelope or gazelle can lie down and get up in a space the length and width of its body. In this case a suitable crate need not be much wider or longer than the body. If a crate is too long, the animal may charge the door, causing injury to the head and neck. If a crate is too wide, the animal may attempt to turn around and get stuck in a dangerous position. The height of a crate for a horned animal should allow the animal to stand with its head in a normal position. If the height of the crate permits the animal to jump up, it may fall over backward.

Many deaths have been caused by animals having too much room. In one instance a fringe-eared oryx walked into a too-large crate, the door was closed behind her, and she became frightened and tried to turn around. In the ensuing struggle she tipped herself upside down with her head pinned beneath her body. Fortunately the crate could be opened quickly and the animal pulled out; if the incident had occurred en route, no one could have saved her.

It is vital that crate construction provide for adequate air exchange, not only to supply sufficient oxygen for respiration, but to remove odors and noxious gases such as ammonia. A black buck, 130 cm long and 80 cm high at the shoulder, placed in a crate with inside dimensions of 180×135 cm high would occupy 25-35% of the volume of the crate, leaving approximately 595 L of air space. If the resting black buck needs 15 L of air per minute at an ambient temperature of 18-21 C, the air in the crate would be exhausted in 39 minutes. If the animal is excited or the temperature is raised, the air could be exhausted in 10-15 minutes. Elevated ambient temperatures increase the air requirement, because the animal breathes more rapidly to assist in cooling [5].

Ventilation ports should be provided near the floor above the bedding and also near the top of the crate to provide a draught effect, drawing air into and out of the crate. However, any opening through which a foot or horn could protrude must be screened. Ventilation ports have the added advantage of permitting visual access to the animal. All openings should be closed with sliding panels rather than hinged doors [5].

Since animals are susceptible to carbon monoxide poisoning, crates should not be placed near exhaust fumes from internal combustion engines. The floors of crates for ungulates are frequently cleated to prevent slipping; this is acceptable if the cleats are short (8 mm high). Higher cleats cause discomfort to a recumbent animal and also have detrimental effects; uneven pressure may cause trauma to localized areas of the limbs or body, especially if the animal becomes weakened and lies down for a long time [5].

Most hoofed animals should be shipped in individual crates. This is of prime importance if the animals are horned. Members of a group may have been companions for a long time, but when they are placed in close confinement, aggressiveness may develop which can result in fatalities to subordinates. The behavior of animals under stress or confined in close quarters is unpredictable.

At no time should animals of varying sizes be placed in the same shipping crate—even mothers and offspring. Small animals are apt to be trampled by the larger. If transporting cannot be delayed until after weaning, the young must be separated by a partition or put into a separate crate. An infant that has been kept away from its mother for an extended period may engorge on milk when they are reunited. If manual handling of the female is possible, it is desirable to partially milk the female before permitting the infant to nurse, so it cannot overeat.

When crates containing large hoofed animals are loaded, they should be oriented with the hind end facing the front of the vehicle. Smaller species will be more comfortable if the crate is loaded sideways. This orientation may prevent serious damage from jolting stops and starts.

The release of an animal from a crate may also be dangerous. Some animals bolt from the crate immediately upon seeing the open door. Others, especially if the interior is dark, accept the crate as a haven and refuse to leave it. The least stressful method of release is to place the crate within an enclosure or in the doorway of the enclosure, open the crate, and let the animal come out on its own to investigate the new surroundings.

CHEMICAL RESTRAINT [1,2,3,7,11,12,16]

Hoofed stock show wide species variation in response to various immobilizing agents. Table 23.3 lists representative species of ungulates and recommended dosages of some restraint drugs. No cut-and-dried formula can be given for the use of chemical immobilizing agents in hoofed animals; there are too many variables. The excitement of the animal, the environmental conditions at the time of immobilization, and other pertinent factors make it mandatory to evaluate each case individually. Indicated dosages can serve only as rough guides.

Large and flighty bovidae can be handled properly only by chemical immobilization (Fig. 23.5). To subject these animals to physical restraint is unwise. When injected, the animal should not be in an area including a moat, a pond, or a wire or net fence into which it might fall or stumble as the drug takes effect.

Approach an immobilized animal cautiously (Fig. 23.6) and maintain control (Fig. 23.7). Keep the head and shoulders of an immobilized animal slightly elevated while transporting and during other manipulative procedures (Fig. 23.8) to minimize the chance of regurgitation. If regurgitation begins, lower the head immediately to allow free flow of ingesta.

Numerous problems are associated with chemical restraint of animals from herds. Separation of a darted

TABLE 23.3. Chemical restraint agent dosages in hoofed stock

		Combination		Combination			Tilazol Combination			Combination		
Animal	Etorphine	Etorphine	Xylazine	Etorphine	Acepromazine	Xylazine	Tiletamine	Zolazepam	Phencyclidine hydrochloride	Phencyclidine	Acepromazine	References
Zebra, ass, wild horse	0.01	2-5 mg total	25-100 mg total	2-5 mg total	5-20 mg total	1-5			1.9-2.2			1,2,7,U
Tapir	0.007						3	3	0.5-2.2			1,3,U
Rhinoceros, black and white	2 mg total	1 mg total	0.265	2 mg total	20 mg total	0.25-0.5						1,7,U
Wild pig	0.02						5-20	5-20	4.2-6.5			1,3,7,U
Peccary	0.01-0.02						5-20	5-20				U
Hippopotamus	0.0033											
Nile	4-8 mg total					1-1.5				0.3-0.5	0.25	1,7
Pygmy	6 mg total									0.5	0.25	1
Camelid	0.007					0.25-0.51	2.4	2.4				1,7,U
Cervid	0.02					1-8	5-20	5-20				1,7,14,U
Giraffe	0.007	0.002	0.02				8	8				1,2,4,U
Pronghorn							10	10				U
Antelope	0.01-0.02					0.5	5-10	5-10				1,7,U
Gazelle						0.5	5-10	5-10				1,7,U
Oryx	0.02			0.011	0.196							1,7
Wild cattle							3-5	3-5				U
Wild sheep, goat	0.03						2.5-7.5	2.5-7.5				1,U
Musk-ox						1-1.5						
						2.4						7,9

Note: Dosage is in mg/kg unless otherwise noted.
U = unpublished data.

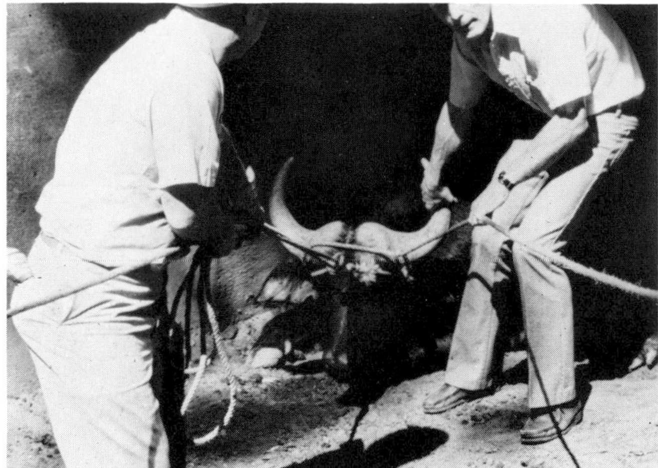

FIG. 23.5. Immobilized African buffalo being moved to a shipping crate.

FIG. 23.6. An immobilized antelope must be approached slowly and steadily, with as little noise as possible and no sudden or jerky movement.

FIG. 23.7. Controlling the head while working on partially immobilized ungulate.

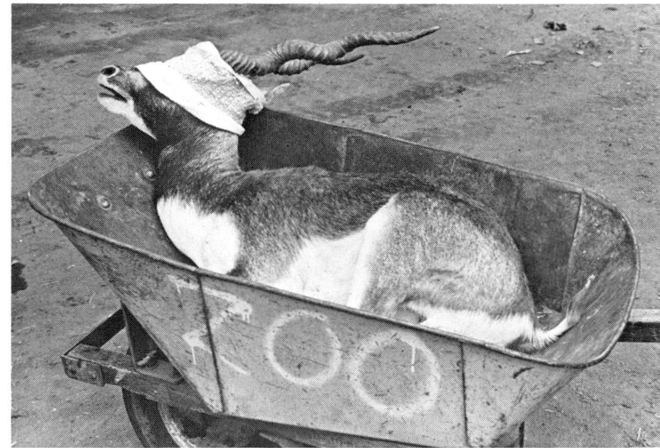

FIG. 23.8. Moving a black buck. Visual stimulation is blocked and head is elevated to reduce chances of regurgitation.

animal from the herd or retrieval after it becomes immobile is difficult. If the enclosure permits, the use of some sort of vehicle may be the most satisfactory method of retrieving the animal.

The use of cap-powered dart syringes in small ungulates is dangerous, because the syringe may be driven clear through the animal. Many species have thin skin that will be penetrated by the dart if the proper weapon is not used or if the proper charge is not selected. Blowguns are recommended for use on small or thin-skinned species when chemical immobilization is necessary.

As a hoofed animal begins to recover from chemical restraint, it requires special protection. Preferably the recovery should take place in a small enclosure containing only the immobilized animal; otherwise, cagemates or penmates may inflict injury while the patient is incapable of protecting itself.

Visual acuity in these animals is diminished with drugs, so prominent sight barriers should be maintained during the recovery process, especially if the immobilizing agent lacks an antidote.

PERISSODACTYLA

Horse, Zebra, Ass

DANGER POTENTIAL. Members of this group are efficient and effective kickers, strikers, and biters. They should not be considered as just slightly wild horses, because they lash out much faster than the domestic horse and are far more likely to kick or strike.

PHYSICAL RESTRAINT. There is no effective way to physically restrain adult animals of this category. Young animals can be roped and stretched to immobilize legs, but adults are so strong that to apply the pressure and stretch necessary to restrain them is to risk injury to the animal.

The twitch, lip chain, or other similar devices customarily employed when restraining domestic equids are not suitable for use on wild counterparts of the horse.

TRANSPORT. Wild equids are moved in individual crates. The crate must be heavily constructed and anchored firmly on the bed of the truck (Fig. 23.9). Removal of the crate with a forklift (Fig. 23.10) and subsequent release must be done gently and quietly. Crates should not be allowed to tip. When an animal is released, place the crate in a gate or doorway, open the door, and allow the animal to emerge of its own accord (Fig. 23.11).

FIG. 23.9. Well-anchored shipping crate.

CHEMICAL RESTRAINT. Zebras, asses, and wild horses can be immobilized using etorphine hydrochloride alone or preferably in combination with either acepromazine maleate or xylazine hydrochloride (Table 23.3). The animal should lie quietly for 10-20 minutes following administration of etorphine, remaining sedated for as long as 45 minutes. Blindfolding prolongs the effect (Fig. 23.12). When the reversal agent is administered, be prepared to quit the stall immediately, because the animal will arise quickly and may become aggressive.

It may be necessary to immobilize a wild equid (preferably with the etorphine-xylazine combination) in order to crate it. Place the crate in front of the immobi-

FIG. 23.10. Removal of crate with a forklift.

FIG. 23.11. Crate placed in a gate for release of an animal.

FIG. 23.12. Blindfolding a zebra following chemical immobilization prolongs sedation.

lized animal, reverse the etorphine, wait 3-4 minutes for the reversal to become effective, and assist the animal to its feet, directing it quickly into the box. A rope looped around the neck may help direct a zebra into the crate and prevent it from rearing and falling backward. A second rope should be attached to the loop to assure removal of the neck loop once the zebra is in the crate. Only occasionally will the animal jump up sooner than anticipated and foil the plan. Close observation following etorphine reversal may obviate this hazard.

Tapir

Tapirs are aquatic to semiaquatic odd-toed ungulates with heavy bodies and short legs. The dental structure of the tapir is similar to that of the horse.

DANGER POTENTIAL. Tapirs bite in defense. The bite wound is similar to that inflicted by the horse. A number of handlers have mistakenly assumed the tapir to be always a gentle, inoffensive animal and have been viciously attacked and maimed.

PHYSICAL RESTRAINT. Tapirs are often docile, allowing casual examination or even injection without any specific restriction (Fig. 23.13). They can be put into squeeze cages, but they often resent being placed in such a device and thrash about wildly. They may push against the top of the squeeze cage and abrade the skin over the withers. This can be obviated by attaching a burlap sack filled with straw or some other type of padding to the top of the squeeze cage before the tapir is admitted. They also have tender feet which can be lacerated during a struggle.

CHEMICAL RESTRAINT. Tapirs can be immobilized with etorphine or combinations of etorphine and xylazine (Table 23.3).

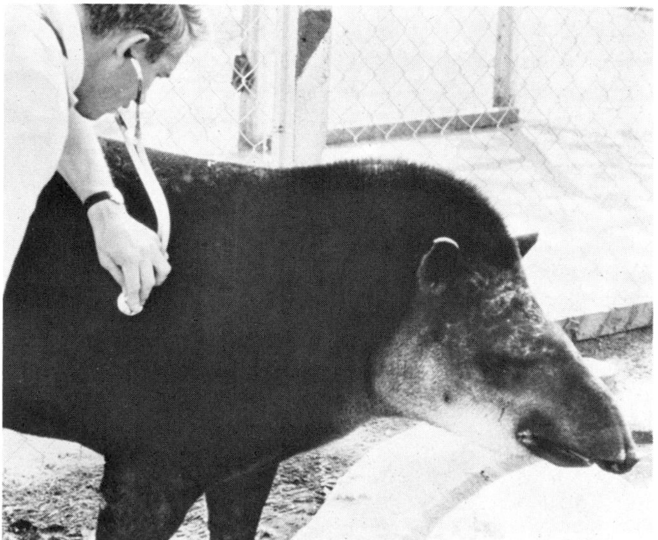

FIG. 23.13. Some tapirs are docile enough to allow physical examination without physical restraint.

Rhinoceros

DANGER POTENTIAL. Rhinoceros are unpredictable large ungulates. The black rhino and Indian rhino are more aggressive than the more phlegmatic white rhino, but all are large and dangerous animals. Rhinos do not kick or strike and rarely bite, but some use their canines as tusks like boar pigs. The primary defense method is ramming with the horn. They may also trample an enemy or crush one against a wall.

PHYSICAL RESTRAINT. Except for young calves, which are handled by methods similar to those applied to other small ungulates, the physical restraint of rhinos is next to impossible. The nearest approach to physical restraint is to entice the animal into a shipping crate with bars in the tailgate, through which a handler can carry out gentle manipulations or administer injections. Some easily agitated animals may traumatize themselves by fighting the crate.

TRANSPORT. Rhinos are transported in large well-constructed shipping crates (Fig. 23.14), which usually must be positioned in a doorway by heavy cranes. The animal is enticed into the crate with food, and the door is closed behind it. Some time is required to fast the animal sufficiently so it will enter the crate for food. Alternatively, a low dose of etorphine can be injected and the rhino led into the crate (Fig. 23.15). Once the animal is in the crate, the crane operator must lift and move the crate carefully, taking pains to avoid tipping it (Fig. 23.16). The crate should be placed on the truck bed or in an airplane with the tail end toward the front. Sudden stops then result in the animal bumping its hindquarters instead of its head against the crate wall.

CHEMICAL RESTRAINT. Etorphine hydrochloride is the drug of choice for immobilizing rhinoceros (Table 23.3). A sedative dose varies from 0.5 mg to 1 mg (total dose). The

FIG. 23.14. A heavily constructed rhino crate left in place so that animal can be fed in the crate for a few days prior to capture.

FIG. 23.15. Wide-lipped rhino, under mild sedation with etorphine hydrochloride, being led to shipping crate.

FIG. 23.16. Rhino crates being moved with heavy equipment.

rhino may remain standing but manageable (see Fig. 23.15). An immobilizing dose is 1-2 mg (total dose). Reversal with diprenorphine is not always complete with a single injection and may have to be repeated once or twice. The heavy epidermis of the rhino would seem to defy penetration by a dart syringe needle, but it is penetrable. A right-angle hit of sufficient force will penetrate the hide on the lateral side of the hind leg or shoulder, but the skin of the medial aspect of the hind leg is much thinner and more easily penetrated, making this area a more desirable target site if it is accessible. The needle should be 4.5-6 cm long [7] and 16 gauge (2 mm ID). A pole syringe can be directed to the medial aspect of the front or hind leg or behind the ear.

ARTIODACTYLA

Wild Pig, Peccary

DANGER POTENTIAL. Members of these two families bite and gore. They neither kick nor strike, and most of them are not heavy enough to crush or trample. They possess sharp teeth, with the canine teeth elongated to form tusks, particularly in the male (Fig. 23.17). These tusks are formidable weapons, used to rend and disembowel opponents. Wild pigs can be aggressive. Do not enter an enclosure confining them without protection.

The peccary is a small ferocious animal with a vicious disposition. It can inflict serious or fatal injury with its tusks and needle-sharp teeth, which can easily penetrate heavy leather gloves. Peccaries are fast and agile and are quick to resent restraint. Even small piglets are obstreperous and hard to subdue (Fig. 23.18).

ANATOMY AND PHYSIOLOGY. Wild pigs and peccaries are heavy-bodied, short-necked, short-legged animals. Physically they resemble domestic pigs. Both have the thick layer of insulating fat common to all species of swine, predisposing them to overheating when subjected to restraint.

PHYSICAL RESTRAINT. Movement can be restricted by enticing the animals into a smaller enclosure with food. Smaller

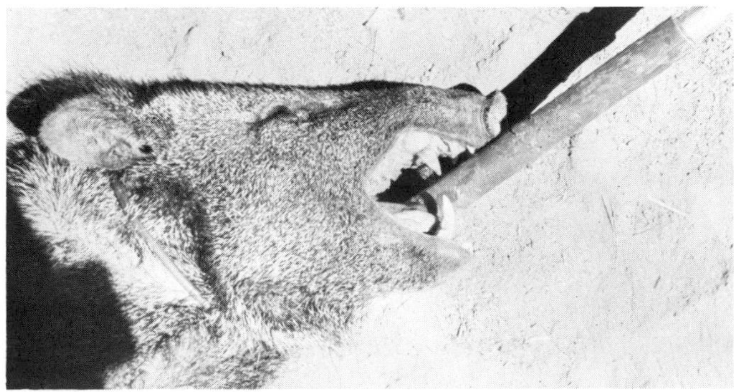

FIG. 23.17. Elongated canine teeth of collared peccary form tusks.

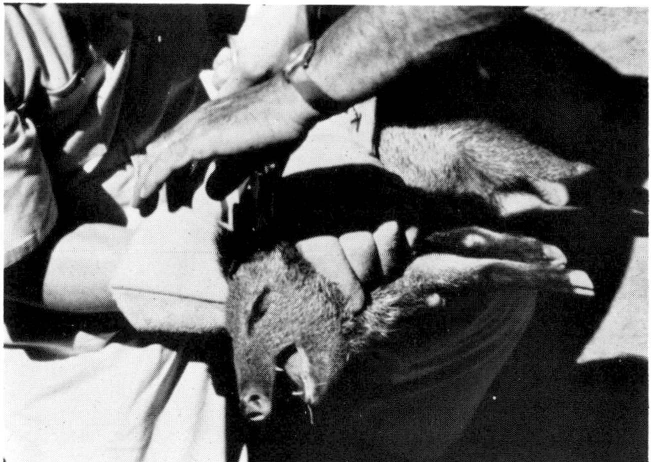

FIG. 23.18. Manual restraint of juvenile collared peccary.

pigs and peccaries can then be snared singly, grasped by the hind legs, and stretched. The come-along must be tightly secured so it does not slip over the narrow head. Gloves offer some protection from teeth.

Peccaries are more agile and aggressive than wild pigs and consequently must be handled with greater care. Peccary piglets must also be snared and stretched, though newborn wild pigs can be handled manually in the same manner as domestic piglets. Large members of this group must be run through a chute or put into a squeeze cage to make examinations, give injections, or obtain laboratory samples.

CHEMICAL RESTRAINT. Wild pigs and peccaries can be immobilized with phencyclidine, ketamine, or xylazine (Table 23.3). Figures 23.19 and 23.20 show a peccary grasped after partial sedation. It would be foolhardy to grab the unsedated animal in this manner.

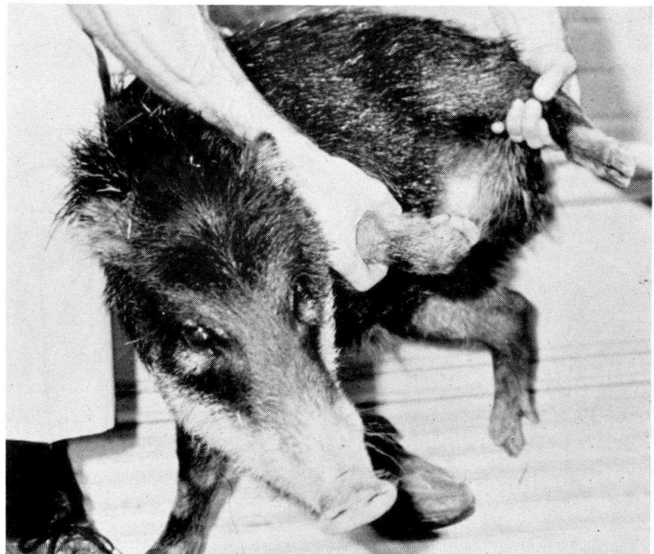

FIG. 23.19. Casting partially immobilized white-lipped peccary.

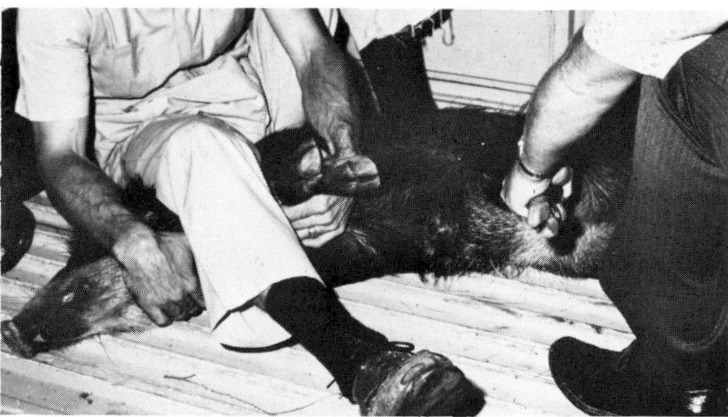

FIG. 23.20. Restraint of partially sedated white-lipped peccary for collecting semen sample.

Hippopotamus

The two species in this family vary greatly in size and temperament. Both are herbivores that spend varying amounts of time in an aquatic environment. The Nile hippopotamus is truly an aquatic species but grazes on land during the night. The pygmy hippopotamus is more terrestrial in habit but is never found far from water.

DANGER POTENTIAL. The Nile hippopotamus is huge; if cornered it can easily trample or crush a person, though the primary offensive and defensive weapons are the tusks, which are elongated canine teeth. The hippopotamus is likely to bite. Its jaws are powerful and the tusks will impale anything taken into the mouth. The much smaller pygmy hippopotamus also has formidable tusks, capable of seriously injuring or killing the unwary person who provokes an attack (Fig. 23.21). It can be guided into a squeeze chute for examination or treatment. The pygmy hippo produces a skin secretion that makes the body slippery. They are impossible to hold by hand when they are excited and are difficult to move when immobilized. The adult Nile hippo cannot be handled manually. Certain objectives can be achieved if the animal can be persuaded through a chute arrangement or restricted in a small pen. Hippos will frequently open their mouths when offered

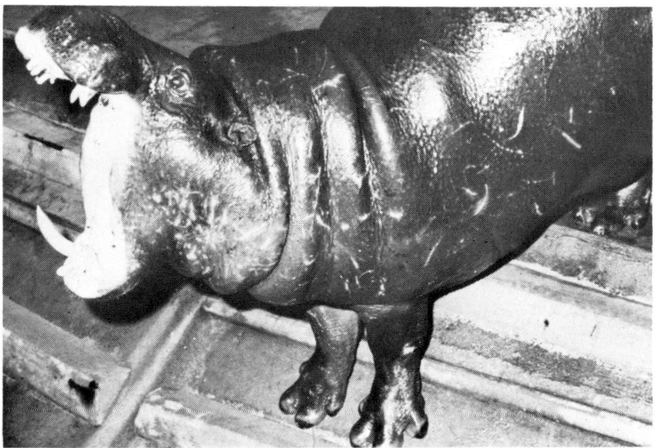

FIG. 23.21. Pygmy hippopotamus.

food, allowing visual examination of dental and other oral structures. Intramuscular injections can be given with a stick syringe.

If restricted activity or extensive examination and/or treatment is necessary, chemical restraint agents must be administered.

TRANSPORT. Hippos are moved in a manner similar to the rhinos.

CHEMICAL RESTRAINT. Etorphine and phencyclidine hydrochloride have been used on the hippo (Table 23.3), using a 7-9 cm, 16 gauge needle without a barb. Hippos dive into available water if frightened or angry, so immobilization must be carried out away from water sources. Once the hippo is immobilized, keep its body surface moistened with water. During warm weather, beads of red-tinged sweat will appear on the skin of the Nile hippo. This is a normal phenomenon. Keep the hippo's head higher than its body after immobilization.

A combination of phencyclidine hydrochloride and promazine hydrochloride (250 mg total dose of each) is used to immobilize an adult pygmy hippo.[1]

Camel, South American Camelids (Llama, Alpaca, Guanaco, Vicuna)

DANGER POTENTIAL. Bactrian and dromedary camels possess enlarged canine teeth and are likely to bite. South American camelids may also bite but are less likely to do so, and their teeth are not so damaging. The camel is also capable of kicking in any direction. It may strike with the foreleg, kick to the side, or kick backward. The speed and strength of a camel's kick is incomprehensible.

On one occasion I was attempting to apply medication to the hind leg of a bactrian camel that was standing next to a 10 cm × 10 cm (4 in. × 4 in.) beam. When the animal felt the medication touch its leg, it lashed out. Fortunately for me the hoof struck the beam, snapping it in two.

In addition to biting, kicking, and striking, the camel and South American camelids have the disagreeable behavioral trait of regurgitating the foul-smelling contents of the rumen onto the handler. The animal may spew the material 6-10 ft.

South American camelids are not likely to kick but use their front legs to attack. Aggressive individuals run at an adversary, rear up on the hind legs, bend the front legs at the knees, and strike down with powerful blows on the enemy. This action may be accompanied by biting and/or spitting.

ANATOMY AND PHYSIOLOGY. The lateral processes of the cervical vertebrae of camelids extend ventrally to encase the trachea and other vital organs. This structure is advantageous for roping these animals; there is little danger of strangling them, since the trachea is protected within the bony canal.

The male camel has a specialized inflatable diverticulum off the soft palate called the gula. When the animal is angered or displaying, this structure protrudes from the mouth, appearing as if the tongue has prolapsed. The animal also emits a rather harsh bellow. This phenomenon is a type of display behavior. The gula will be extruded when the animal is physically restrained and temporarily protrude when the animal is chemically restrained.

PHYSICAL RESTRAINT. South American camelids can usually be handled quite easily (Fig. 23.22). A rope can be tossed over the head to hold it while one or two other handlers approach from the side to further restrict the animal's activity. If more complete immobilization is required, a rope can be placed around the hind legs to stretch the animal (Figs. 23.23, 23.24).

An interesting and hazardous phenomenon sometimes occurs when a llama is roped and stretched in the presence of other llamas: the free individuals may attack the restrained animal.

FIG. 23.22. Llamas are usually docile and can be handled as shown.

FIG. 23.23. Restraining a llama by stretching a hind leg.

1. Personal communication, Dr. M. Bush, National Zoo, Washington, D.C.

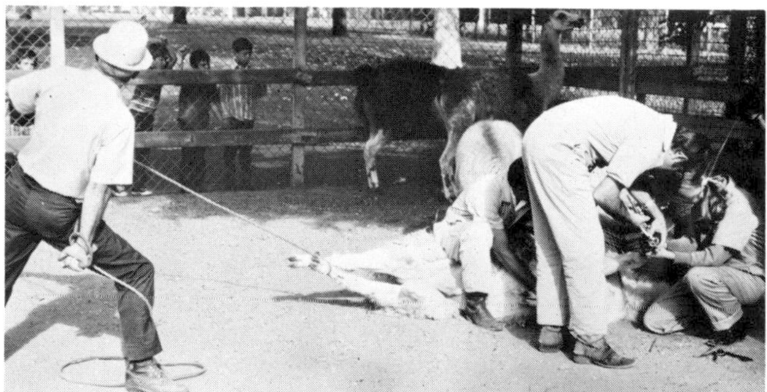

FIG. 23.24. South American camelids seldom struggle once cast and stretched.

Vicunas and guanacos are the wild counterparts of the domesticated llama and alpaca. These animals are flighty and likely to injure themselves if chased or roped. However, in general they can be handled in the same manner as the llama or alpaca.

Camels may also be roped. They can be haltered (Fig. 23.25) and snubbed to a tree or post. With adequate patience and long ropes, handlers can place special hobbles on the hind legs, as illustrated in Figure 23.26. Two people are required, but they must be careful to stay out of range of a kick from the hind legs. The camel may jump around during the process of applying the hobble. Once the hobble is in place, the animal can be more safely approached to take blood samples or make examinations. The hobble should not be tied with a firm knot; it should always be possible to jerk a loop and instantly release the knot if necessary.

Camels can be restricted by running them through chutes. The walls of the chute should be solid, perhaps of plywood, to prevent the animal from sticking a foot through a space and fracturing a leg. Figure 23.27 illustrates a camel in a scale. Notice the opportunity for a leg to be placed through the bars. This camel did just that

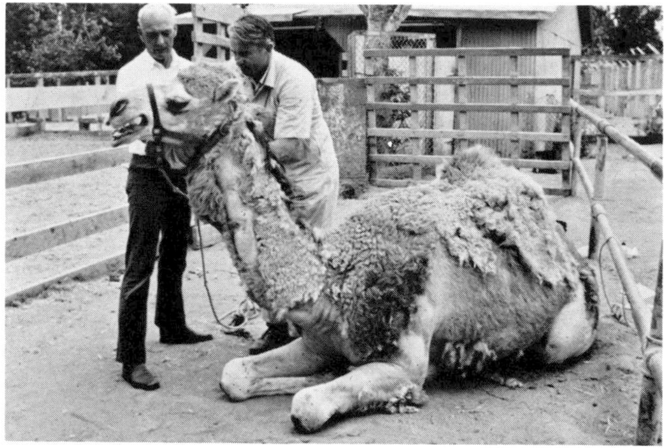

FIG. 23.25. Camels can be haltered for head control.

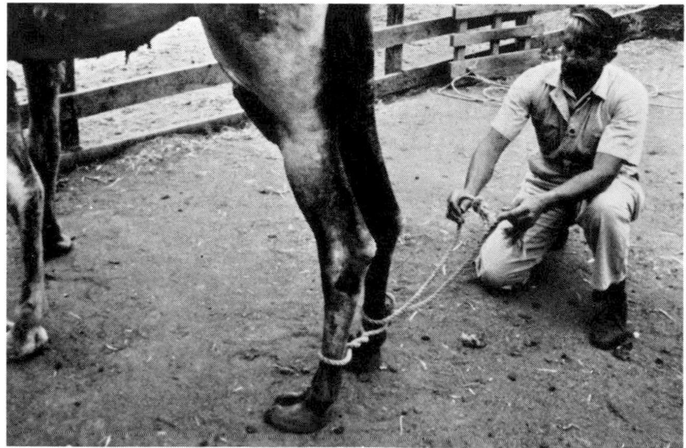

FIG. 23.26. Figure eight rope applied to hind legs of a camel whose head is tied up.

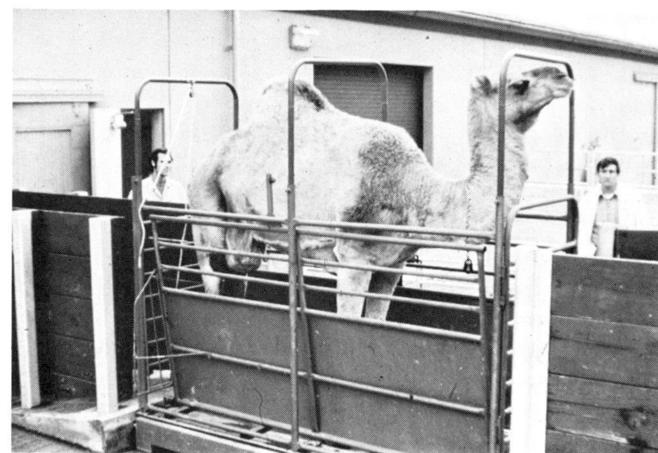

FIG. 23.27. Obtaining accurate body weight is important for determining medication dosage.

FIG. 23.28. Rope sling used to lift seriously ill camel.

when an attempt was made to give an intramuscular injection. As a colleague bent to assist the animal in releasing its foot, the camel reached around and grasped him by the back of the neck. I slapped the animal as hard as I could in the face, and it released its grip; another few seconds might have meant a fractured neck as the animal shook him.

Figure 23.28 illustrates the use of a sling to hoist a sick camel. Calves of all camelids can be handled like calves of domestic cattle.

CHEMICAL RESTRAINT. Xylazine alone or a xylazine and etorphine combination is effective in the camel (Table 23.3).

Deer, Elk, Moose

DANGER POTENTIAL. Small members of these groups bite. The enlarged canine teeth, protruding from the mouth as tusks, are formidable and must be reckoned with during restraint. Other cervids seldom bite. The primary weapons of most cervids are the antlers, used both in display and as weapons. All members of this group strike with their front feet and are capable of inflicting serious wounds with the sharp hoofs.

ANATOMY AND PHYSIOLOGY. Antlers are normally shed and regrown annually. They are specialized bony protuberances stemming from a pedicle on the poll. As antlers develop, they are covered by a highly vascularized velvet. During this period of time, the bone is relatively soft and easily broken. If cervids must be manipulated while in velvet, use extreme caution. Grasp the antlers only at the base. An antler fractured while in velvet will bleed profusely. When the antler matures, the velvet dries and is rubbed off until the antler is highly polished. At this point, the rut begins and the animal becomes more aggressive.

A rutting animal is dangerous not only to other cervids but also to keepers who must enter the enclosure. It may be necessary to amputate the antler of a rutting buck. Leave approximately 4-6 in. of the stump on the skull for a fulcrum against which the animal can rub to remove the base. If it is cut too far distally, the purpose of removing the antler is defeated; if it is cut too close to the head, the stag will fail to shed the small scur, causing abnormal antler growth around the scur the next year.

Antlers are limited to the male of the species except for caribou (reindeer), in which both sexes grow antlers.

Reindeer have been domesticated and utilized for milk and meat in the northern latitudes for many centuries. Their wild counterpart, the caribou, is a relatively calm, phlegmatic animal and can be handled manually. Caribou can be held by the antlers or placed in crates or squeeze cages for manipulation.

Special chutes can be constructed for handling cervids. Figures 23.29-23.31 illustrate a chute arrangement used to handle elk. Either cow elk or deantlered bulls can be herded into the chute area and gradually moved into a funnel-shaped arrangement that leads to a narrow channel, admitting animals to the chute in single file. The animals are herded to the periphery of the corral to begin their movement through the funnel. The sides of the enclosure and chute are high. The inside of the chute should be smooth, covered with plywood or a similar material so animals cannot gain a footing to climb the walls. As the animals progress along the channel, doors can be closed at intervals to separate individuals. A special cattle chute is placed at the end of the channel. Boxes equipped with small doors to allow quick intramuscular injection for chemical immobilization or administration of medication can also be placed at the end of a chute (Fig. 23.32).

FIG. 23.30. Commercial cattle chute modified to accommodate wapiti.

Small cervids without antlers can be handled manually. Figure 23.33 illustrates one method of holding a small mule deer. When grasped in this position, the animal will usually relax, entering a trancelike state. If additional restraint is required, the animal can be placed in lateral recumbency and held as shown in Figure 23.34, or it can be held with the forelimbs brought up behind the head and the hind legs stretched out (Fig. 23.35). Sometimes several people are needed to restrain a cervid (Fig. 23.36). Padding beneath the animal prevents trauma.

FIG. 23.29. Special circular chute designed for handling wapiti.

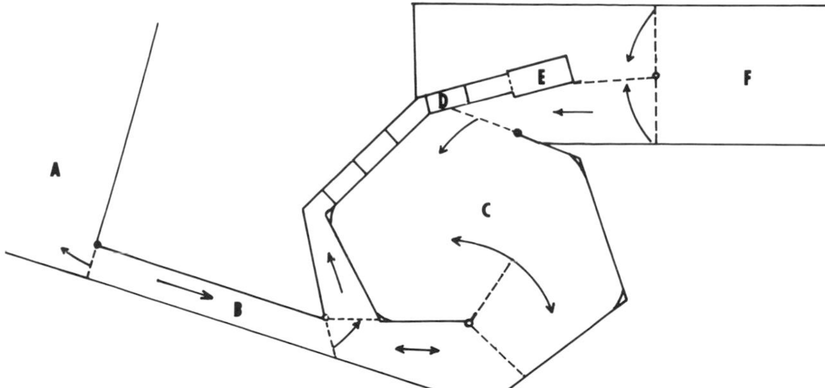

FIG. 23.31. Diagram of holding area and approach to a chute for handling hoofed animals: **A.** General enclosure. **B.** Alley to chute area. **C.** Inner court. **D.** Compartmentalized chute area. **E.** Squeeze chute. **F.** Release area.

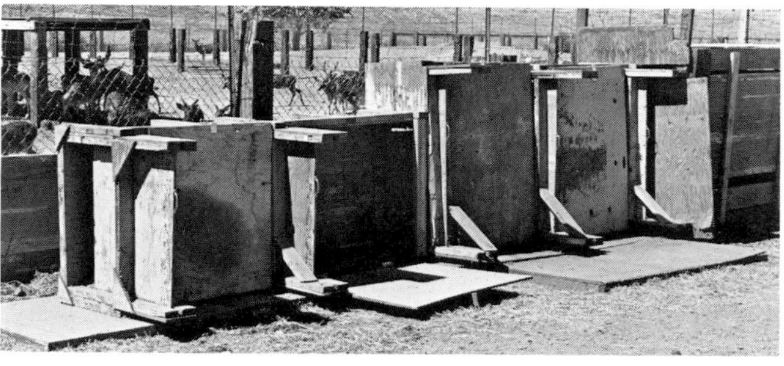

FIG. 23.32. Individual restraint boxes for handling cervids.

FIG. 23.33. Restraint of a small mule deer. Legs are directed away from handler.

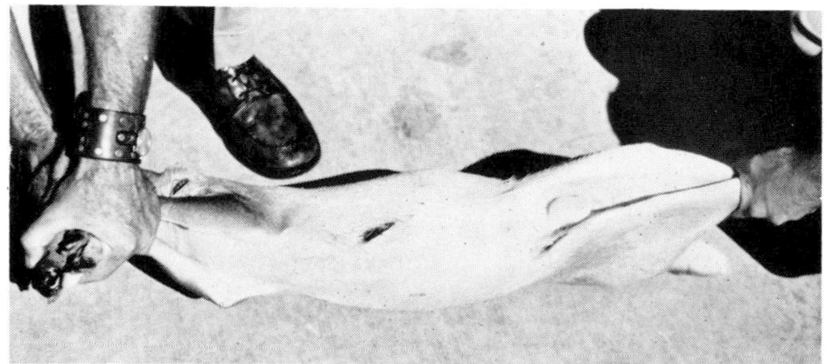

FIG. 23.34. Stretching and holding a muntjac deer.

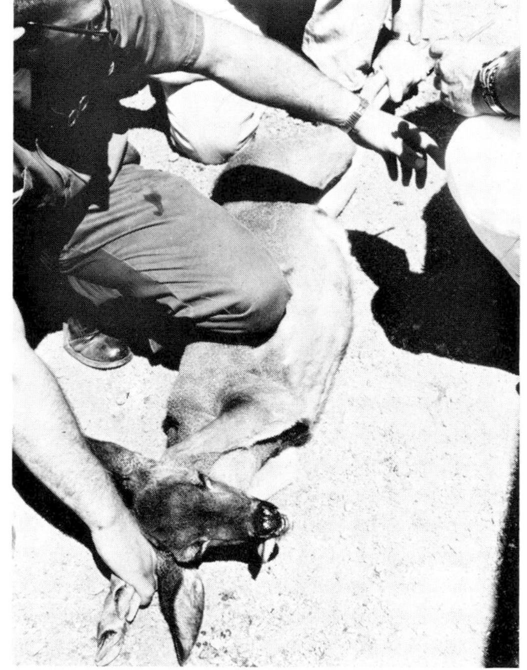

FIG. 23.35. Deer restrained by placing head between extended forelimbs and stretching hind legs.

Cervids can be roped if necessary, but this is a poor form of restraint (Fig. 23.37); they may respond to being caught by charging the roper instead of pulling back to take up the slack. The polished antlers can be used as leverage for restraint if the handler is strong enough to hold the animal (Fig. 23.38). When in velvet, this procedure is not suitable because the antlers are sensitive, highly vascular, and easily fractured.

To capture cervids with a net, lay the net in a path that is likely to be traveled by a group of animals. Anchor one end of the net to a fence or tree. The other end is held by one of the handlers. Carefully haze the animals over the collapsed net. As the animals pass over the net, the selected individual is watched; as it approaches the net, the handler jerks the net up taut. The animal will run into the net. As soon as the animal is entangled, the handlers can subdue it by grasping the front and hind legs and stretching the animal.

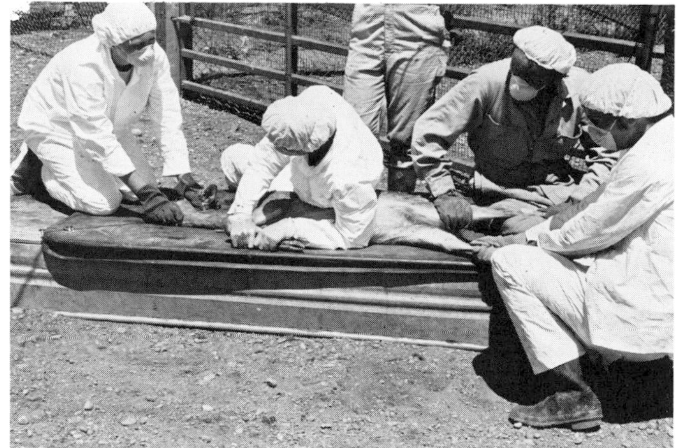

FIG. 23.36. Air mattress or other padding minimizes trauma to animal and handlers during physical restraint. These handlers are wearing protective clothing and masks as special precautions because this deer has tuberculosis.

FIG. 23.37. Roping is a poor form of restraint for cervids.

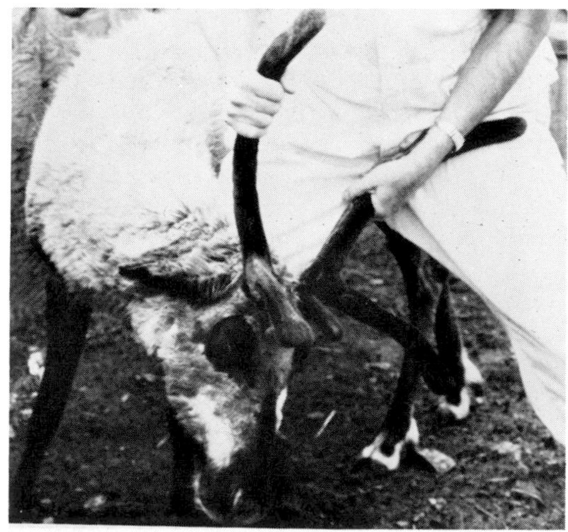

FIG. 23.38. Polished antlers of cervids can be used to restrain some species.

TRANSPORT. Moving large cervids such as elk and moose is a major project. Usually these animals require chemical immobilization. Dragging these heavy-bodied animals from one place to another may severely abrade the skin. Instead, carefully pull the immobilized animal onto a sheet of plywood or a canvas tarp, which can then be dragged or carried to another location. Figure 23.39 shows a wapiti secured to a plywood sheet following immobilization with etorphine hydrochloride. The animal was then lifted out of a moated enclosure (Fig. 23.40).

FIG. 23.39. Chemically immobilized wapiti lashed to sheet of plywood.

FIG. 23.40. Plywood stretcher being lifted out of a moated enclosure.

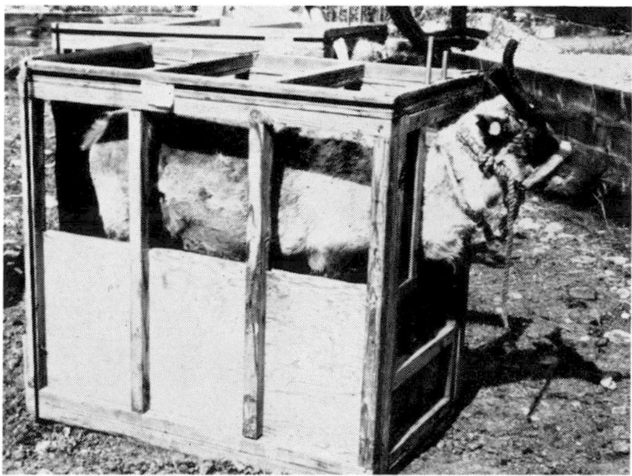

FIG. 23.41. Crate for transporting reindeer.

Individual crates are usually used to transport cervids (Fig. 23.41). A group of animals can be shipped together in a truck if sizes and sexes are segregated.

CHEMICAL RESTRAINT. Etorphine, xylazine, and tiletamine-zolazepam have all been used in various species of cervids. The last seems to be the most satisfactory agent (Table 23.3). Slight elevation of the forequarters will prevent mechanical expulsion of the rumen contents during immobilization. When immobilizing or physically restraining hoofed animals, it is desirable to diminish visual sensations by blindfolding.

Giraffe, Okapi

DANGER POTENTIAL. Giraffe and okapi do not bite, but they can kick and strike in any direction. The extremely long legs are formidable weapons. I once entered a box stall housing a yearling giraffe that was lying in the center of the stall. My entrance startled the animal, and it jumped up and struck at me so fast I was barely able to jump backward out of the stall. On another occasion a colleague was kicked twice on the hand before he could withdraw it. These animals are extraordinarily quick and agile and are capable of inflicting lethal blows with the feet.

The head is also dangerous when used as a powerful battering ram. In normal intraspecific fighting, giraffe stand side by side, head to tail, slapping each other in the body with the side or the back of the head. The same behavioral characteristic is used against people. An animal in a chute (Fig. 23.42) may reach back and knock down a person climbing up the side of the chute.

FIG. 23.43. A giraffe calf can be restrained in the same manner as a horse foal. The grip on the tail can be relaxed except when the calf struggles, otherwise it will slump to the floor.

FIG. 23.42. Giraffe being milked through bars of a timber chute.

FIG. 23.44. Using a mattress to protect handler from the kicking and striking of a giraffe calf.

PHYSICAL RESTRAINT. Giraffe calves weighing up to 111 kg (250 lb) can be grasped and handled with the same techniques used for foals (Fig. 23.43). As the animal grows, it may be desirable to use a shield such as a small mattress as protection from striking or kicking (Fig. 23.44).

Adult giraffe cannot be physically restrained except in specialized chutes. Figure 23.42 illustrates such a chute constructed of wood. Figures 23.45 and 23.46 show a similar arrangement constructed of chain link fencing. For such chutes to be effective, the animals must be forced to pass through them periodically. Ropes can be positioned against a stall wall to make a temporary chute.

Once an animal is in a chute, it should be allowed to relax before beginning general examination, tuberculin testing, or withdrawing blood samples. Any more complicated or painful procedure necessitates chemical immobilization.

FIG. 23.45. Giraffe chute constructed of pipe and chain link fencing. The chute was intercollated into an alley through which the giraffe moved each day to reach the outside paddock.

FIG. 23.46. Inside the giraffe chute.

TRANSPORT. Giraffe are not easy to herd. The most desirable method of separating individuals is to use a device made by placing one wall of the barn on rails so it can be moved to restrict the location of a given animal. Placing an adult giraffe in a shipping crate can be difficult. However, if the animal can be squeezed into a small area, it may walk quietly into a crate (Figs. 23.47, 23.48).

FIG. 23.47. Young giraffe being directed into shipping crate with plywood shields.

FIG. 23.48. Giraffe in a shipping crate.

CHEMICAL RESTRAINT. Of all the ungulates, the giraffe has the poorest history of successful immobilization [4,15]. The anatomical structure of the neck almost precludes chemical immobilization without traumatizing the neck. It is difficult but necessary to support the head in an elevated position. Recovery likewise is hazardous, inasmuch as the return of control of the head and neck does not necessarily coincide with return of the ability to stand up. Etorphine hydrochloride and xylazine hydrochloride have been used, as well as mixtures of the two.

A technique proven successful at the Fresno (California) Zoo for animals requiring hoof trimming begins with an intramuscular injection of either promazine hydrochloride (1 mg/kg) or acepromazine maleate (0.025–0.03 mg/kg).[2] One hour after injection the animal is approached and a casting harness similar to that used to restrain equines is applied. Handlers must take care to remain sufficiently far from the feet to avoid being trampled or struck. The stall should be previously prepared with a deep layer of straw. As tranquilization deepens, the casting harness is gradually tightened until the pull causes the animal to slump to the floor. Sufficient personnel should be on hand to support both head and neck during the procedure. After the feet are trimmed, the casting harness is removed and the animal allowed to lie quietly until it recovers. The tranquilization is light enough that the animal can keep its head erect by the time it is released from the harness.

The okapi is handled in much the same manner as the giraffe. The animals can be restricted in a chute area for general examination, superficial evaluation, or administration of intramuscular injections. If more restrictive restraint is required, chemical immobilization is necessary.

Cattle-Type Animals, Antelope, Sheep, Goat, Pronghorn

DANGER POTENTIAL. Horns are formidable weapons possessed by many members of the Bovidae family. They are found on both sexes, but males have larger horns and are usually more likely to fight than females. The horns may be massive structures such as those of the bighorn sheep or African buffalo (Fig. 23.49), fine and fragile, or any gradation between. The horns may be used strictly for display or continually in mock or serious battle.

In addition to the horns, many members of this family have sharp hoofs. Wild species are particularly likely to lash out with the feet and may cause fractures or other severe injuries with a kick.

ANATOMY AND PHYSIOLOGY. Horns are modified epidermal structures consisting of a cornified outer layer covering a bony core. Horns are permanent, not shed annually like antlers. They are used for display and fighting and in some species function in a thermoregulatory capacity. The horns of some species may be grasped as handles for controlling the head. Horns are easily damaged during restraint, particularly those of young animals in which the bony core has not yet permanently fused to the skull.

BEHAVIOR. There are widely separated extremes of behavior within this group—from docile animals, such as some of

2. Unpublished data.

FIG. 23.49. Cape buffalo is a formidable adversary. Chemical immobilization is the only safe method of restraint.

the sheep that are rarely aggressive, to fierce animals like the gnu, which may attack without provocation.

Introducing a new animal into an established hoofed stock exhibit is dangerous and must be done gradually and carefully. These animals form social groups with hierarchial status. The newly introduced animal is likely to be aggressively harassed. To prevent harassment, the animal should first be released into a small box stall or pen near the other animals, permitting them to become acquainted as they nose each other over gates or through fences. Adequate sight and physical barriers should be provided within the enclosure to allow the new animal to find a refuge.

PHYSICAL RESTRAINT. Smaller members of this family can be manually restrained. The animals should be restricted to a small enclosure, grasped (Fig. 23.50), and the legs stretched and held. Ropes are sometimes used to capture these animals. Once the animal is caught, assistants must quicky grasp the animal, lest it jump towards the roper or injure itself by struggling. A tarp or blanket may prevent injury from flailing feet. Sufficient restraint is usually

FIG. 23.50. If an antelope is to be manually captured within an enclosure, it must be done quickly before the animal traumatizes itself.

achieved by grasping both front and hind feet, controlling the head at the same time.

A special restraint bag for handling small wild ruminants has been designed by personnel of the USDA Animal Import Center at Clifton, New Jersey (Fig. 23.51; see also Fig. 2.5). The fabric used in the construction of the bag is nylon impregnated with a rubber or vinyl base compound (Hypolon). Fabric weighing at least 16–20 oz/sq yd must be used to be sure the bag is capable of withstanding the thrusts of hoofs and horns. The bag is approximately 1 m (48 in.) in diameter and 2 m long. Handholds are attached to either end so the bag can be held firmly to the ends of crates containing the animals. The crate door is opened and the animal encouraged to jump out of the crate into the bag. A foam rubber pad is preplaced beneath the bag. As soon as the animal jumps inside, handlers close the bag from behind and lay the animal on its side. Access to the animal is gained through the ports. The animal can be examined, medicated, vaccinated, or tattooed; blood samples can be obtained with minimal stress.

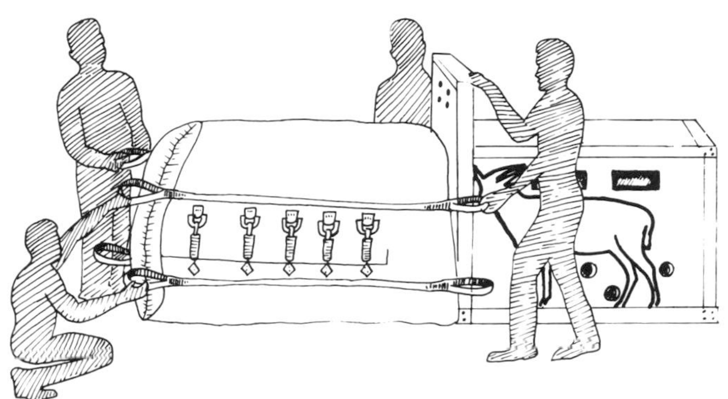

FIG. 23.51. Diagram of use of a restraint bag designed by the USDA Animal Import Center, Clifton, N.J.

Fences adequate to contain quiet hoofed animals may not prevent excited animals from jumping out of an enclosure. Furthermore, when frightened, some hoofed animals may actually climb fences—an act they would probably not attempt if undisturbed. The escape response is intense in these animals, and they will extend themselves to achieve escape in manners beyond general comprehension. I have seen a bison jump or climb a 2.5 m (8 ft) chain link fence to escape capture.

When individuals must be captured from herds of ungulates, it is desirable to restrict the herd to a small enclosure such as a box stall before beginning capture operations. Sometimes it is necessary to construct such a confining area out of plywood sheeting and temporary poles. Even temporary construction should be quite substantial, because the animals will likely bump against it or charge it; if it is flimsy, it will break and cause more serious injuries than may be imagined.

Once an animal group is inside a shed or a small enclosure, rope loops may be tossed onto the heads of selected individuals so they can be dragged out and grasped without alarming others of the herd. I have handled mouflon and aoudad sheep on numerous occasions in this manner without serious injury to animals or staff.

Physical restraint sometimes seems to become a contest between the animal and the restrainer. Various techniques are used to grasp the animal by the horns and twist it down, or to hold it when it is down. In most instances the hind legs must be stretched, requiring more than one person to accomplish the restraint (Figs. 23.52–23.54). When physically restraining animals, be sure the animal's eye is not ground into the earth. A mattress will minimize trauma to both the animal and the handlers who are holding the animal down.

Nets are used to handle a wide variety of hoofed animals; even the lordly musk-ox can be captured with a heavy cargo net. Figures 23.55–23.57 show a net being

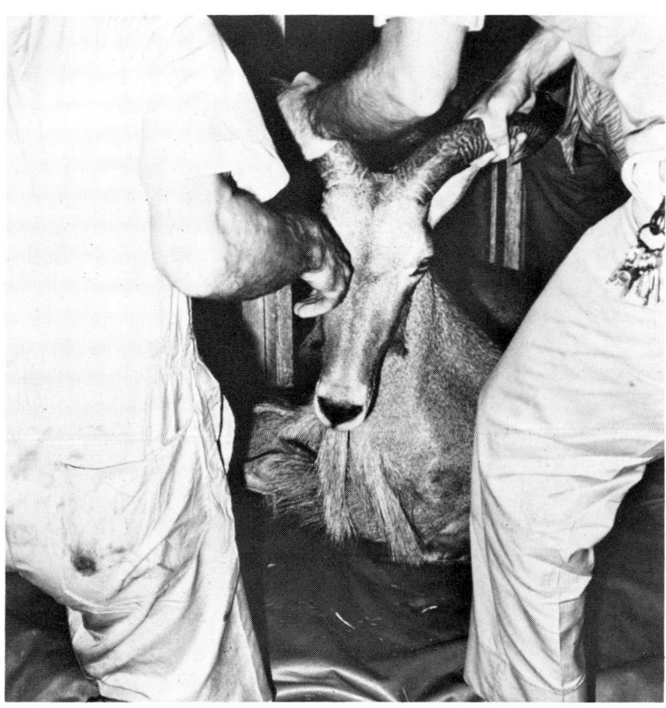

FIG. 23.52. Aoudad ram being removed from shipping crate.

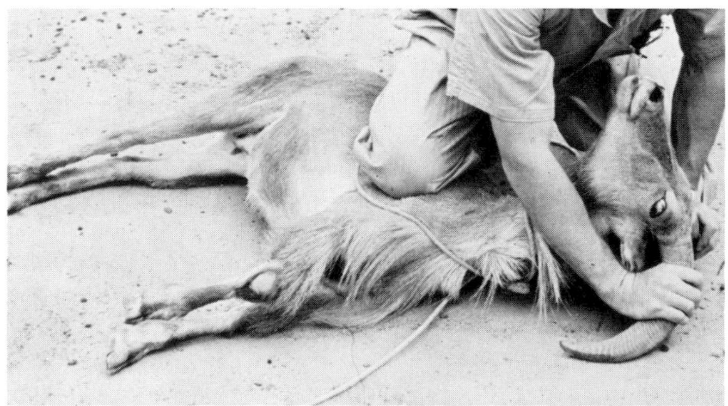

FIG. 23.53. Aoudad being restrained following capture by roping.

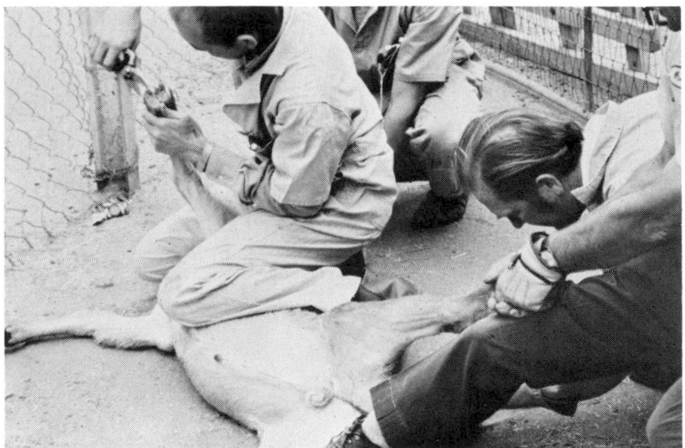

FIG. 23.54. Restraining a wild sheep for hoof trimming and tuberculin testing.

FIG. 23.55. Approaching a Dall sheep with a net.

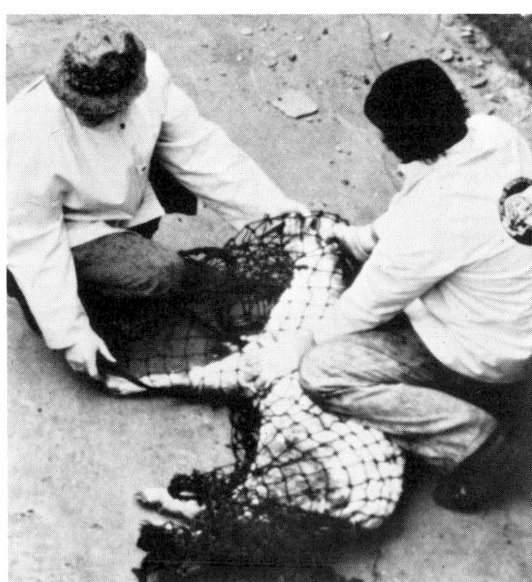

FIG. 23.56. The sheep rushed at the net and became entangled in it.

FIG. 23.57. Using net as a sling to retrieve animal from a moat.

used to capture a Dall sheep. The ewe was in a corner of a moat, being approached by attendants with the stretched net. As the attendants drew near, she bolted to escape and became entangled in the net. In this case the net served as a sling to lift her from the moat.

Most wild bovids will flee on the approach of a person. However, some species such as the gnu, particularly the males, may become very aggressive when approached. When cornered and unable to escape, many otherwise docile animals turn and attack. Plywood shields offer some protection to the person who must enter an enclosure to direct a single animal into another area (Fig. 23.58). Metal

FIG. 23.58. Plywood shield with handhold on the back.

or wooden handles can be attached to the reverse side of the shield for either one or two men to grasp. Shields can be attached to the front of a vehicle if the pen is large enough to allow the operation of a tractor, a skip loader, a truck, or other motorized equipment. If any young animals are in the herd, they should be captured quickly, or the large rams should be removed as rapidly as possible.

Squeeze cages and chutes have been designed for use with hoofed animals. Figure 23.59 shows a squeeze cage used at the San Diego Wild Animal Park. It is adjustable to fit a wide variety of hoofed animals. Extreme caution must be used when putting horned animals into such a device. In many instances, it is wiser to use chemical immobilization on these to preclude possible injury to the horns.

FIG. 23.59. An excellent squeeze chute and transfer crate for hoofed animals.

Occasionally tuberculin testing programs or other activities necessitate the handling of large numbers of hoofed stock. Chutes can be improvised. Railroad ties or other timbers can be sunk into the ground and plywood sheeting applied to the inside, forming a chute. The inside of the chute must be smooth to prevent animals from gaining a foothold to climb out.

In one instance a chute built to handle bison was unsatisfactory because the animals were able to catch their hoofs in the planking on the inside of the 8 ft chute and climb out. When the inside planking was covered with plywood, climbing was minimized. Some of the bison tore off the outer shell of the horn by rubbing their heads up and down against the inside of the chute. Fortunately the horny shells grew back within a few months. However, the shape of the damaged horn was not as symmetrical as the original.

Confining animals in close quarters stimulates aggesssive tendencies, and the animals may fight. Time spent on any procedure requiring close confinement should be kept to a minimum.

Attempting to drive animals into a chute area or box stall or enclosed pen is a frustrating experience (Fig. 23.60). The animals sense a change and are not easily herded into a new enclosure. Sheep will follow a leader, and it may be possible to rope the dominant animal and

FIG. 23.60. Bongo being directed into trailer with plywood shields. Individuals within a species vary greatly as to the degree they will tolerate such manipulation.

pull it into the enclosure. The other animals will then follow him in. A usually successful technique is to feed the animals inside the restraining enclosure for a few days before the procedure is scheduled. Plastic sheeting has been an important contribution to the art of persuading animals to enter a new area.

Most ungulates are diurnal and have limited night vision. If these animals can be confined inside a darkened stall, it is often possible for a keeper or handler to enter such an enclosure, grab a selected individual, and carry it outside without alarming the group. Obviously this may be dangerous if the group includes horned individuals.

If animals in a darkened stall or enclosure can be shifted into a darkened funnel leading into a chute arrangement with doors that can be closed behind each individual, animals of hoofed species can be handled more easily and with little stress.

Commercial cattle chutes can be used with some wild species (Fig. 23.61). Zebu cattle or other cattle-type wild animals can be directed through alleyways into one of the squeeze chutes. Commercial chutes frequently must be modified to prevent wild animals from climbing up the inside. The chute can be lined with plywood sheeting.

Use only those chutes that permit the animal to move its head down if it falls or throws itself in the chute. Chutes must open from the front to allow aid to be given to an animal that falls. The upright posts of the front gate can be padded to minimize trauma to the shoulders of the animal if it lunges against the semiopened gate to escape.

Many wild animal facilities now have built-in chutes or semisqueeze cages to assist in the management of bovids. Sometimes animals will refuse to move through the chute arrangement. Rather than beating or slapping the animals, an electric shock prod is an effective means of hurrying the animal along the alleyway. Electric shock prods may persuade stubborn animals to move, but be cautious

FIG. 23.61. Zebu restrained in commercial cattle chute.

in their use; excessive stimulation may unnecessarily agitate an animal or trigger aggressive behavior.

Most people who have dealt with a large variety of hoofed animals have used rope as a tool of restraint. Although proficiency with a rope can be an invaluable asset, it must be recognized that rope is of limited value for capturing most wild ungulates. The commotion generated by swinging or tossing the rope is much more threatening than the typical wild ungulate can tolerate. It will usually bolt and run headlong and may injure itself.

Several factors should be kept in mind when using ropes on wild ungulates: (1) Consider the general temperament of the animal. The extremely nervous, flighty gazelle and antelope species are likely to be injured during such a capture operation. Wild sheep, on the other hand, can frequently be caught with a rope if necessary. (2) Decide how to control the animal immediately after capture. Many a roper has gloated over a successful catch, only to find the captured animal to be uncontrollable. Instead of the typical pullback response exhibited by a domestic cow, sheep, or goat, many wild animals attack the roper or lunge wildly in every direction, making it impossible to grasp the animal. This is particularly dangerous with animals that have heavy or sharp horns. If roping is used to capture these animals, a dally post is needed to restrict the animal's subsequent activity; otherwise, assistants attempting to grasp the animal can be injured. Large species such as bison, yak, or water buffalo are so strong that even two or three people may be unable to slow or stop them sufficiently to take a dally around a post or a tree. More phlegmatic bovid species such as the yak can usually be roped by the horns, tied or dallied to a post, and pulled into lateral recumbency by another rope tied to the hind legs. (3) Consider whether or not the rope can be quickly released after the manipulation has been completed. Ropes equipped with quick-release hondas can be used, but they lack balance for throwing and are not likely to be the rope of choice for such operations. It may be necessary to tie a small rope to the honda of the lariat so the lariat can be easily pulled free when the animal is released. Keep in mind that the animal may be highly agitated and apt to attack the handlers as it is released. Be sure handlers are protected when working with ropes on wild animals. Keep a sharp knife readily available at all times to cut the rope. Unforeseen circumstances frequently develop that necessitate quick release of an animal.

Any skilled calf roper may be humiliated when attempting to rope antelope. The speed and dodging ability of antelope is phenomenal. Tossing a loop is not a usable approach except in a massed group of animals. This technique is simply not fast enough to ensnare an animal that can jump and dodge as rapidly as an antelope. A swing toss must be used. One must use a small loop and give ample lead to the animal; otherwise it will jump completely through or past the loop.

Any roping operation carried out in the midst of a group of animals is fraught with danger to both roper and animal. The normal flight response sends animals jumping and scurrying here and there throughout the pen or enclosure. The heights to which frightened animals can jump is astounding. I have seen mouflon sheep standing 62.5 cm high at the shoulder jump over the head of an adult man. The behavior of excited and/or cornered animals is totally unpredictable. If sufficient numbers of animals are panicked, they may run over or jump on the roper.

On one occasion of roping mouflon sheep, one ewe attempted to jump over an assistant. The ewe did not quite make it and placed all four feet on the person's forehead to ricochet. Needless to say, this was rather traumatic to the individual involved.

Hoofed animals that have been in captivity for a long time have not been able to maintain their wild athletic condition. When they become excited and run around the pen as a roper attempts to capture one of them, great harm can be done. The animals may overheat or become exhausted. They may injure themselves or others by jumping on top of one another, especially when there are young members in the herd. The young are particularly subject to trampling and exhaustion. I have seen animals drop from exhaustion during such maneuvers and die from pulmonary edema brought on by severe exertion. Such chases may also be the inciting cause for capture myopathy (see Chapter 7).

When shipping horned animals, protect the horns from trauma. The heavy horns of animals like buffalo or eland may shred the inside of a crate. Pieces of hose applied to the tip of the horn can prevent this (Fig. 23.62). Tape the hose onto the horns, bending the tip of the hose over and taping it to protect the tip of the horn and the inside of the crate. Hose applied to an overly aggressive animal will also protect others from injury.

Rubber balls were placed on the tips of the horns of an exceptionally aggressive saiga antelope male to prevent him from traumatizing cagemates (Fig. 23.63).

FIG. 23.62. Animal ready for crating. Horns are protected by placing rubber hosing over the tips.

FIG. 23.63. Rubber balls placed on horns prevent aggressive saiga antelope from goring penmates.

It may be necessary to dehorn or tip the horns of extremely and continuously aggressive animals. The horns of most species can be tipped without causing significant hemorrhage if no more than 1 or 2 in. of the sharp point is removed. Removing more than this will cut the bony core, causing hemorrhage. Cauterizing will stop hemorrhage. Complete dehorning is a surgical procedure that requires anesthesia and hemorrhage control. Dehorning an adult animal will open the cornual sinus, requiring special care to prevent infection and/or fly infestation.

REFERENCES

1. Alford, B. T.; Burkhart, R. L.; and Johnson, W. P. 1974. Etorphine and diprenorphine as immobilizing and reversing agents in captive and free-ranging mammals. J. Am. Vet. Med. Assoc. 164:702-5.
2. Bauditz, R. 1972. Sedation, immobilization and anesthesia with Rompun in captive and free-living wild animals. Vet. Med. Rev. 3:204-26.
3. Beck, C. C. 1972. Chemical restraint of exotic species. J. Zoo Med. 3:3-66.
4. Bush, M.; Ensley, P. K.; Mehren, K.; and Rapley, W. 1976. Giraffe immobilization utilizing xylazine and etorphine hydrochloride. J. Am. Vet. Med. Assoc. 169:884-85.
5. Fowler, M. E. 1974. Veterinary aspects of restraint and transport of wild animals. Int. Zoo Yearb. 14:28-33.
6. Graham-Jones, O. 1974. Some aspects of air transport of animals. Int. Zoo Yearb. 14:34-37.
7. Harthoorn, A. M. 1976. The Chemical Capture of Animals. London: Baillière, Tindall.
8. IATA Live Animal Manual. 2nd ed. 1970. Montreal, Canada: International Air Transport Association.
9. Jones, D. M. 1971. Sedation of a bull musk ox. Int. Zoo Yearb. 11:242-44.
10. Karsten, P. 1974. Safety Manual for Zoo Keepers. Alberta, Canada: Calgary Zoo. (Animal restraint)
11. Klös, H. G., and Lang, E. M. 1976. Zootierkrankheiten. Berlin: Paul Parey.
12. Liscinsky, S. A.; Howard, G. P.; and Waldeisen, R. B. 1969. A new device for injecting powdered drugs. J. Wildl. Manage. 33:1037-38.
13. Oelofse, J. 1970. Plastic for game catching. Oryx 10:306-8.
14. Roughton, R. D. 1975. Xylazine as an immobilizing agent for captive white-tailed deer. J. Am. Vet. Med. Assoc. 167:574.
15. York, W.; Kidder, C.; and Durr, C. 1973. Chemical restraint and castration of an adult giraffe. J. Zoo Anim. Med. 4:17-21.
16. Young, D. E., ed. 1973. The Capture and Care of Wild Animals. Capetown, South Africa: Human and Rousseau.

24 BIRDS

CLASSIFICATION

Flightless Birds
- Order Sphenisciformes: penguin

Superorder Ratitae
- Order Struthioniformes: ostrich
- Order Casuariiformes: cassowary
- Order Apterygiformes: kiwi
- Order Rheiformes: rhea

Water Birds
- Order Anseriformes: duck, goose
- Order Gaviiformes: loon
- Order Podicipediformes: grebe
- Order Pelecaniformes: pelican, cormorant

Shore and Gull-like Birds
- Order Charadriiformes: plover, gull
- Order Procellariiformes: albatross, petrel

Raptors
- Order Falconiformes: hawk, eagle
- Order Strigiformes: owl

Galliformlike Birds
- Order Galliformes: pheasant, grouse
- Order Tinamiformes: tinamou

Long-billed, Long-legged Birds
- Order Ciconiiformes: heron, stork, flamingo
- Order Gruiformes: crane, rail

Large-billed Birds
- Order Coraciiformes: kingfisher, roller, hornbill
- Order Piciformes: woodpecker, toucan
- Order Cuculiformes: cuckoo

Pigeons and Doves
- Order Columbiformes

Psittacine Birds
- Order Psittaciformes: parrot

Hummingbirds and Swifts
- Order Apodiformes

Song, Perching, and Miscellaneous Birds
- Order Caprimulgiformes: frog mouth
- Order Coliiformes: coly
- Order Trogoniformes: trogon
- Order Passeriformes: finch, warbler, crow

GENERALLY the male is called simply a male or a cock, the female is a hen. Newly hatched birds are called chicks. The male falcon is called a tiercel, the female is the falcon. Newly hatched pigeons are squabs. Waterfowl names are the same as those designating domestic species.

Approximately 8,600 species of birds, grouped in twenty-seven orders, are distributed throughout the world. Birds vary in size from a 2 g hummingbird to an ostrich weighing 136 kg (300 lb). Aviculture is a popular hobby and/or serious pursuit of many persons. Husbandry practices necessary to maintain birds in an aviary are well known for some species. Most species of birds have been maintained in either public or private aviaries at one time or another.

It would be impossible to discuss every species or even every family of birds in a book of this type. Fortunately groups of birds with like anatomical or behavioral traits respond similarly to certain types of restraint practices.

ANATOMY AND PHYSIOLOGY

Respiratory function in birds is unique. The lungs are assisted in respiration by a series of air sacs. Birds lack a complete diaphragm, and the lungs are intimately associated with the chest wall. Thus inspiration and expiration are dependent on a bellows system to pull air into the lungs and push it back out. On inspiration the sternum moves downward and forward, expanding the chest and abdominal cavities and drawing air into the respiratory tract (Fig. 24.1A). On expiration the sternum moves backward and upward, compressing the structures and forcing air out through the trachea (Fig. 24.1B). Any manipulation that prevents the proper excursion of the sternum will interfere with respiration. Hands and fingers must not completely encircle the thoracic cavity and immobilize the sternum (Fig. 24.2), and the sternum must be able to move freely when a bird is placed into a plastic tube or encircled with a stockinette.

Cartilaginous rings completely encircle the trachea of birds, minimizing the danger of tracheal collapse during manipulation. However, the extreme mobility of the neck and trachea makes it possible to kink the trachea and inhibit the free flow of air.

The location of the nostrils varies from species to species. Some species are capable of breathing through the mouth, others are not. Before instituting any restraint practice that would cover or interfere with the nostrils, ascertain the specific type of breathing required by that species. Also examine the nares to make certain they are not plugged by feed or exudate.

Feathers cover the entire surface of the body of a bird, forming an efficient insulating layer. The presence of this insulating layer may be detrimental to a bird under restraint. Muscular activity increases heat production, which, coupled with inability to dissipate heat from the body surface, may produce hyperthermia. Birds normally maintain body temperatures up to 41-42 C, higher than those of mammals. Cellular necrosis of all animal tissue begins at body temperatures of 45 C, leaving a bird little margin for increase from normal.

The bones of the wings and legs of birds are constructed for lightness. Some bones are pneumatized, being connected to the air sac system of the respiratory tract; others are hollow. The cortex of the humerus and femur are thin. These modifications assist flight but increase fragility. Fractures are easily induced by rough manipulation.

Both the upper and lower jaws of psittacine birds have movable articulation with the skull.

Birds possess four structures that are used in defense and offense: the beak, the wings, the feet, and the legs. Beaks

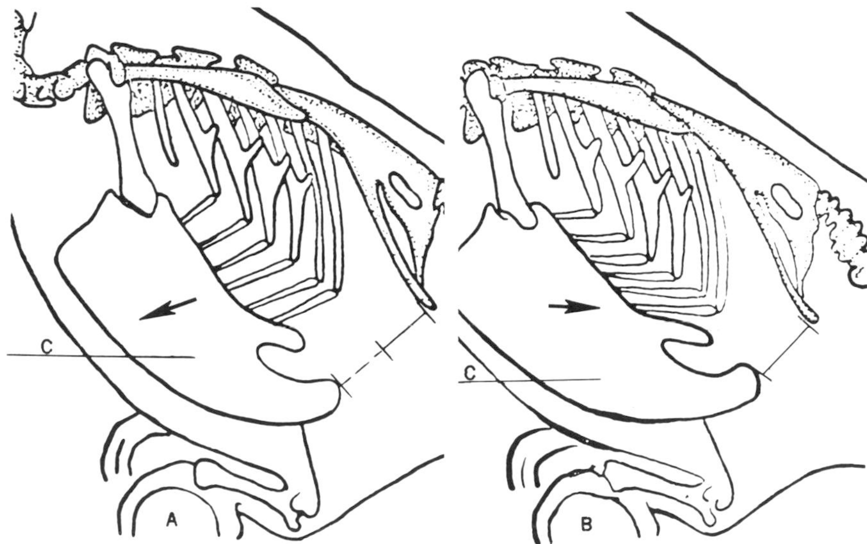

FIG. 24.1. Diagram showing movement of sternum and ribs during respiration: **A.** Inspiration. **B.** Expiration. **C.** Sternum (keel).

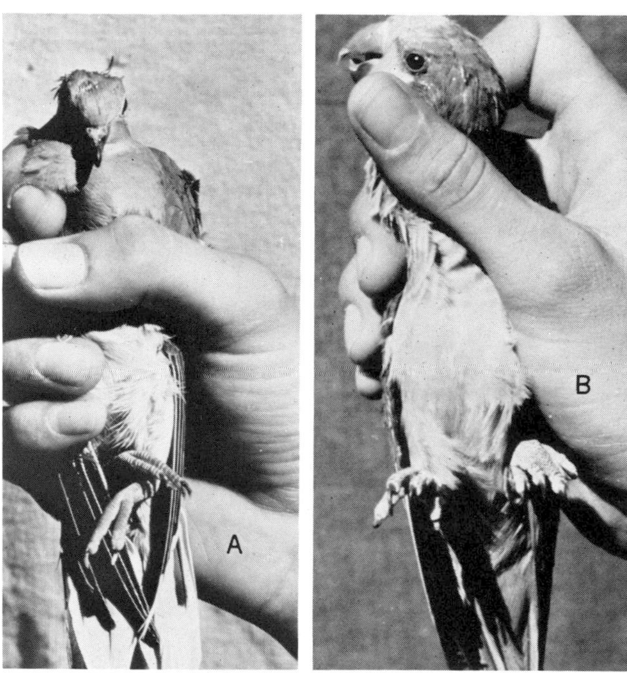

FIG. 24.2. Poor restraint: **A.** Hand completely encircling the body may inhibit respiration. **B.** Twisting the head may restrict air passages.

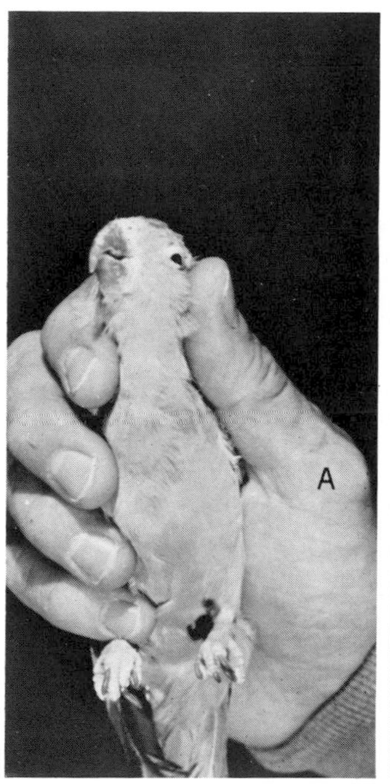

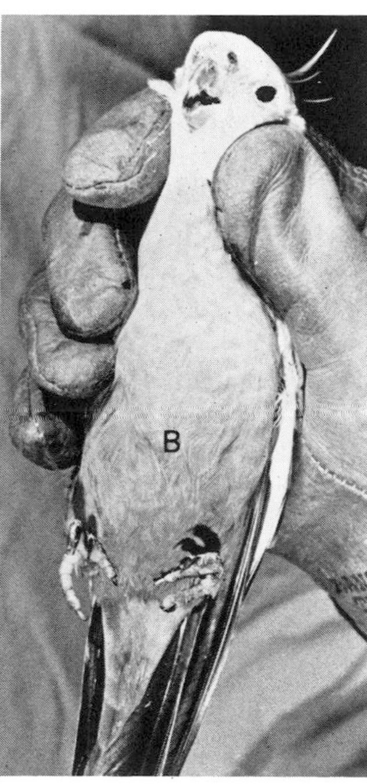

FIG. 24.3. Proper hold for small psittacine birds: **A.** Budgerigar. **B.** Handling a cockatiel may require gloves as protection from a stronger beak.

of birds are highly varied, adapted for the food-gathering habits of individual species. Feet also are adapted for various food-gathering or defensive habits. Only large birds defend themselves with wings or legs.

PHYSICAL RESTRAINT

Some general procedures are applicable for the handling of all types of birds. The head must be controlled. This is accomplished by grasping the bird at the nape. The thumb and forefinger are used to fix the head more firmly if necessary (Fig. 24.3). It is seldom necessary to completely encircle the neck in order to hold the head still. Restraint of most birds is accomplished by approaching the bird from behind, grasping the head and the body, and holding the legs together.

A net is the common tool used in capturing and restraining birds [17]. Many birds are extraordinarily fragile, and a careless handler can easily fracture wings and legs if he slams the hoop into the bird or removes the bird from the net in a rough manner. All birds should be handled gently and carefully. Once the bird is captured, the basic procedure is to grasp the bird by the head at the nape of the neck or by the beak, depending on the species, and carefully remove the net. The legs of raptors must be grasped first, then the head, as described.

The safe release of birds following restraint requires care and attention. To release a bird, turn it right side up and allow it to sit on the floor. Make certain that none of the talons or claws are enmeshed in clothing or gloves and that the bill is free; then give the bird a slight push to release it. Be especially wary of the talons of raptors.

Do not release a bird in midair. The bird has been disoriented during the period of restraint and may have been kept in an abnormal position for a long time. It may need a moment or two to regain its balance and composure.

TRANSPORT

Many specialized crates and cages are used for shipping and moving birds. Hummingbirds can be placed in individual stockings or fabric jackets [22] in front of a source of liquid diet. Antarctic penguins must be kept cool. Heavy-gauge wire must be used for caging large macaws to preclude their chewing their way out.

Unique requirements must be met when transporting each group of birds. Handlers with no experience in transporting birds should contact a private or public aviary or zoo for information.

CHEMICAL RESTRAINT

Ketamine hydrochloride (25-50 mg/kg) is the standard immobilizing and anesthetic agent for use in birds [1,2,4, 6,14,20]. One milligram will sedate a 50 g budgerigar for conducting radiographic examinations and minor surgery. Xylazine hydrochloride has been used effectively and safely in quail, budgerigars, herons, curlews, chickens, and turkeys. Dosages up to 200 mg/kg were tolerated well, but the duration of sedation was prolonged (5-15 hours) [19]. A recommended sedative, anesthetic dose is 10-30 mg/kg. The agents are administered intramuscularly in the breast and thigh (see Fig. 24.60).

Both methoxyflurane and halothane are used extensively to anesthetize birds for prolonged procedures. Other anesthetic agents have also been used successfully [1,3,5, 7,12,13,21].

Various agents have been incorporated into baits for the capture of free-living birds. Alpha chloralose and tribromoethanol (Avertin) have been effective for capturing seed-eating birds. Nontreated grain is fed for a few days, then the bait is provided. The dose of alpha chloralose is 0.5 mg/250 g (1 cup) of grain [25]. Higher concentrations of the drug may cause mortality. Allow 1-2 hours for development of the maximum effect.

Tribromoethanol is added to the grain at the rate of 3 g/250 g. A combination of alpha chloralose (0.5 g) and tribromoethanol (0.5 g/250 g of grain) is also used.

Recovery from anesthesia or chemical immobilization can be made safer by putting the bird in a cotton stockinette. The stockinette is flexible, heat retention is minimal, and the bird cannot thrash and catch wings or legs in the cage as it recovers from anesthesia.

FLIGHTLESS BIRDS [5,9,16]

Danger Potential

Penguin wings are modifed to form flippers used for swimming. The wings must be adequately controlled; otherwise the penguin may beat a handler and inflict bruises or severe injuries, especially if a wing is flipped into the face. Penguins also have sharp beaks that can seriously tear the flesh of a manipulator.

Other flightless birds are referred to as ratites because of their raftlike sternum. The primary defensive structures of the large ratites are the feet and legs. Large ostriches (Fig. 24.4) and cassowarys (Fig. 24.5) are especially dangerous. Never stand in front of one of these birds. An ostrich or a cassowary can disembowel and kill a man with a quick forward thrust of its clawed feet. The power of the blow itself can fracture an arm or leg. Even the smaller rhea or emu can inflict a severe contusion with a kick.

FIG. 24.4. African ostrich.

Do not enter an ostrich or cassowary enclosure without some means of protection; these birds are highly aggressive and could attack, running down and knocking over a handler. As a rule, emus and rheas are not likely to be aggressive, and it is usually safe to enter an enclosure with them.

Physical Restraint

Approach a small penguin from behind and grasp it at the base of both wings. The bird may turn its head and peck at the fingers, so heavy gloves must be used. Large species such as the emperor or king penguin should be hooded; otherwise the hands can be injured even though gloved. Restraint can be obtained by grasping the feet. Some penguins will beat their flippers furiously and may

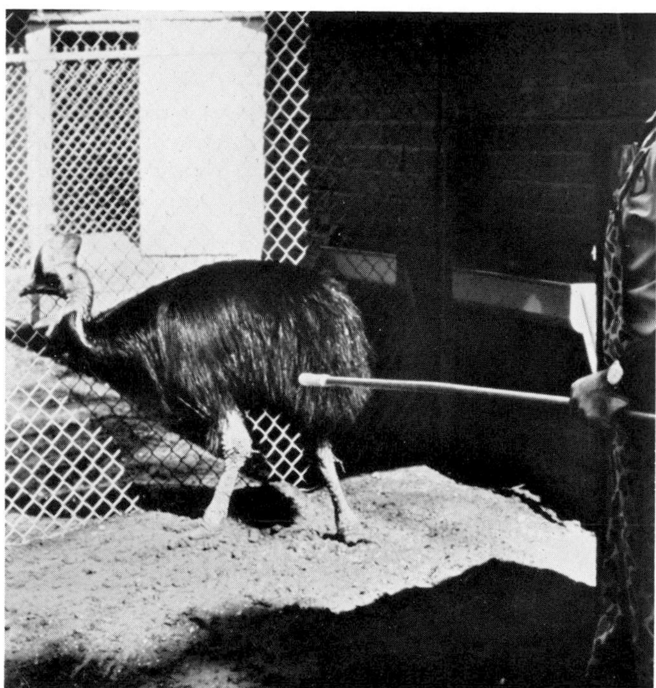

FIG. 24.5. Cassowary being injected with chemical restraint agent via stick syringe. Handler is standing behind plywood shield.

bruise the handler. Another person is needed to restrict movement of the flippers. An excellent method for capturing and handling large penguins is to use an appropriately sized plastic garbage can. Remove the bottom, slowly approach the bird, and place the inverted container over it. The bird's movements are restricted without undue trauma, yet many manipulations can be carried out through the open bottom.

Both large and small species can be captured initially by netting them (Fig. 24.6). A cargo net thrown over the bird is useful for handling ratites.

FIG. 24.7. Obtaining blood sample from penguin by venipuncture of the brachial vein.

Blood samples can be obtained from a penguin by venipuncture of the brachial vein (Fig. 24.7) or by clipping a toenail (Figs. 24.8, 24.9).

A plywood shield generally offers suitable protection from ostrich or cassowary, since it can be moved forward to close proximity with the bird (see Fig. 24.5). Remember though that these birds can jump considerable heights and may jump over a too-short shield. To deal with an extremely aggressive cassowary, attach the shield to the front of a vehicle; a swift kick from one of these birds can knock a person down even though shielded.

A long pole with a branched or forked structure such as bicycle handlebars attached to the end can be used to keep an ostrich or cassowary away from the handler (Fig. 24.10).

Ratites are much easier to handle if they can be hooded first (Figs. 24.11-24.13). Once the hood is placed, the bird can be grasped by the wing stubs from the side or behind (Fig. 24.14). Other means of holding small to medium-sized ratites are illustrated in Figures 24.15-24.17.

FIG. 24.6. **A.** While holding head through the net, regrasp head under the net. **B.** Bird is then freed of the net.

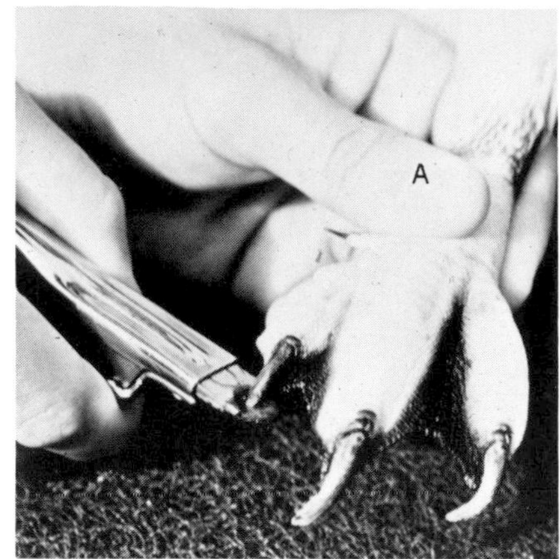

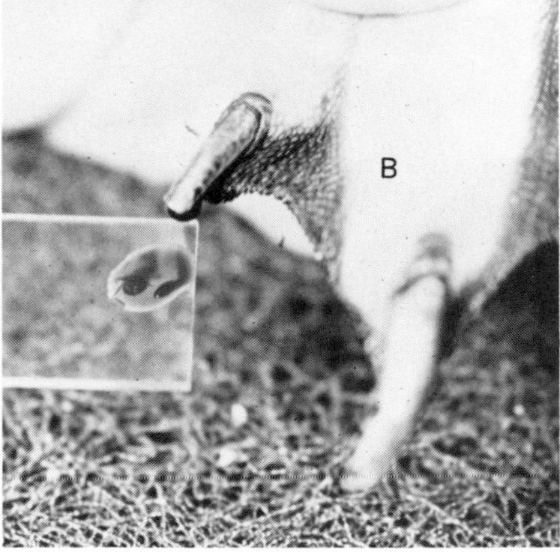

FIG. 24.8. **A.** Blood sample can be obtained from many birds by clipping a toenail. **B.** Blood smear can be made directly.

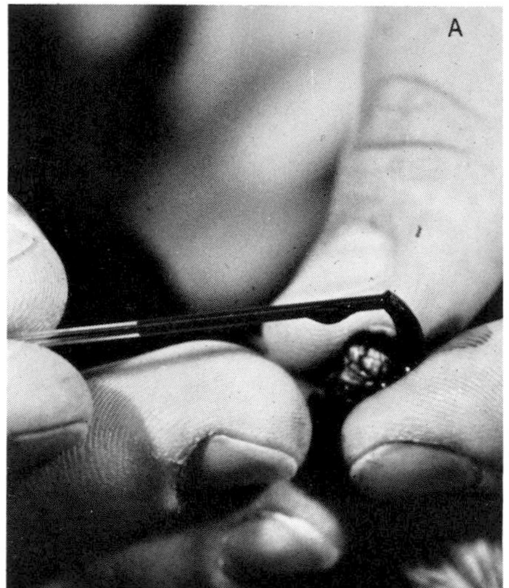

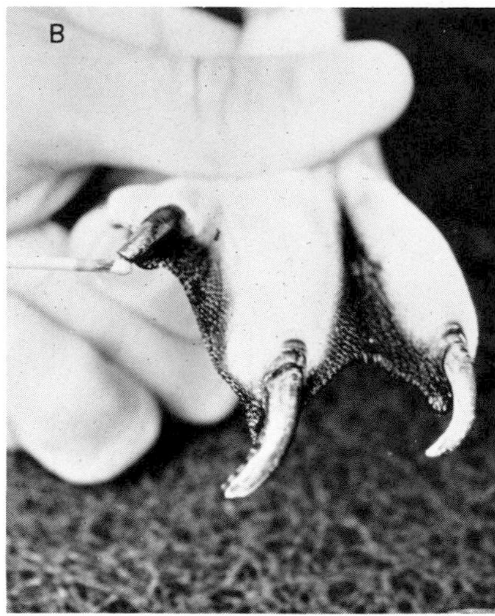

FIG. 24.9. **A.** Small blood sample can be collected in a capillary tube. **B.** Cauterize exposed nail with silver nitrate, as illustrated, or sear wound with hot spatula.

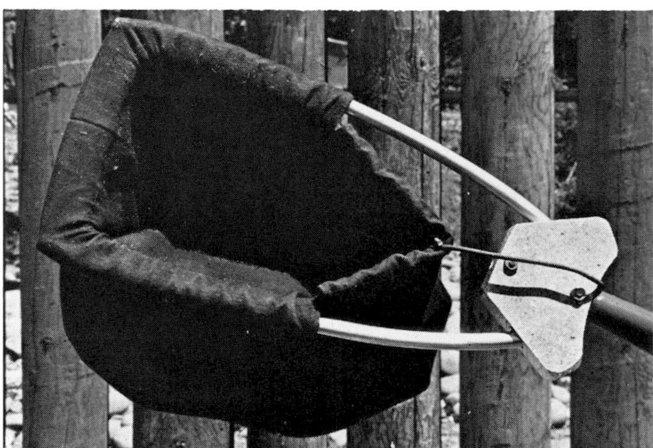

FIG. 24.11. Apparatus for applying hood to a ratite.

FIG. 24.10. Bicycle handlebars adapted as a keep-away.

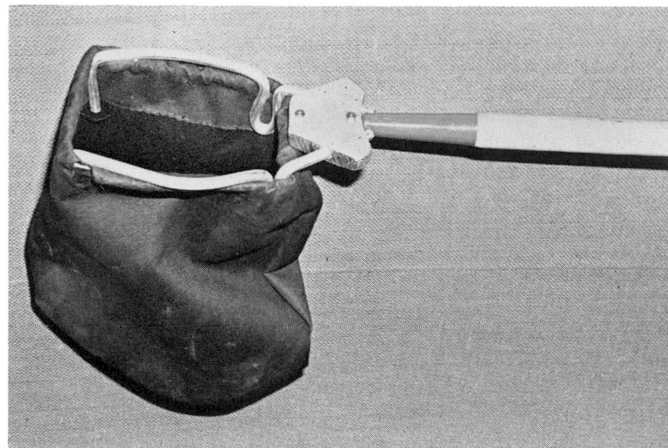

FIG. 24.12. Alternate method of applying hood to a ratite.

FIG. 24.13. Placing hood on an ostrich.

FIG. 24.14. Grasping hooded ostrich by wing stubs.

A sling is useful when treating fractures or supporting debilitated animals (Fig. 24.18). Do not suspend the bird clear of the floor; pressure on the legs is necessary for homeostasis (see Chapter 5).

The kiwi is not difficult to handle. The beak is not sharp, being especially adapted for probing in loose humus for invertebrates, and is not used aggesively against handlers. The bird can easily be picked up and held (Fig. 24.19.

Chemical agents can be administered by blowgun, Cap-Chur projector, or with a stick syringe if the handler can approach the animal behind a shield to jab the syringe into the leg muscle (see Fig. 24.5) [3,5].

Transport small flightless birds in burlap sacks with the head poking through a hole in one corner (Fig. 24.20).

FIG. 24.15. Methods of restraining small to medium-sized ratites.

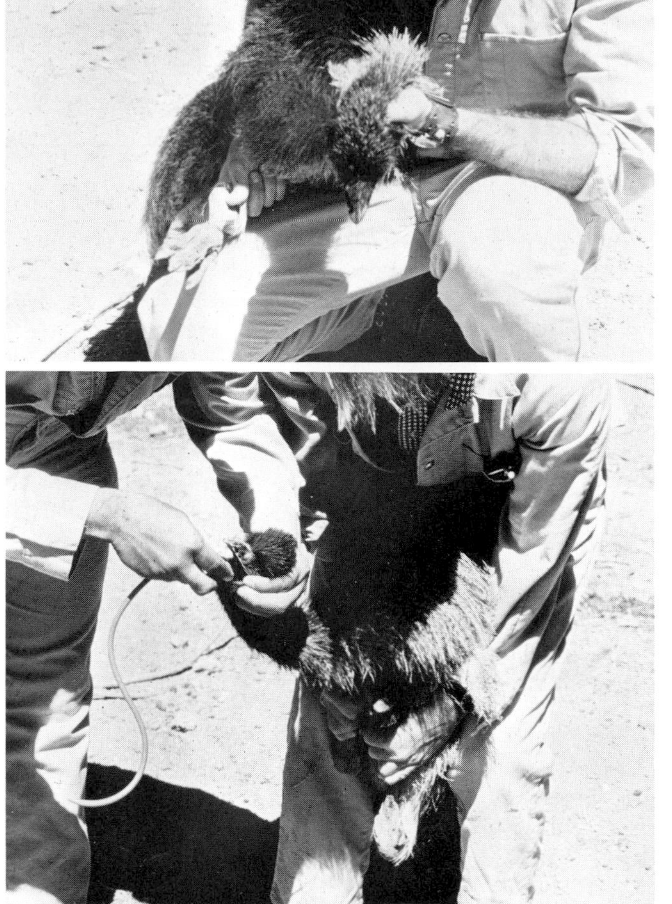

FIG. 24.16. Methods of holding and medicating small ratites.

FIG. 24.18. Improvised sling for an emu.

WATER BIRDS

Danger Potential

Ducks, geese, and swans are not innately aggressive. However, large angry geese and swans may attack and inflict significant injuries.

Most waterfowl have dull bills. Although the bills may not be particularly sharp, many are heavy and strong—able to pinch the handler severely if they peck. A few fish-eating species have a sharp hook on the tip and serrations along the margins of the beak to facilitate grasping slip-

FIG. 24.17. **A.** Emu can be restrained on its back, but its legs are very strong. **B.** Staged picture to show dangerous method of handling an emu. Its legs could reach up and rake handler's face.

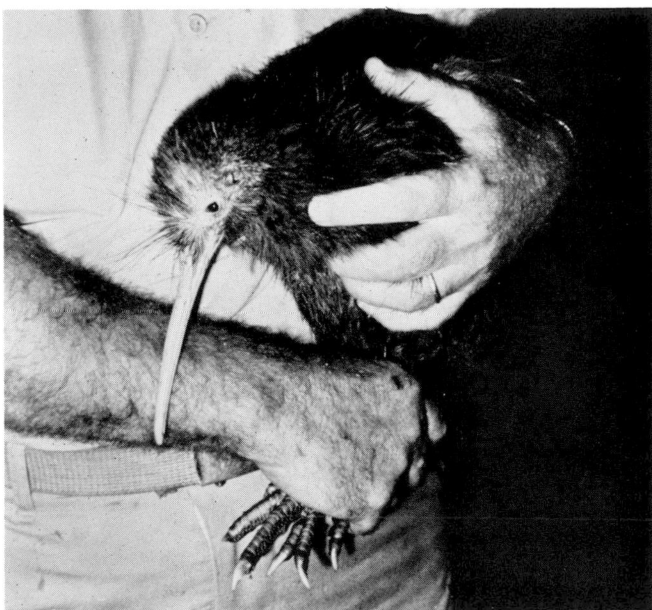

FIG. 24.19. The kiwi, a small ratite from New Zealand, is not aggressive and is easily handled.

FIG. 24.20. Small emu can be transported short distances in a burlap sack.

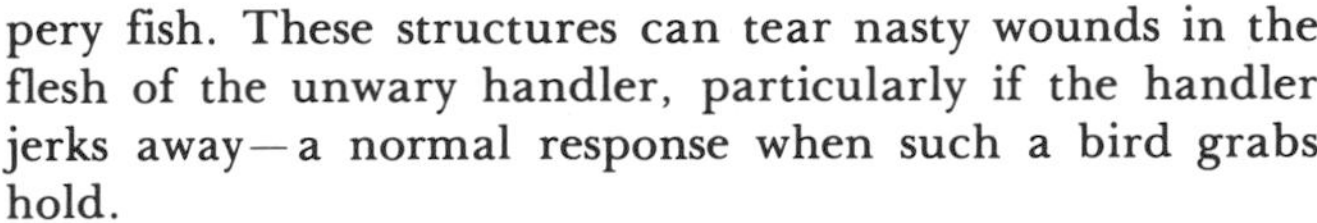

pery fish. These structures can tear nasty wounds in the flesh of the unwary handler, particularly if the handler jerks away—a normal response when such a bird grabs hold.

Large geese and swans can beat the handler vigorously with their wings. They threaten by lowering the head, extending the wings, and rushing toward the enemy, hissing loudly. If the attack is pressed, the person may be pecked and beaten.

Both spur-winged geese and screamers have sharp spurs at the carpus of the wing. These may be deliberately and effectively used against an animal handler. When cornered the screamer may fly at the keeper, flailing its wings.

Restrained waterfowl may flail with the feet, but few have long claws. Some tree-nesting species have claws and will scratch a handler if proper precautions are not taken.

Pelicans are relatively harmless, though a handler should be wary of the sharp hook on the tip of the large beak, since these birds will peck at the eyes. A cornered pelican may fly at a handler, snapping its beak. Grab the beak if threatened. In close quarters the pelican will also beat with its wings.

Grebes and loons have long sharp beaks used for impaling fish. These birds will peck at the face, especially the eyes, of handlers.

Physical Restraint

All waterfowl can be initially captured with nets. The size of the cordage, the mesh, and the hoop should vary according to the size of the bird. Once captured, most species can be extracted from the net and held by the wings and the head in the same manner as domestic waterfowl (see Fig. 15.30).

The hook can also be used on captive wild waterfowl, as illustrated in Figure 15.27. A cannon net is used to capture wild ducks and geese. The net is placed in an area where birds can be baited with food. When the birds congregate and are feeding, the net is projected over the top of the birds, incarcerating them.

Swans and large geese can be caught initially by hand. The birds are cornered, and the keeper grasps them quick-

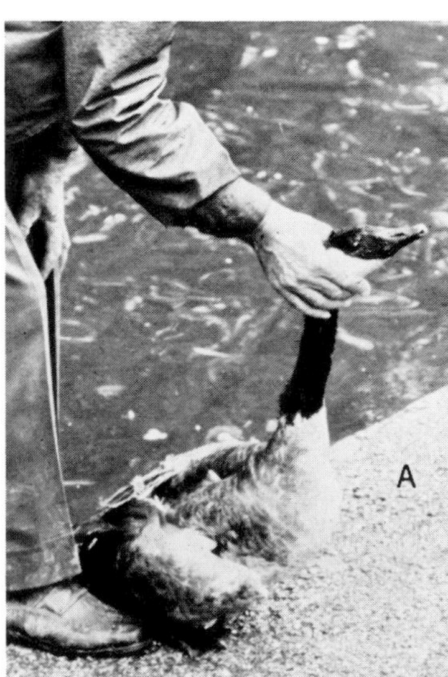

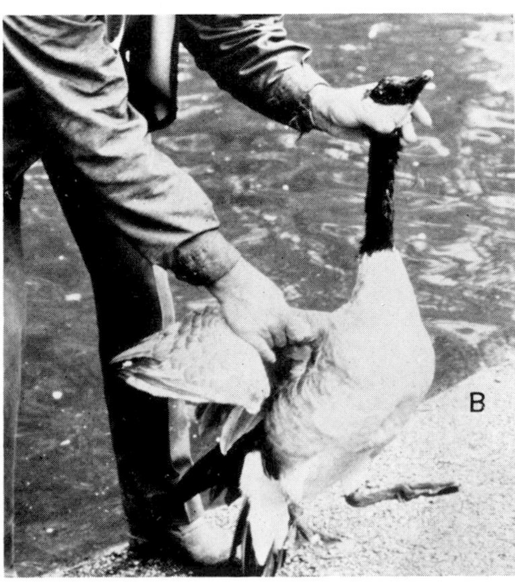

FIG. 24.21. Capturing a goose: **A.** First grasp the neck. **B.** Immobilize the wings.

ly by the neck and then at the base of the wings. The approach should be from the rear (Fig. 24.21). A special restraint jacket can also be used [10].

When attempting to capture a screamer or a spur-winged goose, approach from the rear, using either a net or a broom to ward off the spurred wings. The spurs of birds in an aviary can be dulled or clipped to prevent damage to other birds in the collection or to the keepers. Once captured, these birds are manipulated in the same manner as other waterfowl (see Fig. 15.33).

Control the beak at all times when handling grebes and loons or place a cork or other blunt object over the sharp tip of the beak.

SHORE AND GULL-LIKE BIRDS

The numerous species of small to medium-sized shore birds with long legs and short to long beaks share many characteristics important for restraint. Most of these birds have tiny fragile legs that are easily injured by the hoops of nets or by too severe pressure applied during capturing or handling.

Danger Potential

Some shore birds have long sharp bills, but most are not highly aggressive. Nevertheless, care should be taken in protecting one's face from the beak of any bird.

Gulls and terns are equipped with more formidable beaks than other shore birds. Protect your face when handling these birds, because they often peck at the eyes. Gulls and terns will also peck at the handler's fingers and arms if not prevented.

Physical Restraint

Shore birds are not difficult to handle. All can be captured easily by careful placement of a hoop net. Gently extricate them from the net, holding small species by the body with fingers at the back of the head. Hold larger birds by the body and legs. Place a shore bird in a stockinette or a section of lady's hose for prolonged restraint.

RAPTORS [4,11]

Danger Potential

All raptors are carnivorous. The diet of smaller species may consist primarily of insects, but all have beaks and talons adapted for grasping and rending flesh. The sharp tearing beaks are capable of wounding severely. Eagles or large owls can sever a finger from the hand. The claws or talons of raptors are long, strong, and sharp—well adapted for grasping and penetrating prey species.

Vultures employ the beak as a defensive weapon, seldom using talons. Other raptorial species rarely defend themselves with the beak, relying on the talons for protection, but hawks and owls will peck when being restrained.

A handler who is grasped by the powerful talons of a large eagle or owl may find it impossible to get free without assistance, unless the bird chooses to let go.

FIG. 24.22. Hooded falcon.

Physical Restraint

It is advisable to wear heavy leather gloves when working with birds of prey even though gloves do not afford full protection. The beak and talons of raptors can cut through leather without much difficulty.

The tools of the falconer are useful for restraining all raptorial species. The hood is an important tool for decreasing stress and facilitating restraint (Fig. 24.22). If leather hoods are not available, a cloth hood is simple to construct. The falconer places leather bands called jesses around the metatarsi. They are customarily attached to a swivel and thence to a tether, which secures the bird to a perch or a block. It may take some time for the bird to become accustomed to being restrained in this manner. A few birds refuse to accept this type of restraint and continually fly at the falconer or away from the perch to the limit of the tether. This is called "bating." The bird can injure itself if bating behavior is persistent, but usually after one or two attempts it will not fly from the perch unless frightened.

Vultures are easily captured with a net. The head must be controlled at all times. The hand holding the legs should also keep the wings folded against the bird's body to obviate flapping. If the wings are free to flap, they may be injured and also would certainly interfere with any procedure. Metomidate (3-4 mg/kg) has been reported to be an effective sedative agent [8,15].

Tamed captive falcons and hawks are easily handled. To pick up a tame raptor from a perch, first hood the bird. Then approach it from behind and place both hands over the back and wings. Control the feet by clasping the fingers over the legs. For more security separate the little finger

FIG. 24.23. Capturing owl by throwing a towel over it.

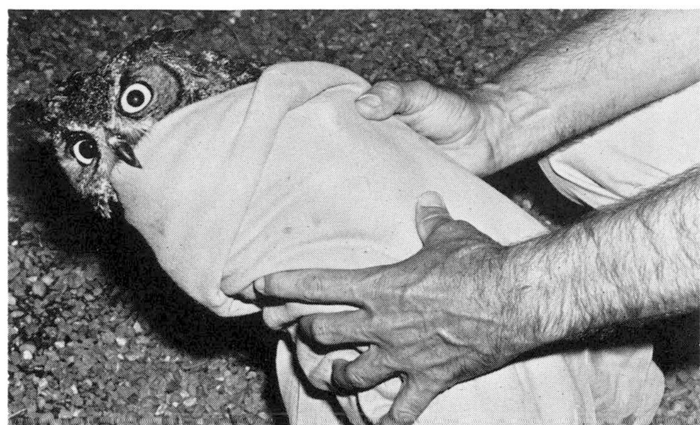

FIG. 24.24. Restraining owl by wrapping it in a towel.

from the ring finger, placing them on either side of the legs.

Raptors can be captured by throwing a towel, laboratory coat, or other cloth over them (Fig. 24.23) and wrapping them in the fabric (Fig. 24.24). This technique is most successful if the bird is on the floor of the enclosure when captured.

Once the bird is in hand, control it by grasping the feet and the head in various manners dictated by what is to be done (Figs. 24.25-24.27). Although gloves are desirable for initial capture, they should be removed when holding the bird to better sense the degree of pressure being applied.

To grasp a bird presented in a cardboard box, place both hands on the back over the wings and legs. Press the bird to the floor of the box and direct the fingers of both hands around and underneath the body to grasp the legs and control the feet before lifting the bird.

Owls often throw themselves on their backs and direct a flailing set of formidable talons toward anyone attempting to pick them up (Fig. 24.28). A great horned owl can drive a talon completely through the heaviest leather glove, so

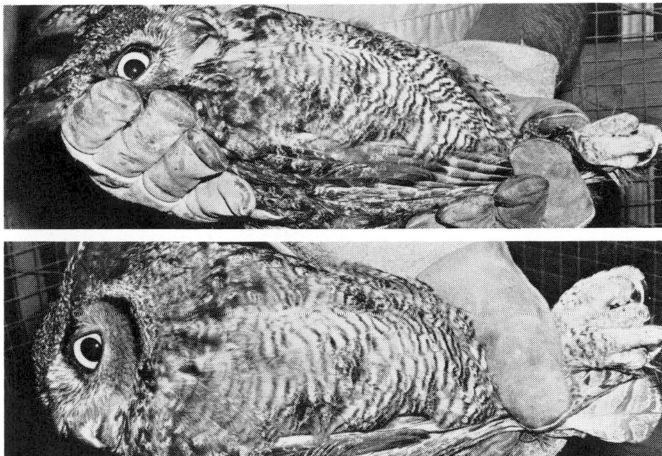

FIG. 24.25. Alternate methods of holding an owl following initial capture.

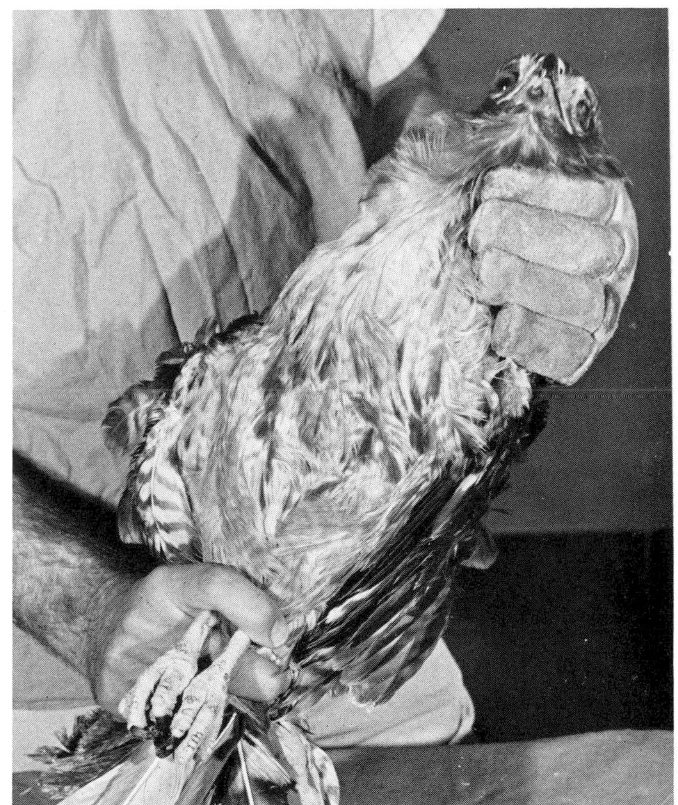

FIG. 24.26. Properly secured hawk.

approach this bird with caution. Dangle an empty glove above the bird and, while the talons are attached to the glove, grab the legs with the other hand (Fig. 24.29). Young owls can sometimes be captured from this prone position by presenting a towel or small piece of cloth for the talons to grasp. Lift the bird into the air upside down by the towel, reaching beneath it to clutch the legs with the other hand.

If a raptor is perched, the approach can be from either

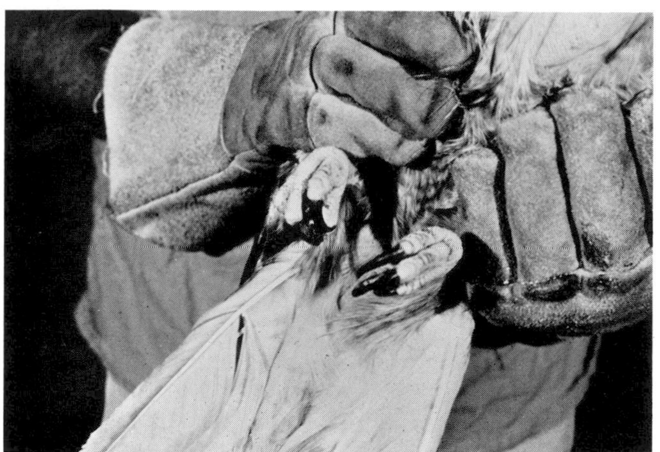

FIG. 24.27. Alternate method of securing feet and wings of a hawk.

FIG. 24.28. Owl flipped on its back to defend itself with its talons.

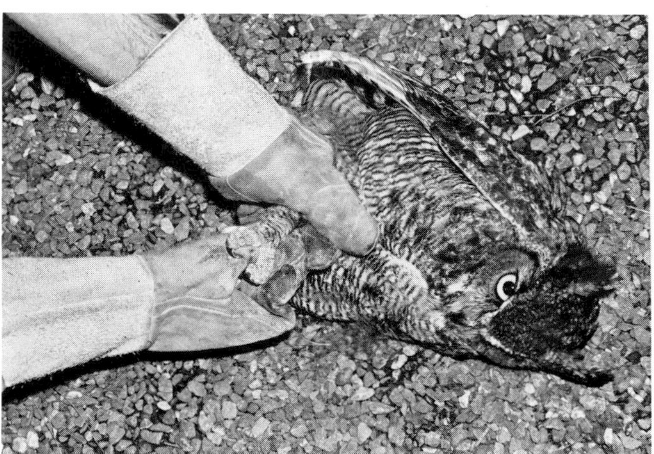

FIG. 24.29. Owl's legs can be grasped while it is intent on clawing at another object such as a dangling glove or towel.

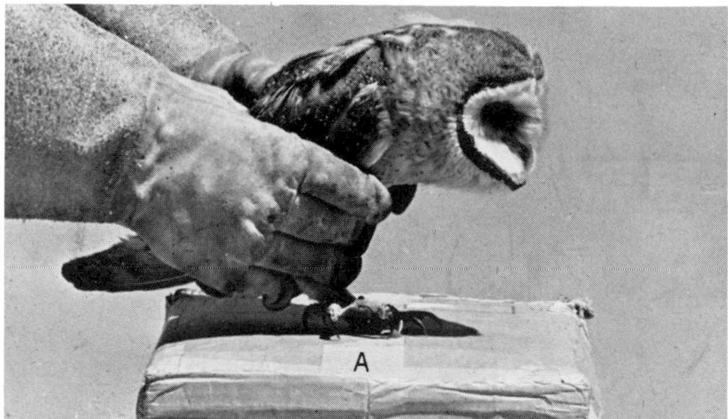

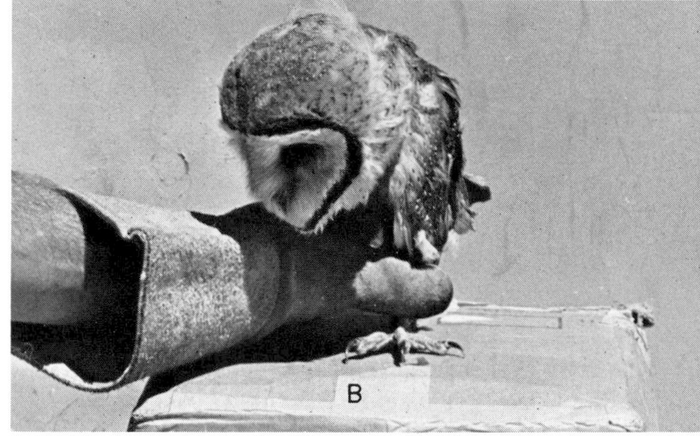

FIG. 24.30. Approaching a perched raptor: **A.** From rear. **B.** From front.

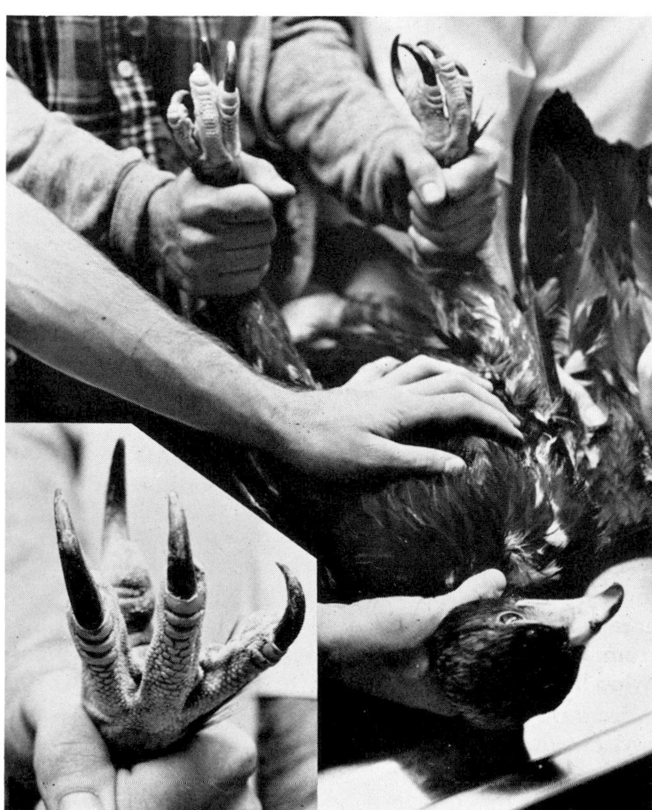

FIG. 24.31. Golden eagle restrained on a table. Talons must be carefully controlled.

the back or front. From the back, grasp the wings, body, and legs together (Fig. 24.30A). When approaching from the front, grasp the legs first (Fig. 24.30B).

If a bird impales a handler with a talon during the capturing process, the bird should be released and allowed to move away. If the bird is held and continues to struggle against capture, its grip will be maintained or enhanced. An impaled talon can be released by straightening the leg at the tarsal-metatarsal articulation, relaxing the tendon-tightening mechanism that operates to maintain the grip when the leg is flexed. A second person may be needed to force the leg to straighten and release the talons.

It is impossible for an unassisted person to get free from the talons of a bird such as the golden or bald eagle (Figs. 24.31, 24.32) unless the bird is killed first. I have attempted to remove a burlap sack from the clutches of a golden eagle. It took one person on each talon to force the bird to relinquish its grip on the sack.

FIG. 24.32. Restraining an eagle. (Eagles and hawks do not usually bite, but they may. Be cautious.)

Medium-sized raptors are effectively restrained after initial capture by placing them in a stockinette or nylon hose (Fig. 24.33). The stockinette serves as a hood to diminish visual stimulation and restricts wing action. To examine a wing or a leg, cut a hole through the stockinette and extract the limb (Fig. 24.34). Birds restrained in nylon hose must be watched carefully, since nylon retains heat to a greater degree than cotton stockinette. Birds have died from hyperthermia as a result of prolonged restraint in nylon hose.

Examination of the mouth of a large raptor can be carried out quite easily with a dowel speculum between the upper and lower beak (Fig. 24.35). Collars to prevent self-mutilation can be placed on quiet raptors (Fig. 24.36). The bird must not be able to reach the margin of the collar with the tip of its beak.

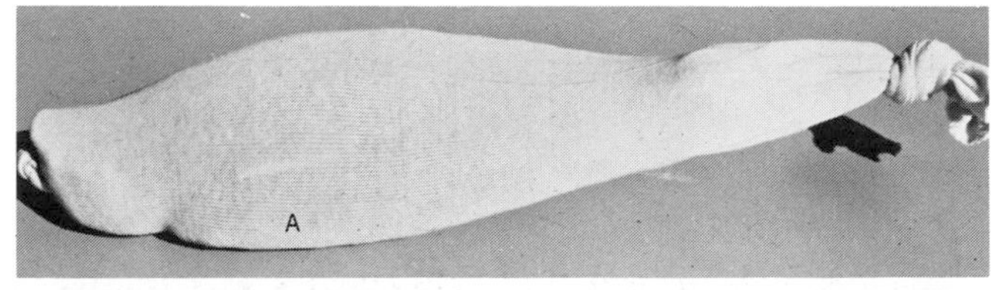

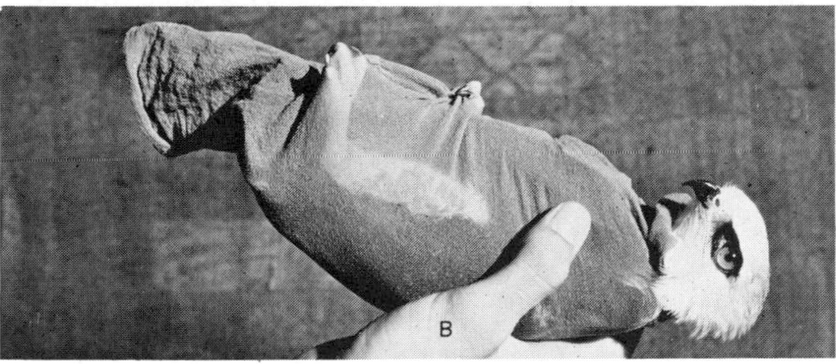

FIG. 24.33. **A.** Stockinette restraint. **B.** Nylon hose used to restrict movement of a raptor.

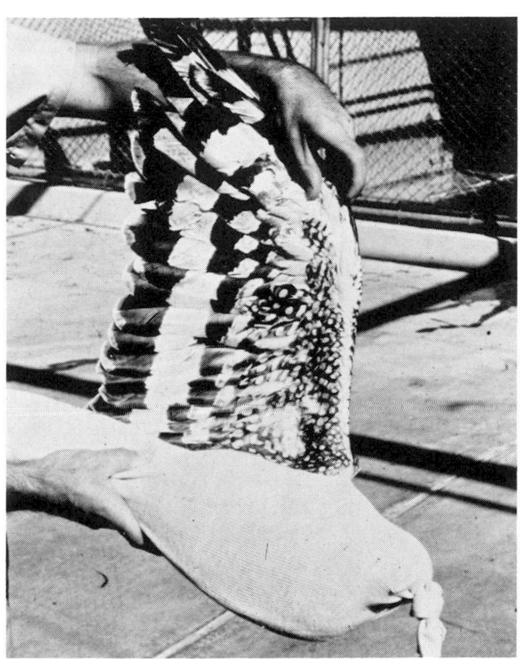

FIG. 24.34. Using a stockinette to control a raptor while exposing a wing for examination.

A raptor can be examined for external parasites as illustrated in Figure 24.37.

GALLIFORM BIRDS

Galliforms are heavy-bodied, short-legged, slow-flying birds. These birds spend most of the time on the ground searching for seeds, insects, or herbs.

Danger Potential

Most birds in this group are inoffensive docile birds, which can be handled without danger of being pecked or scratched. However, all have claws and may scratch when excited, though none are capable of causing serious injury. Males in this group may exhibit large tarsal spurs, primari-

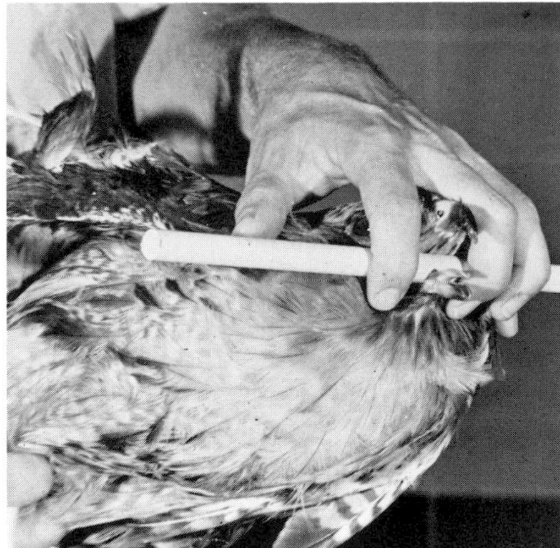

FIG. 24.35. Dowel used as speculum for a hawk.

FIG. 24.36. Cardboard collar used to prevent self-mutilation during wound healing.

ly used in fighting among themselves but which may also be used defensively when capture is imminent.

Do not grab these birds by the feathers, particularly the tail feathers. Pheasants, crowned pigeons, and turacos release their feathers readily when captured. Although not of long-term significance, lost tail feathers disfigure species in which the tail feathers are important for exhibition.

Although generally mild mannered, some pheasant species occasionally become aggressive, particularly during the breeding season. A cock may fly at an intruder with feet outstretched, attacking with tarsal spurs and wings [16]. Protect your face when attempting to capture male birds.

Physical Restraint

Once the bird is captured, hold the wings close to the bird's body and control the legs (Figs. 24.38–24.41). The beak is rarely used for defense.

FIG. 24.37. Collecting external parasites from a hawk by placing it in a plastic bag containing chloroform-impregnated pledget of cotton, making sure the head is kept out of the bag.

FIG. 24.38. Controlling legs and wings of pheasant.

LONG-BILLED, LONG-LEGGED BIRDS [21]

Danger Potential

The primary defense of long-beaked birds is pecking. Herons, storks, cranes, and other birds with large or long sharp-pointed bills may peck at the face and eyes of handlers, inflicting serious injury. Flamingos frequently attempt to peck captors and can inflict rather nasty

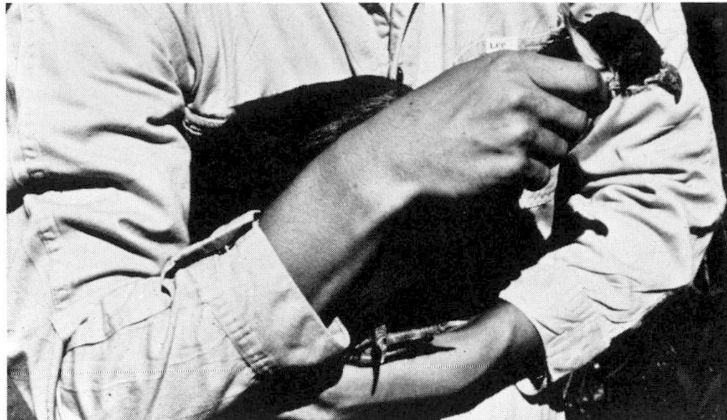

FIG. 24.39. Restraining pheasant by holding it next to the body.

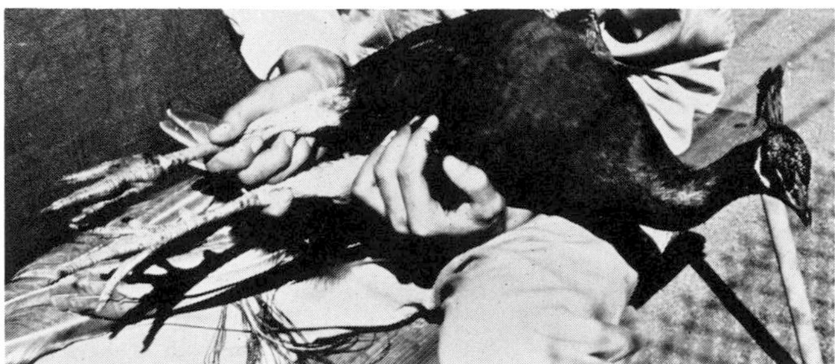

FIG. 24.40. Restraining peafowl by controlling the legs.

FIG. 24.41. Holding peafowl for temporary transport.

wounds with the serrated margins of the blunt recurved beak. Do not peek into a crate or wire enclosure containing one of these birds, lest it peck at your eyes. When capturing long-billed birds, control the head first by grasping either the neck or the bill.

Members of this group of birds will attempt to scratch with their long legs, but scratches are seldom severe. The African crowned crane is an exception. This bird will claw much as does a hawk. Guard against injury from the feet, the beating wings, and the beak of this crane.

Physical Restraint

The long thin legs of these birds are easily broken by injudicious handling. Restraint should be applied gently, exerting minimum pressure.

A person should not enter an enclosure confining large cranes without some means of protection such as a broom, a net, or a stick to hold the bird away and keep it from pecking. Cranes can be hooded, as is done with ostriches, to eliminate visual stimuli. A combination hood and net may cover the head of a crane until the neck can be grasped and the wings controlled.

Cranes can usually be herded into a corner with a sheet of plastic, a shield, or a fence panel. If a fence panel is used, the birds can be squeezed into a corner, enabling the handler to grasp an individual by the beak or the neck. Further restraint is imposed by controlling the wings (Figs. 24.42, 24.43). When prolonged restraint of a crane or stork is necessary, tape the bill shut and impale a blunt object such as a cork or rubber stopper on the tip of the beak to prevent jabbing injuries (Fig. 24.44). Large cranes can be moved by placing a soft rope or a sock around the base

FIG. 24.42. Grasp a crane by the neck first to prevent personal injury from the beak, then grab the wings.

FIG. 24.43. A crane can be held as illustrated, but the legs will flail and may scratch the handler.

FIG. 24.44. Protecting from sharp bill by placing a cork over the taped beak, leaving nostrils uncovered.

of both wings and controlling the beak (Fig. 24.45) [16]. If cranes and storks must be handled frequently, wear a metal face mask to protect the eyes and face, as shown in Figure 24.48. It is not desirable to capture cranes or storks with a net. The fine bones are fragile and easily fractured or injured if struck with the hoop or entangled in the mesh.

Flamingos can usually be slowly herded into a corner, permitting handlers to enter the enclosure to capture them, fastening them by the neck before grabbing the base of the wings. Do not clutch flamingos by their lower legs or attempt to net them, lest the long spindly legs be injured. To hold a flamingo, grasp the neck just below the head with one hand and the base of the legs with the other (Fig. 24.46).

Long-legged birds present special problems during recovery from anesthesia or immobilization. If left to their own devices while awakening, they are likely to stagger and fall, injuring themselves. A satisfactory means of controlling the bird during recovery is to place it in a burlap sack with the head exposed; this prevents it from standing until completely over the effects of the anesthetic (Fig. 24.47). These birds are subject to capture myopathy (Chapter 7), so leg activity should not be severely restricted [26].

FIG. 24.45. Moving a saurus crane by controlling head and wings.

FIG. 24.46. When restraining a flamingo, restrict head movement and grasp legs close to the bird's body.

LARGE-BILLED BIRDS

Danger Potential

Toucans, hornbills, and others of this group will peck the face and hands with their massive bills. If you must enter an enclosure with one of these birds, use a mask or a plastic shield (Fig. 24.48).

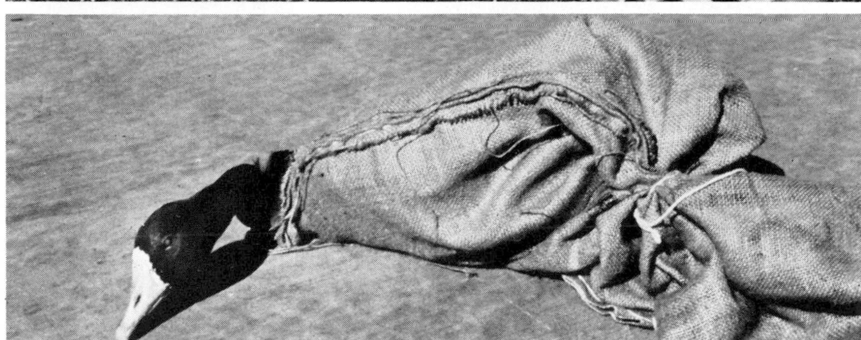

FIG. 24.47. Crane and goose placed in bags for temporary restraint.

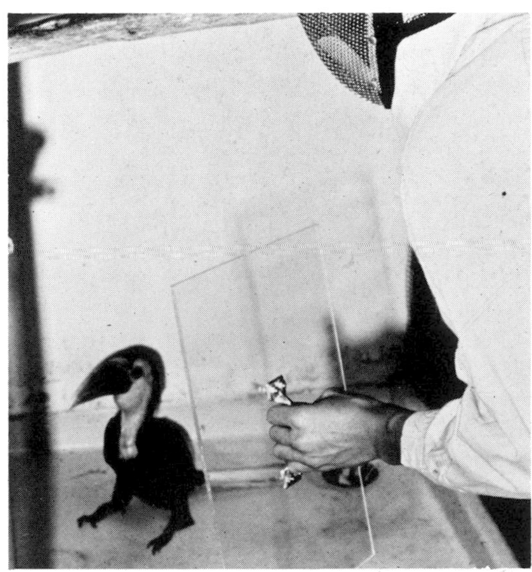

FIG. 24.48. Capturing a hornbill. Handler is protected by fencer's face mask and plastic shield.

Physical Restraint

Birds in this group should be initially captured with a net (Fig. 24.49). The bill should be clasped and held shut while the bird is extracted from the net and during subsequent procedures. The bill can be taped shut and a cork impaled on the tip if prolonged restraint is necessary. Once in hand, the beak and wings can be held together close to the body of the bird; or the wings can be grasped as shown in Figures 24.50 and 24.51.

PIGEONS AND DOVES [18]

Doves are perhaps the most inoffensive of wild birds to handle. They do not scratch; they are not likely to peck; they have no wing spurs; and they have mild, docile dispositions.

Wild pigeons are handled by the same methods as domestic species. Basic handling procedure is to grasp from above and behind, pressing the wings close to the bird's body. Control of pigeons and doves is shown in Figures 24.52–24.55. Birds of these species are also easily netted.

The large crowned pigeon will shed its feathers if the feathers are grasped improperly during the restraint procedure. Do not grasp the tail, or you will pull out the tail feathers.

PSITTACINE BIRDS—PARROTS, PARAKEETS, LORYS

Danger Potential

Psittacines all have large heavy bills and strong jaws. The diet of these birds consists primarily of nuts and seeds, though some are fruit eaters. The bills of all are capable of seriously injuring an unwary handler. The beak of a large parrot or macaw is a formidable weapon. A large macaw such as a hyacinth can easily crush the bone of a finger. A glove affords little protection from crushing.

Some parrots are well adapted to climbing and clinging to branches. These have sharp claws that can injure. However, light gloves afford adequate protection from scratching by the claws of psittacine birds.

The kea parrot is a member of the psittacine group, but its habits resemble those of raptors. It is carnivorous and uses its beak and talons in the same manner and for the

FIG. 24.49. Capturing and restraining an aracari.

FIG. 24.50. Proper restraint for an aracari.

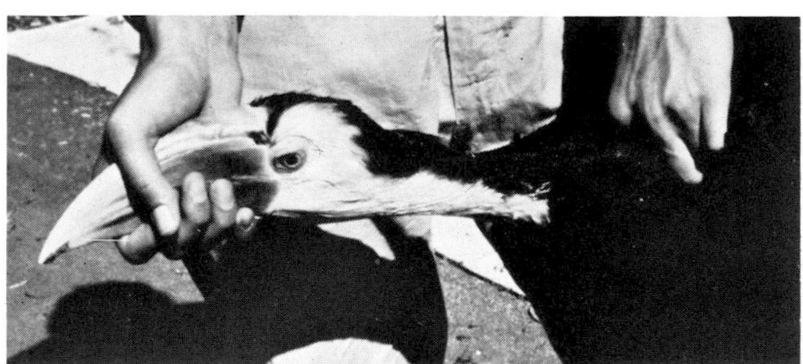

FIG. 24.51. Proper restraint for a hornbill.

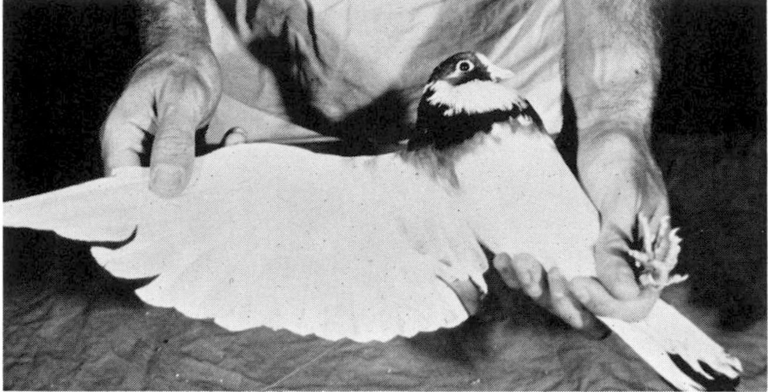

FIG. 24.53. One person can easily restrain and examine a pigeon.

same purposes as do raptors. It should be handled as if it were a raptor.

Physical Restraint

Darkening the room has a sedative effect on diurnal birds and facilitates capture of psittacines from cages or aviaries. The basic procedure when capturing and handling all psittacine birds is to control the head. A light glove may be worn, but it is usually possible to safely grasp small psittacine birds without gloves.

Parakeets (budgerigars) are routinely kept in small cages. If confined in an aviary, they should be captured with a small net. If in a small cage, remove all obstructions such as perches, mirrors, or other dangling objects against which an excited bird may fly and injure itself. Corner the bird against the cage wall or the floor. Approach it from behind and above, placing the thumb and forefinger on each side of the head (Fig. 24.56). Grasp it firmly, but do

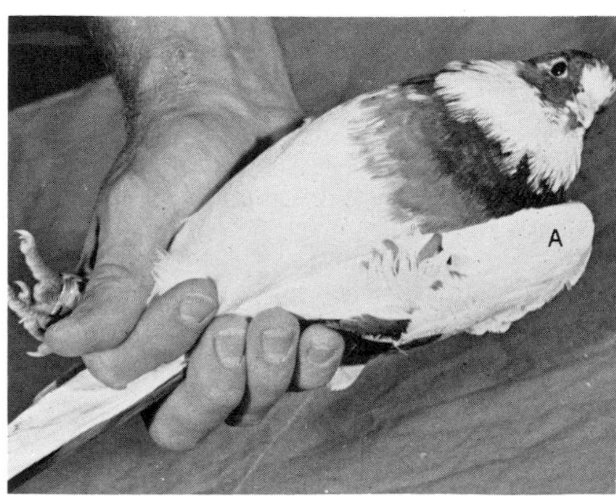

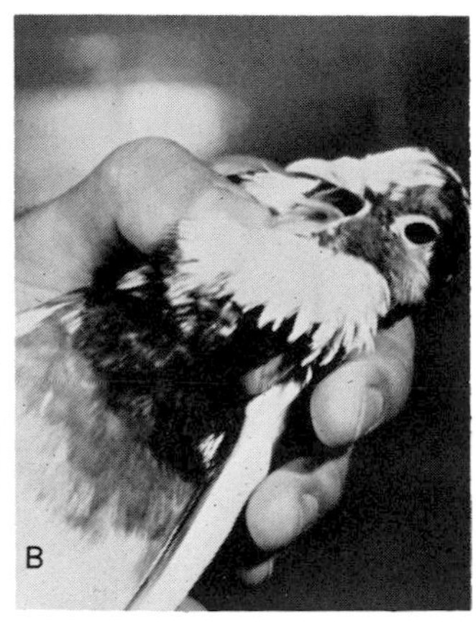

FIG. 24.52. **A.** Proper pigeon handling. **B.** Examining mouth of a pigeon.

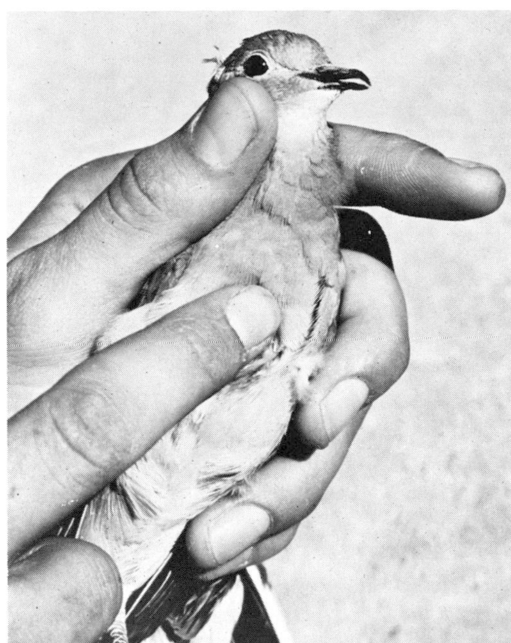

FIG. 24.54. Small dove can be handled like a budgerigar or finch.

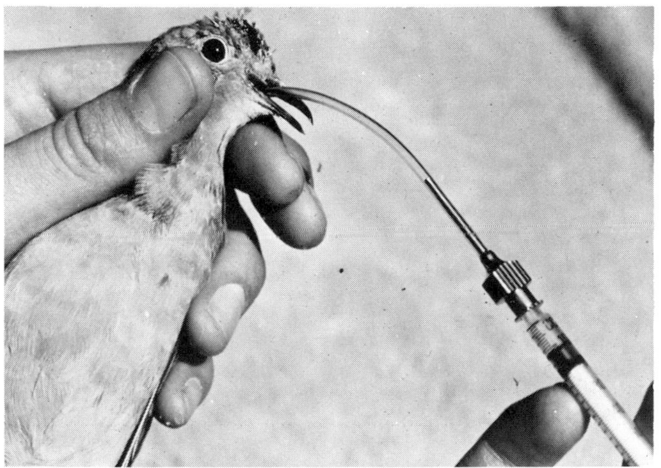

FIG. 24.55. Stomach tube placement in a dove (no speculum is necessary).

not crush it. Remove the bird from the cage, fixing the grip to position the bird on its back in the cupped hand with the head controlled by thumb and forefinger (see Fig. 24.3A).

Larger members of this group are handled in much the same way as parakeets except that as the size of the bird increases, heavier gloves must be used to protect the hands (Fig. 24.57). If gloves are not available, use a towel to capture the medium-sized bird (Fig. 24.58).

Gloves offer insufficient protection from large macaws and parrots. Divert the attention of the bird before attempting capture. The bird can be brought out of the cage and the head pressed with a stick to divert attention. Some large birds that have been caught previously with gloves

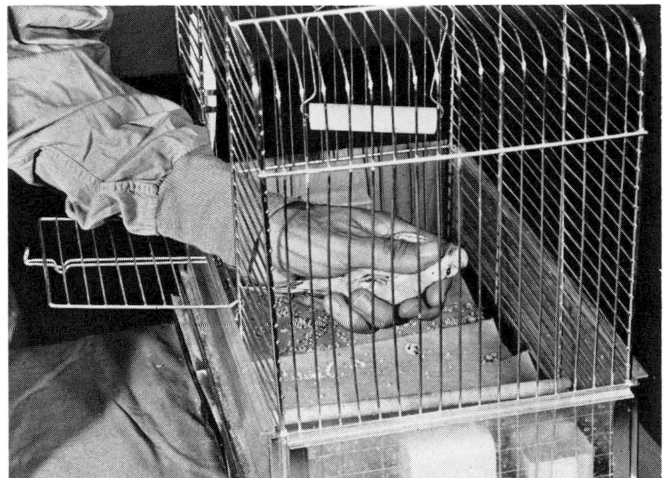

FIG. 24.56. Capturing budgerigar in a cage by grabbing it from behind.

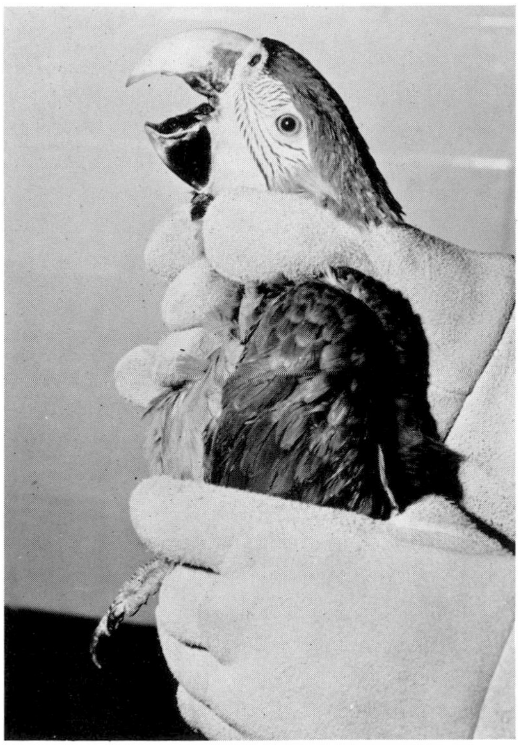

FIG. 24.57. Proper restraint of a large psittacine bird.

may set up a terrible racket and fight capture. Darkening the room and/or covering the bird with a towel may reduce excitement and obstreperous behavior. Large psittacines are usually grasped first behind the head at the nape; then the tail and limbs are held to prevent the bird from flying forward. Once the bird is in hand, the grip can be maintained more satisfactorily with the bare hand, so the gloves should be removed.

If it is necessary to transfer a captured psittacine bird to another person, the second person places a hand over the

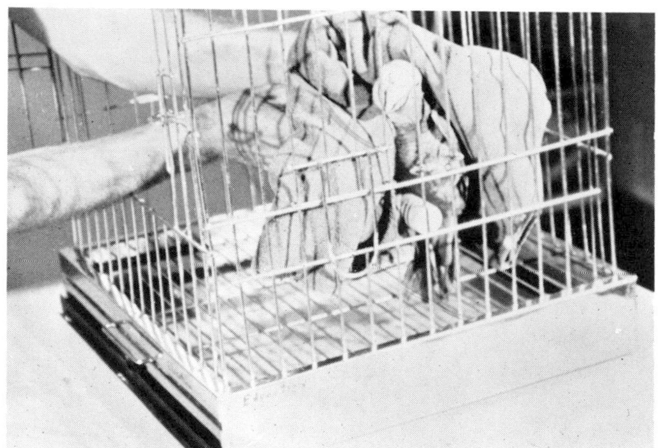

FIG. 24.58. Capturing parrot by surrounding it with a towel.

hand holding the head, imposing the grip as the first person releases it.

When holding the head of a psittacine bird, be certain that thumb and forefinger are kept close to the mandible (Fig. 24.59A). If the bird is able to achieve even slight mobility of the head, it will be able to manipulate the mandible into position to grab the finger (Fig. 24.59B). Even a gloved finger can be bruised by the strong beak of a macaw. To restrict the head more securely, place thumb and forefinger on the lateral surface of the mandible.

Open the mouth of a psittacine bird by inserting various specula (Figs. 24.60, 24.61). A stomach tube can be inserted into the esophagus quite easily when the mouth is held open. The speculum must be held firmly in the mouth. If it should loosen, the bird may clamp down on the stomach tube, cut the tube in half, and swallow the severed segment.

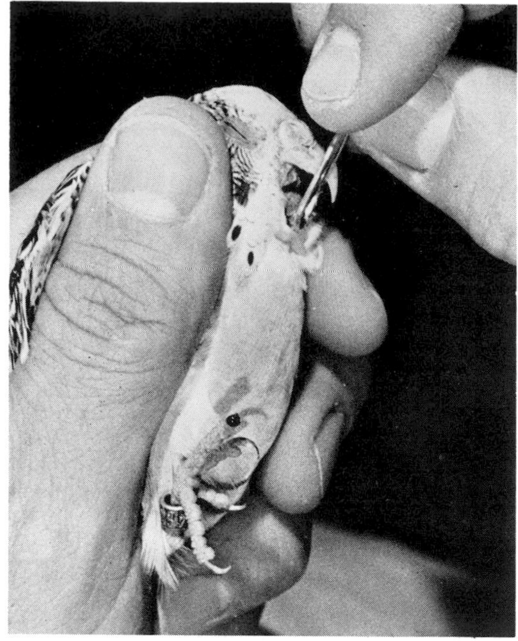

FIG. 24.60. Using paper clip as oral speculum for a small psittacine.

A budgerigar may be intubated without the use of a speculum by inserting the tube gently through the commissure of the mouth. At the commissure the horny beak cannot cut the tube. The tube can be inserted, held in this position while medication is administered, and withdrawn without causing a great deal of discomfort.

Discarded plastic syringe holders or other plastic tubes may be used to control small birds for short periods (Figs. 24.62, 24.63). Tubes must be of sufficient diameter to permit respiration.

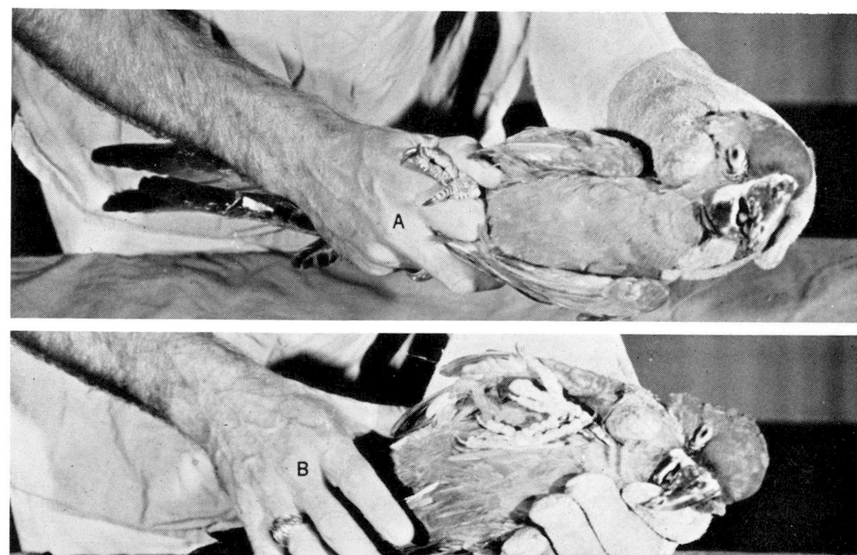

FIG. 24.59. **A.** Proper placement of fingers when holding a large macaw. **B.** Improper placement may result in the beak grasping a finger.

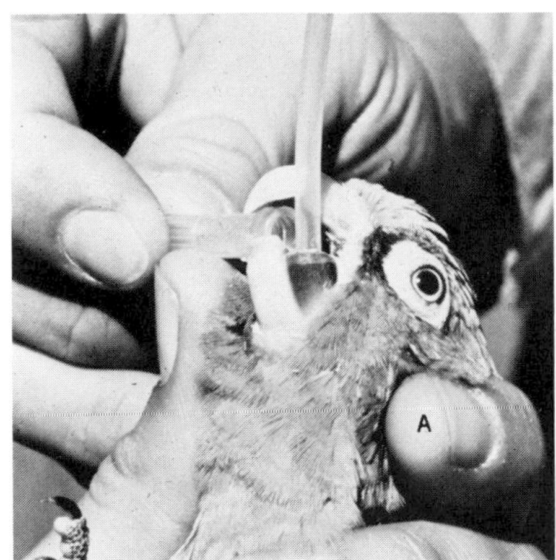

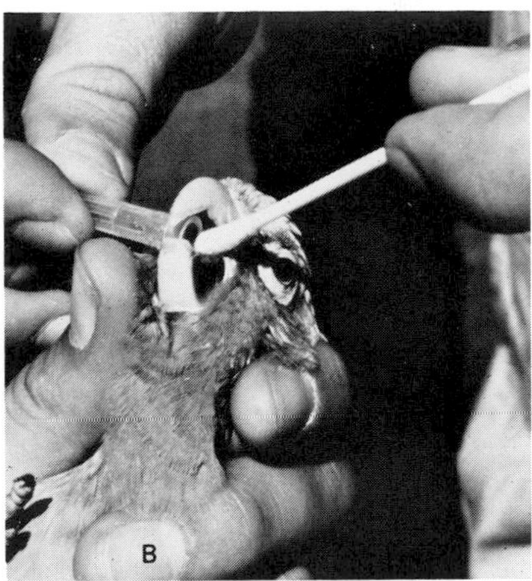

FIG. 24.61. Using plastic needle holder as a speculum to: **A.** Pass a stomach tube. **B.** Swab the oral cavity.

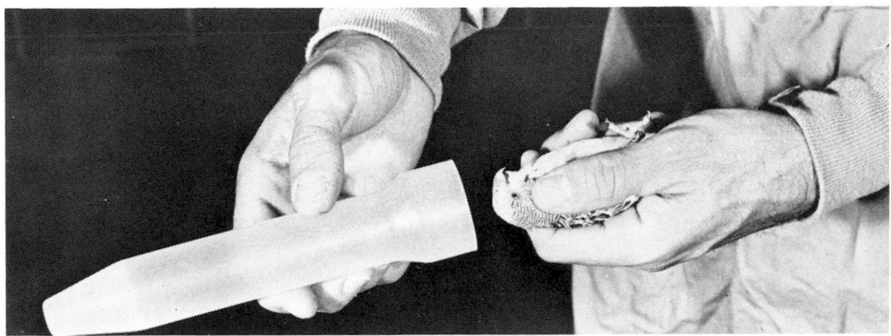

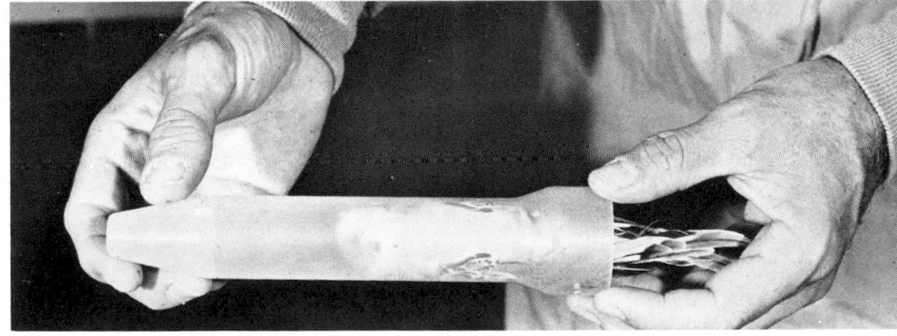

FIG. 24.62. Plastic syringe holder used to restrain a budgerigar.

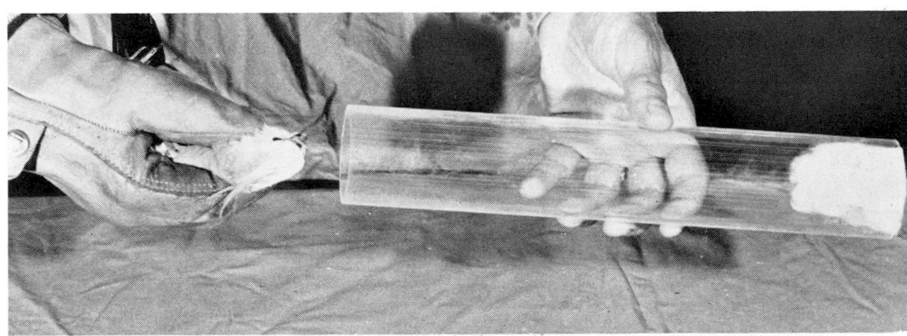

FIG. 24.63. Plastic tubes can be selected to restrain any small bird.

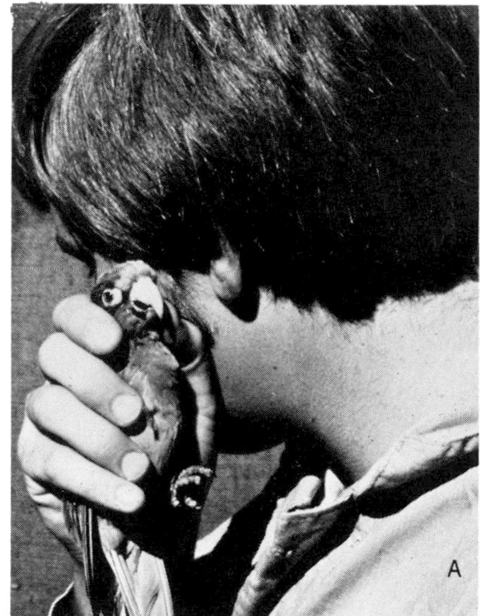

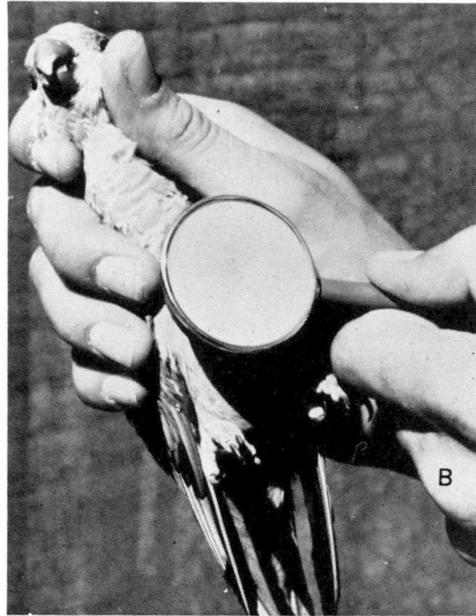

FIG. 24.64 Evaluation of respiratory sounds: **A.** Direct listening. **B.** Using a stethoscope.

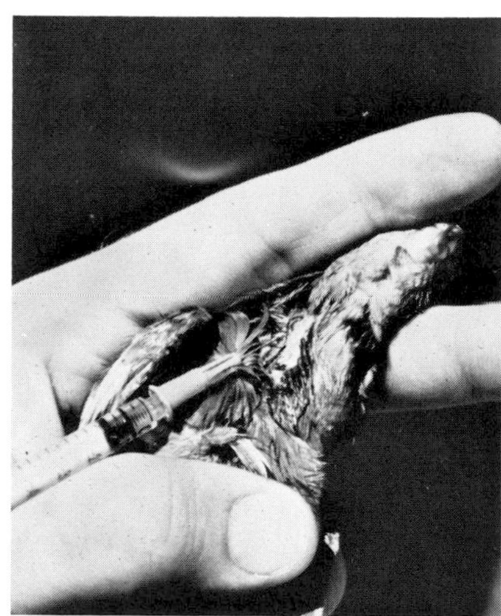

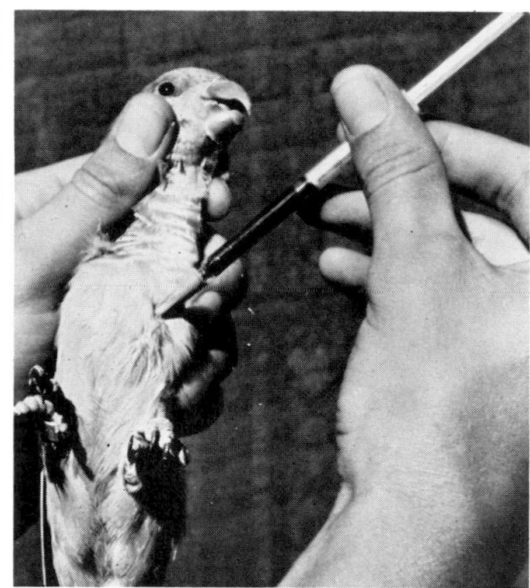

FIG. 24.65. Bleeding a small bird from the jugular vein *(left)*. Intramuscular injection into the breast muscle *(right)*.

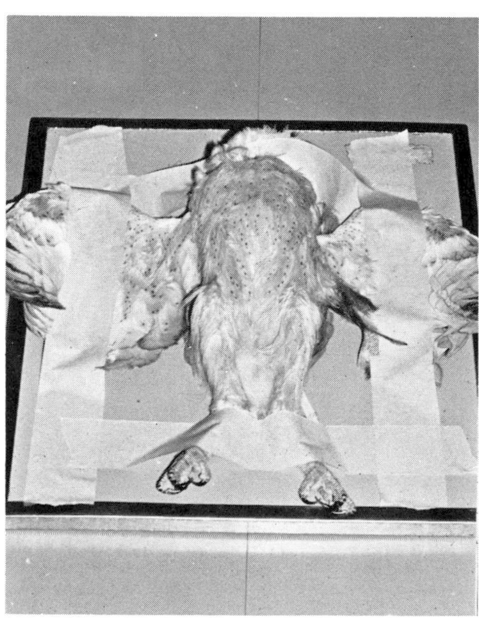

FIG. 24.66. Positioning for avian radiography [23,24]: Dorsoventral *(left)*. Lateral *(right)*.

Miscellaneous examinations and other techniques are illustrated in Figures 24.64 through 24.66. Collars are used to prevent self-mutilation (Figs. 24.67–24.69). When examining small caged birds, it is a good policy to have an oxygen jar available (Fig. 24.70) to alleviate respiratory distress.

HUMMINGBIRDS AND SWIFTS

It is extremely difficult to capture the swift-flying hummingbird, which is capable of reversing flight instantaneously. Mist nets are used to capture hummingbirds and other small rapid-flying birds.

After initial capture, these birds can be restrained and handled by cupping them gently in the hand (Fig. 24.71). None are aggressive or capable of injuring the handler.

SONG, PERCHING, AND MISCELLANEOUS BIRDS

The small delicate finches and warblers, crows, and hundreds of other species of birds of various sizes are handled in much the same way. Most are inoffensive, and little protection is needed for the hands.

Physical Restraint

In an aviary the birds are captured with a net, carefully removed, and held cupped in the hand with the hand around the base of the head (Fig. 24.72). Small birds rarely injure a handler when they peck, but the habit of securing the head properly should become deeply ingrained. The feet may require controlling, though most of these birds will not scratch.

When holding a bird in the cupped hand, do not completely surround the sternum and interfere with respiration. Tiny finches and warblers are extremely difficult to hold safely. Suggested techniques are illustrated in Figure 24.73.

Darken the room where caged birds are kept to diminish activity and fright before removing them from the cage. Approach birds from behind. A light glove may be required for handling mynah birds and some of the heavier birds such as crows, ravens, and jays, which may resist capture and restraint. Follow the same procedure described

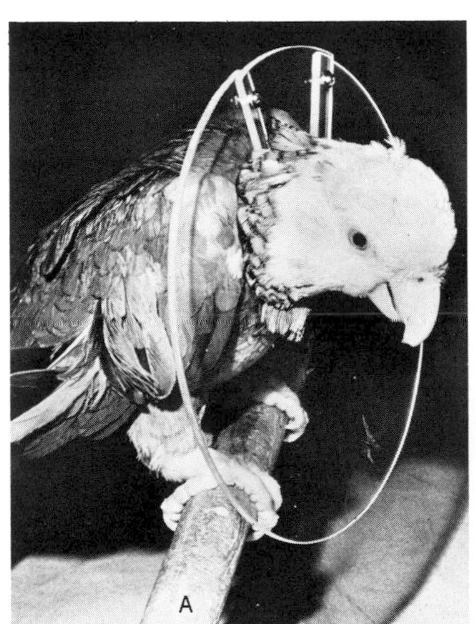

FIG. 24.67. **A.** Collars inhibit self-mutilation. **B.** But some birds are clever.

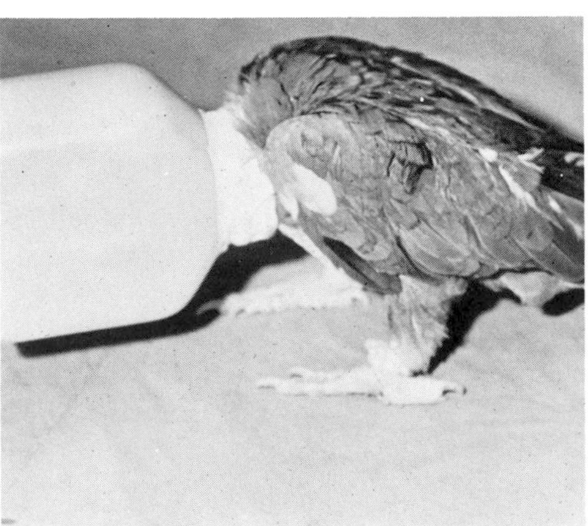

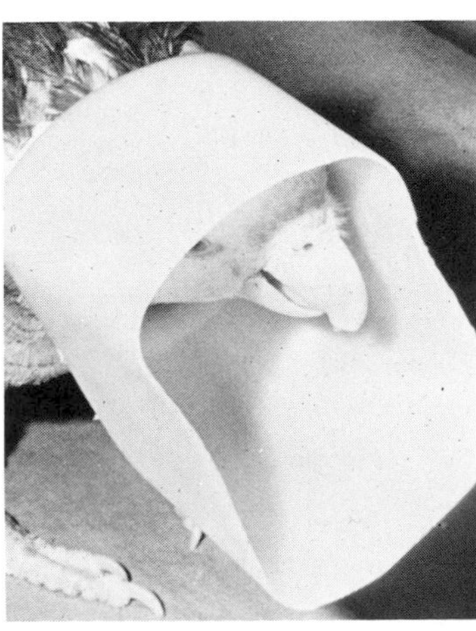

FIG. 24.68. Discarded plastic drug bottle serves as collar. Special arrangements for food and water must be provided.

FIG. 24.69. Improvised collars to prevent self-mutilation: **A.** Cardboard on a macaw. **B.** Plastic playing card on a budgerigar.

FIG. 24.70. An oxygen jar should be readily available when working with small birds.

for capturing the budgerigar. The beak can be taped shut to eliminate pecking (Fig. 24.74).

When examining small birds, be sure that any fans in the room are turned off. Exhaust fans can suck a small bird into the blades.

REFERENCES

1. Amand, W. B. 1974. Avian anesthesia. In R. W. Kirk, ed. Current Veterinary Therapy V. Philadelphia: W. B. Saunders.
2. Beck, C. C. 1972. Chemical restraint of exotic species. J. Zoo Anim. Med. 3:1-67.
3. Beck, M. W. R. 1969. Emu capture in the field using an immobilizing drug and Cap-Chur apparatus. Aust. Wildl. Res. 14:195-97.
4. Berger, D. D., and Mueller, H. C. 1959. The bal-chatri: A trap for the birds of prey. Bird Banding 30:18-26.
5. Blackshow, G. D., and Wakeman, B. 1971. Immobilization of an adult ostrich for surgery. J. Zoo Anim. Med. 2:11-12.
6. Borzio, F. 1973. Ketamine hydrochloride as an anesthetic for wildfowl. Vet. Med. Small Anim. Clin. 68:1364-65.
7. Cline, D. R., and Greenwood, R. J. 1972. Effects of certain anesthetic agents on mallard ducks. J. Am. Vet. Med. Assoc. 161:624-33.
8. Cooper, J. 1973. Use of the drug metomidate to facilitate the handling of vultures. J. Zoo Anim. Med. 4:6.
9. Dilbone, R. P. 1968. Experiences with anesthesia in the penguin. J. Small Anim. Pract. 9:138-39.
10. Evans, M., and Kear, J. 1976. A jacket for holding large birds. Int. Zoo Yearb. 15:191-93.
11. Fuller, M. R. 1975. A technique for holding and handling raptors. J. Wildl. Manage. 39:824-25.
12. Gandal, C. P. 1962. Avian anesthesia. Int. Zoo Yearb. 4:141.
13. Graham-Jones, O. 1965. Restraint and anesthesia of small cage birds. J. Small Anim. Pract. 6:31-39.
14. Harthoorn, A. M. 1976. The Chemical Capture of Animals, pp. 118, 238. London: Baillière, Tindall.
15. Houston, D. C., and Cooper, J. E. 1973. Use of the drug metomidate to facilitate the handling of vultures. Int. Zoo Yearb. 13:269-70.
16. Karsten, P. 1974. Safety Manual for Zoo Keepers. Alberta, Canada: Calgary Zoo.
17. Kerlin, R. E., and Susman, O. 1963. Capture, processing and venipuncture of wild birds. Proc. Annu. Meet. Am. Vet. Med. Assoc., p. 111.
18. Levi, W. M. 1957. The Pigeon. Sumter, S.C.: Levi.
19. Levinger, I. M.; Kedem, J.; and Abram, M. 1973. A new anesthetic agent for birds. Br. Vet. J. 129:296-300.
20. Mandelker, L. 1972. Ketamine hydrochloride as an anesthetic for parakeets. Vet. Med. Small Anim. Clin. 67:55-56.
21. Pomeroy, D., and Woodford, M. H. 1976. Drug immobilization of marabou storks. J. Wildl. Manage. 40:177-79.
22. Rutger, A., and Norris, K. A., eds., 1970, 1972. Encyclopedia of Aviculture, vols. 1, 2. London: Blandford Press.
23. Silverman, S. 1975. Avian radiographic technique. In J. W. Ticer, ed. Radiographic Techniques in Small Animal Practice. Philadelphia: W. B. Saunders.
24. Whittow, G. E., and Ossorio, N. 1970. A new technique for anesthetizing birds. Lab. Anim. Care 20:651-56.
25. Williams, L. E., and Phillips, R. W. 1973. Capturing sandhill cranes with alpha chloralose. J. Wildl. Manage. 37:94-97.
26. Young, E. 1966. Leg paralysis in the greater flamingo and lesser flamingo following capture and transportation. Int. Zoo Yearb. 6:226-27.

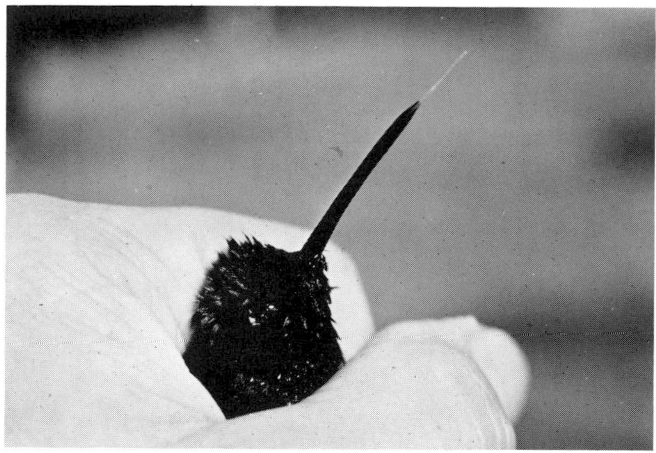

FIG. 24.71. Tiny hummingbird is difficult to hold without crushing it.

FIG. 24.74. Taping beak to prevent bothersome pecking.

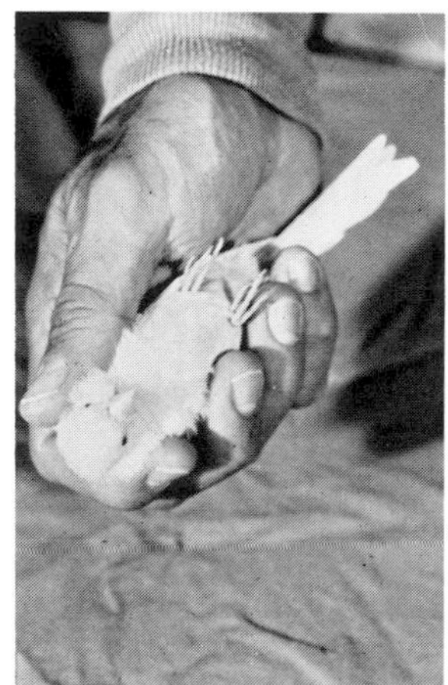

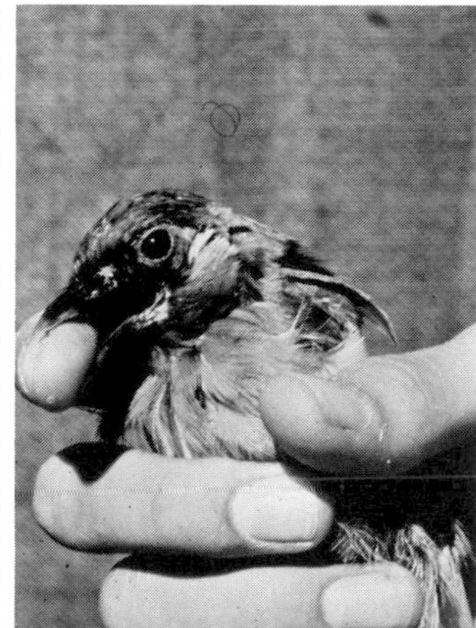

FIG. 24.72. Proper restraint of a canary *(left)*. Proper restraint of a passerine bird *(center)*. Peck of this small bird will not injure the finger *(right)*.

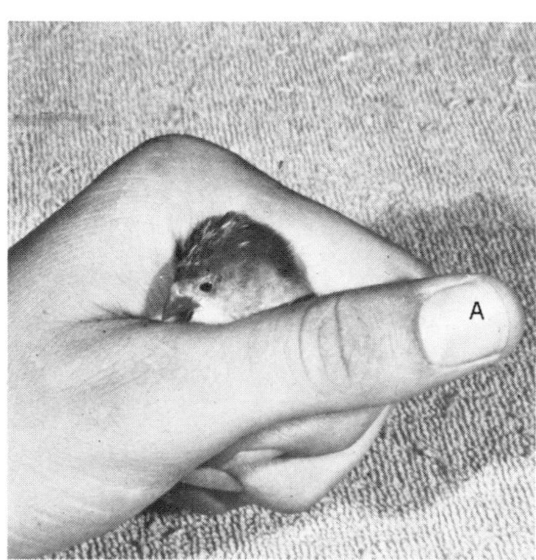

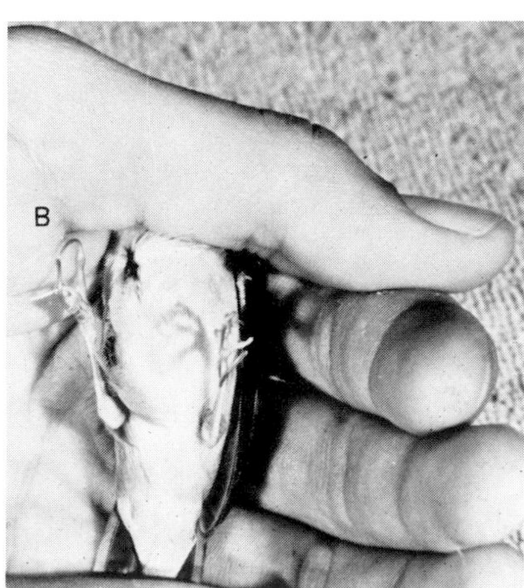

FIG. 24.73. **A.** A tiny finch may require holding as illustrated.
B. Examination of legs and body can be accomplished as shown.

25 REPTILES

CLASSIFICATION

Class Reptilia
- Order Crocodilia (2 families, 8 genera, 21 species)
  - Family Gavialidae: gavial
  - Family Crocodilidae: alligator, crocodile, caiman
- Order Chelonia (Testudines) (8 families, 56 genera, 219 species): turtle, tortoise, sea turtle, snapping turtle
- Order Rhynchocephalia (1 family, 1 genus, 1 species)
  - Family Sphenodontidae: tuatara
- Order Squamata
  - Suborder Lacertilia (19 families, 360 genera, 2,839 species): gecko, monitor, iguana, chameleon, Gila monster, legless lizard
  - Suborder Ophidia (9 families, 416 genera, 2,005 species)
    - Family Boidae: giant constrictor
    - Family Elapidae: cobra, krait
    - Family Hydrophidae: sea snake
    - Family Viperidae: rattlesnake, viper
    - Family Colubridae: nonpoisonous species

GENERALLY adults are called males and females. A male crocodile is called a bull and the female a cow. Newly hatched or newborns are called hatchlings or juveniles.

Although members of this class share sufficient morphological characteristics to warrant close zoological classification, the requirements for restraint and handling vary widely with the species; therefore, for discussion they will be divided into groups that can be handled with similar techniques.

All reptiles continue to grow throughout life, although the rate of growth slows markedly as they age. Size is an indication of age, but other factors such as nutritional level and environmental temperatures also determine size.

CROCODILIANS

Crocodilians include alligators, crocodiles, gavials, and caimans. Approximately twenty-one species are scattered throughout tropical and subtropical areas of the world.

Danger Potential

All crocodilians are carnivores, capable of inflicting serious bites even in the newly hatched stage. Crocodilians lurk until a prey species nears, then quickly grasp it and tear off flesh by flinging the head about. They exhibit the same behavior when they bite a person; they grasp and then flip the head, tearing out chunks of tissue.

In all crocodilians the muscles that close the mouth are very strong, but those that open it are weak, so a person's hands can easily hold the mouth closed.

The tail of crocodilians is used to propel them through the water. During restraint vicious lashing with the tail is a principal method of attack, particularly by larger species.

Physiology and Behavior

Crocodilians are primarily aquatic species which come ashore and bask in the sun to absorb heat. Although they must surface to breathe, they are able to stay submerged for many minutes.

During colder months of the year some crocodilians become anorectic and enter a torpid state. During this time they subsist on energy reserves stored in the body. Captivity may adversely modify the ability to properly prepare for the torpid state; thus captive animals may be marginally hypoglycemic [11]. Handling one of these animals while it is in the torpid state may send it into hypoglycemic shock by stimulating a too-sudden return to activity, for which the animal is incapable of mobilizing sufficient glucose to meet energy needs. The reaction of torpid crocodilians to chemical restraint agents is extremely unpredictable.

Physical Restraint

Small specimens up to 0.6 m (2 ft) in length can be manually handled without difficulty (Figs. 25.1, 25.2). It may be necessary to pin the head partially to allow a safe approach (Fig. 25.3). These animals may scratch, so provide suitable protection in the form of gloves or clothing. The tail must be restrained at all times; otherwise even a small animal may slap the handler in the face or injure itself by wildly flailing the tail.

Specimens up to 2 m (6 ft) in length can be handled with a snare. Nooses vary in design from swiveled snares, available commercially, to homemade cable snares. With either type, once the noose is around the neck, the animal is likely to try to twirl on the snare; unless the handler is prepared to twirl the snare with the animal, strangulation

FIG. 25.1. Grasping small crocodilian by the neck and tail.

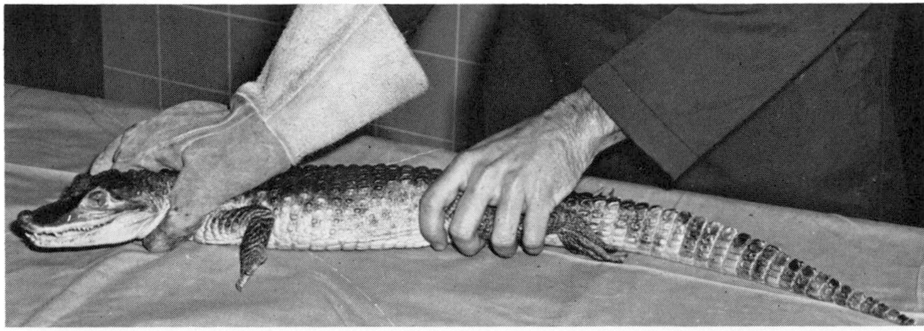

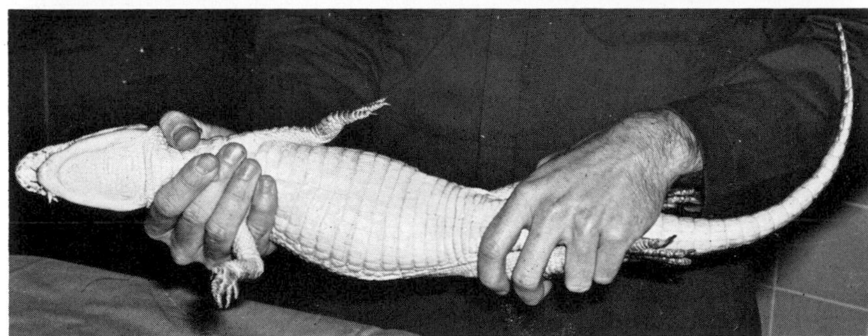

FIG. 25.2. Holding small caiman to prevent biting and minimize scratching. Notice that hind feet are restrained also.

FIG. 25.3. Controlling small caiman by partially pinning its head.

or neck injuries may occur. Immediately grasping the tail as soon as a snare is placed around the neck will inhibit twirling on the snare and prevent flailing (Fig. 25.4). Even with a snare around the neck, the head of a crocodile is able to flip rapidly from side to side.

Manual handling of large crocodilians is both difficult and hazardous. Desirable methods of approaching large crocodilians vary with the species. An alligator or caiman can be manhandled by two or three persons who jump on it simultaneously, one grasping the front legs and controlling the body and another the tail (Fig. 25.5). Speed, agility, and timing are important. *Do not* attempt this with a crocodile; it is much faster and more aggressive than alligators or caimans. Always use nets or special squeeze cages to manipulate any species of crocodile. Large crocodilians can be restrained by the use of heavy cargo nets or by ropes placed around the mouth, legs, and tail. Once the animal is secured, the mouth can be taped shut with electricians' tape or duct tape or tied shut with small ropes (Fig. 25.6).

Large crocodilians are capable of knocking a person down by merely flipping the head or tail from side to side. Be cautious. Some species of crocodilians have teeth protruding outside the mouth. If the head is permitted to flail from side to side, these teeth can lacerate anyone nearby. The tail must always be held by one or two individuals or secured by ropes or heavy nets; otherwise the flipping of the tail can inflict lethal injuries. When the animal is under control, it can be turned over or manipulated in any manner necessary for examination, treatment, or obtaining laboratory samples.

A rope noose can be tossed over the head and snout of a crocodilian. If two ropes are used, the animal can be dragged into a crate for shipment. The ropes can be wrapped around the animal to completely truss it up, or it can be lashed to a plank. The plank can serve as a stretcher, or a human stretcher can be used to carry specimens up to 2 m long.

The staff at the London Zoo has developed a special funnel bag to control crocodilians [1]. The heavy canvas bag is placed over the jaws and head. Then ropes are lashed around the bag to keep the animal's mouth shut.

Crocodilians dissipate heat by evaporation from the mucous membrane of the mouth. A struggling alligator with its mouth lashed shut may rapidly overheat. Be prepared to monitor the animal and cool it if necessary.

Persons working in zoos or other exhibit facilities often

FIG. 25.4. Controlling small caiman with a snare.

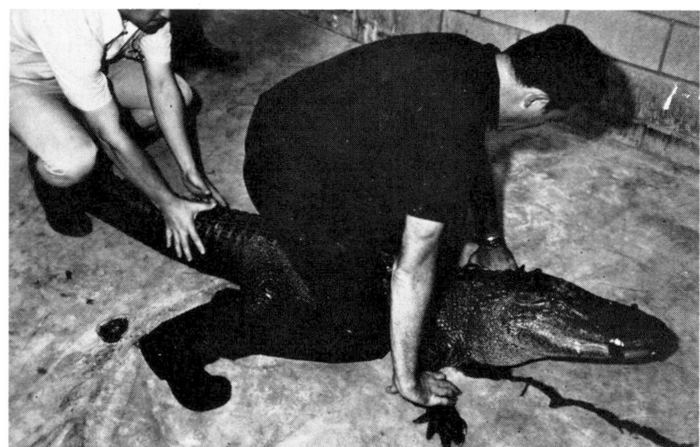

FIG. 25.5. A large alligator can be manually restrained by an experienced, confident team.

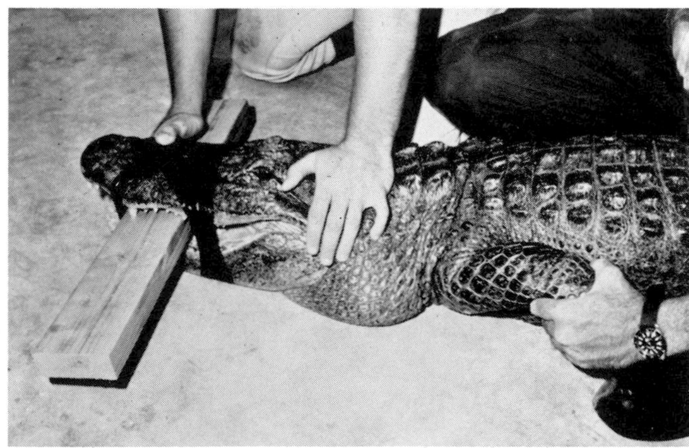

FIG. 25.6. Once alligator is in hand, the mouth can easily be kept closed by taping it. The board in this instance holds mouth open for passage of a stomach tube.

guide alligators or caimans by grasping the tails (Fig. 25.7). This is not without some danger. It is safer if a colleague keeps the animal from flipping around to the side with a broom or a stick. Both handlers should remain on the same side of the animal.

Squeeze cages for crocodilians are commercially available, or they can be constructed (Fig. 25.8). Squeeze cages should be immersible. The animal may be enticed to enter the submerged cage for food, or the cage may be placed in the area of the pond where the animal habitually lurks.

Small blood samples sufficient for a routine hemogram can be collected by clipping a toenail of any clawed reptile.

Transport

Crocodilians can be moved or shipped in dog cages as long as facilities are provided to maintain the proper ambient temperature. They will probably not feed while out of water, but this is of no consequence.

Chemical Restraint

Drug reactions of the poikilothermic crocodilians are so variable that precise recommendations are difficult. Etorphine hydrochloride (0.5 mg/kg) has been suggested as a restraint agent [10,11]. Ketamine hydrochloride (50 mg/kg) has also been used.[1]

1. Unpublished data.

FIG. 25.7. Pulling large alligator from a pool.

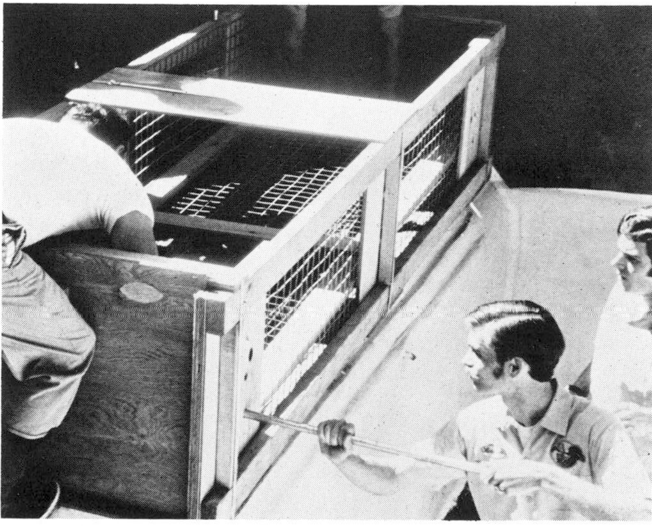

FIG. 25.8. Portable squeeze cage for capturing and transporting a crocodilian. Cage is lowered into a pool and animal baited inside. Upon removal, upper movable wall can be pressed over animal to restrain it for examination or treatment.

CHELONIANS

The terminology used to describe chelonians varies. Generally aquatic species are called "turtles"; "tortoises" are terrestrial species found in arid parts of the world; and "terrapins" are species that may have both terrestrial and aquatic phases. However, some freshwater turtles leave the water to bask in the sun and air; other species such as the sea turtles are totally aquatic, living their entire lives in the sea except for short periods when females deposit eggs on sandy beaches.

Danger Potential

The major weapons of turtles and tortoises are the hard bony plates that replace teeth in both upper and lower jaws. There is marked variation in the aggressiveness of different species. Snapping turtles are notorious biters, and occasionally a large tortoise may nip at a person. Some soft-shelled turtles and a few of the side-necked turtles occasionally bite. The side-necked turtle scratches by continually raking with the legs as long as it is grasped, but the wounds inflicted are usually superficial. The western pond turtle and some other species are also persistent scratchers, constantly attempting to win freedom when grasped.

Anatomy

Both turtles and tortoises have an exoskeleton composed of the carapace and the plastron. The carapace supplies a convenient handle for restraint, but it is difficult to examine the animal if the head and limbs are retracted tightly into the shell. In some of the larger species it is impossible to extract the head and legs without injuring the animal unless chemical immobilizers are administered.

Physical Restraint

To pick up a small to medium-sized turtle or tortoise, grasp the sides of the carapace (Fig. 25.9). From this posi-

FIG. 25.9. Holding a small tortoise.

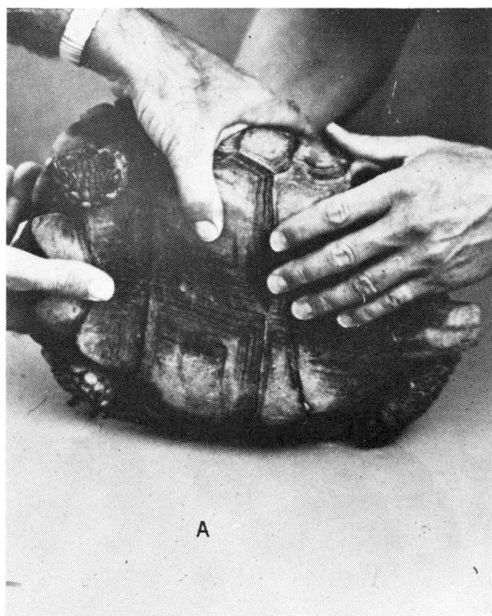

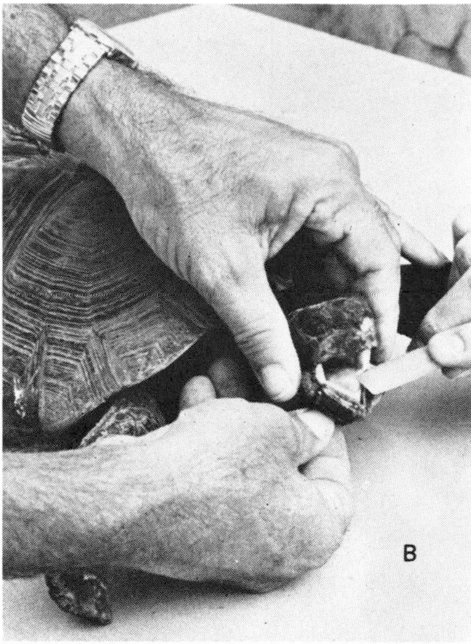

FIG. 25.10. **A.** Examining plastron of a tortoise. **B.** Fixing head for examining mouth of tortoise.

tion it can be turned over or around as needed (Fig. 25.10).

Do not turn the animal over and right it so rapidly that a loop of intestine folds on itself and obstructs the bowel. Do not hold the animal with the head down for long periods, since this may interfere with respiration.

If a limb must be examined, grasp the foot and apply gentle, steady traction to withdraw the leg from the protective cover of the carapace and plastron. Do not jerk on the limb, lest injury occur to small bones and other structures of the leg. An unsedated large tortoise may be strong enough to defy attempts to pull out the leg.

The head is more difficult to extract, but if you wait until the head is extended voluntarily, you can gently put your fingers behind the head and hold it out. Considerable effort may be made by the turtle to retract the head, and judgment must be made as to the degree of force that can appropriately be exerted to keep the head out. Forceps can be used to pull out the head of a small to medium-sized chelonian (Figs. 25.11, 25.12). A word of caution: If a tortoise with a carapace length of 25 cm pulls its legs and head with forceps attached back under the carapace, it may be extremely difficult to retrieve the forceps (Fig. 25.13).

Soft-shell turtles require careful handling, since they tend to be slightly aggressive and will bite and scratch. The soft carapace makes it difficult to grasp the animal firmly, and heavy-handed restraint practices can impair respiration or damage internal organs. Soft-shells can be netted to lift them out of the water. Wear light gloves to gently grasp the animal.

Snapping turtles, weighing up to 100 kg (220 lb), are

FIG. 25.11. Extracting head of tortoise from its shell.

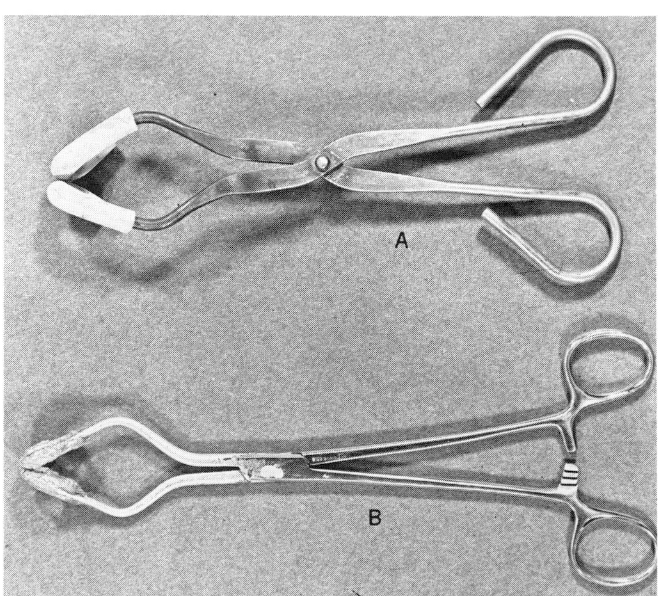

FIG. 25.12. Forceps modified for extraction of head of tortoises and turtles: **A.** Crucible tongs covered with rubber tubing. **B.** Tongue forceps, bent and covered with silicone.

FIG. 25.13. Forceps caught by a retracting tortoise.

particularly hazardous to handle. They can easily bite off a finger or inflict other serious injury. A snapper is best handled by grasping the tail and lifting it off the ground or out of the water. Do not allow the head to dangle close to your leg, lest the animal reach out and bite (Figs. 25.14, 25.15). Once the tail has been grasped, take a firmer hold on the animal by carefully running the hand closely over the top of the carapace and grasping the carapace just above the head (Fig. 25.16).

Another technique for lifting the snapping turtle is to grasp it quickly on either side of the carapace (Fig. 25.17). Two persons may be required to lift a large snapper (Fig. 25.18).

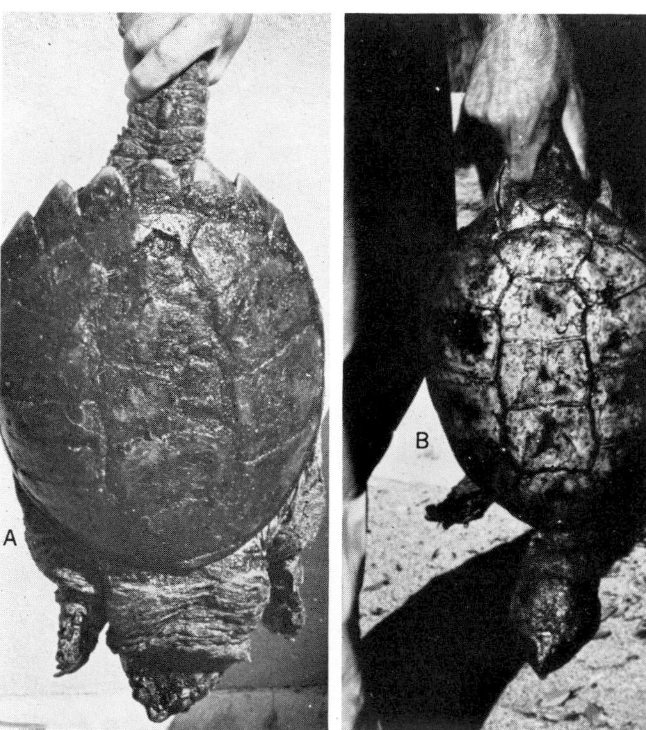

FIG. 25.14. **A.** Holding small snapping turtle by the tail. **B.** Do not allow head of snapper to approach the leg.

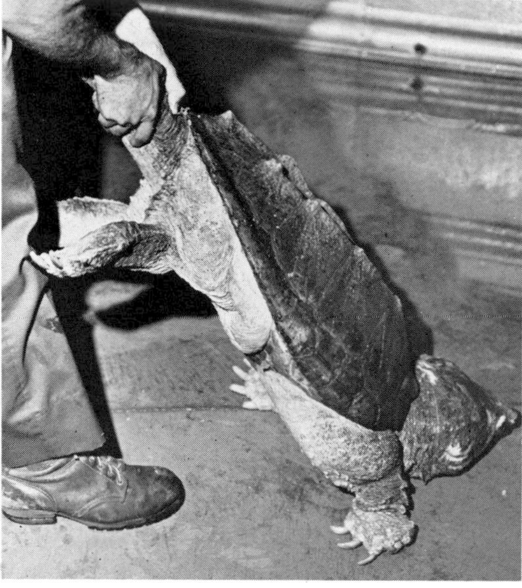

FIG. 25.15. One person can move a large snapping turtle by the tail, but it is difficult to safely lift and hold it off the ground by the tail.

Snapping turtles can be turned over onto the carapace. They usually struggle for a moment, then relax. Hold the mouth shut by pressing the lower jaw against the upper jaw. Keep the hands in a position that permits quick withdrawal if the turtle should succeed in righting itself.

A blood sample can be obtained from a snapping turtle

FIG. 25.16. One-person carry for large snapping turtle. The jaws of this animal are capable of amputating a digit and severely injuring a limb.

FIG. 25.17. Picking up large snapping turtle by the carapace.

by cutting a toenail or through venipuncture of the tail vein. Straighten the tail and cleanse the ventral surface just distal to the vent. Insert a 22 gauge, 2 cm (0.8 in.) needle perpendicular to the longitudinal axis until the vertebral body is touched. Establish a negative pressure in the syringe and then slowly withdraw the needle until blood wells up into the syringe.

Sea turtles weigh up to 600 kg (1,320 lb). They are herbivorous and generally not aggressive. Small specimens can be grasped directly from a tank or netted. Drain the tank

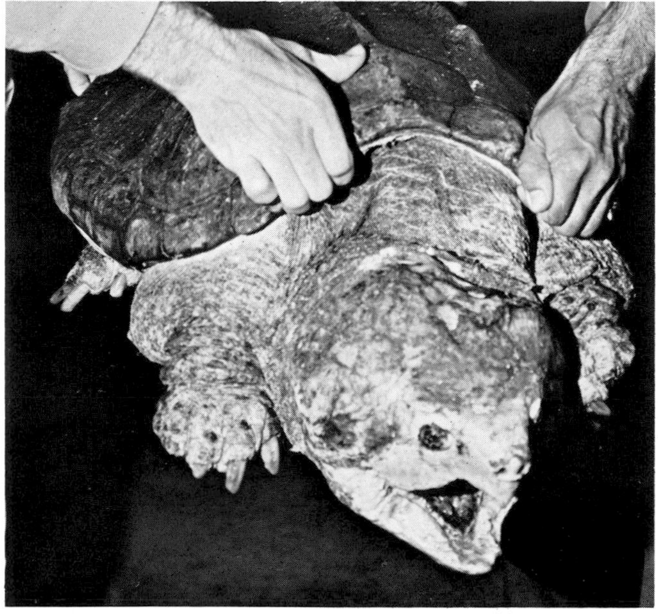

FIG. 25.18. Two-person carry for large snapping turtle.

to drydock large specimens. The flippers are strong and may strike a handler if the turtle is kept right side up. If a turtle is tipped upside down, it tends to relax and quietly allow examination and minor surgery.

Turtles and tortoises can be bled from the jugular vein (Fig. 25.19). The animal should be slightly sedated with ketamine hydrochloride. Extend the head and apply pressure at the base of the neck. The vein will become visible on the ventrolateral aspect of the neck. The vein is very mobile, so no definite landmarks can be stated.

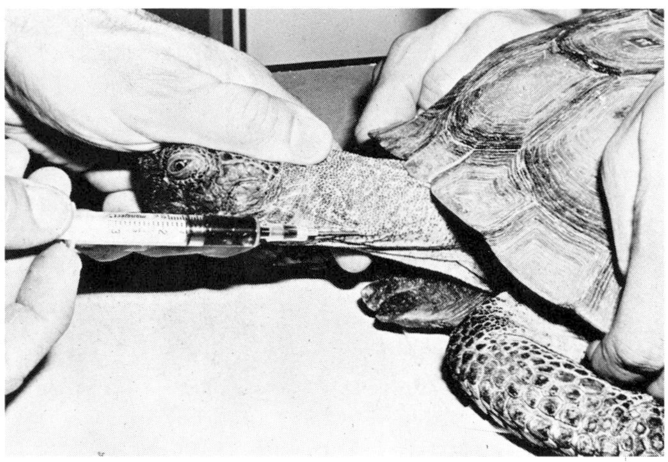

FIG. 25.19. Jugular venipuncture in a sedated tortoise.

Direct intracardial penetration can be made through the plastron[3]. Anesthetize and clean the tortoise for a sterile puncture. Penetrate on the midline at the junction of the pectoral and abdominal shield (Fig. 25.20).

The mouth of a chelonian can be opened with wooden or plastic wedges (see Fig. 2.17). Dowels or sheep and swine specula can be used on large species (Fig. 25.21).

Giant tortoises that are weakened or paralyzed can be supported as illustrated in Figure 25.22.

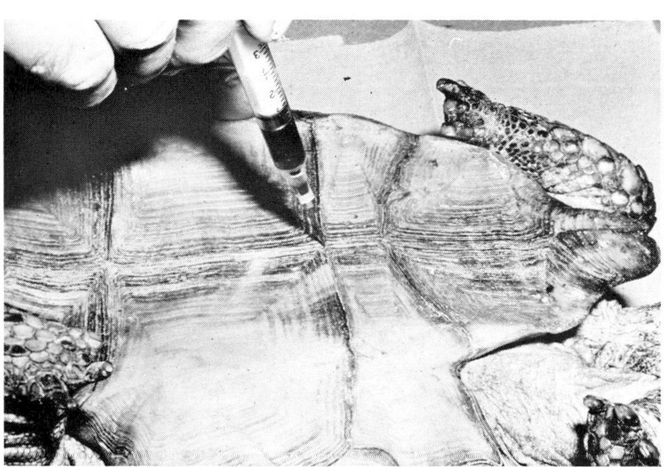

FIG. 25.20. Collecting intracardial blood sample from tortoise.

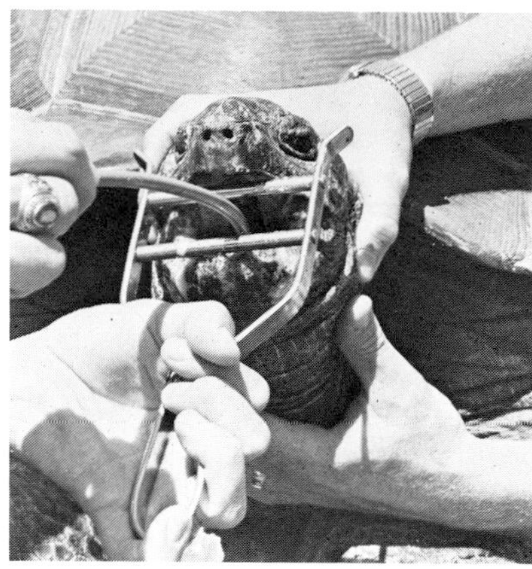

FIG. 25.21. Mouth speculum in place on a giant tortoise. Large animal must be depressed or sedated in order to hold the head out.

FIG. 25.22. Special dolly allows partially paralyzed tortoise to move about.

Transport

Terrestrial species are easily moved in small cages. Aquatic species must be kept damp but need not be immersed. Even sea turtles can tolerate being out of water for hours if they are kept cool and moist. Foam rubber soaked with water provides some moisture for shipping aquatic turtles or amphibians.

Chemical Restraint

Chemical restraint of tortoises and turtles is easily carried out using ketamine hydrochloride (15-60 mg/kg)[2] or Tilazol (3-5 mg/kg)[7]. In all instances the animal should be weighed and 30% of the weight deducted to exclude the weights of carapace and plastron. Care should be taken to place chemically immobilized chelonians in a recovery environment maintained at a temperature of 28-32 C (83-90 F). If the ambient air temperature is low, the recovery period will be dangerously prolonged. If the temperature is too high, the reptile may develop hyperthermia. Chelonians can be placed in an incubator or on a heating pad if supplemental heat is needed. Make certain that the temperature does not rise above 32 C (90 F). A chemically immobilized tortoise or turtle is incapable of behavioral or physical thermoregulation.

Recovery from chemical immobilization may require several hours. The animal should be carefully monitored until it recovers. Do not allow aquatic species access to water for at least 12 hours after completion of the procedure.

2. Unpublished data.

LIZARDS

Lizards vary in size from tiny skinks weighing a few grams to monitor lizards such as the Komodo dragon weighing 90 kg (200 lb). Obviously restraint techniques suitable for use on such a variety of species must differ.

Danger Potential

Most lizards are carnivorous, preying on insects, other reptiles, small mammals, or birds. Dental structures vary from hard bony plates to sharp teeth. The tails of certain species are weapons that can be used as whips. This is particularly true of the common green iguana, whose tail exceeds 3 ft in length at maturity. This tail can inflict a painful if not serious injury to the unwary handler. Many lizards are agile climbers with long, sharp, grasping claws which can inflict nasty scratches.

There are only two species of poisonous lizards—the Gila monster and the Mexican beaded lizard. Both of these species occur in the arid deserts of southwestern United States and Sonora, Mexico. They are rather phlegmatic lizards, not significantly aggressive unless molested. If annoyed, they are capable of grasping firmly and refusing to let go. A continual chewing motion allows venom, extruded onto the teeth of the lower jaw, to work its way into the punctured skin of the victim.

Physical Restraint

Tiny lizards can be hand held (Figs. 25.23, 25.24A). Some of them are pugnacious and will attempt to bite. Thin gloves will protect the hands (Figs. 25.25, 25.26). Small lizards can also be placed into plastic tubes (Fig. 25.24B) for radiographic studies, medication, or inhalation anesthesia. If the tube is small and a snug fit, it can be turned over for examination of the ventral aspect.

FIG. 25.23. Many small lizards can be manually restrained.

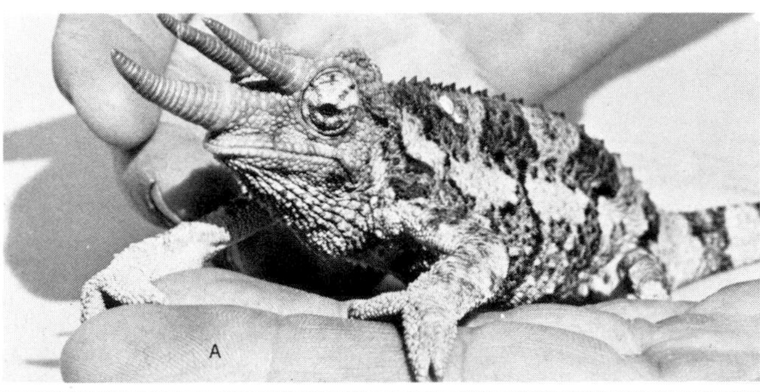

FIG. 25.24. **A.** Chameleon is docile and can be handled with minimal restraint. **B.** Small lizard or gecko placed in plastic tube for close inspection.

FIG. 25.26. Using gloves to grasp rough-scaled sungazer lizard.

A lizard should not be caught by the tail. Some species are able to discard the tail as a device to distract predators. Although the tail will eventually regrow, the individual will be a poor exhibit for many weeks.

Larger lizards such as iguanas can be presented for examination in a sack. Determine the location of the head and grasp the animal quickly behind the head through the sack, controlling it until you can regrab the animal as the sack is carefully removed. At all times keep the tail of larger species restrained (Fig. 25.27).

If one must handle a lizard to expose the ventral aspect, a long-sleeved shirt or wide-gauntleted gloves should be worn to protect from scratches. Once the animal is in hand, it can be situated in a variety of positions, as indicated in Figure 25.27. The mouth can be opened for ex-

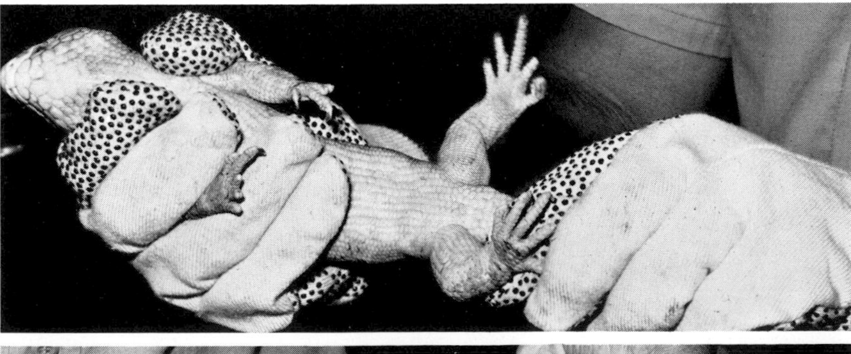

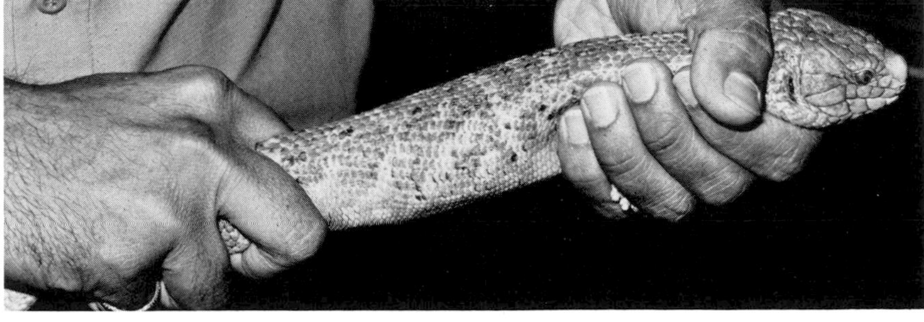

FIG. 25.25. Proper method of holding medium-sized lizard. Initial capture with a gloved hand may prevent biting and scratching. After capture, lizard can be safely held bare-handed.

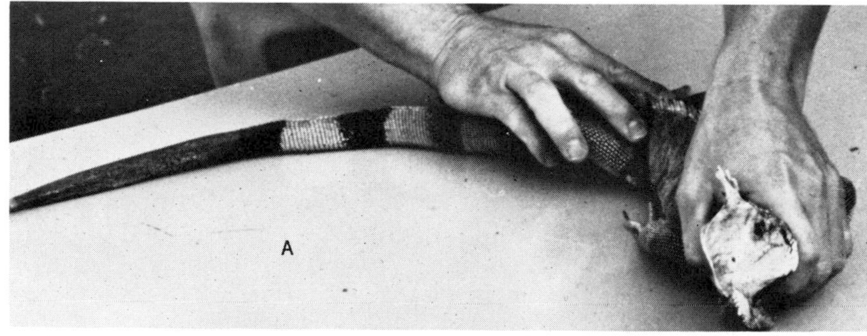

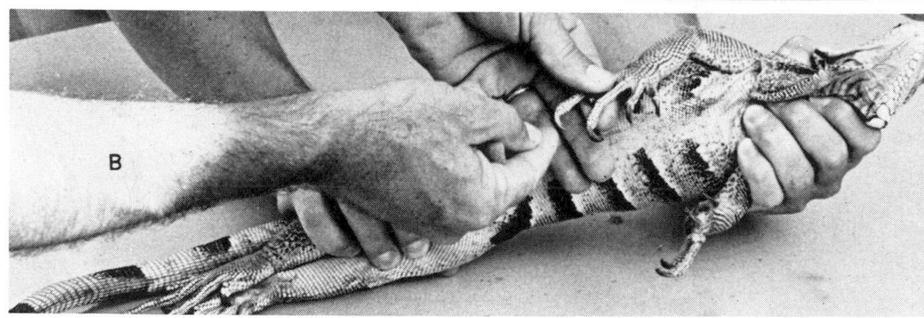

FIG. 25.27. **A.** Aggressive iguana is grasped from above, securing head and whiplike tail at the same time. **B.** Once animal is grasped, include hind legs in the tail hold to preclude scratching.

FIG. 25.28. Mouth of iguana can be opened as shown.

FIG. 25.29. Large pet iguana is handled by controlling head and tail.

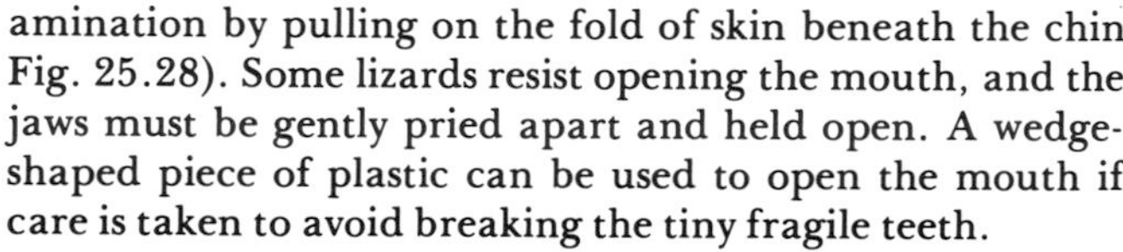

amination by pulling on the fold of skin beneath the chin Fig. 25.28). Some lizards resist opening the mouth, and the jaws must be gently pried apart and held open. A wedge-shaped piece of plastic can be used to open the mouth if care is taken to avoid breaking the tiny fragile teeth.

Some pet iguanas can be held without grasping the nape of the neck (Fig. 25.29). However, they frequently bite when held in this fashion by a strange person who manipulates the animal in an unaccustomed manner.

Large lizards must be handled either with nets or snares (Fig. 25.30). Take care to avoid injury to the neck when tightening a snare. Snares, cables, or ropes covered with plastic or rubber tubing are safest. As soon as the snare has been placed on the neck, grasp the tail, as with crocodilians. If manipulation is to be prolonged, tape the hind legs to the tail, as shown in Figure 25.30. The ensnared animal can be transported from one place to another with additional support beneath the body. Do not allow the weight to be suspended solely from the neck and tail.

An experienced individual can handle poisonous lizards by grasping them behind the head and neck, as indicated in Figures 25.31 and 25.32. They can also be grasped by the tail and lifted with a hook (Fig. 25.33 upper). They are unlikely to climb up their own tails onto the handler's hand. The inexperienced or nervous handler should control the head with a snake hook. These lizards can be persuaded to walk into a plastic tube, where they can be manipulated and examined as desired (Fig. 25.33 lower).

Lizards have a vein on the ventral side of the tail. To collect a blood sample, place the lizard on its back. Locate one of the ventral spinous processes; then insert a 20 gauge needle, directing it forward at a 45° angle parallel with the ventral spinous process. Insert the needle until the tip

FIG. 25.30. Snare can be used to control large monitors. Note tape on hind legs to prevent scratching.

FIG. 25.32. Proper hold for a Gila monster.

strikes the vertebral body. Withdraw the plunger to establish slight negative pressure in the syringe, then pull the needle slowly back 1-2 mm until blood wells into the syringe [2].

Toenails are easily accessible on lizards. A blood sample can be withdrawn into a capillary tube by cutting a toenail. Following collection the tip of the nail should be cauterized with silver nitrate or a hot spatula.

SNAKES

Snakes exhibit extreme diversity in morphology, physiology, activity, food habits, and other biological parameters.

FIG. 25.31. Gila monster approached from above and behind.

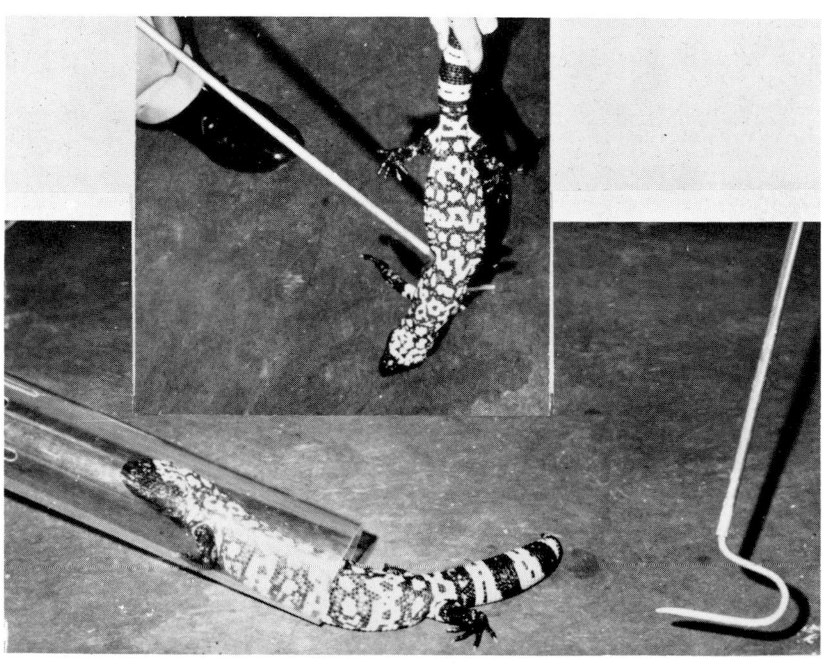

FIG. 25.33. Gila monster handled by the tail; hook gives additional support *(upper)*. Gila crawling into a plastic tube *(lower)*.

Danger Potential

All snakes can bite. Small specimens may be unable to open the mouth widely enough to be a hazard to human beings. Others can inflict serious injury or death. Venomous species require special handling techniques.

Constrictors kill their prey by coiling tightly around the body and preventing respiration, thus suffocating the victim. They do not crush the victim's bones. A 3 m (10 ft) boa allowed to completely encircle the body or neck of a restrainer is capable of causing death if assistance is not rendered quickly (Fig. 25.34).

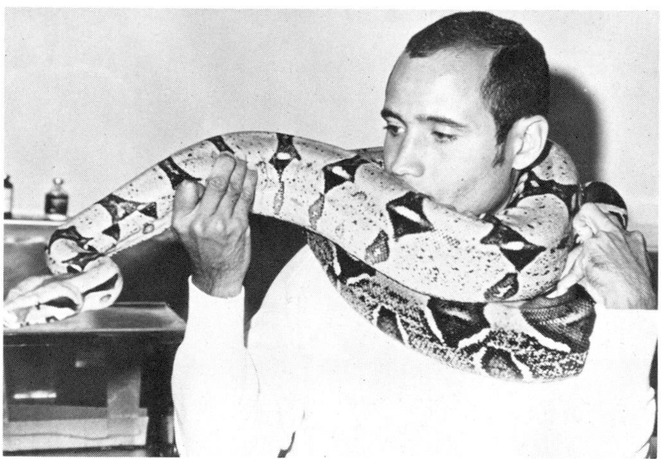

FIG. 25.34. Staged picture illustrates a very dangerous practice.

Nonpoisonous Snakes

Many nonpoisonous snakes are inoffensive (Fig. 25.35) and will not bite unless tormented unmercifully. The California boa can be manipulated without danger of being bitten. Some of the larger nonpoisonous snakes require control of the head, particularly when manipulating them for examination. When holding a snake it is important to support the body (Fig. 25.36). An unsupported snake becomes insecure and restless and may thrash about (Fig. 25.37). If the body is left dangling, a vigorous snake may thrash until its neck is dislocated or fractured. One cribo held by the neck (Fig. 25.38) thrashed so vigorously that the vertebral column was fractured.

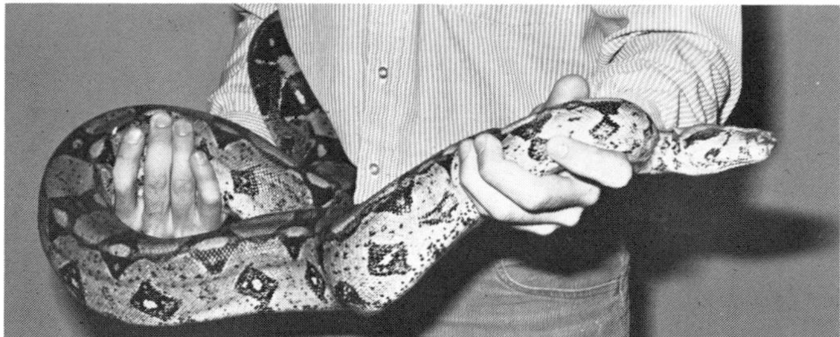

FIG. 25.36. Proper support of a medium-sized snake.

FIG. 25.37. Improper method of holding a medium-sized snake.

The mouth of a properly held snake can be opened either by pulling on the loose fold of skin between the lower jaws or by gently inserting a plastic spatula, covered forceps, or tongue depressor into the mouth, taking care not to damage the teeth (Figs. 25.39, 25.40).

Snake hooks are fundamental tools for working with reptiles (Fig. 25.41). Hooks can be used for directing movement, lifting snakes from containers, and a variety of other restraint procedures. A snake hook can be used to pin the head of any snake to the ground, allowing the handler to safely grasp it (Fig. 25.42). Only sufficient pressure to hold the snake should be exerted; too much pressure on the

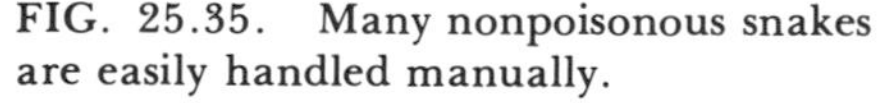

FIG. 25.35. Many nonpoisonous snakes are easily handled manually.

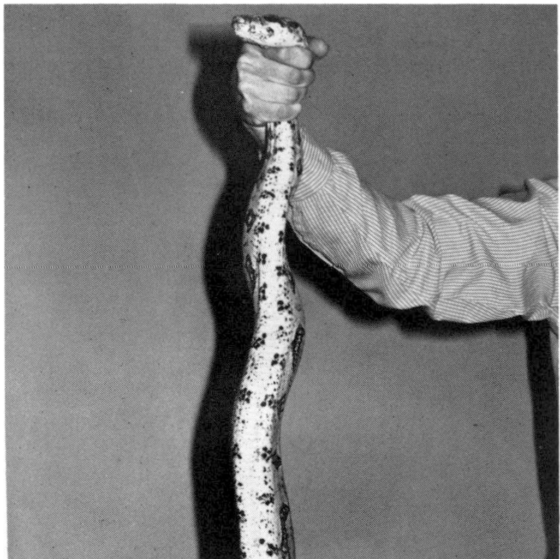

FIG. 25.38. Snake improperly lifted by the neck.

neck can seriously injure the spine or dislocate the head. Furthermore, if a manipulation is rough, a snake may subsequently refuse to eat, even to the point of starvation.

Be very cautious when dealing with large constrictors. Remove them from a cage with a large hook. If a constrictor is known to be docile, an experienced handler may be able to reach in carefully and grasp the animal. Remember that a snake in its own cage may behave in a territorial manner and is likely to be more aggressive than if it is removed from the cage into strange territory. Once the animal is out of the cage, it can be placed on the floor and the head gently pinned until it is grasped (Fig. 25.42).

Do not allow any large snake to throw a loop around your neck or body. It is natural for a snake to coil. A snake coiled around an arm will feel comfortable, and the arm will suffer no harm (Fig. 25.43).

Some of the more agile nonpoisonous snakes are difficult to handle. If pinned they thrash and often injure themselves. The experienced handler can usually pin and grab them quickly, but other techniques are more suitable, especially for novices. The snake loop or noose is a more effective tool for these species (Fig. 25.44).

FIG. 25.40. Opening mouth of snake with plastic wedge.

The Pilston snake tong is not a suitable tool for direct handling of snakes (Fig. 25.41), since it is likely to cause injury—particularly in the hands of a novice. It is useful for removing dishes, for feeding, or for holding plastic tubes.

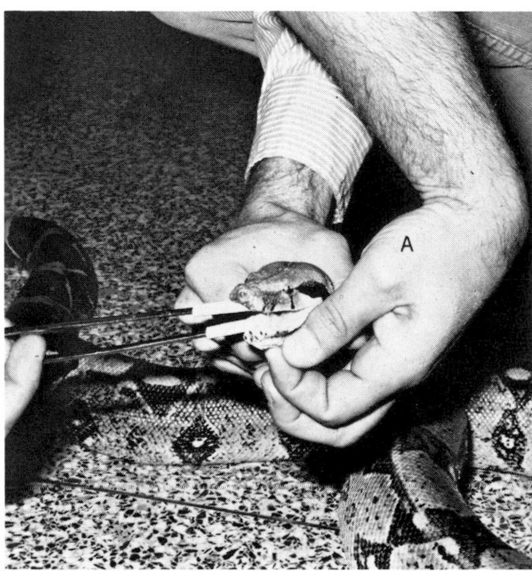

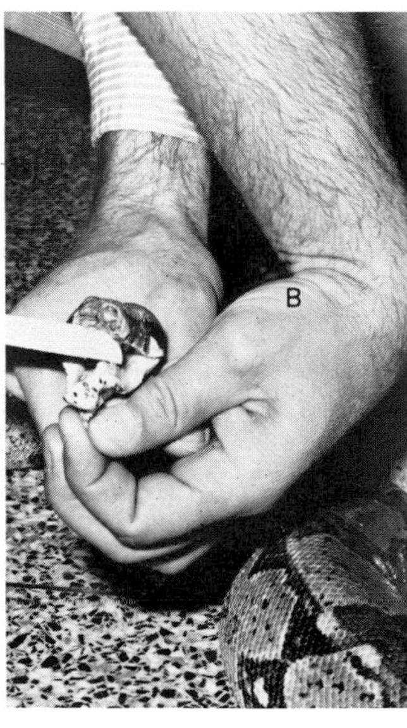

FIG. 25.39. Opening mouth of large snake: **A.** Covered thumb forceps. **B.** Tongue depressor.

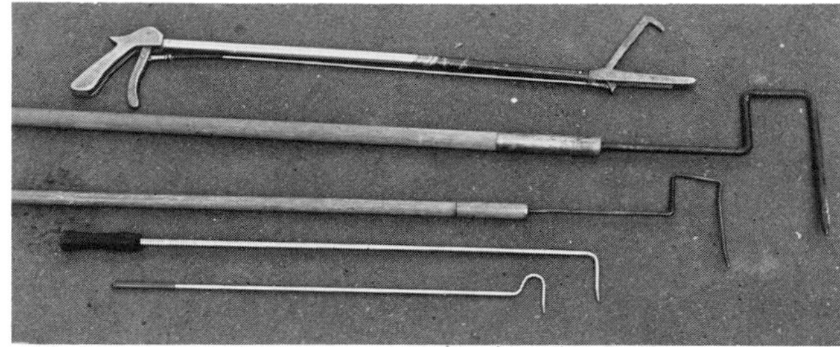

FIG. 25.41. Pilston snake tong and snake hooks.

The degree of agitation, aggressiveness, or nervousness exhibited by a snake may depend on the temperature at which it is handled and/or the amount of excitement it has experienced immediately prior to the manipulative procedure. An excited snake should be allowed to rest for a time. A can may be put over the snake to allow it to settle in darkness without the stimulation of outside influences (Fig. 25.45).

Mild restraint can be applied by cooling the animal. I vividly recall a client extracting a very docile python from a walk-in refrigerator when I arrived to examine the snake.

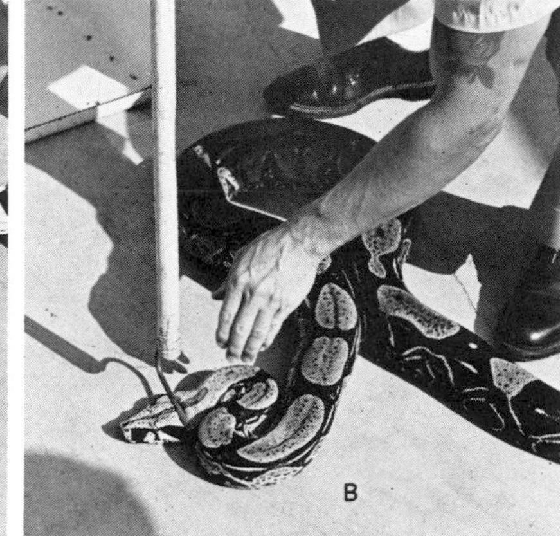

FIG. 25.42. **A.** Using snake hook to remove snake from cage. **B.** Gently pinning snake to aid in grasping the head.

FIG. 25.43. Supporting large constrictor snake by coiling it around an arm.

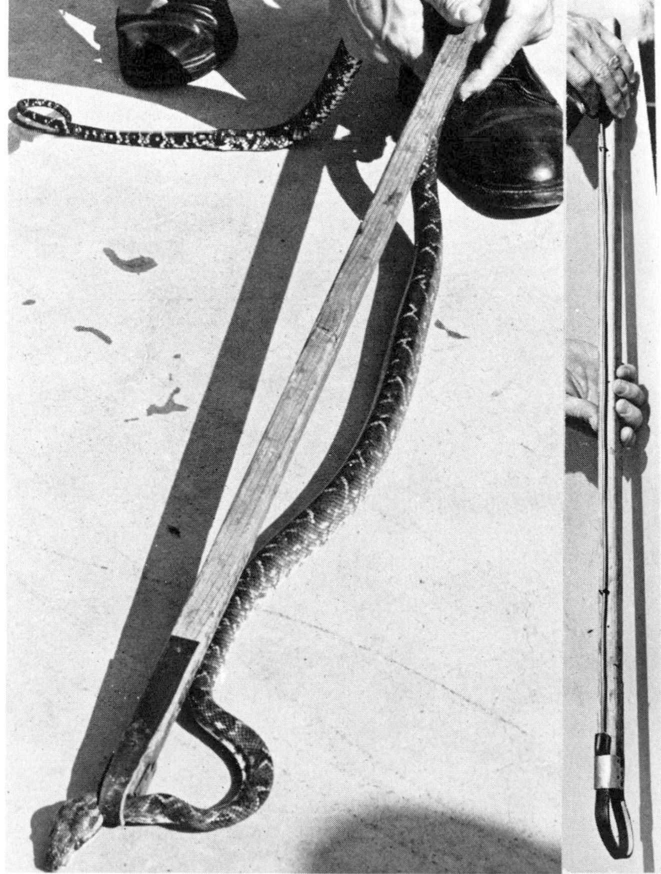

FIG. 25.44. Strap snake loop—used at the San Diego Zoo.

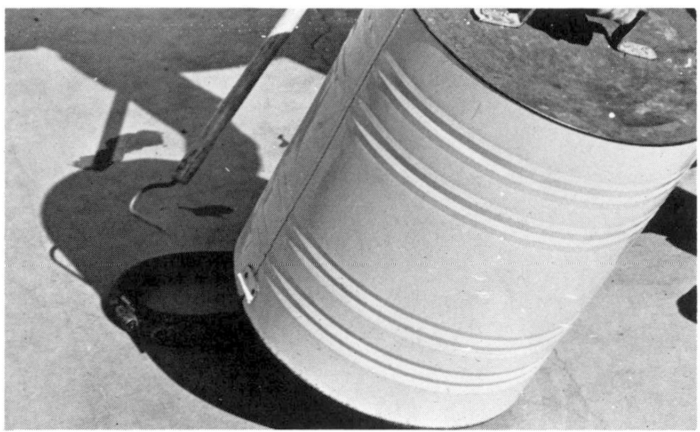

FIG. 25.45. Large tin can temporarily restricts the activity of a snake.

The desirability of cooling a snake is somewhat questionable, however, because of the likelihood that respiratory ailments will ensue from prolonged chilling. Cooling as a restraint technique should be administered with caution and should never be relied on to control poisonous snakes. If a household refrigerator is used to cool a snake, be sure the temperature does not fall below 3 C (38 F). The animal can be packed in an ice bath to maintain cooling during surgery (Fig. 25.46). The body must be continually surrounded with ice. Make certain that no foreign substances are allowed into the ice water, since this may lower the temperature to below 0 C (32 F) and cause permanent frost damage to the snake. After cooling, warm the snake slowly to prevent respiratory infection.

A small plastic shield is illustrated in Figure 25.47. This can be used to capture a large slightly aggressive nonpoisonous snake. The shield allows constant sight of the head; by applying gentle pressure, the shield can immobilize the snake sufficiently to enable the handler to grasp it behind the head.

Another capture technique is to allow a snake to begin to engulf prey, usually a rodent (Fig. 25.48). Then grasp the snake behind the head. This technique is less desirable than others because of the danger of regurgitation if a snake is handled soon after eating. Since regurgitation is undesirable, it is generally wise to refrain from manipulation of snakes other than for emergencies during the first two or three days after consumption of food.

Sexing snakes is a routine procedure. The male snake

FIG. 25.46. Anesthesia for nonvenomous reptiles—placing it in ice bath.

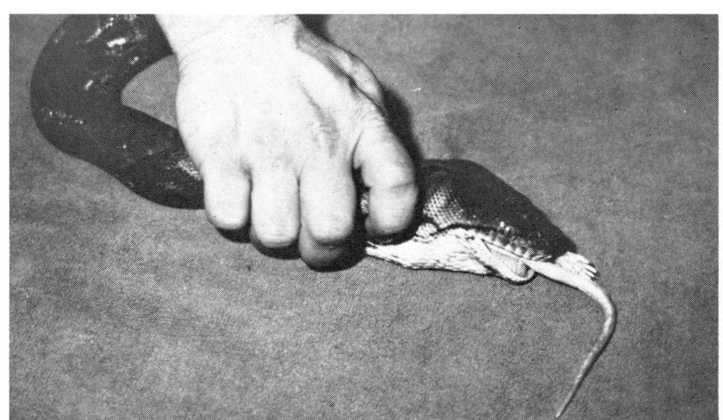

FIG. 25.48. Snake grasped just after it has started to engulf prey.

FIG. 25.47. Use of plastic shield to approach and capture aggressive nonpoisonous snake.

has paired hemipenes recessed into a diverticulum posterior or caudal to the vent. The sex of the animal can be determined by gently inserting a probe into the diverticulum (Figs. 25.49, 25.50). In the male the probe can be inserted to a considerable depth, depending on the size of the snake. Inserting the probe for a distance of 3 cm in a 60 cm (24 in.) male snake is not unusual, whereas in the female the probe can be inserted only 0.6 cm (0.25 in.).

The length of the probe to be used is determined by the size of the snake (Fig. 25.51). Probes are available commercially, or they can be improvised. It is important that the tip to be inserted into the diverticulum be a tiny ball and not a sharp point. In numerous instances straightened-out paper clips inserted to determine sex have penetrated the diverticulum, resulting in abscesses and frequently in the subsequent death of the snake.

Intramuscular injections can be given to a snake in the large muscles that parallel the vertebral column. These muscles directly overlie the ribs, so take care to avoid penetration of the abdominal or thoracic cavity and the subsequent injection of material into the lung or viscera. Hold the snake in the manner illustrated in Figure 25.52. Use the free hand to insert the needle into the muscle in a diagonal direction.

Subcutaneous infusion in the snake can be carried out at the lateral side of the back muscles. There is a slight groove in this area, and the skin here is relatively free of underlying tissue, allowing more space for distribution of the fluid.

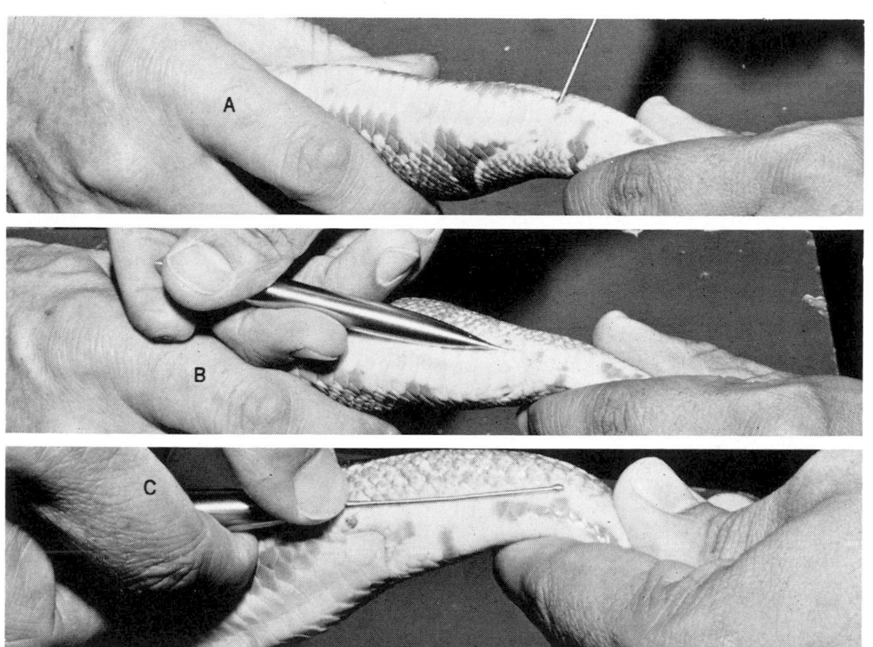

FIG. 25.49. Sexing a snake: **A.** Insert probe into lateral aspect of the vent. **B.** Direct probe posteriorly into bursa of the male. **C.** Evaluate depth of insertion.

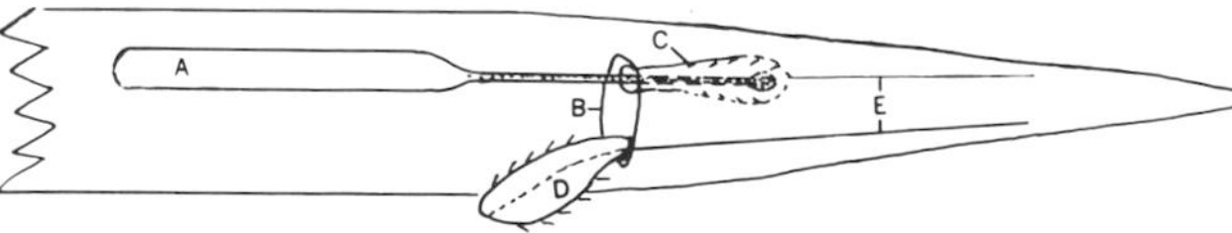

FIG. 25.50. Diagram of the hemipenes of a male snake: **A.** Probe. **B.** Vent. **C.** Bursa of the hemipenis. **D.** Erect hemipenis. **E.** Retractor muscles of the hemipenes.

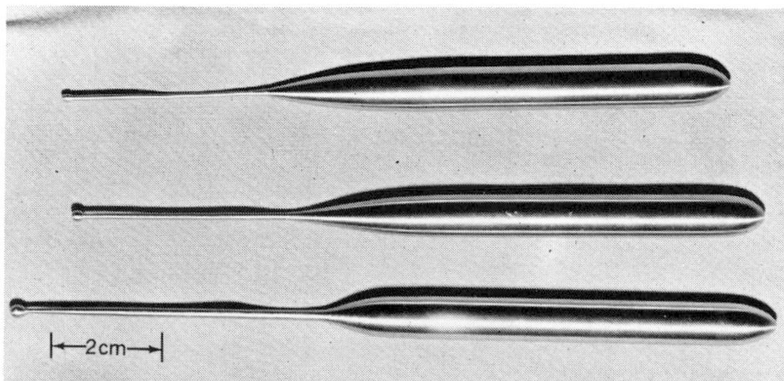

FIG. 25.51. Probes used to determine sex in snakes.

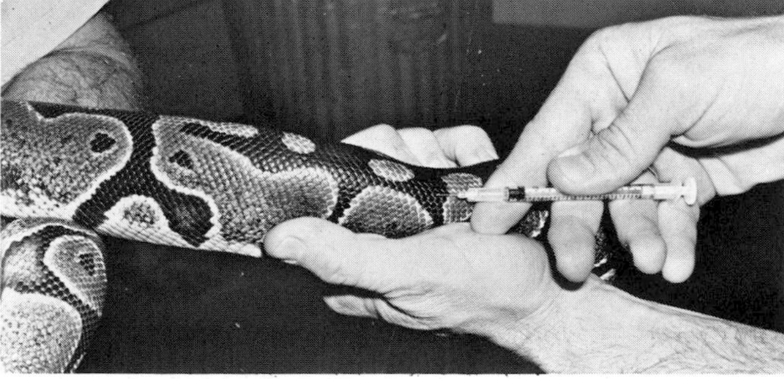

FIG. 25.52. Intramuscular injection can be given on either side of the vertebral column.

Obtaining blood samples from a snake is not easy. Some collectors advocate snipping off the tip of the tail, a technique that not only is disfiguring but may fail to yield a significant quantity of blood. A direct cardiac puncture is more satisfactory for obtaining a blood specimen. Although the procedure can be carried out on an unanesthetized snake, it is safer to anesthetize the animal. After immobilization the snake is placed on its back. The heart can usually be seen pulsating at the junction of the anterior and middle portions of the snake's body. The heart is mobile and can be moved forward or backward, so it must be fixed between the thumb and the finger before proceeding. A 20 guauge, 3.8 cm (1.5 in.) needle is inserted between the scales over the heart. Insert the needle slowly until the heart is reached, a point defined by movement of the needle in the hand; then penetrate the ventricle with a quick jab. When the sample has been obtained and the needle withdrawn, apply slight pressure on the heart for a moment to assist in sealing the puncture.

Unextruded caps over the eyes are a common problem of reptiles in captivity. The cap is an epithelial structure over the cornea of the eye, continuous with the skin, and should be shed with the skin. For various reasons the caps may fail to shed. As many as five or six may stack up if successive caps fail to shed. As the snake prepares to shed, the corneas of the eyes usually become opaque, making the snake relatively blind until the skin is shed. A snake is unlikely to feed during this period; if the caps are retained indefinitely, the snake may starve.

Reptile keepers usually try to loosen caps by soaking the animal, hoping the caps will come off without further treatment. If they do not come off, the caps must be removed manually. The snake should be restrained, either physically or with chemical immobilizers. Poisonous snakes must be partially anesthetized. The cap is then gently removed with fine forceps.

Poisonous Snakes

Many species of poisonous snakes are found throughout the world. Each species has different characteristics, degree of agility, and method of striking; but all are characterized by the presence of sacs from which the venom is extruded into fangs for envenomation of prey species or enemies. It is unwise to restrain any poisonous snake unless antivenin is at hand. Antivenin for a given species of snake must usually be obtained from the native country of the snake. Maintaining a stock of antivenins is costly, but the bite of many of these snakes is lethal unless such protective agents are available within minutes. If a bite occurs during a manipulative procedure and antivenin is unavailable, contact the nearest large reptile collection—either zoo or private facility.

Various groups of poisonous snakes differ in behavioral traits sufficiently to require the development of specialized restraint and handling techniques. No one should handle poisonous snakes without first developing expertise and confidence by practicing the techniques on nonpoisonous snakes. It it important to be confident that you can complete the procedure before beginning it. There may not be a second chance.

Vipers and pit vipers are usually somewhat phlegmatic heavy-bodied snakes. They arrange their bodies into a series of undulating folds from which position they can strike in any direction. The maximum striking distance is approximately two-thirds the length of the body. No snake flies through the air when it strikes. There are nearly as many techniques for handling vipers as there are handlers. Some of the more heroic involve direct catching with the bare hands, a technique that should be left to exhibitionists. Small vipers can be held for intramuscular injections by pressing them with wire screen (Fig. 25.53).

Pinning a snake is a common procedure, but it should be used on venomous species only by the experienced snake handler. Figure 25.54 shows the sequence of proper pinning. The hook is gently pressed behind the head; then the thumb and forefinger are used to grasp just behind the jaws. A firm hold must be kept until the snake is released. An alternate hold is with the thumb and second finger, the index finger being placed over the top of the head in the manner illustrated in Figure 25.54C.

FIG. 25.53. Small vipers can be pressed with a wire screen.

It is extremely difficult to pin one of the agile elapid snakes without injuring it. Neither is it wise to attempt to pin a massive snake such as a gaboon viper or an African puff adder. The snake noose or loop is more effective to control both types.

Plastic tubes of various sizes make excellent tools for handling many species of poisonous snakes [8]. They are now being used extensively by snake handlers in the United States. The plastic tubes can be capped on one end or left open. Slots in the sides of the tubes permit various procedures to be successfully completed in relative safety for both person and snake (Fig. 25.55).

The plastic tube should be sized so that the thickest portion of the body of the snake can barely pass through it. Otherwise the snake may turn around and come out. To tube the snake, place it on the floor with a hook. Hold the tube with a tong or, with docile or slow-moving species, by hand (Figs. 25.56, 25.57). The tube may also be placed along a wall. When the snake has crawled into the tube one-third of its length, very slowly and deliberately reach down and grasp with one hand both snake and tube at the point they adjoin (Fig. 25.58). Maintain this grasp continually until the snake is released. Never hold the tube

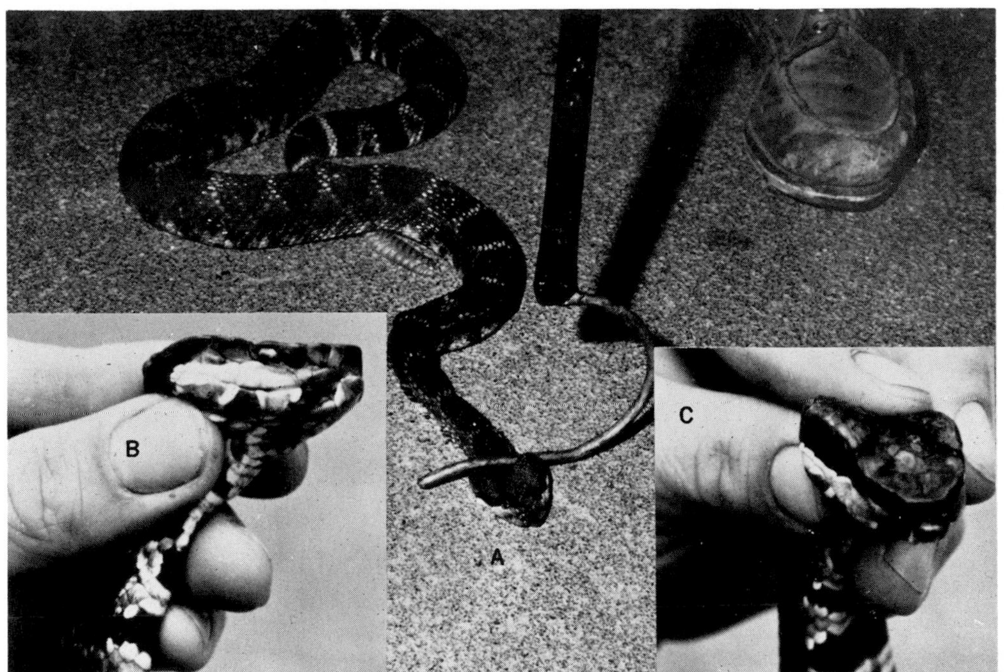

FIG. 25.54. **A.** Pinning a snake. **B, C.** Alternate ways of holding a venomous snake.

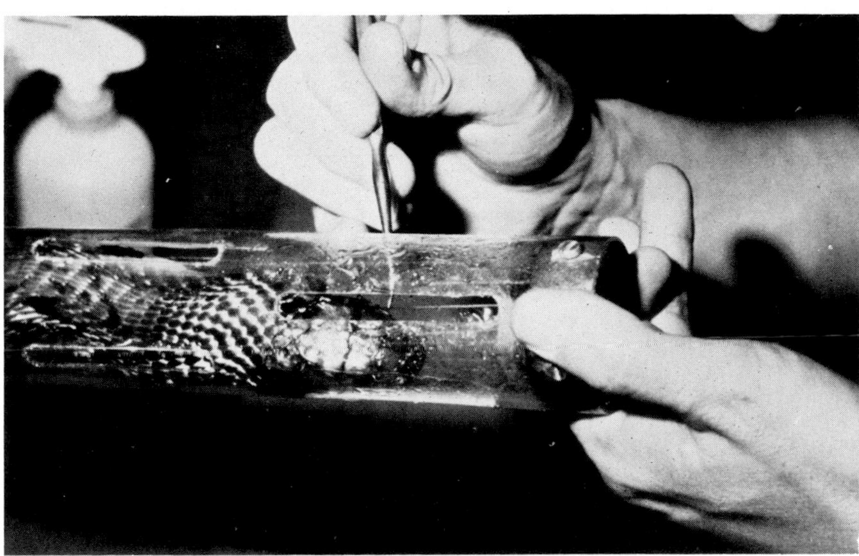

FIG. 25.55. Removing retained eye caps from cobra through slots in plastic tube.

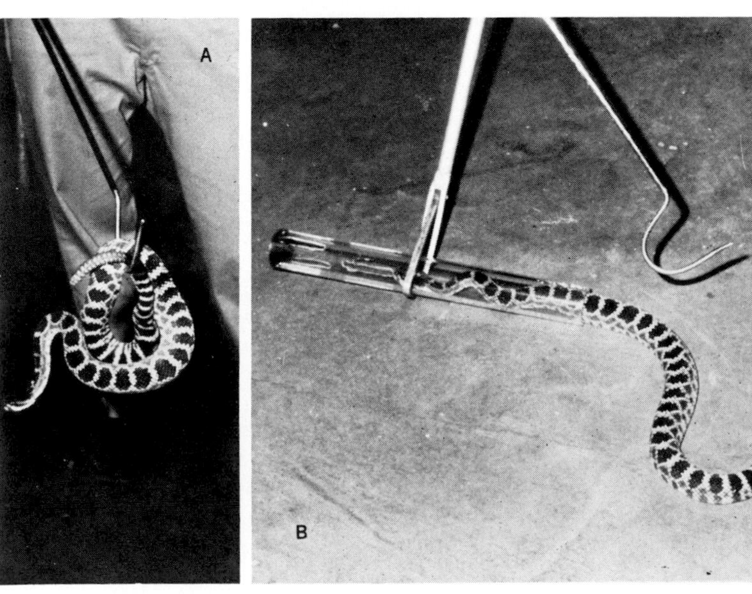

FIG. 25.56. Plastic tube restraint for snakes: **A.** Removal of snake to the floor with a hook. **B.** Holding plastic tube with Pilston snake tong while snake crawls in.

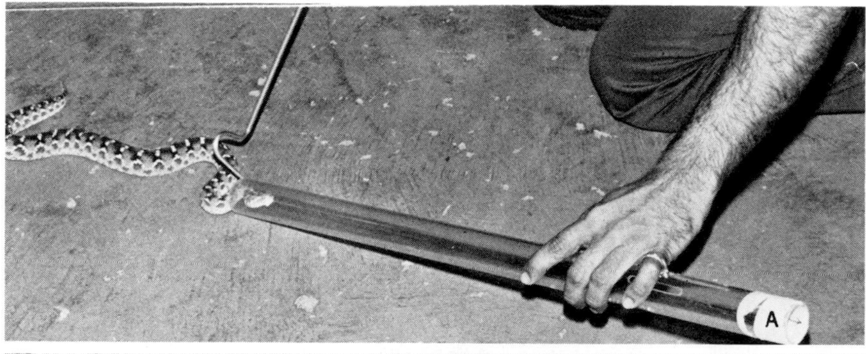

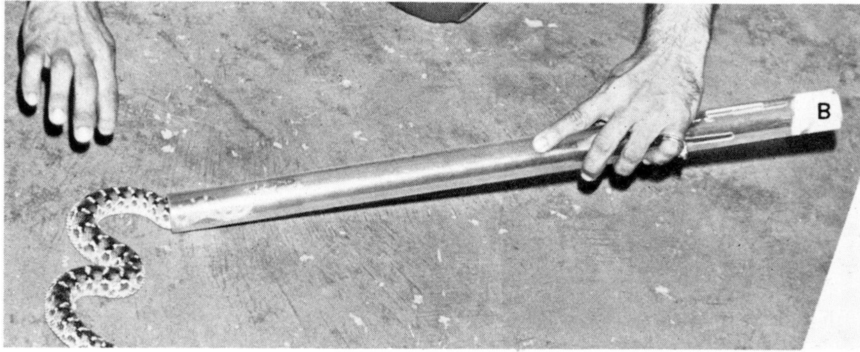

FIG. 25.57. **A.** Guiding snake into plastic tube with a hook. **B.** Preparing to grasp snake and tube at the same point.

FIG. 25.58. As soon as snake and tube are grasped, unit can be manipulated into any position.

FIG. 25.59. Knowledge of cobra's behavior allows experienced keeper to touch the head while snake is hooded. Cobra strikes forward and downward.

with one hand and the snake with the other; the snake might back out of the tube and bite.

Various manipulative procedures can be carried out with the snake in a tube. Examination of both ventral and dorsal aspects, intramuscular injections, forced sheddings, sexing, and removal of caps from the eyes can be carried out with a great degree of safety on venomous snakes in tubes.

It is sometimes difficult to induce some flighty snakes such as cobras and other elapids to enter a tube, but with patience most of them can be successfully tubed. There are some exceptions. This technique is dangerous and unsuitable for handling large, extremely aggressive, and fast-moving snakes such as the boomslang or the king cobra. With all species of elapids it is essential to hold the plastic tube with long forceps such as the Pilston tong.

The elapidae (cobra) family, in addition to being generally more aggressive and equipped with a more toxic venom, are flightier in temperament than vipers and consequently more dangerous to manipulate. The cobra's defensive posture is to raise the body to a vertical position with hood up. These snakes strike forward and downward from that position (Fig. 25.59). Other venomous snakes have different striking patterns.

Squeeze boxes (Fig. 25.60) are more suitable than tubes for handling large, swift, and aggressive elapid snakes. Squeeze boxes can be incorporated directly into the permanent cage. This is particularly important for a snake such as the king cobra. Covering the squeeze cage with a solid top darkens it and creates a refuge for the snake, enticing it to crawl inside. Then a trap door is closed and the snake is contained. Removing the solid top permits pressing the snake with the screen squeeze, as illustrated in Figures 25.61 and 25.62. The screen permits the removal of eye caps and administration of intramuscular injections.

FIG. 25.60. Special snake squeeze box for handling cobras.

A versatile squeeze cage can be constructed with a removable top and slotted sides. The snake is hooked into the open box and a plastic or screen squeeze is inserted into a slot to press the animal (Fig. 25.63). The squeeze is held in place with removable adjustable rods (Fig. 25.64). To examine anterior surfaces, tip the cage upside down, release the squeeze slightly to let the snake flip over, then retighten the squeeze (Fig. 25.64B).

Another type of restraint cage is constructed with a small side door. The hole must be too small to accommodate the snake's head while the body is blocking the hole, or the snake will escape. The snake hook is inserted through the small opening and the snake pulled over against the hole (Fig. 25.65). Intramuscular or subcutaneous injections can be given. This box is not suitable for the removal of caps, to alleviate poor shedding, or to carry out other elaborate procedures.

FIG. 25.61. When solid top is removed, screen squeeze can be pressed onto the snake.

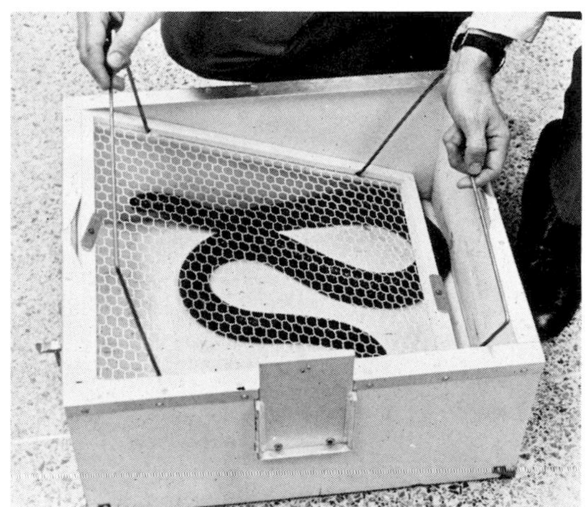

FIG. 25.62. Screen squeeze must be pressed straight down, otherwise snake may escape.

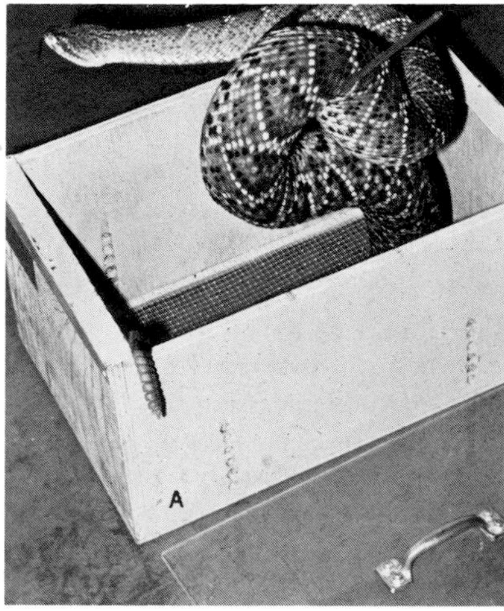

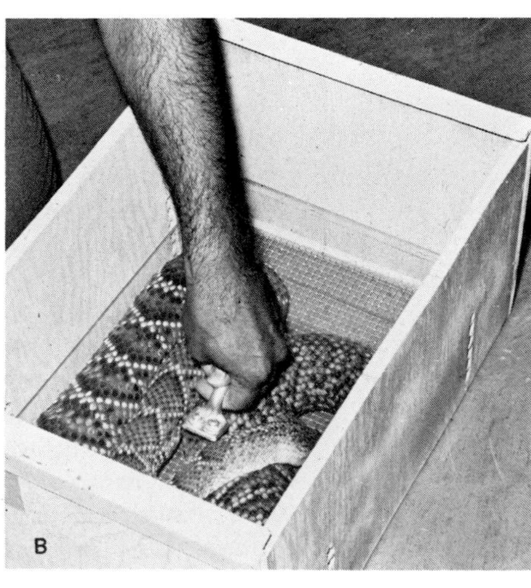

FIG. 25.63. **A.** Placing rattlesnake in homemade squeeze. **B.** Pressing snake against screen bottom with fitted sheet of plastic.

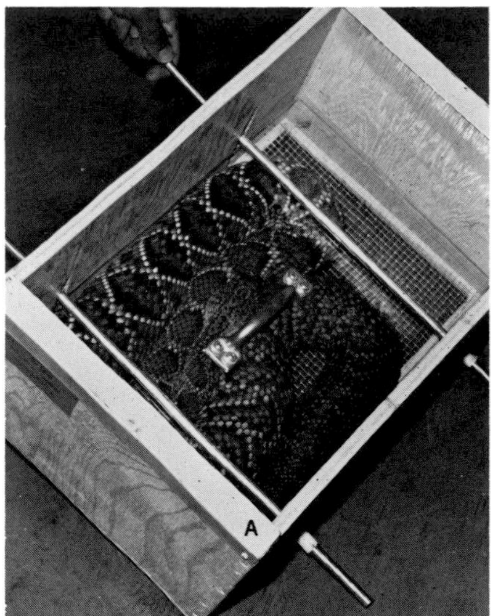

FIG. 25.64. **A.** Holding the squeeze with aluminum rods. **B.** Cage is turned over to allow examination and/or intramuscular injections through screen.

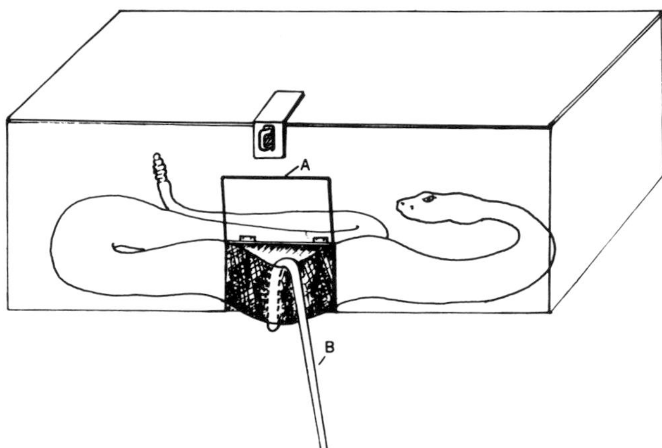

FIG. 25.65. Diagram of restraint box containing a side door. **A.** Door. **B.** Snake hook.

FIG. 25.66. Face shield for working with a spitting cobra.

A limited number of elapids are capable of forcibly ejecting venom from the fangs at the eyes of handlers, temporarily or permanently blinding the victim. Handlers of spitting cobras must use specialized equipment—usually a plastic shield or goggles—to protect the eyes (Fig. 25.66). Except for this equipment, spitting cobras are handled in the same manner as other cobra-type snakes.

Sea snakes present particularly difficult restraint problems. They have very short fangs, but the venom of these serpents is the most highly toxic known. Usually they are handled with small nets or squeeze cages as illustrated for other elapids.

Venomous snakes can be rendered nonvenomous by incision and ligation of the duct emptying the venom gland [6]. Some handlers have advocated the removal of the entire venom gland, but this requires radical surgery and is dangerously traumatic for the snake. Removal of the fangs is of no value in rendering a venomous snake harmless; new fangs will appear within a few days.

Snakebite

The bite from a nonpoisonous snake is rarely serious, except that the mouth of any snake is likely to harbor potentially dangerous bacteria. Any snakebite should be treated as a puncture wound. Initial bleeding should be encouraged to cleanse the tracts, and the area should be thoroughly washed with soap and water.

Small colubrid and boid snakes may bite and persist in their grasp. It is a mistake to attempt to tear away from them since this may lacerate the skin. Grasp the head and force the mouth open to disengage the snake.

Snakes that prey on birds frequently have long teeth to penetrate the feather layer and reach the bird's body. A bite from a green tree boa or an anaconda can result in a serious laceration. Prompt medical attention should be sought.

Those who handle poisonous snakes risk being bitten and must be prepared to apply first aid procedures to themselves or associates. The likelihood of being bitten is probably directly correlated with the number of snakes handled and the care exercised in working with the animals. The emotional response of a victim to a venomous snakebite may complicate first aid treatment. The bitten person may become hysterical, faint, hyperventilate, or suffer from neurogenic shock in addition to suffering from the effects of the venom.

The managers of most large reptile collections follow a planned protocol when dealing with snakebite emergencies. Following is an example of such a protocol.

1. Return the snake to its own enclosure.
2. If returning the snake within 30 seconds seems improbable, kill the snake by a quick blow to the head with a hook or any other nearby object that will enable you to reach out without risking another bite.
3. Sound any alarm system in use or call for help.
4. If an identification slip is on the cage, remove this and keep it on your person.
5. Remove appropriate antivenin from refrigerated storage and place it nearby.
6. Lie down and rest until help arrives.
7. If others are present in the immediate area at the time of the bite, they should carry out steps 1-5 and the victim should lie down and rest.

Russell has outlined appropriate steps for first aid treatment of any venomous snakebite [9]. These steps are summarized below:

Step 1. Apply a constricting band. In the case of a bite by a pit viper the band should be placed 4-8 cm proximal to the bite and tightened only enough to occlude lymphatic and venous return. It is neither necessary nor safe to hamper arterial flow. The constricting band should be loosened for 90 seconds every 10 minutes.

If the bite is from an elapid, the value of constriction is questionable. If used, a light tourniquet should be applied and left in place until antivenin is given.

Step 2. Incision and suction are indicated for viper and pit viper bites but are apparently valueless for elapid bites. Incisions should be 2-5 cm long over each fang mark and only deep enough to penetrate the skin. Incision and suction is of little value if delayed over 30 minutes. Suction should be continued for at least 1 hour.

Step 3. Immobilize the bite area, splinting limbs if possible. Keep the bite area below the heart level. Have the patient sit or lie down and avoid exertion. Fear and excitement can be alleviated by calm actions and reassurance on the part of the person giving the first aid.

Step 4. Transport the victim to a hospital for intensive care by a physician.

Step 5. Provide the physician with information about the species of snake involved and the time interval since the bite, point out constricting bands or tourniquets, report any unusual signs, and give details of any treatment given.

Step 6. Antivenin should be administered only by trained persons with equipment and medication available to cope with anaphylaxis should the victim prove to be sensitive to horse serum.

Transport

Snakes are adept at escaping through tiny openings. Any shipping crate must be checked carefully for loose screens or doors. Glass or plastic cages can be used for short local trips, but wooden cages must be used for interzoo shipment. Cages for venomous species must be lockable.

Reptiles must be protected from extreme temperature variations. Styrofoam iceboxes make excellent temporary carrying cages for local trips, or foam rubber or styrofoam can be laminated onto the inside of wooden crates.

Snakes are commonly transported in sacks. A hoop sack (Fig. 25.67) is particularly desirable for transporting poisonous snakes, since a snake can be hooked into the sack while the handler remains at a safe distance. Once the snake is inside, the hoop is flipped over so the snake cannot crawl out. Then a cord is tied around the sack. If the sack is constructed with a double bottom, it can be grasped to tip the snake out. The double bottom eliminates the danger of being bitten through the canvas sack.

FIG. 25.67. **A.** Placing snake on a hook into snake sack. **B.** False bottom sewn into sack allows grasping without danger of being bitten through the sack.

Tie a sack firmly at the top. Snakes easily force themselves through very small openings. Furthermore, it is important to check sacks to be certain there are no tiny holes that could be enlarged by a snake forcing its way through.

Docile nonvenomous snakes can be placed directly in the sack (Fig. 25.68 left). With more aggressive nonvenomous

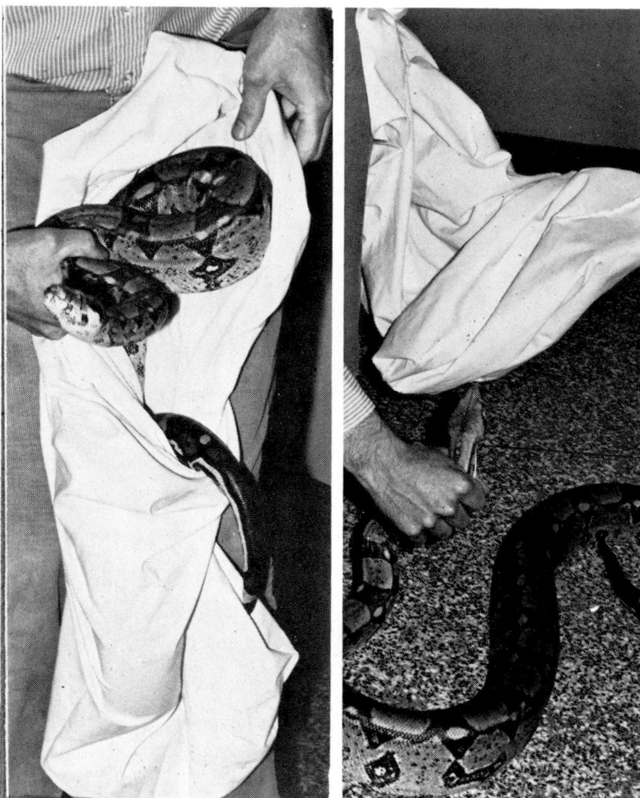

FIG. 25.68. Putting docile nonpoisonous snake into sack *(left)*. Sacking aggressive nonpoisonous snake: Turn sack inside out, keeping the hand in the inverted sack *(right)*.

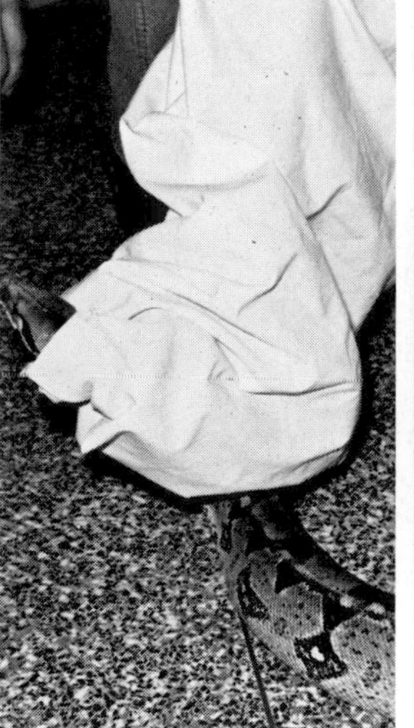

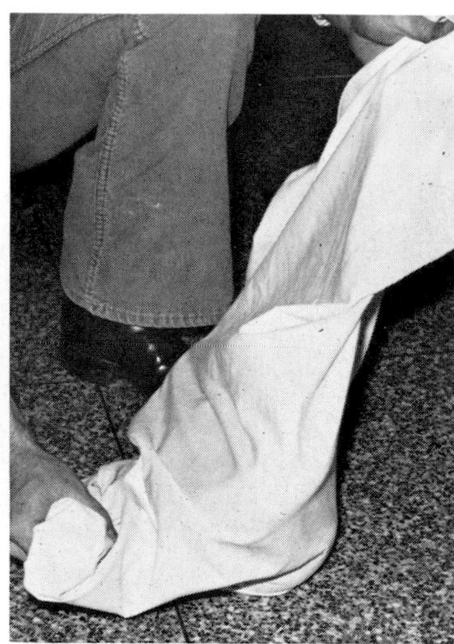

FIG. 25.69. Sacking aggressive nonpoisonous snake *(cont.)*: Regrasp snake behind its head with the inverted sack *(left)*. Sack is pulled over body of the snake *(right)*.

FIG. 25.70. Open-mouth stance typical of ketamine hydrochloride anesthesia in a snake.

snakes, turn the sack inside out and grasp the head through it, everting the sack over the snake (Figs. 25.68 right, 25.69).

Be observant when placing a sacked snake in a strange place. Be sure that it will not get too hot or too cold. Additionally, do not place the sack on a chair or any other place where someone might unknowingly place a hand on it or sit on it. In one instance, a rattlesnake in a sack was brought into my office in my absence and placed on a chair. When I returned, I did not notice the sack and, during the course of casual conversation, started to sit down. Fortunately those nearby alerted me before I sat on the snake.

Chemical Restraint

Intramuscular ketamine hydrochloride has proved to be an efficient, effective immobilizing and anesthetizing agent for use in all species of snakes on which it has been tested [4,5,7]. The required dosage (55-88 mg/kg) is higher than that required for mammals of comparable weights. The drug produces mild sedation or profound anesthesia depending on the dose used [5]. One of the first signs of a snake's impending immobilization after intramuscular injection of ketamine is a characteristic elevation of the head in a peculiar stargazing manner, with the mouth held partially open (Fig. 25.70).

All snake immobilization procedures should be carried out with the snake on a heating pad or in a warm environment—not on a cold stainless steel table. Aftercare must include monitoring the environmental temperature to maintain sufficient body heat to allow the animal to metabolize the drug.

Several other methods are used for anesthetizing venomous snakes. One method requires a box or glass jar. The anesthetic gas is passed into the box or jar (Fig. 25.71), or a pledget of cotton soaked with anesthetic can be placed inside the box or jar containing the snake. The progression of anesthesia is determined by inverting the jar. If the snake is unable to right itself, it is probably anesthetized (Fig. 25.72).

Since a snake can hold its breath for 15-20 minutes, inhalant anesthesia may become a prolonged procedure. For this reason apnea, a common concern of the mammalian restrainer, is not a serious problem of restrained reptiles. In fact it is sometimes difficult to ascertain whether a snake is actually alive during anesthetic procedures. However, it is easy to insert a tube past the glottis through the trachea and respire a distressed animal either manually, using mouth-tube respiration, or with inhalation equipment.

Another technique for anesthesia can be used by persons who are confident in working with snakes. Catch the snake with a loop and grasp it behind the head. Place a gauze roll in the mouth to expose the glottis (Fig. 25.73) and intubate the animal intratracheally, permitting anesthetic to be forced into the lungs (Fig. 25.74).

Once the snake is anesthetized, tape it to a board (Figs. 25.75, 25.76). Masking tape can be used on small snakes to minimize scale damage. Use only adhesive tape on large vipers.

FIG. 25.71. Desiccating jar used to administer volatile anesthesia to a snake.

FIG. 25.72. Anesthesia is reached when snake fails to right itself if inverted.

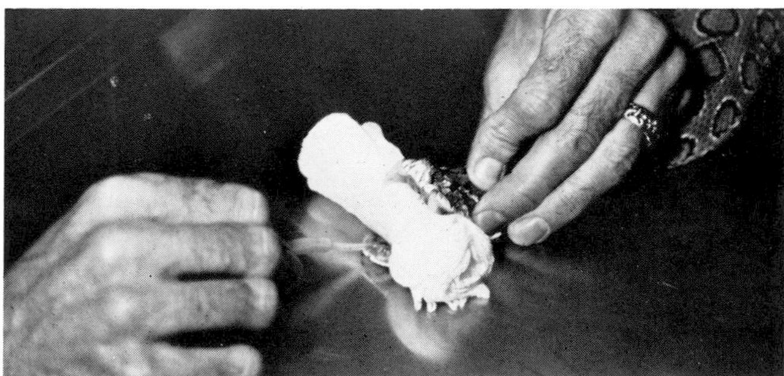

FIG. 25.73. Gauze roll holds mouth of Russell's viper open while inserting trachea tube for anesthesia.

REFERENCES

1. Ball, D. J. 1974. Handling and restraint of reptiles. Int. Zoo Yearb. 14:138-40.
2. Esra, G. N.; Benirschke, K.; and Griner, L. A. 1975. Blood collecting technique in lizards. J. Am. Vet. Med. Assoc. 167:555.
3. Gandal, C. P. 1958. A practical method of obtaining blood from anesthetized turtles by means of cardiac puncture. Zoologica 43:93-94.

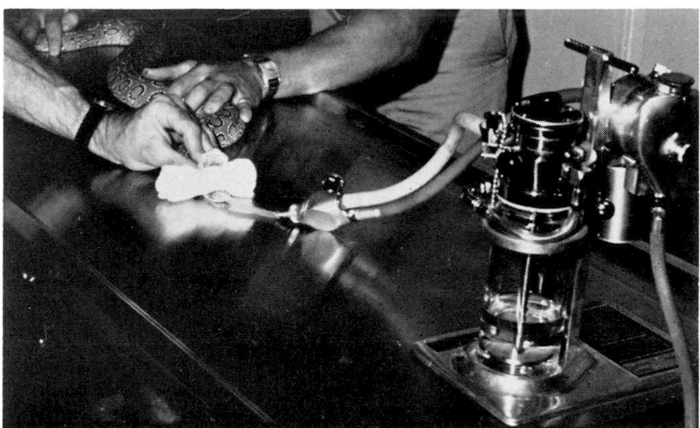

FIG. 25.74. Anesthetizing Russell's viper with forced breathing of volatile anesthetic.

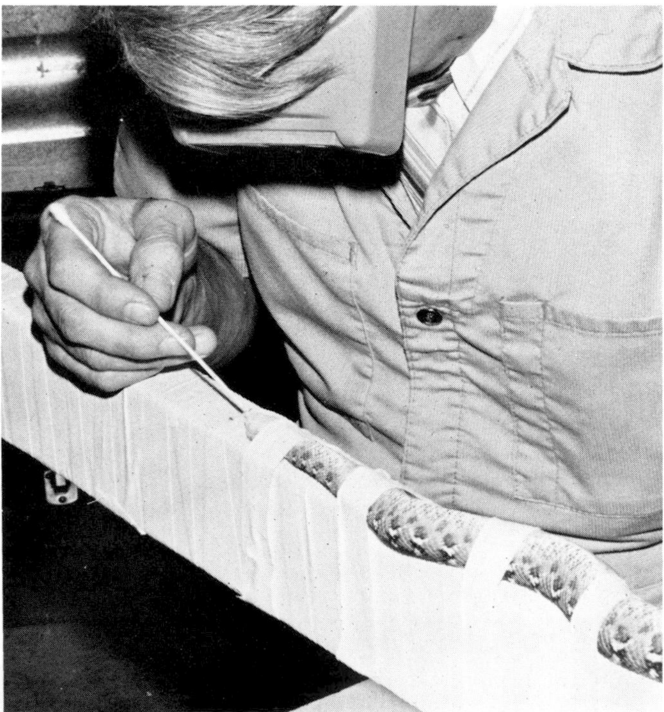

FIG. 25.75. Snake taped to a board.

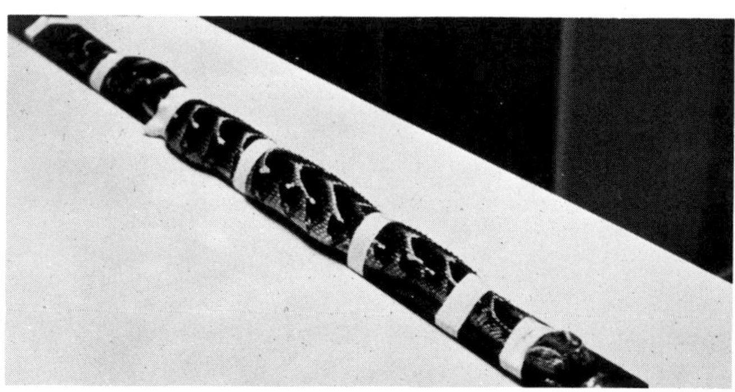

FIG. 25.76. Venomous snake sedated and immobilized.

4. Glenn, J. L.; Straight, R.; and Synder, C. C. 1972. Clinical use of ketamine hydrochloride as an anesthetic agent for snakes. Am. J. Vet. Res. 33:1091-1103.
5. ______. 1972. Ketalar: A new anesthetic for use in snakes. Int. Zoo Yearb. 12:224-26.
6. ______. 1973. Surgical technique for isolation of the main venom gland of viperid, crotalid and elapid snakes. Toxicon 11:231-33.
7. Klide, A. M., and Klein, L. V. 1973. Chemical restraint of three reptilian species. J. Zoo Anim. Med. 4:8-11.
8. Murphy, J. B. 1971. A method for immobilizing venomous snakes at Dallas Zoo. Int. Zoo Yearb. 11:233.
9. Russell, F. E. 1967. First aid for snake venom poisoning. Toxicon 4:285-89.
10. Wallach, J. D., and Hoessle, C. 1970. M99 as an immobilizing agent in poikilotherms. Vet. Med.Small Anim. Clin. 65:163-67.
11. Wallach, J. D.; Hoessle, C.; and Bennett, J. 1967. Hypoglycemic shock in captive alligators. J. Am. Vet. Med. Assoc. 151:893-96.

SUPPLEMENTAL READING

Caras, R. 1974. Venomous Animals of the World. Englewood Cliffs, N.J.: Prentice-Hall.

Moore, G. M., ed. 1965. Poisonous Snakes of the World, p. 5099. Washington, D.C.: U.S. Government Printing Office.

Oehme, F. W.; Brown, J. F.; and Fowler, M. E. 1975. Toxins of animal origin. In J. J. Casarett and J. Doull, eds. Toxicology, pp. 570-90. New York: Macmillan.

Porter, K. R. 1972. Herpetology. Philadelphia: W. B. Saunders.

Romer, A. S. 1966. Vertebrate Paleontology, 3rd ed. Chicago: Univ. of Chicago Press.

26 AMPHIBIANS AND FISH

CLASSIFICATION

Class Amphibia (over 2,500 species)
  Order Anura (Salienta): frogs, toads
  Order Caudata (Urodela): salamanders, sirenids
  Order Gymnophonia (Caecilia): caecilians
Class Pisces (over 25,000 species ranging in length from a few millimeters to 18 m)

AMPHIBIANS

Danger Potential

Although toothless, large salamanders and a few of the large toads are capable of inflicting a painful bite with hardened cornified plates similar to those of turtles. A large amphibian's jaws are strong enough to give a sharp pinch.

No amphibian has a venomous bite. The secretion of the parotid gland of certain toads (Colorado River and marine) is toxic to dogs if ingested (Fig. 26.1). The material also irritates the human eye if it gets into the conjunctival sac. Hands must be thoroughly washed after handling a venomous toad lest a hand should inadvertently touch the mouth or rub an eye. In one unique instance a Colombian giant toad actually projected the secretion from the parotid gland into the eye of a keeper who was holding the toad (Fig. 26.2). The keeper experienced an immediate burning sensation and quickly rinsed the secretion from his eyes, but the irritation persisted for an hour. No additional toxic manifestations were noted.

Anatomy and Physiology

Amphibians are obligate aquatic animals for at least the reproductive phase of the life cycle. Some species have developed the capacity to lay eggs that develop and metamorphose in minimal quantities of water, such as the amount that accumulates in the junctions of leaves and stems of plants in a rain forest.

The skin glands of amphibians produce a secretion that protects their skin from the deleterious effects of water. These secretions prevent desiccation, inhibit the growth of microorganisms on the skin, and in specialized cases discourage predators [1]. The secretions may cause problems for the animal restrainer, because they make the animal slippery and difficult to grasp and hold. All species of toads secrete substances (parotid) that are repulsive to animals that bite or mouth them. Puppies may bite toads once but usually resist the impulse a second time.

Frogs of the genus *Dendrobates* and others produce a potent biotoxin, with effects much like those of a cardioactive glycoside. The skins of these frogs were used by indigenous peoples to make the substance used to coat arrowheads for hunting wild animals in Central and South America.

The skin of some amphibians functions as a supplementary respiratory organ.

Physical Restraint [2]

Frogs can be captured from the water or while on a dry surface. Small species can be netted or grasped bare-handed (Fig. 26.3), but some of the larger may bite, so take

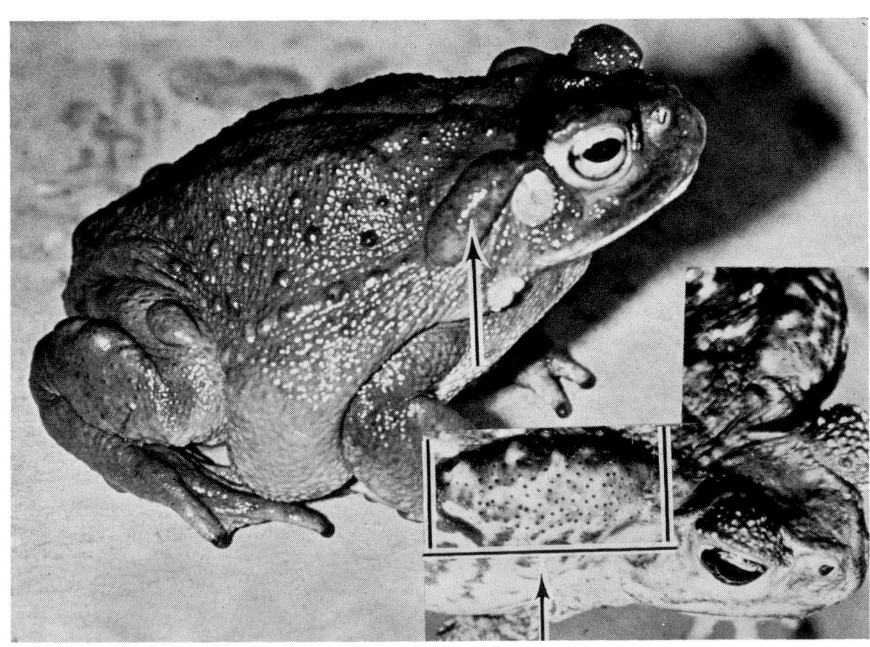

FIG. 26.1. Colorado River toad. Arrow points to parotid gland.

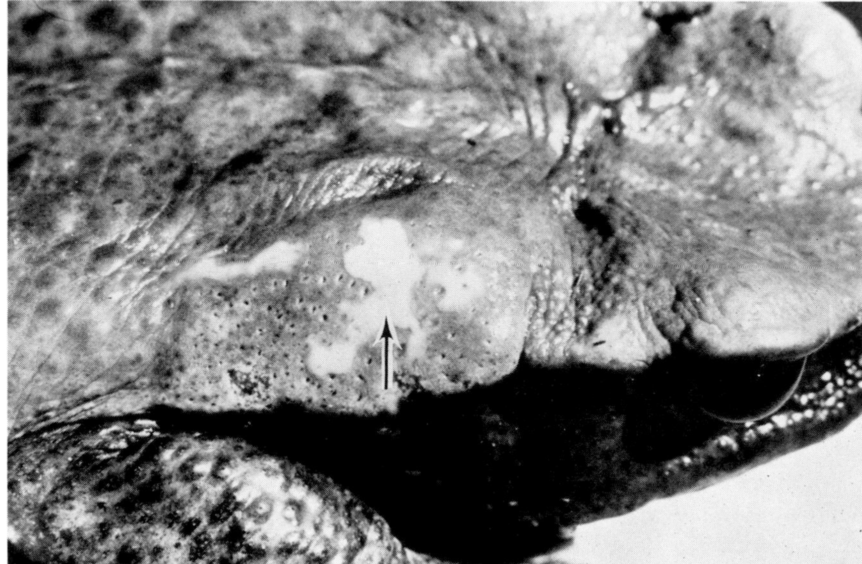

FIG. 26.2. Secretion from parotid gland of Colombian giant toad.

precautions. Since the skin surface is slippery, the hand must surround the frog (Fig. 26.4). If the animal is to be held for more than a few seconds, wrap the legs with gauze to assist in control (Figs. 26.5, 26.6).

Small toads do not bite and can be handled like frogs (Fig. 26.7). The skin is usually not as slippery as the frog's. Large frogs and toads are picked up and held by the hind legs, with support of the body (Fig. 26.8).

Smaller species of salamanders can be grasped barehanded. One or both hands may be used. Moisten the hands first (Fig. 26.9). Aquatic species are more slippery. Large salamanders like the hellbender or giant salamander should be handled more cautiously. The initial grasp should be made quickly over the back, as one would grasp a crocodilian.

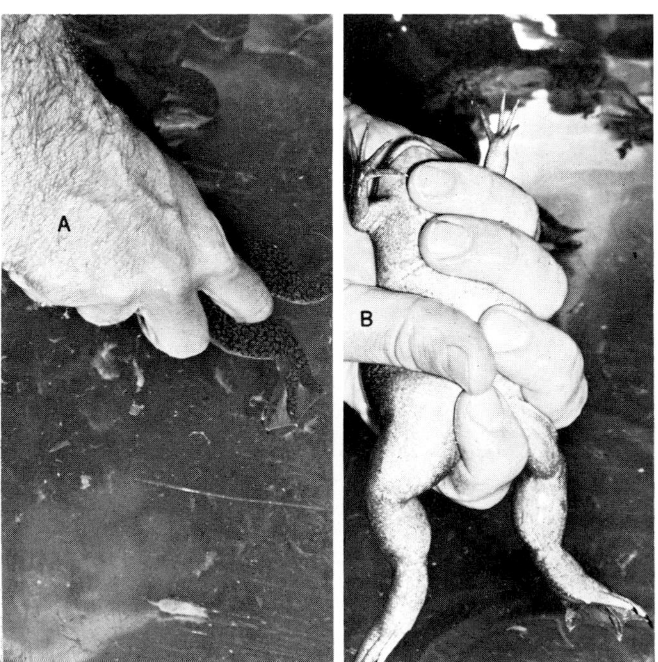

FIG. 26.3. **A.** Grasping clawed toad in a tank. **B.** One-hand method of holding toad.

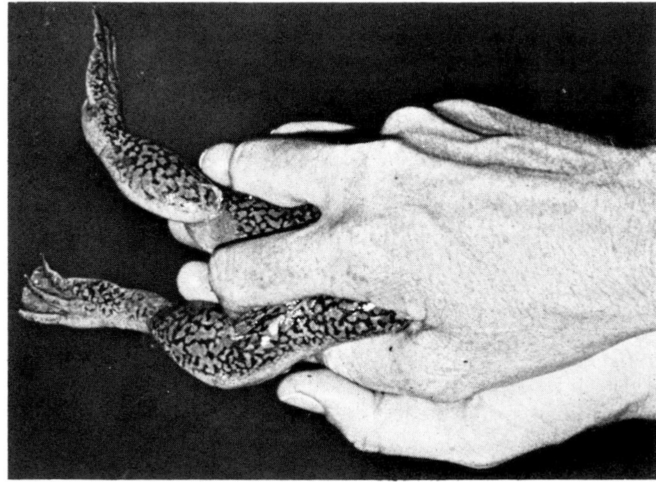

FIG. 26.4. Two-hand method for holding frog.

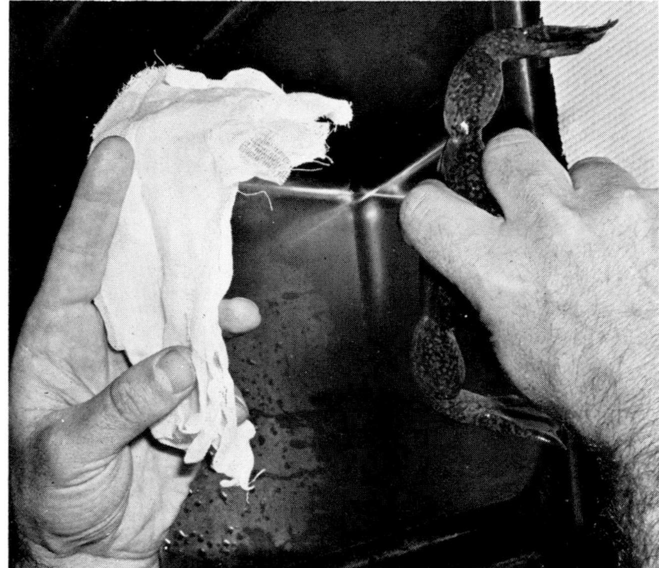

FIG. 26.5. Placing gauze on the legs to provide firmer grip.

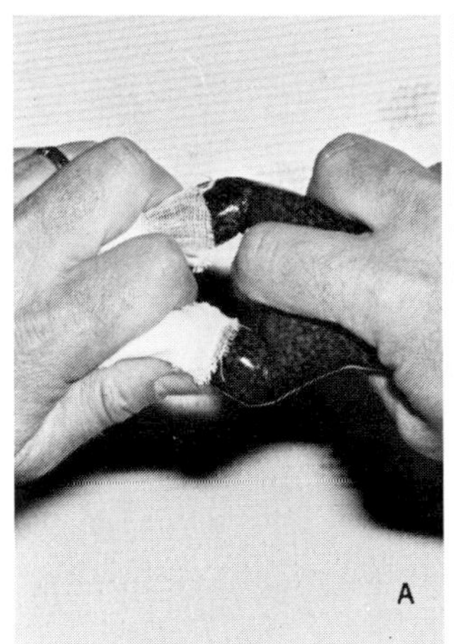

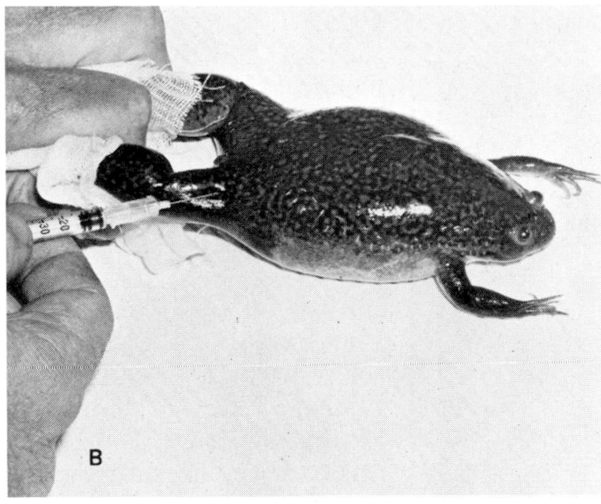

FIG. 26.6. **A.** Using gauze to provide firmer grip. **B.** Subcutaneous injection of a frog.

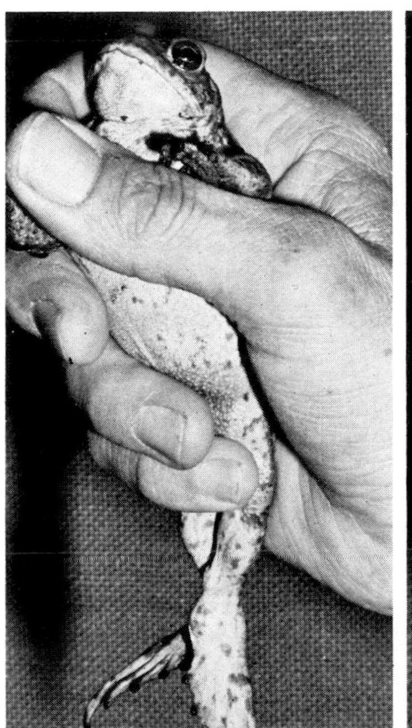
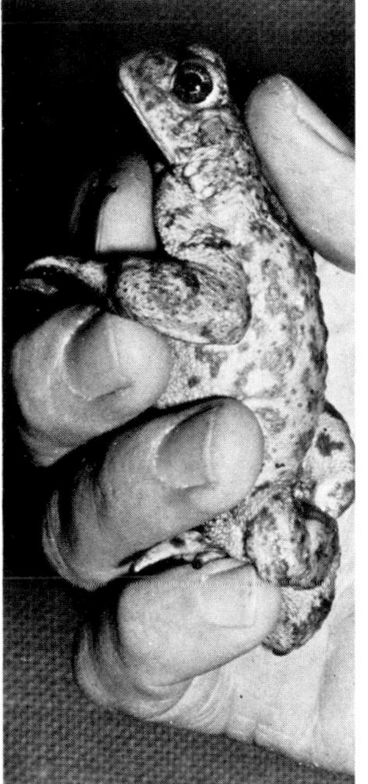

FIG. 26.7. Methods for holding small toad.

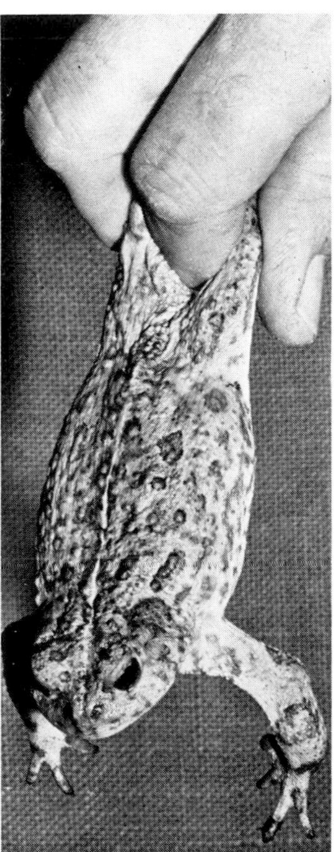
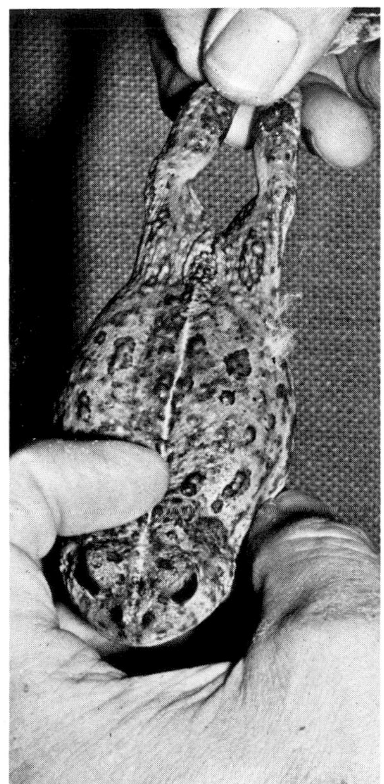

FIG. 26.8. Picking up and holding toad by hind legs.

Caecilians are legless burrowing amphibians. They are seldom seen in the wild and are poor exhibit animals. Caecilians are docile and can be handled like a snake.

Chemical Restraint

The skin of amphibians is sufficiently permeable to allow absorption of anesthetic agents (Fig. 26.10). Frogs and toads can be immersed in a 1-2% solution of ethyl carbamate (urethane), a 10% solution of ethyl alcohol, or a 0.1% solution of tricaine methanesulfonate (MS-222, Sandoz) [5]. Intravascular injections are administered to frogs and toads through the dorsal lymph sacs. The sacs are actually paired lymph hearts, located dorsally on either side of the last vertebrae (urostyle) [5]. Locate the sacs by observing the rhythmic beating under bright overhead illumination, and use a 25 gauge, 2.5 cm needle [5].

Anesthetics administered intravascularly include hexobarbital (120 mg/kg), pentobarbital (60 mg/kg), 1-2 ml

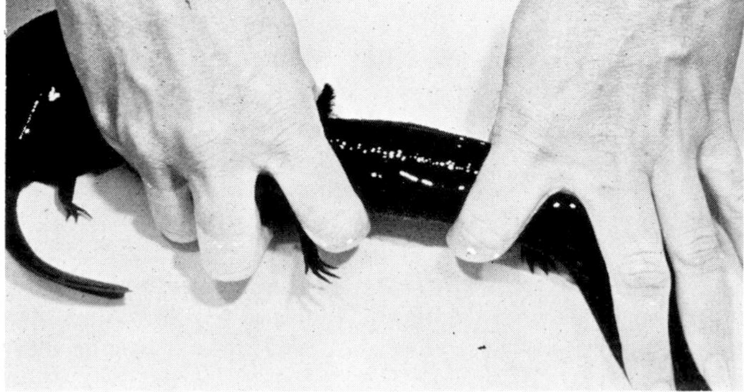

FIG. 26.9. Grasping axolotl salamander with moistened hands.

FIG. 26.10. Toad placed in anesthetic solution.

of a 10% solution of chloral hydrate, and 5% ethyl carbamate (0.04-0.12 ml/g) [5].

Salamanders are anesthetized by immersion in a solution of tricaine methanesulfonate (1:3,000) or 1-2% ethyl carbamate. Sedation is not needed for handling caecilians.

FISH

Fish culture for food and as a hobby has become a multimillion dollar business. The industry is not new; food fish have been kept in home ponds or tanks for over 4,000 years [8,9]. The keeping of fish as pets dates back to the ancient Roman and Chinese empires.

Anatomy and Physiology

Streamlined fish with no demarcation between head and body are difficult to grasp or hold. The surface of the body is coated with a protective mucus which further complicates handling.

Fish breathe via gills. They can live out of water for only a few minutes. Any handling procedure prolonged beyond this limit must provide oxygenated water to bathe the gills. Cold water contains more oxygen than warm water. If fish are held in small aquaria or transport tanks, the water must be aerated or the fish will die—especially if the water is warm.

Danger Potential

Fish can bite, sting, and abrade the skin of handlers. Their bite is not venomous, but carnivorous species are efficient predators with numerous teeth designed for grasping, shearing, and tearing flesh. Free-living sharks are notorious for their gruesome attacks on people. In captivity injuries are not as likely to occur, but these animals can and do bite if mishandled.

Smaller species may be equally aggressive. The diminutive piranha can severely injure. A barracuda or a moray eel can inflict serious damage. The list of fish that may bite is long. It is important to keep hands clear of the mouths of fish.

Sharks have extremely rough skin that can abrade the skin of a handler.

Certain species of fish have developed the capacity to generate powerful electrical charges. Electric rays are found in three families distributed throughout the warm oceans of the world. Electrical discharges of over 200 volts and 2,000 watts have been measured from one of these rays [6]. This electricity is generated in one or more specialized organs. Electrical shock is used to procure food as well as to serve as an effective deterrent to predators.

Electric organs are also present in the electric catfish (Africa), stargazer (USA), and the electric eel (South America) [6]. The electric eel can develop a charge of 550 volts. Animals coming within the electrical field will be stunned. Touching the animal while it is discharging may be fatal to persons. Touching the eel simultaneously at two sites increases electrical conductivity and hence the hazard [6].

Many species of fish have venomous spines which inject secretions of varying degrees of toxicity. Very few venomous fish have been studied. Probably only a fraction of the actual number of venomous species are known [1]. Halstead [3] lists 4 shark, 58 stingray, 47 catfish, 4 weeverfish, 57 scorpion fish, 15 toadfish, 3 stargazer, 8 rabbitfish, and 8 surgeonfish species as venomous. The spines are usually associated with one or more of the fins, although in the case of the surgeonfish the lancetlike spine is on the lateral surface of the base of the tail.

Tropical fish enthusiasts may be at risk of envenomation from small specimens of poisonous fish. One aquarist was injured while netting a small specimen of oriental catfish from a tank. The fish flopped from the net and he automatically reached out to grab it. A dorsal spine was driven into his hand. Pain was instantaneous and violent, and he was incapacitated within seconds. Sedation and general nursing care were required to effect recovery within two days.

The syndrome of envenomation by most stingray fish includes severe pain, inflammation of the wound, and temporary paralysis. There are no antivenins available.

Fish spines need not be venomous to inflict injury. The cartilaginous rays which keep the fins expanded may puncture a hand that grabs them. Those who must handle catfish repeatedly may wear light cotton gloves to minimize injuries from the spines. The value of the gloves as protection to the handler's hands must be balanced against the increased abrasion and trauma inflicted on the fish.

Being drenched is one unique hazard of working with certain fish. A small shark once swam up to the side of the tank and spit water all over my shirt. I was told this behavior is not unexpected for that species of shark.

Physical Restraint

Nets are the primary tools for handling fish. The size of the net and the mesh vary from the lightweight, fine-meshed small nets used to transfer tropical fish (Fig. 26.11) to heavy, coarse commercial nets for harvesting tunafish. All shapes and sizes in between are employed (Figs. 26.12-26.14). Plastic and wooden panels can be used to direct fish into a limited area to permit netting or into a smaller capture tank submerged in a large tank.

Some fish have loosely adhered scales which can be easily scraped off by harsh handling. Mucous secretions on the

FIG. 26.11. Netting tropical fish.

FIG. 26.12. Netting fish from a fish hatchery raceway.

FIG. 26.13. Using a hoop net to collect samples of fish.

FIG. 26.14. Another type of fish net.

surface of the skin plus the scales protect the epithelium from microbial infection. Avoid rubbing off the mucus or scales since this may allow penetration of pathogenic bacteria. Harsh, prolonged, or repeated handling of fish can lead to heavy death loss as a result of stress and reduced resistance to bacterial infection.

Although nets are essential fish-handling tools, there are hazards associated with their use. Scales can be abraded and fins entangled in the mesh. The lidless eyes of fish usually bulge from the surface of the body and are thus susceptible to abrasions if the net rubs tightly against the fish. Fish can also see the approaching net and may become traumatized as they dart about frantically to elude capture. Each net should be restricted to use in a single tank or be disinfected between tanks to avoid spreading infections or parasites from tank to tank.

Some tropical fish enthusiasts recommend the use of polyurethane plastic bags for fish capture instead of nets [8]. The bag is transparent, thus the fish experience less fear and are not able to elude capture so readily. Another

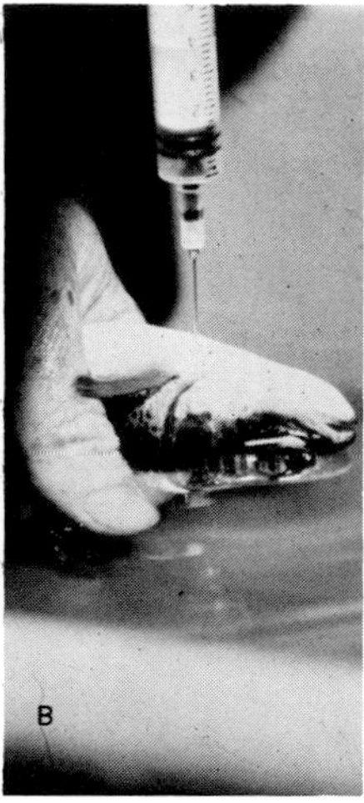

FIG. 26.15. **A.** Anesthetized fish. **B.** Obtaining intracardial blood sample from a trout.

important benefit is that the disposable plastic bag is germ- and parasite-free, unlike nets used repeatedly between tanks.

Chemical Restraint

Numerous chemical agents have been dissolved in water to anesthetize fish (Fig. 26.15) [4,7]. A few are listed in Table 26.1.

TABLE 26.1. Anesthetic agents for fish

Anesthetic	Concentration per Liter H_2O	Anesthetic Qualities		
		Induction	Maintenance	Recovery
Ether	10-50 ml	2-3 min.	fair	5-30 min
Sodium secobarbital	35 mg	30-60 min.	good	60+ min.
Ethyl carbamate (urethan)	5-40 mg	2-3 min.	good	10-15 min.
Chloral hydrate	0.8-0.9 g	8-10 min.	poor	20-30 min.
Tricaine methanesulfonate (Finquel, MS-222)	25-100 mg	1-3 min.	excellent	3-15 min.
Quinaldine	0.01-0.03 mg	1-3 min.	fair	5-20 min.

Source: Adapted from Klontz [7].

REFERENCES

1. Caras, R. 1974. Venomous Animals of the World, p. 103. Englewood Cliffs, N.J.: Prentice-Hall.
2. Frazer, J. F. D. 1967. Frogs and toads. In The UFAW Handbook on the Care and Management of Laboratory Animals, 3rd ed. London: E & S Livingstone.
3. Halstead, B. W., and Cowville, D. A. 1965-1970. Poisonous and Venomous Marine Animals of the World, 3 vols. Washington, D.C.: U.S. Government Printing Office.
4. Healy, E. G. 1964. Anesthesia in fishes. In O. Graham-Jones, ed. Small Animal Anesthesia. Elmsford, N.Y.: Pergamon Press. .
5. Kaplan, H. M. 1969. Anesthesia in amphibians and reptiles. Fed. Proc. 28:1541-46.
6. Klausewitz, W. 1973. The cartilaginous fishes. In B. Grzimek, ed. Grzimek's Animal Life Encyclopedia, pp. 121, 296. New York: Van Nostrand Reinhold.
7. Klontz, G. W. 1964. Anesthesia of fishes. In Anesthesia in Experimental Animals. (Proceedings of a symposium) Brooks Air Force Base, Texas.
8. Spotte, S. 1970. Fish and Invertebrate Culture. New York: Wiley-Interscience.
9. ______. 1973. Marine Aquarium Keeping. New York: John Wiley & Sons.

APPENDIXES

APPENDIX A. DOMESTIC ANIMALS

Common Name	Scientific Name
Mammals:	
Alpaca	*Llama glama pacos*
Ass (donkey)	*Equus asinus*
Banteng	*Bibos banteng*
Buffalo, water	*Bubalus bubalis*
Camel, bactrian	*Camelus bactrianus*
Camel, dromedary	*Camelus dromedarius*
Cat	*Felis catus*
Cattle, European	*Bos taurus*
Cattle, zebu	*Bos indicus*
Dog	*Canis familiaris*
Elephant, Asian	*Elephas maximus*
Ferret	*Mustela putorius furo*
Fox	*Vulpes fulva*
Gayal	*Bos guarus frontalis*
Goat	*Capra hircus*
Guinea pig	*Cavia porcellanus*
Hamster, golden	*Mesocricetus auratus*
Horse	*Equus caballus*
Kouprey	*Bos sauveli*
Llama	*Llama glama*
Mink	*Mustela vison*
Mouse	*Mus musculus*
Mule	*Equus* sp.
Musk-ox	*Ovibos moschatus*
Rabbit	*Oryctolagus cuniculus*
Rat	*Rattus norvegicus*
Reindeer	*Rangifer tarandus*
Sheep	*Ovis aries*
Swine	*Sus scrofa*
Yak	*Bos mutus grunniens*
Birds:	
Budgerigar	*Melopsitticus undulatus*
Canary	*Serinus canarius*
Chicken	*Gallus gallus*
Duck, Muscovy	*Cairina moschata*
Duck, Pekin	*Anas platyrhyncos*
Goose	*Anser anser*
Goose, Canada	*Branta canadensis*
Guinea fowl	*Numida meleagris*
Peafowl	*Pavo cristatus*
Pheasant, ring-necked	*Phasianus colchicus torquatus*
Pigeon	*Columba livia*
Quail, coturnix	*Coturnix coturnix*
Swan, mute	*Cygnus olor*
Turkey	*Meleagris gallopavo*

APPENDIX B. WILD ANIMALS (Scientific names of animals specifically mentioned in the text)[1]

Common Name	Scientific Name
Monotremes and Marsupials	
Bandicoot	*Perameles* spp.
Echidna	*Tachyglossus aculeatus*
Glider, sugar	*Petaurus breviceps*
Kangaroo, gray	*Macropus giganteus*
Kangaroo, red	*Macropus rufus*
Koala	*Phascolarctos cinereus*
Opossum, Virginia	*Didelphis virginiana*
Phalanger, brush-tailed	*Trichosurus vulpecula*
Platypus	*Ornithorhynchus anatinus*
Tasmanian devil	*Sarcophilus harrisii*
Wallaby, agile	*Macropus agilis*
Wallaby, red-necked	*Macropus rufogrisea*
Wallaroo	*Macropus robustus*
Wombat, common	*Wombatus ursinus*
Small Mammals	
Aardvark	*Orycteropus afer*
Agouti	*Dasyprocta* spp.
Anteater, giant	*Myrmecophaga tridactyla*
Armadillo, three-banded	*Tolypeutes tricinctus*
Armadillo, nine-banded	*Dasypus novemcinctus*
Armadillo, giant	*Priodontes giganteus*
Bat, fruit	Megachiroptera (suborder)
Bat, insectivorous	Microchiroptera (suborder)
Bat, vampire	*Desmodus rotundus*
Beaver	*Castor canadensis*
Capybara	*Hydrochoerus capybara*
Chinchilla	*Chinchilla laniger*
Flying lemur	*Cynocephalus* sp.
Golden mole	Family Chyrsochloridae
Hedgehog	*Erinaceus europaeus*
Hyrax, rock	*Procavia capensis*
Kangaroo rat	*Dipodomys* spp.
Mouse, deer	*Peromyscus* spp.
Muskrat	*Ondatra zibethica*
Pangolin	*Manis* spp.
Pika	*Ochotona* spp.
Porcupine, African crested	*Hystrix cristata*
Porcupine, Brazilian tree	*Coendou prehensilis*
Porcupine, North American	*Erethizon dorsatum*
Shrew, American short-tailed	*Blarina brevicauda*
Shrew, bicolored water	*Neomys fodiens bicolor*
Shrew, elephant	*Elephantulus* spp.
Shrew, European water	*Neomys fodiens fodiens*
Shrew, masked	*Sorex cinereus*
Sloth, 3-toed	*Bradypus tridactylus*
Solenodon, Haitian	*Solenodon paradoxus*
Springhaas	*Pedetes* sp.
Tamandua	*Tamandua tetradactyla*
Tenrec	*Tenrec ecaudatus*
Woodchuck	*Marmota* sp.

1. References for nomenclature follow.

Common Name	Scientific Name
Carnivores	
Aardwolf	*Proteles cristatus*
Bear, American black	*Ursus americanus*
Bear, grizzly	*Ursus arctus horribilus*
Bear, polar	*Ursus maritimus*
Bear, sun	*Helarctos malayanus*
Binturong	*Arctictis binturong*
Bobcat	*Felis rufus*
Cacomistle	*Bassariscus astutus*
Cat, leopard	*Felis bengalensis*
Cheetah	*Acinonyx jubatus*
Civet cat	*Viverra civetta*
Coatimundi	*Nasua nasua*
Coyote	*Canis latrans*
Fennec	*Fennecus zerda*
Fox, gray	*Urocyon cinereoargenzeus*
Fox, red	*Vulpes vulpes*
Grison	*Galictis vittata*
Hyena, spotted	*Crocuta crocuta*
Kinkajou	*Potos flavus*
Leopard, clouded	*Neofelis nebulosa*
Lion	*Panthera leo*
Lion, mountain	*Felis concolor*
Mink	*Mustela vison*
Mongoose	Family Vivveridae (order Carnivora)
Ocelot	*Felis pardalis*
Otter, North American	*Lutra canadensis*
Otter, sea	*Enhydra lutris*
Otter, small clawed	*Amblonyx cinerea*
Panda, giant	*Ailuropoda melanoleuca*
Panda, lesser	*Ailurus fulgens*
Raccoon, North American	*Procyon lotor*
Skunk, striped	*Mephitis mephitis*
Tiger	*Panthera tigris*
Weasel	*Mustela* spp.
Wolf, gray	*Canis lupus*
Wolverine	*Gulo gulo*
Primates	
Capuchin	*Cebus capucinus*
Chimpanzee	*Pan troglodytes*
Gibbon, white-handed	*Hylobates lar*
Gorilla	*Gorilla gorilla*
Langur	*Presbytis* spp.
Loris, slow	*Nycticebus coucang*
Macaque, Philippine	*Macaca philippinenon*
Macaque, rhesus	*Macaca mulatta*
Marmoset	*Callithrix jacchus*
Marmoset, pygmy	*Callithrix pygmaea*
Monkey, colobus	*Colobus* spp.
Monkey, green	*Cercopithecus aethiops*
Monkey, roloway	*Cercopithecus diana roloway*
Monkey, spider	*Ateles* spp.
Monkey, squirrel	*Saimiri sciurea*
Monkey, woolly	*Lagothrix* spp.
Orangutan	*Pongo pygmaeus*
Tamarin, cotton-topped	*Saguinus oedipus*
Tarsier	*Tarsius* spp.
Tree shrew	*Tupaia* sp.
Marine Mammals	
Dolphin, Atlantic bottle-nosed	*Tursiops truncatus*
Dugong	*Dugong dugong*

Common Name	Scientific Name
Manatee	*Trichechus manatus*
Seal, elephant	*Mirounga angustirostris*
Seal, harbor	*Phoca vitulina*
Seal, weddel	*Leptonychotes weddelli*
Sea lion, California	*Zalophus californianus*
Walrus	*Odobenus rosmarus*
Whale, blue	*Balaenoptera musculus*
Whale, killer	*Orcinus orca*
Elephants	
Elephant, African	*Loxodonta africana*
Elephant, Asian	*Elephas maximus*
Hoofed Stock	
Perissodactylids:	
Horse, Przewalski's	*Equus przewalskii*
Rhinoceros, black	*Diceros bicornis*
Rhinoceros, Indian	*Rhinoceros unicornis*
Rhinoceros, white	*Ceratotherium simus*
Tapir, Brazilian	*Tapirus terrestris*
Tapir, Malayan	*Tapirus indicus*
Tapir, mountain	*Tapirus pinchaque*
Zebra, Grant's	*Equus burchelli*
Artiodactylids:	
Alpaca	*Lama glama pacos*
Aoudad	*Ammotragus lervia*
Antelope, roan	*Hippotragus equinus*
Antelope, sable	*Hippotragus niger*
Antelope, saiga	*Saiga tatarica*
Banteng	*Bibos javanicus*
Bison, American	*Bison bison*
Black buck	*Antilope cervicapra*
Bongo	*Taurotragus euryceros*
Buffalo, Cape	*Syncerus caffer*
Buffalo, water	*Bubalus arnee*
Bushbuck	*Tragelaphus scriptus*
Camel, bactrian	*Camelus bactrianus*
Camel, dromedary	*Camelus dromedarius*
Caribou	*Rangifer tarandus caribou*
Deer, fallow	*Cervus dama*
Deer, mule	*Odocoileus hemionus*
Deer, muntjac	*Muntiacus muntjak*
Deer, sika	*Cervus nippon*
Deer, white-tail	*Odocoileus virginianus*
Eland	*Taurotragus oryx*
Elk (wapiti)	*Cervus elaphus canadensis*
Gazelle, Dorcas	*Gazella dorcas*
Gazelle, Thompson	*Gazella thomsoni*
Giraffe	*Giraffa camelopardalis*
Gnu, white-tail	*Connochaetus gnu*
Goat, Rocky Mountain	*Oreamnos americanus*
Guanaco	*Lama glama guanicoe*
Hippopotamus, Nile	*Hippopotamus amphibius*
Hippopotamus, pygmy	*Choeropsis liberiensis*
Ibex	*Capra ibex*
Impala	*Aepyceros melampus*
Llama	*Lama glama*
Moose	*Alces alces*
Musk-ox	*Ovibos moschatus*
Nyala	*Tragelaphus angasi*
Okapi	*Okapia johnstoni*
Oryx, fringe-eared	*Oryx gazella callotis*
Peccary, collared	*Tayassu tajacu*
Peccary, white-lipped	*Tayassu albinostris*
Pronghorn	*Antilocapra americana*

Common Name	Scientific Name
Reindeer	*Rangifer tarandus*
Sheep, bighorn	*Ovis canadensis*
Sheep, Dall	*Ovis canadensis dalli*
Sheep, mouflon	*Ovis musimon*
Tahr	*Hemitragus jemlahicus*
Vicuna	*Lama vicugna*
Warthog	*Phacochoerus aethiopicus*
Yak	*Bos grunniens*

Birds

Common Name	Scientific Name
Aracari	*Pteroglossus* sp.
Bald eagle	*Haliaeetus leucocephalus*
Budgerigar (parakeet)	*Melopsittacus undulatus*
Canada goose	*Branta canadensis*
Cassowary, double-wattled	*Casuarius casuarius*
Cockatiel	*Nymphicus hollandicus*
Coly	*Colius* spp.
Crow	*Corvus* spp.
Crowned crane	*Balearica pavonina*
Crowned pigeon	*Goura cristata*
Duck, mallard	*Anas platyrhyncos*
Emu	*Dromaius novae-hollandiae*
Flamingo, American	*Phoenicopterus ruber*
Goose, spur-winged	*Plectropterus gambensis*
Golden eagle	*Aquila chrysaetos*
Great horned owl	*Bubo virginianus*
Hornbill	*Buceros* spp.
Kiwi	*Apteryx australis*
Macaw, hyacinth	*Anodorhynchus hyacinthinus*
Ostrich	*Struthio camelus*
Parrot, kea	*Nestor notabilis*
Pelican, American white	*Pelecanus erythrorhynchos*
Penguin, Humboldt	*Spheniscus humboldti*
Rhea, greater	*Rhea americana*
Saurus crane	*Grus antigone*
Screamer, horned	*Anhima cornuta*
Toucan	*Ramphastos* spp.
Trogan	*Trogon* spp.
Turaco	*Muso phagidae*

Reptiles

Common Name	Scientific Name
African puff adder	*Bitis arietans*
Anaconda	*Eunectes murinus*
Boa, California	*Lichanura roseofusca*
Boa, common	*Constrictor constrictor*
Boa, green tree	*Corallus caninus*
Boomslang	*Dispholidus typus*
Caiman	*Caiman* spp.
Cribo	*Spilotes pullatus*
Crocodile	*Crocodylus* spp.
Gila monster	*Heloderma suspectum*
Iguana, common green	*Iguana iguana*
King cobra	*Ophiophagus hannah*
Komodo dragon	*Varanus komodoensis*
Lizard, Mexican beaded	*Heloderma mexicana*
Lizard, sungazer	*Cordylus giganteus*
Rattlesnake	*Crotalus* spp.
Turtle, side-necked	Plerodira (suborder)
Turtle, snapping	*Chelydra serpentina*
Turtle, western pond	*Clemmys marmorata*
Viper, Russell's	*Vipera russellii*

Amphibians and Fish

Common Name	Scientific Name
Axolotl	*Ambystoma mexicanum*
Barracuda	*Sphyraenoidei*
Catfish, electric	*Malapterurus electricus*
Catfish, oriental	*Plotosus lineatus*
Eel, electric	*Electrophorous electricus*
Eel, moray	Family Muraenidae
Hellbender	*Cryptobranchus alleganiensis*
Rabbit fish	*Chimaera* sp.
Salamander, giant	*Andrias davidianus*
Scorpion fish	Scorpaenoidei (suborder)
Stargazer	*Astroscopus* spp.
Stingray	*Dasyatis* spp.
Surgeonfish	*Acanthuridae*
Toad, marine	*Bufo marinus*
Toad, Colombian giant	*Bufo blombergi*
Toad, Colorado River	*Bufo alvarius*
Toad, clawed	*Xenopus laevis*
Toadfish	*Opsatus beta*
Tunafish	*Thunnus* spp.
Weeverfish	Trachinoidei (suborder)

REFERENCES

Clements, J. F. 1974. Birds of the World: A Checklist. New York: Two Continents.

Grzimek, B., ed. 1975. Grzimek's Animal Life Encyclopedia. New York: Van Nostrand Reinhold.

Mammalian Taxonomic Directory. International Species Inventory System. Minnesota Zoological Garden, St. Paul, Minnesota, July 1974. (This is a system adapted to computers that combines taxonomic data from detailed monographs on specific orders.)

APPENDIX C. GENERIC NAMES, COMMON OR TRADE NAMES, AND SOURCES OF DRUGS MENTIONED IN THE TEXT

Generic	Common or Trade	Source[1]
Acetylpromazine maleate	Acepromazine	2
Alpha-chloralose	chloralose	17
	alpha-chloralose	11
Antivenin	antivenin	29
Atropine sulfate	atropine	9,12,13,30
Calcium borogluconate	calcium borogluconate	30
Calcium gluconate	calcium gluconate	25,30
Chloral hydrate	Equithesin	14
	chloral hydrate	25
	Equised	5
Chlorpromazine hydrochloride	Thorazine	21
Desoxycorticosterone acetate	DOCA	18
Dexamethasone	Azium	24
Diazepam	Valium, Tranimal	22
Diprenorphine hydrochloride	M50-50	8
Droperidol	Inapsine	21
Droperidol + fentanyl citrate	Innovar-Vet	21
d-tubocurarine chloride	*d*-tubocurarine	1
Epinephrine	adrenalin	20,30
Ether, diethyl	ether	25,30
Ethyl carbamate	urethan	15
Etorphine hydrochloride	M-99	8
Fentanyl citrate	fentanyl, Fentanest, Sublimaze	21
Glycerol guaiacolate	glycerol guaiacolate Glycodex	5
Halothane	Fluothane	2
Hydrocortisone sodium succinate	Solu-Cortef	27
Ketamine hydrochloride	Ketaject, Ketalar, Ketanest, Ketaset, Vetalar	4,20
Lactated Ringer's solution	lactated Ringer's	30
Methoxyflurane	Metafane	21
	Penthrane	1
Nalorphine hydrobromide	Lethidrone	6
Nalorphine hydrochloride	Nalline	16
Naloxone hydrochloride	Narcan	10
Nicotine alkaloid	Capchur solution	19
Ouabain	ouabain	12
Pentobarbital, sodium	Nembutal	1
Phencyclidine hydrochloride	Sernyl, Sernylan	3
Phenobarbital, sodium	phenobarbital	30
Phenylbutazone	Butazolidin	14
Prednisolone sodium succinate	Solu-Delta-Cortef	27
Quinaldine	quinaldine	15
Secobarbital, sodium	Seconal	1
Sodium bicarbonate	sodium bicarbonate	30
Sodium hypochlorite	bleach (household)	30
Succinylcholine chloride	Anectine	6
	Quelicin	1
	Sucostrin	25
Thiamyl sodium	Surital	20
Tiletamine hydrochloride	CI-634, Zolazepam	20
Tribromoethanol	Avertin	28
Tricaine methanesulfonate	Finquel, MS-222	23
Tuberculin: mammalian, avian	tuberculin	26
Xylazine hydrochloride	BAY 1470, Rompun	7

1. Listed in Appendix D.

APPENDIX D. FIRMS SUPPLYING DRUGS MENTIONED IN THE TEXT

1. Abbott Laboratories
   Veterinary Division
   P.O. Box 68
   Abbott Park
   North Chicago, IL 60064
   (312) 688-5109
2. Ayerst Laboratories
   685 Third Avenue
   New York, NY 10017
   (212) 986-1000 X777
3. Bio-Ceutics Laboratories
   St. Joseph, MO 64502
   (816) 233-2804
4. Bristol Laboratories
   Veterinary Products Division
   P.O. Box 657
   Syracuse, NY 13201
   (315) 470-2753
5. Burns-Biotec
   7711 Oakport Street
   Oakland, CA 94621
   (415) 562-0117
6. Burroughs Wellcome Co.
   3030 Cornwallis Road
   Research Triangle Park, NC 27709
   (919) 549-8371
7. Chemagro Agricultural Chemicals
   P.O. Box 4913
   Kansas City, MO 64120
   (913) 681-2451
8. D-M Pharmaceuticals, Inc.
   P.O. Box 1584
   Rockville, MD 20850
   (301) 762-3113
9. Eli Lilly & Co.
   P.O. Box 618
   Indianapolis, IN 46206
   (317) 636-2211
10. Endo Laboratories
    1000 Stewart Avenue
    Garden City, NY 11530
    (516) 832-2210
11. Fisher Scientific Co.
    711 Forbes Avenue
    Pittsburgh, PA 15219
    (412) 562-8300
12. Fort Dodge Laboratories
    P.O. Box 518
    Fort Dodge, IA 50501
    (515) 573-3131
13. Haver-Lockhart Laboratories
    P.O. Box 390
    Shawnee Mission, KS 66201
    (913) 631-4800
14. Jensen Salsbery Laboratories
    520 West 21st Street
    Kansas City, MO 64141
    (816) 321-1070
15. Matheson Coleman & Bell
    P.O. Box 7203
    Los Angeles, CA 90022
    (213) 685-5280
16. Merck & Co.
    Merck Chem. Division
    Rahway, NJ 07065
    (201) 574-4000
17. Nutritional Biochemicals Corp.
    26201 Miles Road
    Cleveland, OH 44128
    (216) 831-3000
18. Organon, Inc.
    375 Mt. Pleasant Avenue
    West Orange, NJ 07052
    (201) 325-4500
19. Palmer Chemical & Equipment Co.
    Atlanta, GA 30300
    (404) 942-4395
20. Parke, Davis & Co.
    Joseph Campan at the River
    Detroit, MI 48232
    (313) 567-5300
21. Pitman-Moore Co.
    P.O. Box 344
    Washington Crossing, NJ 08560
    (609) 737-3700
22. Roche Laboratories
    Nutley, NJ 07110
    (201) 235-5000
23. Sandoz Pharmaceuticals
    Route 10
    East Hanover, NJ 07936
    (201)386-1000
24. Schering Corp.
    Bloomfield, NJ 07003
    (201) 743-6000
25. E. R. Squibb & Sons
    P.O. Box 4000
    Princeton, NJ 08540
    (609) 921-4688
26. United States Department of Agriculture
    APHIS-VSL
    Ames, IA 50010
    (515) 232-0250
27. Upjohn Co.
    301 Henrietta Street
    Kalamazoo, MI 49001
    (616) 323-4000
28. Winthrop Laboratories
    90 Park Avenue
    New York, NY 10016
    (212) 972-4141
29. Wyeth Laboratories
    P.O. Box 8299
    Philadelphia, PA 19101
    (215) 688-4400
30. Various pharmaceutical and chemical supply companies

APPENDIX E. SOURCES OF RESTRAINT EQUIPMENT AND SUPPLIES

General Equipment

Nasco
901 Janesville Avenue
Fort Atkinson, WI 53538
(414) 563-2446
or
1524 Princeton Avenue
Modesto, CA 95352
(209) 529-6957

McMasters-Carr, wholesaler
P.O. Box 54960
Los Angeles, CA 90054
(213) 945-1311
or
P.O. Box 4355
Chicago, IL 60680
(312) 281-1010

Large hardware stores

Large mail order firms (ask for farm or tool catalogs)
Montgomery Ward
Sears Roebuck

Chemical Restraint Equipment

Syringes
Pneu-Dart, Inc.
Williamsport, PA 17701
(717) 323-2710
Pole syringes
Kay Research Products
1525 E. 53rd Street, Suite 503
Chicago, IL 60615
(312) 643-9044
Weapons
Palmer Chemical Equipment Co., Inc.
Palmer Village
P.O. Box 867
Douglasville, GA 30134
(404) 942-4397

Paxarms Ltd.
P.O. Box 317
Timaru, New Zealand

Donjoy Industries Pty. Ltd.
Salisbury, Rhodesia

Gloves

Routine animal-handling gloves
Ketch-all Co.
Department VMA
2537 University Avenue
San Diego, CA 92104
(714) 297-1953
Special order primate gloves
Lithgow Services
1205 S. Railroad Avenue
San Mateo, CA 94402
(415) 349-2310
Chain mail gloves
Butcher supply firms

Hooks

For elephants
John Beery Co.
2415 Webster Street
Alameda, CA 94501
(415) 769-8200

For reptiles
FurMont Reptile Hooks
Fuhrman Diversified
1212 W. Flamingo
Seabrook, TX 77586
(713) 272-4832

Horse Equipment

Colorado Saddlery
1411 Market Street
Denver, CO 80202

H. Kaufman & Sons
Saddlery Co.
139-141 E. 24th Street
New York, NY 10010

Nets

Hill & Hill Custom Veterinary Supplies
324 E. Shamrock
Rialto, CA 92376
(805) 268-1037

Beckman Net Co.
3729 Ross Street
Madison, WI 53707
(608) 233-6991
Attn: Milo Beckman

Flexi-Nets
Fuhrman Diversified (FurMont Reptile Hooks also)
1212 W. Flamingo
Seabrook, TX 77586

West Coast Netting, Inc.
14929 Clark Avenue
City of Industry, CA 91745
(213) 330-3207

Ropes and Chains

ACCO
American Chain Division
454 E. Princess Street
York, PA 17403
(717) 741-0847

The Cordage Group
Division of Columbian Rope Co.
Auburn, NY 13021
(315) 253-3221

Tubbs Cordage Co.
200 Bush Street
San Francisco, CA 94104
(415) 495-7155

Large hardware stores

Snares

Ketch-all Co.
Department VMA
2537 University Avenue
San Diego, CA 92104
(714) 297-1953

Squeeze Cages (for primates, carnivores, marine mammals)

Research Equipment Company, Inc.
P.O. Box 1151
Bryan, TX 77801
(713) 779-4459

Veterinary Equipment

Haver-Lockhart Laboratories
Box 390
Shawnee Mission, KS 66201
(913) 631-4800

Jensen-Salsbery Laboratories
P.O. Box 167
Kansas City, MO 64141
(816) 321-1070

Western States Veterinary Supply Co.
5530 West Colfax Avenue
Lakewood, CO 80214
(303) 233-1212

APPENDIX F. ABBREVIATIONS USED IN THIS BOOK

U.S. Customary

av—avoirdupois
ft—foot, feet
gal—gallon(s)
in.—inch(es)
lb—pound(s)
mi—mile(s)
oz—ounce(s)
qt—quart(s)
sq—square
yd—yard(s)
ac—acre
apoth—apothecaries' weight (pharmaceutical)
Sp.—species
Spp.—species (plural)
B.W.—body weight
wt—weight

Metric

cc—cubic centimeter(s) = ml
cm—centimeter(s)
cu—cubic
g—gram(s)
ha—hectare(s)
kcal—kilocalorie(s)
kg—kilogram(s)
L—liter
m—meter(s)
mEq—milliequivalent
ml—milliliter = cc
mg—milligram
mm—millimeter(s)
t—metric ton(s)
μl—microliter
dl—deciliter = 100ml

APPENDIX G. CONVERSION TABLES

Linear

1 millimeter = 0.039 inch
1 meter = 3.281 feet
1 meter = 1.094 yards
1 kilometer = 0.621 mile

1 inch = 25.4 millimeters
1 foot = 0.305 meter
1 yard = 0.914 meter
1 mile = 1.609 kilometers

Volume

1 liter = 33.815 fluid ounces
1 liter = 1.057 quarts
1 liter = 0.264 gallon

1 fluid ounce = 29.573 milliliters
1 fluid ounce = 0.03 liter
1 pint = 0.473 liter
1 quart = 0.946 liter
1 U.S. gallon = 3.785 liters
= 0.83 Br. imperial gal
1 British imperial gallon = 4.545 liters
= 1.2 U.S. gal

Area

1 hectare = 0.004 square mile
1 hectare = 107,639.1 square feet
1 hectare (10,000 sq m) = 2.47 acres

1 acre (43,560 sq ft) = 0.405 hectare
1 acre = 4046.86 square meters

Mass

1 milligram = 1/60 grain (apoth)
1 gram = 0.035 ounce
1 gram = 15.432 grains (apoth)
1 kilogram = 2.2 pounds
1 metric ton (1,000 kg) = 1.102 tons
1 mg/kg = 0.454 mg/lb

1 grain (apoth) = 60 milligrams
1 ounce (av) = 28.35 grams
1 pound = 0.454 kilogram (454 g)
1 ton (2,000 lb) = 0.907 metric ton (1,000 kg)
1 mg/lb = 2.2 mg/kg

Temperature (degrees Celsius to degrees Fahrenheit)

C	F	C	F	C	F	C	F
25	77.0	33	91.4	38.5	101.3	43	109.4
26	78.8	34	93.2	39	102.2	44	111.2
27	80.6	35	95.0	39.5	103.1	45	113.0
28	82.4	36	96.8	40	104.0	46	114.8
29	84.2	36.5	97.7	40.5	104.9	47	116.8
30	86.0	37	98.6	41	105.8	48	118.4
31	87.8	37.5	99.5	41.5	106.7	49	120.2
32	89.6	38	100.4	42	107.6	50	122.0

INDEX

Aardvark, 189, 199
Aardwolf, 201, 204, 208. *See also* Carnivores
Abrasion, 75
Acepromazine maleate, 49, 51, 112, 155, 169, 188, 191, 238, 243, 245, 256
Acetylcholine (ACh), 58
Acidosis
  metabolic, 40-41, 81
  respiratory, 81
Adaptation, physiological, 54
  syndrome, 53
Addison's disease, 60, 84
Adenohypophyseal hormones, 58
Adenohypophysis, 58
Adrenal cortex, 58
Adrenal medulla, 55, 56, 58
Adrenocortical insufficiency, 84
Agalactia, 89, 90
Agouti, 189
Alarm response, 53, 56, 58, 87
Albatross, 262
Aldosterone, 58
Alkalosis, 81
Allergies, 91
Alligator. *See* Crocodilians
Alpaca. *See* Camelids
Alpha chloralose, 264
Amphibia, 311
Amphibians
  anatomy and physiology, 311
  chemical restraint, 313-14
  classification, 311
  danger potential, 311
  physical restraint, 311-13
Analgesia, analgesic, 51
Anal sacs, 205
Anectine. *See* Succinylcholine chloride
Anesthesia. *See also* Restraint, chemical
  definition, 51
  jar, for snakes, 309
Animal Import Center, Clifton, New Jersey, 11, 257
Animals' rights, 5
Anorexia, 87, 89
Anoxia, 73, 81-82, 89, 90
Anseriformes, 181, 262
Anteater. *See also* Edentates
  giant, 55, 191, 192
  tamandua, 191, 192
Antelope, 7, 11, 14, 77, 89, 243, 244, 256, 260, 261. *See also* Bovids, wild
Anterior pituitary hormones, 58
Antlers, 77, 91, 251, 253
Aoudad, 257
Ape, 9, 214, 219. *See also* Primates
Apodiformes, 262
Aracari, 278
Armadillo, 189, 191. *See also* Edentates
  giant, nine-banded, three-banded, 192, 193
Artiodactyla, 113, 131, 139, 181, 240, 247-61
Ass. *See* Horse, domestic; Equids, wild
Ataractic, ataraxia, 51
Atropine sulfate, 48, 51, 87, 213
Avertin. *See* Tribromoethanol
Azaperone, 147
Azium. *See* Dexamethasone

Baboon. *See* Primates
Bag. *See also* Sack
  ambu resuscitation, 42
  canvas, tapering, 186
  cat, 10, 158
  funnel, 287
  special restraint, 11, 186, 257
Bandicoot, 183, 184. *See also* Marsupials
Banteng, 113
Barrier, 6, 13
  bale of straw, 14
  bars, 15
  blanket, 14, 256
  burlap, 240
  fence, 240
  gates, solid, 14
  mattress, 14, 187, 254
  plastic sheeting, 14, 186, 240, 275
  poles, 15
  wire panels, 12, 14
Basal metabolism, 63, 71
Bat
  anatomy and behavior, 190
  chemical restraint, 190-91
  classification, 189
  fruit, 190
  insectivorous, 190
  physical restraint, 190
  vampire, 190
Bating, 270
BAY 1470. *See* Xylazine
Bear, 13. *See also* Carnivores
  chemical restraint, 204, 205
  classification, 201
  physical restraint, 202-4
  polar, 203, 204, 205, 223
Beaver, 189, 190. *See also* Rodents
Behavior. *See also* individual species
  aspects, 4-5
  characteristics, 4
  patterns, 6-8
Birds, 8, 9
  domestic. *See* Poultry and waterfowl, domestic
  wild. *See* Birds, wild
Birds, wild
  anatomy and physiology, 262
  chemical restraint, 264
  classification, 262
  collars, 274, 283, 284
  danger potential, 8, 9. *See also* individual species or group
  feather damage, 81
  flightless, 264-68
  galliform, 273-74, 275, 277
  hummingbirds and swifts, 264, 283
  large-billed, 276-77, 278
  long-billed, long-legged, 274-76
  physical restraint, 9, 11, 13, 14, 15, 263. *See also* Poultry and waterfowl, domestic
  pigeons and doves, 277, 278
  psittacine, 277-83, 284
  radiography, 282
  raptors, 270-73, 274
  releasing, after restraint, 264
  shore, 270
  song, perching, and miscellaneous, 283-85
  transport, 264, 269, 277. *See also* Poultry and waterfowl, domestic
  water, 268-70, 277. *See also* Waterfowl, domestic
Birds, wild, flightless. *See also* Birds, wild
  danger potential, 264
  chemical restraint, 264, 265
  obtaining blood samples, 265, 266
  physical restraint, 264-68, 269
Birds, wild, galliform. *See also* Birds, wild; Poultry and waterfowl, domestic
  danger potential, 273-74
  physical restraint, 274, 275
Birds, wild, large-billed. *See also* Birds, wild
  danger potential, 276
  physical restraint, 277
Birds, wild, long-billed, long-legged. *See also* Birds, wild
  capture myopathy, 276
  danger potential, 274-76
  physical restraint, 275-76, 277
  recovery, from anesthesia, 276
Birds, wild, psittacine, 11. *See also* Birds, wild
  capturing, from cage, 278, 279, 280
  collars, 283, 284
  danger potential, 277-78, 279
  intubation, 280
  obtaining blood samples, 282
  physical restraint, 278-83, 284
  specula, 280, 281
Birds, wild, raptors. *See also* Birds, wild
  danger potential, 270
  examination, for parasites, 274
  physical restraint, 270-73, 274
Birds, wild, shore and gull-like. *See also* Birds, wild
  danger potential, 270
  physical restraint, 270
Birds, wild, song, perching, and miscellaneous. *See also* Birds, wild
  obtaining blood samples, 282
  physical restraint, 283-84, 285
Birds, wild, water. *See also* Birds, wild; Waterfowl, domestic
  danger potential, 268-69
  physical restraint, 269-70
Bison, 5, 260. *See also* Bovids, wild
  American, 77, 259
Biting, 8, 9, 75, 286
Black buck, 9, 91, 242, 244. *See also* Bovids, wild
Blindfold, 9, 10
  cetacean, 224, 225
  crane, 275
  hoofed stock, 244, 245
  horse, 108
  raptor, 270
  ratite, 265-67
Bloat, 89
Block, 31, 32
Block and tackle, 22, 30, 32, 237
Blood
  cross matching, 74
  samples, obtaining. *See* individual species
  volume, in vertebrates, 74
Blowgun, 37, 38, 39, 244
Boa, California, 297. *See also* Snakes
Body temperature, 64, 90
Bongo, 259
Bovids, wild. *See also* Hoofed stock; Cattle, domestic; Sheep and goats
  anatomy and physiology, 77, 256
  behavior, 8, 256
  chemical restraint, 242-44
  classification, 240
  danger potential, 256
  dehorning, 261
  physical restraint, 256-61

Bovids, wild *(continued)*
  restraint bag, 257
  roping, 260
  transport, 241-42, 244, 260
Box. *See also* Cage; Crate
  Santini, 228
  snake, squeeze, 305, 306
Bradycardia, 84, 86
Breathing, rapid. *See* Hypoxia
Bruise. *See* Contusion
Budgerigar. *See* Parakeets.
Buffalo
  African or Cape, 89, 244, 256. *See also* Bovids, wild
  Asiatic or water, 113, 181, 260. *See also* Cattle, domestic
Bull lead, 13, 116, 117
Burro. *See* Horse, domestic
Bush baby, 214

Cable. *See* Rope; Snare
Cacomistle, 206
Caecilians, 311, 313
Cage. *See also* Box; Crate
  special primate, 214
  squeeze, 10, 186, 212, 306
  transfer, 10, 204, 207, 213, 219, 259
Cage, squeeze
  carnivore, 11, 203, 211, 212
  commercial, 147, 211
  crocodilian, 289
  hoofed stock, 259
  mammal, small, 10
  pinniped, 228
  primate, 211, 219
  rodent, 195
  snake, 305
  swine, 145, 147
Caiman. *See* Crocodilians
Calcium gluconate, 82, 83
Calf
  casting, 123, 124, 125
  restraint, 123-26
Camel, 9, 240, 249-51. *See also* Camelids
Camelids, 9, 88. *See also* Hoofed stock
  anatomy and physiology, 249
  chemical restraint, 243, 251
  classification, 240
  danger potential, 249
  physical restraint, 31, 249-51
Canary, 285. *See also* Birds, wild, song
Canids, wild, 202. *See also* Carnivores; Dog, domestic
Caprimulgiformes, 262
Capybara, 189, 194. *See also* Rodents
Cardiac tamponade, 87
Cardioactive glycoside, 35
Caribou. *See* Cervids
Carnivora, 148, 156, 201
Carnivores. *See also* Dog, domestic; Cat, domestic
  canids, 202
  chemical restraint, 204, 213
  classification, 201
  danger potential, 9, 201
  felids, 208-12, 213
  hyaenids, 208
  mustelids, 205-8
  names of gender, 201
  procyonids, 204-5
  transport, 75, 212, 213
  ursids, 202-4
  viverrids, 208
  weights, 201
Cassowary, 262, 264, 265. *See also* Birds, wild, flightless
Casting
  calf, 123-26
  cattle, 126-28
  crisscross rope, 127, 133
  elephant, 237, 238
  foal, 102, 103
  horse, 103-5
  swine, 146-47
Cat, domestic, 11. *See also* Cat, wild
  anatomy, physiology, and behavior, 156
  bags, 158
  chemical restraint, 160
  danger potential, 156
  names of gender, 156
  nets, 159
  physical restraint, 156-59
  transport, 160
  weights, 156
Cat, wild. *See also* Carnivores
  behavior, 7, 208
  chemical restraint, 213
  danger potential, 208
  infant care and restraint, 208, 210
  obtaining blood samples, 211-12
  physical restraint, 208-13
  snares, chains, nets, 209, 210, 211
  squeeze cages, 211-12
  transport, 212-13
Catalepsy, 52, 86
Catecholamine, 40, 55, 59, 81, 87
Catfish, electric. *See* Fish
Cattle, domestic, 5, 81, 88, 89. *See also* Hoofed stock
  casting, 126-28
  chemical restraint, 129-30
  chutes, 128-29, 259
  clamps, 118, 121, 122
  classification, 113
  danger potential, 113-14
  electric prods, 15
  foot hitch, 30
  halters, 30, 31, 114, 115, 116
  hobbles, 117, 118, 120
  manipulation of feet and legs, 121-23
  manipulation of tail, 122, 123
  names of gender, 113
  nose ring, 116, 117
  nose tongs, 115, 116
  physical restraint, 114-29
  slinging, 128
  specula, 119, 120
  tail tie, 29, 123
  transport, 129
  weights, 113
Cavy, 189
Cervids. *See also* Hoofed stock
  anatomy and physiology, 251
  chemical restraint, 243, 253, 254
  danger potential, 251
  physical restraint, 9, 251-52, 253
  rut, 4, 251
  transport, 75, 241, 253
Cetacea, 223
Cetaceans,
  anesthesia, 230
  danger potential, 223
  obtaining blood samples, 223, 225
  physical restraint, 223, 224, 225
  release, 225
Chain
  cat, 13, 209, 210
  elephant, 234, 235
  goat, 137
Chameleon, 286, 294. *See also* Lizards
Charadriiformes, 262
Cheetah, 7-8, 212. *See also* Cats, wild
Chelonians
  anatomy, 289
  chemical restraint, 293
  classification, 286
  danger potential, 289
  obtaining blood samples, 291-92
  physical restraint, 289-92
  specula, 293
  transport, 293
Chemical restraint. *See* Restraint, chemical
Chevrotain, 240
Chicken, 7, 60, 86, 171. *See also* Poultry and waterfowl, domestic
  danger potential, 171
  obtaining blood samples, 173, 175
  physical restraint, 172-74, 175
Chimpanzee, 214, 219. *See also* Primates
Chinchilla. *See also* Rodents
  behavior, 196
  dental specula, 198
  fur examination, 196, 197
  physical restraint, 196, 197, 198
Chiroptera. *See* Bat
Chloral hydrate, 111, 112, 314, 316
Cholinergic bradycardia, 86-87
Chute
  camel, 250
  cattle, 128-29, 259
  cervid, 251, 252
  giraffe, 255
  hoofed stock, 259
  tip, 129
CI 744. *See* Tiletamine hydrochloride-zolazepam hydrochloride
Ciconiiformes, 262
Civet cat, 201, 208
Claws, 8, 9
  injury, 80
Clifton quarantine station, 10-11, 257
Coatimundi, 204
Cobra, 286, 303, 304. *See also* Snakes
  king, 305
  spitting, 9, 306
Coliiformes, 262
Collar
  dog, 149
  Elizabethan, 152, 153, 205, 206, 274, 283, 284
  goat, 137
Columbiformes, 262
Coly, 262
Concussion, 76
Conduction, 71
Confinement, 6, 10-12, 15. *See also* Cage; Chute
Constrictor, 298. *See also* Snakes
Controlled Substances Act of 1970, 50-51
Contusion, 75-76
  brain, 76-77
Convection, 71
Convulsions, 69, 77, 81, 82, 83, 90
Cooling, 10
  mechanisms, 64-67, 68
  morphological adaptations, 68
Coraciiformes, 262
Cormorant, 262
Cortisol, 40, 60
  biological effects, 59
  releasing factor, 58
Countercurrent heat exchange system, 68, 71
Cow. *See* Cattle, domestic
Cradle
  horse, 108
  pig, 145, 147
Crane, 9, 14, 262, 275, 276, 277. *See also* Birds, wild, long-billed, long-legged
  African crowned, 275
  saurus, 276
Crate. *See also* Box; Cage
  equid, wild, 245
  giraffe, 255
  nebulization, 228
  shipping, 4, 10, 60, 204, 242, 245
  rhinoceros, 246, 247

wheeled, 229
Cribo, 297. *See also* Snakes
Crocodile. *See* Crocodilians
Crocodilians
chemical restraint, 288
classification, 286
danger potential, 286
funnel bag for, 287
hypnosis, 7
physical restraint, 286-88, 289
physiology and behavior, 286
squeeze cage, 289
torpidity, 83, 286
transport, 288
Crossbow, 38, 39
Crow. *See* Birds, wild, song, perching, miscellaneous
Cuckoo, 262
Cuculiformes, 262
Curare, 35, 50, 51
Cushing's syndrome, 59
Cut. *See* Laceration

Dally, 28, 32
Danger potential. *See* individual species or group
Dart. *See* Syringe
Death
cause, 73
feint, 86
Deer. *See also* Cervids
mule, 240, 251, 252
muntjac, 9, 252
Defense mechanisms, 8, 9. *See also* danger potential of individual species or group
Dehydration, 69, 71, 81, 84-85
Dermoptera. *See* Flying lemur
Desoxycorticosterone acetate (DOCA), 84
Dexamethasone, 85
Dextrose, 83, 84
Diazepam, 44, 49, 51
Diminishing sense perception, 6
sight, 9, 278, 283. *See also* Blindfold
sound, 10
Diprenorphine, 45, 50, 51, 238
Disseminated intravascular coagulation, 69
Dog, domestic. *See also* Carnivores
behavior, 148
chemical restraint, 155
cholinergic bradycardia, 86
collars, 149, 152, 153
danger potential, 148
glucocorticosteroid therapy, 85
leather harness, 149
lolling tongue, related to hyperthermia, 151
medication, 151-52, 153
muzzling, 150-51
names of gender, 148
obtaining blood samples, 153, 154
panting, related to hyperthermia, 68, 151
physical restraint, 149-54
removal from cage, 149-50
sheep, 131
specula, 152, 153
stress response, 58, 59, 60
territoriality, 5
transport, 155
weighing, 153, 154
weights, 148
Dog, wild, 202, 203. *See also* Carnivores; Dog, domestic
Dolphin, 223-25. *See also* Cetaceans
Domestication, 181-82
Dominance, over animals, 7
Donkey. *See* Horse, domestic
Dove. *See also* Birds, wild
classification, 181, 262
intubation, 279
physical restraint, 277, 278
Droperidol. *See* Fentanyl-droperidol
Drugs, restraint, 43-52
Duck
domestic, 171, 177-79
Muscovy, 171
Pekin, 171
wild, 268-70
Dugong. *See* Cirenians

Eagle, 270, 272, 273. *See also* Birds, wild, raptors
Echidna, 183, 184. *See also* Monotremes
Eclampsia. *See* Hypocalcemia
Edentata, 189
Edentates
chemical restraint, 192
classification, 189
danger potential, 191
obtaining blood samples, 191
physical restraint, 191-92
Eel, electric. *See* Fish
Elephant
African, 68, 231
anatomy and physiology, 231
Asian, 231
behavior, 232-33
casting, 237, 238
chaining, 234-35
chemical restraint, 44, 237-38
classification, 231
cooling, 68
danger potential, 9, 91, 233
earing, 236
foot trimming, 236
hook, 233, 234
hyperthermia, 232
physical restraint, 233-38
special needles for, 41, 237
transport, 238
trunk, 231-33
weights, 231
Elk. *See* Cervids
Emphysema, 82
Emu, 10, 268. *See also* Birds, wild, flightless
Endotherm. *See* Homeotherm
Envenomation
fish, 314
monotremes, 183
reptiles, 293, 302, 306-7
Epinephrine, 58, 83, 87
Equids, wild. *See also* Hoofed stock; Horse, domestic
chemical restraint, 245
classification, 240
danger potential, 245
physical restraint, 245
transport, 245
Equipment failure, 41
Estrus, 4
Ether, 316
Ethyl carbamate, 313, 314, 316
Etorphine hydrochloride, 44, 51, 112, 188, 204, 207, 237, 243, 245, 246, 249, 253, 254, 288

Fainting. *See* Cholinergic bradycardia
Falcon, 270. *See also* Birds, wild, raptors
Falconiformes, 262
Fatal syncope. *See* Cholinergic bradycardia
Fear, 6
mastery, 7
Felids. *See* Cats, wild
Fentanyl. *See* Fentanyl-droperidol
Fentanyl-droperidol, 45, 51, 147, 170, 200, 204
Finch, 262, 283-84, 285
Finquel. *See* Tricaine methanesulfonate
Fire extinguisher, 15
Fish
anatomy and physiology, 314
catfish, electric, 314
chemical restraint, 316
classification, 311
danger potential, 314
eel, electric, 314
netting, 315
physical restraint, 314-15
ray, electric, 314
scale damage, 81
shark, 314
stargazer, 314
stingray, 314
venomous, 314
Flamingo, 262, 274, 276. *See also* Birds, wild, long-billed, long-legged
Flanking, 123-24
Flight distance, 7, 8
Fluid requirement, 84
Flying lemur, 189, 190
Foal, 78
capturing, holding, 102
Founder. *See* Laminitis
Fowl, domestic. *See* Poultry and waterfowl, domestic
Fox, 13, 201, 202. *See also* Canids, wild
bat-eared, 68
Fracture
limb, 78-79
skull, 77
Frog. *See* Amphibians
Frogmouth, 262
Frostbite, 71

Galago, 215. *See also* Prosimians
Galliformes, 171
Gastric dilatation, 85, 89
Gavial. *See* Crocodilians
Gaviiformes, 262
Cayal, 113
Gazelle, 7-8, 241, 260. *See also* Hoofed stock
Gecko, 286
General adaptation syndrome, 53
Gibbon, 214. *See also* Primates
Gila monster, 293, 296. *See also* Lizards
Giraffe, 14, 89, 240, 241. *See also* Hoofed stock
chemical restraint, 243, 256
danger potential, 9, 254
physical restraint, 7, 255
transport, 255
Gloves, 14
chain-mail, 14, 219
cotton, 14
gauntleted, double thickness leather, 14, 217, 218
gauntleted, heavy leather, 14, 207, 209, 270-73, 279, 294
Gnu, 86. *See also* Bovids, wild
Goat, domestic. *See also* Hoofed stock
angora, 137
chemical restraint, 138
classification, 131
danger potential, 136
haltering, 137
milking, 137
names of gender, 131
neck chain or collar, 137
physical restraint, 136-38
scent glands, 135-36
transport, 138
weights, 131
Goat, wild. *See* Bovids, wild
Goose, 171, 177, 179, 262, 268-70
spur-winged, 269, 270

Gopher, 189
Gorilla, 214, 219, 221. *See also* Primates
Grebe, 262, 269, 270
Grisson, 206
Grouse, 262
Gruiformes, 262
Guanaco. *See* Camilids
Guinea pig, 161, 167-68, 169. *See also* Laboratory animals; Rodents
Gull, 68, 262, 270

Halothane, 179, 227, 230, 264
Halter
  goat, 137
  leather, 94
  rope, 30, 31, 114, 115
  rope thong, 116
  shanks, 94, 95
  sheep, 133
  tie, 24
Hamster, 161, 168, 170. *See also* Laboratory animals; Rodents
Hare. *See* Rabbit
Harness
  casting, 103-5
  rope, 141
Hawk, 271, 272, 274. *See also* Birds, wild, raptors
Heart flutter. *See* Ventricular fibrillation
Heat. *See also* Thermoregulation
  conservation, 63
  cramps, 69, 71
  dissipation, 68
  exchange, 68
  exhaustion, 69, 71
  production, 63
  stroke or sunstroke, 71
Hedgehog, 189, 190
Hellbender, 312
Hematoma, 74
Hemorrhage, 74-75
Heron, 262, 274
Hexabarbital, 313
Hibernators, 83, 190
Hierarchial social structure, 4, 215, 256
Hinny, 93
Hippopotamus, 232, 240. *See also* Hoofed stock
  chemical restraint, 44, 249
  danger potential, 8, 9, 248
  Nile, 248, 249
  pygmy, 248
  thermoregulation, 69
  transport, 30, 249
Hitch
  anchor, 29, 30
  clove, 23-24
  double half, 24
  foot, 29, 30
  half, 19, 23, 24, 126
  lark's head, 126
  trucker's, 29
Hobble
  breeding, 106-7
  burlap, 107
  casting, 103
  chain, 117, 120
  hock, 120
  leather, 106
  rope, 31, 117, 120
Homeostasis, 53, 71
Homeotherm, 63, 71
Hoofed stock, wild. *See also* Cattle, domestic; Horse, domestic; Sheep and goats, domestic
  bovids, 256-61
  camelids, 249-51
  cervids, 251-54
  chemical restraint, 242-44. *See also* individual species or group
  classification, 240
  equids, 245
  giraffe, okapi, 254-56
  hippopotamus, 248-49
  names of gender, 240
  physical restraint, 240-41. *See also* individual species or group
  rhinoceros, 246-47
  swine, 247-48
  tapir, 246
  transport, 241-42. *See also* individual species or group
  weights, 240
Hoofs, injuries, 75, 79, 80
Hook
  Brummel, 234-35
  elephant, 233, 234
  reptile, 295, 297, 299, 303, 304
  waterfowl, 177, 178, 269
Hornbill, 14, 262, 276, 277, 278
Horns
  danger potential, 8, 91, 113, 256
  injuries, 77-78, 256
  protection, 260
  structure, 77, 256
Horse, domestic, 5, 10
  blindfolding, 9, 108
  breeding hobbles, 106, 107
  casting, 103-5
  catching, 108-9
  chemical restraint, 111-12
  classification, 93
  cradle, 108
  cross tie, 109
  danger potential, 93
  earing, 98, 99
  haltering, 95
  halter tie, 24, 25
  hobbles, 106-7, 108
  hoof damage, 75, 79
  loading, 110-11
  manipulating feet and legs, 99-101
  myopathy, 88
  names of gender, 93
  physical restraint, 12, 15, 93-109
  roping, 25
  sideline, 105-6
  slings, 107-8
  tail tie, 28-29
  transport, 109-11
  twitches, 96-98
Horse, wild. *See* Equids, wild
Hoven. *See* Bloat
Humane considerations, 5
Hummingbird, 262, 264, 283, 285
Hydrocortisone sodium succinate, 85
Hyena, 201, 208. *See also* Carnivores
Hyperadrenalism, 81
Hyperadrenocorticism, 59
Hypercalcemia, 69
Hyperphagia, 56
Hypersexuality, 57
Hyperthermia, 56, 58
  cetaceans, 224
  dogs, 151
  elephants, 232
  mechanisms for coping with, 64-67, 68-69
  muzzled crocodilians, 287
  pinnipeds, 229
  swine, 140
Hypnosis, 7, 52, 86, 161
Hypoadrenocorticism, 59, 60
Hypocalcemia, 82
Hypoglycemia, 58, 83, 90
Hypokalemia, 88
Hypokinesia, 3
Hypophysis, 58
Hyposexuality, 57
Hypothalamic adenohypophyseal adrenal pathway, 54, 55, 58
Hypothalamic adenohypophyseal adrenocortical pathway, 56
Hypothalamus, 58
Hypothermia, 70-71
Hypoxemia, 69
Hypoxia, 56, 81-82, 90
Hyracoidea, 189
Hyrax, 189, 200

Ibex, 240
Idiopathic ataxia. *See* Wobbler syndrome
Iguana, 286, 293, 294, 295. *See also* Lizards
Immobilon. *See* Etorphine hydrochloride
Immobilization. *See* Restraint, chemical; Restraint, physical
Impala, 240
Inapsine. *See* Fentanyl-droperidol
Incubator, 222
Injections, suitable sites, 42
Innovar-Vet. *See* Fentanyl-droperidol
Insectivora, 189
Insectivores
  chemical immobilization, 190
  classification, 189
  danger potential, 189
  sex determination, 190
Insulin shock, 83
International Air Transport Association, 242
Intubation
  endotracheal, 43, 82, 85, 205
  gastric, 84

Jesses, 270

Kangaroo, 183, 184, 187, 188. *See also* Marsupials
Kangaroo rat, 189
Ketaject. *See* Ketamine hydrochloride
Ketalar. *See* Ketamine hydrochloride
Ketamine hydrochloride, 45-46, 51, 147, 160, 170, 179, 188, 191, 192, 193, 199, 204, 206, 207, 222, 230, 243, 248, 264, 288, 293, 308
Ketanest. *See* Ketamine hydrochloride
Ketaset. *See* Ketamine hydrochloride
Kingfisher, 262
Kinkajou, 201, 205. *See also* Procyonids
Kiwi, 262, 267, 269. *See also* Birds, wild, flightless
Knot
  bowline, 22, 23
  double bow, 22, 23
  honda, 24, 26
  Matthew Walker, 21
  overhand, 19
  sheet bend, 22, 23
  single bow, 22, 23
  square, 22, 23
  stopper, 20
  wall, 20
Koala, 183, 185, 186. *See also* Marsupials
Komodo dragon. *See* Lizards
Kouprey, 113
Krait, 286

Laboratory animals. *See also* Rodents
  behavior, 161
  chemical restraint, 169-70
  classification, 161
  danger potential, 161
  guinea pig, 167-68, 169
  hamster, 168, 170

miscellaneous species, 168
mouse, 163, 165, 166
physical restraint, 161-68
post orbital sinus bleeding, of small rodents, 166
rabbit, 161, 162, 163, 164, 165
rat, 166, 167, 168
stress response, 60
transport, 169
weights, 161
Laceration, 75
Lactated Ringers solution, 70, 85
Lagomorpha. *See* Rabbit
Lamb. *See* Sheep
Laminitis, 79
Langur, 55, 214. *See also* Primates
Lariat, 33. *See also* Roping
Laryngoscope, 43. *See also* Intubation
Leash, 149
Lemming, 189, 194. *See also* Rodents
Lemur, 214, 216. *See also* Primates
Leopard. *See* Cat, wild
Leopard, clouded, 208. *See also* Cat, wild
Leopard cat. *See* Cat, wild
Limbic system, 58
Limb injuries, 78
Lion, 7, 10. *See also* Cat, wild
Lizards
classification, 286
danger potential, 293
monitor, 293, 296
obtaining blood samples, 295-96
physical restraint, 11, 293-96
poisonous, 9, 293, 295, 296
sungazer, 294
Llama, 240, 249, 250. *See also* Camelids
Loon, 262, 270. *See also* Birds, wild, water
Loris, 214
Lory. *See* Birds, wild, psittacine

M50-50. *See* Diprenorphine
M99. *See* Etorphine hydrochloride
Macaque, 14, 214, 217. *See also* Primates
Philippine, 4
Macaw, 8, 14, 280. *See also* Birds, wild, psittacine
Macropods, 186-88. *See also* Marsupials
Mammals, domestic, scientific names, 181
Mammals, marine
anatomy and physiology, 223
chemical restraint, 227, 229-30
classification, 223
danger potential, 223
names of gender, 223
physical restraint, 223-28
transport, 228-29
Mammals, small, wild, 189-200
Manatee. *See* Sirenians
Marmoset, 214, 216. *See also* Prosimians
Marmot, 189, 194. *See also* Rodents
Marsupialia, 183
Marsupials
anatomy and behavior, 184
chemical restraint, 188
classification, 183
cooling, 68
danger potential, 184
names of gender, 183
obtaining blood samples, 188
physical restraint, 184-88
transport, 188
Medical problems during restraint. *See* Restraint, medical problems
Mesmerism. *See* Hypnosis
Methoxyflurane, 179, 264
Metomidate, 270
Milk fever. *See* Hypocalcemia
Mineralocorticoids, 58
Mink, 181, 206, 207. *See also* Mustelids
Mole. *See also* Insectivors; Monotremes
golden, 189
marsupial, 183
Mongoose, 201, 208
Monkey. *See also* Primates
cholinergic response, 86
colobus, 55
effects of chronic stress, 60
obtaining blood samples, 219
physical restraint, 217-19
roloway, 70
squirrel, 55
Monotremata, 183
Monotremes
classification, 183
danger potential, 183
physical restraint, 183
transport, 188
Moose. *See* Cervids
Mouse, 163, 165-66. *See also* Laboratory animals; Rodents
marsupial, 183
MS-222. *See* Tricaine methanesulfonate
Mule. *See* Horse, domestic
Muscular necrosis, 88
Musk-ox, 240, 243, 257. *See also* Bovids, wild
Muskrat, 195, 196. *See also* Rodents
Mustelids. *See also* Carnivores
chemical restraint, 206
classification, 181, 201
danger potential, 205
physical restraint, 206-8
skunk musk removal, 206
Musth, 233
Muzzle
crocodilian, 287, 288
dog, 69, 150, 151
leather, 151
wild canid, 202
Myoglobinuria
equine, 88
hyperthermic, 69
paroxysmal paralytic, 88
Myopathy
capture, 87-88, 89, 276
nutritional, 88

Nalline. *See* Nalorphine hydrochloride
Nalorphine hydrochloride, 50, 51
Naloxone hydrochloride, 45, 50, 51
Narcan. *See* Naloxone hydrochloride
Narcosis, 52
Narcotic antagonist, 50
Nerve injury, 80
Neuroleptic, 52
Neuroleptoanalgesia, 52
Net, 269, 283, 287, 314
cannon, 13, 269
cargo, 12, 257, 265
cotton, 13
drop, 13
fine mesh, 187, 217
hoop, 12, 186, 210, 217, 227, 228, 263, 265, 270, 315
manila, 13
mist, 13
nylon, 13
rectangular, 13, 258, 315
wire fence, 240
Nicotine alkaloid (sulfate), 50, 51
Nonspecific response, 54
Norepinephrine, 58, 87
Numbat, 183

Ocelot, 211. *See also* Cat, wild
Octopus, 9
Okapi. *See* Giraffe
Ophidiophobia, 91
Opposum. *See also* Marsupials
classification, 183
physical restraint, 184
torpidity, 7, 40
Virginia, 184, 188
Orangutan. *See* Ape
Ostrich, 10, 262, 264, 265, 266, 267. *See also* Birds, wild, flightless
Otter, 201, 205. *See also* Mustelids
Malayan, 4
river, 207
sea, 207, 223
Overheating. *See* Hyperthermia
Overstraining disease. *See* Myopathy, capture
Owl, 262, 271, 272. *See also* Birds, wild, raptors

Pachyderm, 232
Palmer Cap-Chur equipment, 36, 37, 38, 41
Panda, 201
Pangolin, 9, 189, 192-93
Panting
dog, 68, 151
importance, in thermoregulation, 69, 151, 202
wild canid, 202
Parakeet, 9, 14, 277. *See also* Birds, wild, psittacine
capturing from cage, 278-79
Paralysis
brachial, 80
delayed, 80
facial, 78
neck, from stretching, 78
perineal, 80
radial, 80
Parbuckle, 238
Parrot, 277. *See also* Birds, wild, psittacine
capturing from cage, 280
kea, 277
Passeriformes, 262
Peafowl, 275. *See also* Birds, wild, galliform
Peccary, 240, 247, 248. *See also* Swine, wild
Pelecaniformes, 262
Pelican, 9. *See also* Birds, wild, water
Penguin, 262, 264, 265. *See also* Birds, wild, flightless
Pentobarbital sodium, 191, 313
Perissodactyla, 93, 181, 240, 245-47
Petrel, 262
Phalanger, 183, 185. *See also* Marsupials
Pheasant. *See* Birds, wild, galliform
Phencyclidine hydrochloride, 44, 46-47, 51, 147, 188, 192, 200, 204, 222, 243, 248, 249
Pholidota, 189
Physiological saline, 84, 85
Piciformes, 262
Pig, domestic. *See* Swine, domestic
Pig, wild. *See* Swine, wild
Pigeon. *See* Dove
Pika. *See* Rabbit
Piloerection, 68
Pinnipedia, 223
Pinnipeds. *See also* Mammals, marine
capture, 226-27
chemical restraint, 227, 230
neck strap, 226
obtaining blood samples, 228
physical restraint, 226-28
squeeze cages, 228
Pisces. *See* Fish
Platypus, 183. *See also* Monotremes
Plover, 262
Pneumonia, 82
gangrenous, 88

Podicipediformes, 262
Poikilotherm, 63, 71
Pony. *See* Horse, domestic
Porcupine. *See also* Rodents
  Brazilian tree, 199
  classification, 189
  crested, Old World, 197
  danger potential, 198
  North American, 197, 198
  obtaining blood samples, 199
  physical restraint, 197-200
  physiology and behavior, 196-97
Portland (Oregon) Zoo, 233
Postorbital sinus bleeding, 166
Postrestraint complications, 89
Postures, defensive or protective, 55
Poultry and waterfowl, domestic
  behavior and physiology, 171
  chemical restraint, 179
  classification, 171
  danger potential, 171
  feather damage, 81
  names of gender, 171
  physical restraint, 172-79, 277
  releasing, after restraint, 177
  transport, 178
Prairie dog, 195. *See also* Rodents
Prednisolone, 84
Primates
  anatomy, physiology, and behavior, 214-15
  chemical restraint, 220-21, 222
  classification, 214
  danger potential, 214
  intubation, 220, 222
  obtaining blood samples, 219, 220
  physical restraint, 11, 215-21
  transport, 221
  tuberculin testing, 220
  weights, 214
Proboscidea, 231
Procellariiformes, 262
Procyonids, 201, 205
Prods, electric and battery-operated, 15, 234, 259
Promazine hydrochloride, 112, 192, 249
Pronghorn. *See* Bovids, wild
Prosimians. *See also* Primates
  chemical restraint, 217
  danger potential, 217
  physical restraint, 215-17
Psittaciformes, 262
Puff adder, African, 302. *See also* Snakes, poisonous
Pulmonary edema, 82

Quelicin. *See* Succinylcholine chloride
Quinaldine, 216

Rabbit
  behavior, 161, 193
  chemical restraint, 169-70
  chronic stress syndrome, 60
  classification, 161, 181, 189
  danger potential, 161, 193
  obtaining blood samples, 163, 164, 165
  physical restraint, 161-63, 164, 193
  torpid state, 161
Raccoon, 201, 205. *See also* Procyonids
Radiation, 71
Rail, 262
Raptors. *See* Birds, wild, raptors
Rat, 161, 166-67, 168. *See also* Laboratory animals; Rodents
  kangaroo, 189
Ratitae, 262
Ratite, 266, 267, 268. *See also* Birds, wild, flightless
Rattlesnake, 305. *See also* Snakes, poisonous

Ray. *See* Fish
Receptors, 57, 63
Regurgitation, 9, 82, 88, 90
Reindeer, 240, 253. *See also* Cervids
Renal shut down, 69
Reptiles
  chelonians, 289-93
  classification, 286
  crocodilians, 286-89
  lizards, 293-96
  scale damage, 81
  snakes, 296-310
Reptilia, 286
Restraint. *See also* Restraint, chemical; Restraint, physical
  adverse effects, 41-43
  behavioral aspect, 4
  chronic stress syndrome, 60-61, 89
  environmental considerations, 4
  etiology of signs observed, 90
  factors determining technique selection, 3, 4
  general concepts, 3
  health status, 4
  hierarchial status, 4
  humane considerations, 5
  human injury, 90
  medical problems, 73-91
  physiological and psychological effects, 3, 4
  planning, 4, 5, 73
  postrestraint complications, 89
  psychological, 6-8
  purposes, 3
  territoriality, 5
  tools, 4, 6-16
Restraint, chemical. *See also* individual species or group
  causes of failure, 41-43
  definition of terms, 51-52
  delivery systems, 36-40
  drugs, 43-50
  factors affecting, 40-41, 43
  historical development, 35
  ideal drug, 35-36
  importance, 4
  injection sites, 42
  legal aspects, 50-51
  tranquilizers, 49-50
Restraint, medical problems
  bloat, 89
  etiology of signs, 90
  feather and scale damage, 81
  foot injuries, 79, 80
  head and neck injuries, 76-78
  human injury, 90-91
  limb injuries, 78-80
  metabolic conditions, 81-89
  nerve injuries, 80
  postrestraint complications, 89
  problems of infant animals, 89
  regurgitation, 88
  trauma, 74-81
Restraint, physical, 9-15. *See also* individual species or group
Restraint, tools. *See also* individual tools
  arm extension, 6, 12-13, 38-40
  barriers, 6, 8, 13-14
  boards, 11, 221, 309, 310
  broom, house, 15, 111, 185, 199
  chemical agents, 15
  confinement, 6, 10-12
  diminishing sense perception, 6, 9-10
  hands, 10, 14
  physical force, 6, 14-15
  psychological, 6-8
  shovel, scoop, 15
  special techniques, 15-16
  towels, 11, 271
  voice, 4, 6, 10

Reverence for life, 5
Rhabdomyolysis. *See* Muscular necrosis
Rhea. *See* Birds, wild, flightless
Rhinoceros, 30, 41, 44, 232, 240. *See also* Hoofed stock
  chemical restraint, 246-47
  danger potential, 246
  physical restraint, 246
  transport, 246, 247
Rodentia, 161, 189
Rodents. *See also* Laboratory animals
  classification, 189
  danger potential, 193
  physical restaint, 194-200
  physiology and behavior, 193-94
  stress response, 60
Rompun. *See* Xylazine
Rope
  burn, 76
  cable, 17
  care, 17
  cleaning, 17
  coiling, 25-27
  construction, 17
  crocheted loop for storage, 22
  fibers used, 17, 18
  hanking, 21-22
  laying, 33
  splicing, 19-21
  strand, 17
  strength comparison, 19
  terms, 32-34
  throwing, 24
  whipping, 20
Rope work
  bamboo pole and loop, 12, 28
  basic, 19
  block and tackle, 30, 31, 32
  bracing, 28, 29
  coiling, 25-27
  dally, 28-32
  drag toss, 27
  halter, 30, 31, 114, 115
  halter tie, 24, 25
  hitches, 23-24, 29-30
  hobbles, 31
  honda, 24, 25, 26
  knots, 22-24, 25, 26
  piggin' string, 124
  sideline, 105, 106
  sling, 30-31, 33
  swing toss, 27-28
  tail tie, 28-29

Sack
  burlap, 11
  hoop, 307
  jute, 11
  snake, double-bottomed, 307
Salamander, 311
  axolotl, 313
  giant, 312
Saline, physiological. *See* Dehydration
San Diego Zoo, 216
Schweitzer, Albert, 5
Scoline. *See* Succinylcholine chloride
Screamer, 269, 270. *See also* Birds, wild, water
Seal, 226. *See also* Pinnipeds
  eared, 223
  elephant, 230
  Weddell, 230
Sea lion, 226, 227, 228. *See also* Pinnipeds
  California, 61, 230
Sea snake, 286
Sedation, 52. *See also* Tranquilizers
Seizures. *See* Convulsions
Sernyl. *See* Phencyclidine hydrochloride
Sernylan. *See* Phencyclidine hydrochloride

Seroma, 74
Shackle
  elephant, 234
  hog, 145, 146
Shark, 9, 314. *See also* Fish
Sheep, domestic
  behavior and physiology, 5, 131
  chemical restraint, 138
  classification, 131
  crisscross casting, 133
  crook, use, 133, 134
  danger potential, 131
  dog, 131
  handling lambs, 133-34
  medication techniques, 133-35
  names of gender, 131
  physical restraint, 88, 131-35
  setting up, 132-33
  transport, 138
  weights, 131
Sheep, wild. *See also* Bovids, wild
  bighorn, 77
  Dall, 258
  mouflon, 260
Sheeting
  canvas, heavy, 75, 212, 213, 253
  plastic, opaque, 14, 186, 240-41, 275
Shield
  face mask, 14, 217, 276, 306
  fence panel, 14, 275
  garbage can lid, 214
  plastic, 13, 14, 276, 300
  plywood, 13, 16, 213, 227, 258, 265, 275
Shock
  alarm response, 58
  hemorrhagic, 74
  hypoglycemic, 83, 286
  insulin, 83
  irreversible, 85
  neurogenic, 85
  psychogenic, 85
Shrew, poisonous, 189
Simians
  danger potential, 217
  obtaining blood samples, 219
  physical restraint, 217-18
Sirenia, 223
Sirenians, 223, 226, 229. *See also* Mammals, marine
Skunk. *See also* Mustelids
  spotted, 205
  musk, 206
Sling
  canvas, 107-8, 213
  hip, 128
  ratite, 267, 268
  rope, 15, 30, 108, 250
  stretcher, 223-24, 228-29
Sloth. *See* edentates
Snakebite, 306-7
Snakes
  anesthesia, 300, 308-9
  chemical restraint, 308-10
  classification, 286
  cooling, 10, 299-300
  danger potential, 297, 306-7
  eye cap removal, 302, 303
  fear of, 91
  hook, 13, 297, 299, 302, 303, 304
  loop, 299, 302
  nonpoisonous, 297-302
  noose, 302
  obtaining blood samples, 301
  physical restraint, 297-308
  pinning, 297, 299, 302, 303
  plastic tubes, 302, 303, 304
  poisonous, 9, 302-6
  probes, for sexing, 302
  sexing, 300-301
  squeeze box, 305, 306
  tong, Pilston, 298, 299, 303
  transport, 307-8
Snare
  bear, 203
  cable, 140, 142
  cat, 209, 210
  commercial, 12
  dog, 150
  reptile, 286, 288, 295, 296
  snout, swine, 142, 143
Snatch block, 32, 33
Social distance, 7-8
Sodium bicarbonate, 81, 88
Sodium secobarbital, 316
Solenodon, 189
  Haitian, 189
Solitary confinement, 3
Solu-delta-cortef. *See* Prednisolone
Specific response, 54
Speculum
  broom handle, 16
  canine, 152
  cattle, 118, 119, 120
  chinchilla, 198
  dental, 16
  Frick, 120
  hardwood dowel, 15-16, 220, 274, 292, 293
  Hauptner dental wedge, 118, 119
  needle holder, plastic, 281
  paper clip, 280
  sheep, 133, 292, 293
  wooden or plastic, 16, 292, 298
Sphenisciformes, 262
Splice
  back, 20, 21
  eye, 20-21, 22
  long, 20, 21
  short, 19, 21
Splicing, 19, 34
Sprain, 79
Springhaas, 189
Squeeze cage. *See* Cage, squeeze
Squirrel, 189, 194. *See also* Rodents
Stimuli, noxious, 53, 54, 55
Stockinette, 11, 216, 273
Stocks, 114, 122, 129, 260. *See also* Chute
Stork, 9, 14, 262, 274, 276. *See also* Birds, wild, long-billed, long-legged
Stress
  adaptation syndrome, 53
  alarm response, 58
  basic concepts, 54-55
  body response, 55-57
  chronic restraint syndrome, 60-61, 89
  cold, 70
  definition of terms, 53
  flight distance, 8
  hierarchial, 4
  homeostasis, 53, 71
  pathophysiology, 57-61
  role, in disease, 61, 87, 89
  transport, 4
Stressor
  behavioral, 54, 55
  miscellaneous, 54, 55
  neuroendocrine pathways, 55-57
  psychological, 54
  somatic, 54
  touch, 10
Stretcher
  cetacean, 223-24, 229
  crocodilian, 287
  primate, 221
Strigiformes, 262
Sublimaze. *See* Fentanyl-droperidol
Succinylcholine chloride, 47, 50, 51, 89, 112, 130
Sucostrin. *See* Succinylcholine chloride
Suffocation, 14. *See also* Physical restraint of individual species
Sugar glider, 184. *See also* Marsupials
Sunstroke, 69, 71
Surital. *See* Thiamyl sodium
Swan, 171, 181, 262, 268, 269. *See also* Birds, wild, water; Poultry and waterfowl, domestic
Swift, 262, 283. *See also* Birds, wild
Swine, domestic. *See also* Swine, wild
  anatomy, physiology, and behavior, 6, 139
  casting, 144-46
  chemical restraint, 147
  classification, 139, 181
  danger potential, 139
  handling piglets, 142, 144
  harness, rope, 141
  hierarchy, 4, 139
  hyperthermia, 140
  neck tongs, 13, 142
  physical restraint, 139-47
  shackle, 145, 146
  snare, 13, 142-43
  snout rope, 140-42
  squeeze cage, 145, 147
  transport, 147
  trough, cradle, 145, 147
Swine, wild. *See also* Swine, domestic
  anatomy and physiology, 247
  chemical restraint, 248
  classification, 240
  danger potential, 247
  physical restraint, 247-48
Sympathetic nervous system, 56
Syncope, fatal. *See* Cholinergic bradycardia
Syringe
  Haigh-designed blowgun, 39
  hand-held, 36, 37
  Palmer Cap-Chur, 37, 38, 244
  pole or stick, 38-40, 265
  projected, 36-39
  Reudi-designed blowgun, 39
  Telinject blowgun, 39

Talons, 8, 270, 272, 273. *See also* Birds, wild
Tamandua, 191, 192. *See also* Anteater
Tamarin. *See* Prosimians
Tapir. *See also* Hoofed stock
  chemical restraint, 243, 246
  classification, 240
  danger potential, 246
  physical restraint, 246
  Tarsier, 215, 217. *See also* Prosimians
  Tasmanian devil, 183, 185. *See also* Marsupials
  Teeth, 8, 139
  Temperature, 63
  Tenrec, 189. *See also* Insectivores
  Tern. *See* Birds, wild, shore and gull-like
  Terrapin, 289. *See also* Chelonians
  Territoriality, 5
  Tetany 90
  hypocalcemic, 82
Therapeutic index, 35
Thermoneutral zone, 71
Thermoregulation, 4, 56
  definition of terms, 71
  heat and moisture conservation, 63
  heat production, 63
  mechanisms for cooling, 64-67, 68
  medical problems, 69-71
  morphological adaptations for cooling, 68-69
  physiology, 63
Thiamyl sodium, 230
Tiger, 10, 14, 201, 208, 211. *See also* Cats, wild
  cub, 4, 209, 210

Tilazol. *See* Tiletamine hydrochloride-zolazepam hydrochloride.
Tiletamine hydrochloride-zolazepam hydrochloride, 48, 138, 147, 155, 170, 192, 193, 204, 230, 243, 254, 293
Tinamiformes, 262
Tinamou, 262
Toad. *See also* Amphibians
  clawed, 312
  Colombian giant, 311, 312
  Colorado River, 311
  marine, 311
  parotid gland secretion, 311, 312
Toenails, injury, 80
Tongs
  neck, 13, 142-43, 202
  nose, 13, 115-16
  Pilston snake, 298-99
Torpid state, 7, 40, 161, 192, 286
Tortoise, 286, 289, 290, 291, 292, 293. *See also* Chelonians
Toucan. *See* Birds, wild, large-billed
Tracheotomy, 85
Tranquilizers
  acepromazine maleate, 49
  diazepam, 49-50
Transporting animals. *See also* individual species
  effects, 4
  ventilation requirements, 242
Trauma, 74-76
Tribromoethanol, 264
Tricaine methanesulfonate, 313, 316
Trogon, 262
Trogoniformes, 262
Trough, 145, 147
Tuatara, 286
Tube
  endotracheal, 230
  gila monster, 296
  Kingman, stomach, 119
  plastic, 11, 221, 280, 281, 296, 303, 304
  psittacine, small, 280
  stomach, 218, 220, 280, 281
Tuberculin testing, primate, 220
Tubulidentata, 189
Turaco, 274. *See also* Birds, wild, galliform
Turkey. *See also* Poultry and waterfowl, domestic
  artificial insemination, 176
  obtaining blood samples, 176
  oral medication, 177
  physical restraint, 174-77
Turtle. *See also* Chelonians
  sea, 286, 289, 292
  side-necked, 289
  snapping, 286, 289-91
  soft-shell, 290
  western pond, 289
Tusks
  cervid, 251
  elephant, 232, 233
  hippopotamus, 248
  swine, 139, 247
Twitch, 94, 96-98
Tying-up syndrome, 88
Tympany. *See* Bloat

Unconsciousness, causes, 73
Ungulate. *See* Hoofed stock
Urethan. *See* Ethylcarbamate
Urination, 9, 90
Ursidae, 201. *See* Bear

Vagal bradycardia. *See* Cholinergic bradycardia
Vagal reflex. *See* Cholinergic bradycardia
"Vangspier sindroom." *See* Myopathy, capture
Ventricular fibrillation, 40, 58, 81, 87
Vetalar. *See* Ketamine hydrochloride
Vicuna, 240, 249, 250. *See also* Camelids
Viper, 286, 303. *See also* Snakes, poisonous
Voluntary motor response, 55

Wallaby, 183, 186. *See also* Marsupials
Wallaroo, 9, 183, 186. *See also* Marsupials
Walrus, 223, 226. *See also* Pinnipeds
Wapiti, 14, 240, 251, 253. *See also* Cervids
Warbler. *See* Birds, wild, song
Waterfowl, domestic, 177-79. *See also* Poultry and waterfowl, domestic; Birds, wild, water
Weapons, of animals, 8-9. *See also* danger potential, individual species or group
Weasel, 201, 205. *See also* Mustelids
Whale, 223. *See also* Cetaceans
  killer, 7, 223, 225, 229
Whip, 15, 111, 233, 241
White muscle disease. *See* Myopathy, nutritional
White muscle stress syndrome. *See* Myopathy, capture
Wobbler syndrome, 78
Wolf, 9, 201, 202. *See also* Carnivores
Wolverine, 201, 207. *See also* Mustelids
Wombat, 183, 185, 186. *See also* Marsupials
Woodchuck, 194, 195. *See also* Rodents
Woodpecker, 262
Wound. *See* Laceration

Xylazine, 48, 51, 112, 130, 138, 155, 188, 204, 238, 243, 245, 248, 251, 254, 264

Yak, 5, 113, 181, 260. *See also* Bovids, wild; Cattle, domestic

Zebra, 7, 240, 241, 245. *See also* Equids, wild; Hoofed stock
Zebu, 113, 181, 259, 260. *See also* Bovids, wild; Cattle, domestic